P9-CJE-413

Algebra $\frac{1}{2}$

An Incremental Development

Second Edition

Algebra $\frac{1}{2}$

An Incremental Development

Second Edition

JOHN H. SAXON, JR.

SAXON PUBLISHERS, INC.

Algebra $\frac{1}{2}$: An Incremental Development
Second Edition

Copyright © 1997 by Saxon Publishers, Inc.

Printed in the United States of America.

ISBN: 0-939798-45-X

Editor and production supervisor: Nancy Warren
Compositor: Black Dot Graphics

Twelfth printing: May 1999

Printed on recycled paper

┌─ *Reaching us via the Internet* ─┐

WWW: http://www.saxonpub.com

E-mail: info@saxonpub.com
└──────────────────────────────────┘

Saxon Publishers, Inc.
2450 John Saxon Blvd.
Norman, OK 73071

Contents

Contents

Preface

This is the second edition of a transitional math book designed to permit the student to move from the concrete concepts of arithmetic to the abstract concepts of algebra. The research of Dr. Benjamin Bloom has shown that long-term practice beyond mastery can lead to a state that he calls "automaticity." When automaticity is attained at one conceptual level, the student is freed from the constraints of the mechanics of problem solving at that level and can consider the problems at a higher conceptual level. Thus, this book concentrates on automating the concepts and skills of arithmetic as the abstract concepts of algebra are slowly introduced. The use of every concept previously introduced is required in every problem set thereafter. **This permits students to work on attaining speed and accuracy at every conceptual level.** Students often resist this practice because they feel that if they have already mastered a concept, no further practice is required. They do not realize that being able to work the problem slowly is not sufficient. They need to be reminded that mathematics is like other disciplines. For example, playing a musical instrument well requires long-term practice of the fundamentals. Playing football, golf, tennis, or any other sport well requires long-term practice and automation of fundamentals. Mathematics also requires this long-term practice.

This book has 27 more lessons than the first edition had so that concept presentations could be smoothed out. The development of skills necessary to solve word problems has been much improved, with special attention being paid to the concepts required for the solution of rate problems.

Testing of pre-algebra students has revealed that more than a few students still have difficulty in using fractions, mixed numbers, and decimal numbers in the four basic arithmetic operations. Some students have difficulty in only one area, while others have difficulty in more than one area. Thus, the homework problem sets provide practice in every one of these skill areas for the entire year. The students who are deficient will be afforded the opportunity to learn, and the practice afforded the others will increase and solidify their abilities in these basic areas and permit them to develop speed and accuracy in problem-solving techniques.

Special emphasis is given to reading numbers written in numerical form and to translating numbers from numerical form to word form. Practice problems in these skills are provided in homework problem sets for over half the book. This translation is difficult, and lack of emphasis and long-term practice at the proper point is the reason that many older students are without this skill.

Conceptualization of area, volume, and perimeter is given considerable emphasis. These abstractions are introduced early, and problems involving at least one of these topics appear in every lesson beginning with Lesson 9. Since the words *area* and *volume* designate abstractions, students are encouraged to associate floor tiles with area and sugar cubes with volume. Reification of abstractions facilitates comprehension and leads to long-term retention.

Word problems are given special attention. Every problem set in the book begins with three or four word problems. At first, the problems are straightforward problems that require only reading and writing numbers, addition, subtraction, division, multiplication, or finding the average. The problems are designed to give students practice in reading word problems and deciding what is given and what procedure is required so that the question asked may be answered. This ability to read and to translate is a skill that must be mastered before the solution of more difficult problems can be attempted. Problems about rate are begun in Lesson 36, and time and distance problems are presented in Lesson 61. Later, simple two-step word problems are introduced.

Major emphasis is placed on the most fundamental problem of mathematics —which deals with the fractional part of a number. This problem is introduced early and practiced continually until its companion, the percent problem, is introduced. Then both problems appear regularly until the end of the book. Both problems are approached conceptually, and students are encouraged to draw diagrams of the completed problems. Some students believe that somehow, somewhere, a trick can be found that will allow the answers to percent problems to be obtained without understanding of the concept involved. Since no such trick exists, understanding must occur, and a picture of the problem has proved to be very helpful. This approach will be continued in the next two books in the Saxon series, *Algebra 1* and *Algebra 2*. Students can no longer be permitted to finish mathematics without total mastery of percent and complete understanding of the relationship of percent to the fractional-part-of-a-number problem.

Simple concepts in algebra are introduced early and practiced for the rest of the book. Variables are introduced in Lesson 34, where the numbers of arithmetic are used as replacements for variables in algebraic expressions. Simple equations are introduced in Lesson 39, and equations with fractions are introduced in Lesson 40. These problems allow practice in adding, subtracting, and dividing fractions at the same time that the two basic rules for solving equations are being introduced. This early introduction of simple equations permits the introduction of ratio problems in Lesson 57 and elementary ratio word problems in Lesson 69. These problems are practiced in the problem sets until the end of the book.

Integral exponents and integral roots are introduced in Lesson 46 and are practiced gently thereafter. This early introduction of exponents and roots is necessary because many students need long-term practice in order to understand these notations. Negative numbers are introduced in Lesson 72, and simplification problems that contain elementary combinations of positive and negative numbers will appear in the next 65 problem sets. This will provide excellent preparation for the more complicated expressions that will be encountered in algebra 1.

A study of the homework problem sets will show how this book provides comprehensive review and continued practice of the skills of arithmetic while basic facets of more advanced topics are introduced. The early introduction of these topics will ease the transition to algebra and will ensure higher success percentages for the students who attempt algebra. Thus, this book can be used successfully as a pre-algebra book for gifted and average students and can be used as the first year's course in a two-year algebra 1 sequence for other students. It can also be used as a general mathematics book for students who will not attempt algebra. The only difference will be in the age of the students when they use this book.

The book was written to provide continued practice in skills for an entire year. **To gain maximum advantage from the use of this book, it is necessary that all students work all the problems. The book was designed with the understanding that every**

problem would be worked by every student. In this book, the learning is spread out rather than being concentrated.

I am indebted to Tom Brodsky for his help in revising this book. I thank Shirley McQuade Davis for her ideas on teaching thinking patterns. I thank Joan Coleman for supervising the preparation of the manuscript and for managing production details, and David Pond, Scott Kirby, Chad Threet, and John Chitwood for their artwork.

John Saxon
Norman, Oklahoma

LESSON 1 *Whole number place value · Reading and writing whole numbers · Addition*

1.A
whole number place value

We use the **Hindu-Arabic** system to write our numbers. This system is a base 10 system and has 10 different symbols. The symbols are called **digits,** and they are

$$0, 1, 2, 3, 4, 5, 6, 7, 8, 9$$

When we write whole numbers, we can write the decimal point at the end of the number, or we can leave it off. Thus, both of these

$$427. \qquad 427$$

represent the same number. In the right-hand number, the decimal point is assumed to be after the 7. The value of a digit in a number depends on where the digit appears in the number. The 5 in the left-hand number below

$$415,623 \qquad 701,586 \qquad 731,235$$

has a value of 5000 because it is in the thousands' place. The value of the 5 in the center number is 500 because it is in the hundreds' place. The value of the 5 in the right-hand number is 5 because it is in the units' (ones') place. The first place to the left of the decimal point is the ones' place. We also call this place the **units' place.** The next place is the **tens' place.** The next place is the **hundreds' place.** The next place is the **thousands' place.** Each place to the left has one more zero.

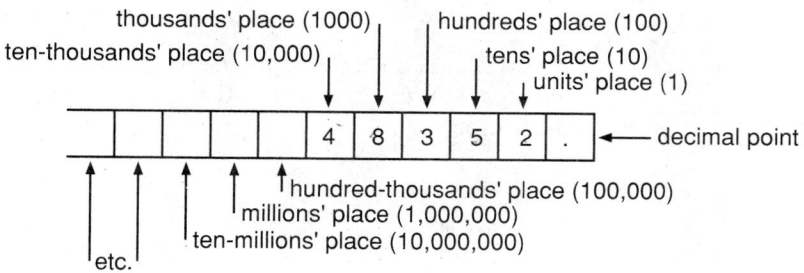

example 1.1   In the number 46,235:
  (a)   What is the value of the digit 5?
  (b)   What is the value of the digit 2?
  (c)   What is the value of the digit 4?

solution   First we write the decimal point at the end.

$$46,235.$$

1

(a)   The 5 is one place to the left of the decimal point. This is the units' place. This digit has a value of 5 × 1, or **5.**

(b)   The 2 is three digits to the left of the decimal point. This is the hundreds' place. This digit has a value of 2 × 100, or **200.**

(c)   The 4 is five digits to the left of the decimal point. This is the ten-thousands' place. This digit has a value of 4 × 10,000, or **40,000.**

1.B
reading and writing whole numbers

We begin by noting that all numbers between 20 and 100 that do not end in zero are hyphenated words when we write them out.

23 is written twenty-three	64 is written sixty-four
35 is written thirty-five	79 is written seventy-nine
42 is written forty-two	86 is written eighty-six
51 is written fifty-one	98 is written ninety-eight

The word **and** is not used when we write whole numbers by using words.

501	is written	five hundred one
	not	five hundred and one
370	is written	three hundred seventy
	not	three hundred and seventy
422	is written	four hundred twenty-two
	not	four hundred and twenty-two

The hyphen is also used in whole numbers when the whole number is used as a modifier. The words

ten thousand

are not hyphenated. But when we use these words as a modifier, as when we say

ten-thousands' place

the words are hyphenated. Other examples of this rule are

hundred-millions' digit
ten-billions' place
hundred-thousands' place

Before we read whole numbers, we begin at the decimal point and move to the left, placing a comma after every three digits. These commas divide the digits into groups of three digits.

Etc.	Trillions			Billions			Millions			Thousands			Units (Ones)			· Decimal point
	Hundreds	Tens	Ones	Hundreds	Tens	Ones	Hundreds	Tens	Ones	Hundreds	Tens	Ones	Hundreds	Tens	Ones	

To read the number 4125678942, we begin on the right end and separate the number into groups of three by writing commas.

4,125,678,942

Then we read the number from left to right, beginning with the leftmost group. First we read the number in the group, and then we read the name of the group. Then we move to the right and repeat the procedure.

example 1.2   Use words to write this number: 51723642

solution   We write the decimal point on the right-hand end. Then we move to the left and place a comma after each group of three digits.

51,723,642.

The leftmost group is the millions' group. We read it as

fifty-one million,

and write the comma after the word *million.* The next three-digit group is the thousands' group. We read it as

seven hundred twenty-three thousand,

and we write the comma after the word *thousand.* The last three-digit group is the units' group. We do not say "units" but just read the three-digit number as

six hundred forty-two.

Note that the words **fifty-one, twenty-three,** and **forty-two** are hyphenated words. Also note that we do not use the word *and* between the groups. Now we put the parts together and read the number as

fifty-one million, seven hundred twenty-three thousand, six hundred forty-two

We note that the commas appear in the same places the commas appeared when we used digits to write the number.

example 1.3   Use digits to write the number fifty-one billion, twenty-seven thousand, five hundred twenty.

solution   The first group is the billions' group. It contains the number fifty-one.

51,   ,   ,

All the groups after the first group must have three digits. There are no millions, so we use three 0s.

51,000,

There are twenty-seven thousands. **So we write 027 in the next group so that the group will contain three digits.**

51,000,027,

Now we finish by writing 520 in the last group.

51,000,027,520

1.C
expanded form

Writing a number in expanded form is a good way to practice the idea of place value. When we write numbers in expanded form, we do not consider the groups of three digits as we do when we write a number in words. Instead we consider the value of every digit in the number. To write a number in expanded form, we write each of the nonzero digits multiplied by the place value of the digit. We use parentheses to enclose each of these multiplications. We put a plus sign between each set of parentheses.

To write 5020 in expanded form, we write

$$(5 \times 1000) + (2 \times 10)$$

because this number contains 5 thousands and 2 tens.

example 1.4 Write the following number in standard form:

$$(4 \times 10,000) + (6 \times 100) + (5 \times 1)$$

solution The number has 4 ten-thousands, **no thousands,** 6 hundreds, **no tens,** and 5 ones. The number is **40,605.**

example 1.5 Write the number 6,305,126 in expanded form.

solution There are 6 millions,   and 3 hundred thousands,
$(6 \times 1,000,000)$ $(3 \times 100,000)$

and 5 thousands,   and 1 hundred,   and 2 tens,   and 6 ones.
(5×1000) (1×100) (2×10) (6×1)

If we put them all together, we get

$$(6 \times 1,000,000) + (3 \times 100,000) + (5 \times 1000) + (1 \times 100) + (2 \times 10) + (6 \times 1)$$

1.D

addition When we add numbers, we call each of the numbers **addends,** and we call the answer a **sum.**

$$
\begin{array}{rl}
523 & \text{addend} \\
619 & \text{addend} \\
+\ 512 & \text{addend} \\
\hline
\mathbf{1654} & \text{sum}
\end{array}
$$

To add whole numbers, we write the numbers so that the last digits of the numbers are aligned vertically.

example 1.6 Add: $4 + 407 + 3526$

solution We write the numbers so that the last digits are aligned vertically. Then we add.

$$
\begin{array}{r}
4 \\
407 \\
+\ 3526 \\
\hline
\mathbf{3937}
\end{array}
$$

practice **a.** In the number 152068, what is the value of the 2?

b. Write in standard form: $(6 \times 1000) + (4 \times 10) + (3 \times 1)$

c. Write 85,020 in expanded form.

d. Use digits to write this number: ten billion, two hundred five million, forty-one thousand, five hundred

e. Use words to write this number: 36025103

problem set 1

1. In the number 5062973, what is the value of each of these digits?
   (a) 6   (b) 9   (c) 3

2. Write the six-digit number that has the digit 4 in the thousands' place, with the remaining digits being 3.

3. Write the seven-digit number that has the digit 3 in the millions' place and the digit 7 in the hundreds' place, with the remaining digits being 6.

4. A number has eight digits. Every digit is 9 except the ten millions' digit, which is 3; the ten-thousands' digit, which is 5; and the units' digit, which is 2. Use digits to write the number.

5. Use digits to write this number: forty-one billion, two hundred thousand, five hundred twenty

6. Use digits to write this number: five hundred seven billion, six hundred forty million, ninety thousand, forty-two

7. Use digits to write this number: four hundred seven trillion, ninety million, seven hundred forty-two thousand, seventy-two

8. Use digits to write this number: nine hundred eighty million, four hundred seventy

Use words to write each number:

9. 517236428
10. 90807060
11. 32000000652
12. 3250000652
13. 6040000
14. 99019900

Write each number in standard form:

15. $(3 \times 100{,}000) + (4 \times 1{,}000) + (2 \times 10)$

16. $(7 \times 10{,}000) + (8 \times 100) + (6 \times 10)$

17. $(9 \times 1000) + (4 \times 100) + (5 \times 1)$

18. $(7 \times 1{,}000{,}000) + (2 \times 10{,}000) + (6 \times 1000)$

Write each number in expanded form:

19. 5280
20. 408
21. 70,600
22. 21,000
23. 4005
24. 9080

Add. Do not use a calculator.

25.
```
   43
   76
   84
 + 91
```

26.
```
  4628
  5734
+ 8416
```

27.
```
  9056
  4708
+ 9076
```

28.
```
  432
  846
  943
+ 721
```

29.
```
  856
  943
  784
+ 947
```

30.
```
  555
  666
  765
+ 567
```

LESSON 2 *The number ray · Ordering · Rounding whole numbers*

2.A

the number ray

A number ray is an arrow that points from left to right. We associate the number zero with the left-hand end of the ray. The left-hand end of the ray is called the **origin.** We divide the number ray into segments whose lengths are equal.

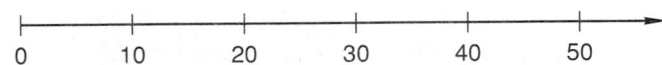

Every positive number can be paired with a point on the number ray. When we put a dot on a number ray to mark the location of a number, we call the dot the **graph** of the number. The number associated with a point on the number ray is called the **coordinate** of the point. Any group of positive numbers can be arranged in order from the least to the greatest. If we graph the numbers 34, 45, and 7 on our number ray, we get this figure.

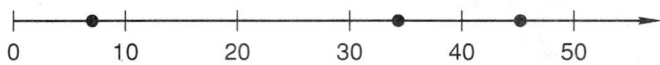

When we graph these numbers, we see that the smallest number is on the left and the greatest number is on the right.

example 2.1 Use a number ray to arrange these numbers in order: 5, 14, 17, 2, and 6

solution The distance between marks on the number ray can be any distance so long as they are all the same. We decide to use spaces 5 units long.

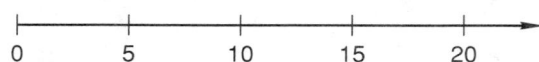

When we graph the numbers, we see that we have arranged them in order.

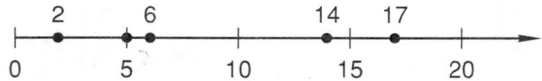

So if we **order** the numbers 5, 14, 17, 2, and 6 we get

$$2, 5, 6, 14, 17$$

example 2.2 Arrange the following numbers in order from least to greatest:
(a) 465   (b) 654   (c) 456   (d) 564

solution The smallest number is the one with the smallest first (leftmost) digit. We have two numbers that begin with 4.

$$465 \quad \text{and} \quad 456$$

Now we look at the second digits and see that the number on the right is the lesser of the two numbers because 5 is less than 6. So the two smallest numbers are

$$456 \quad \text{and} \quad 465$$

This leaves 654 and 564. Of course, 564 is less than 654 because it has the smallest first digit. So our answer is

$$456, 465, 564, 654$$

If we graphed these numbers on a number ray, this is the order in which they would appear.

We often use the **rounded** form of a number. If the distance to the barn is 209 yards, the farmer would probably say that it is 200 yards to the barn. The farmer has rounded 209 to the nearest hundred because 209 yards is closer to 200 yards than it is to 300 yards.

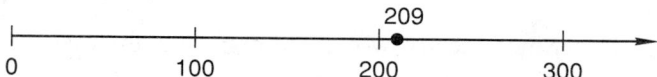

If it is 246 yards to the barn, the farmer might say that it is 250 yards to the barn. If so, the farmer has rounded 246 to the nearest 10 because 246 is closer to 250 than it is to 240.

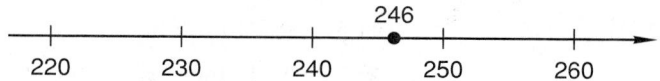

We often round numbers to the nearest ten, the nearest hundred, the nearest thousand, the nearest ten thousand, etc. Rather than draw a number ray, we will use a circle and an arrow to help us round numbers. To demonstrate, we will round 24,374 to the nearest thousand. We will use three steps.

1. Circle the digit in the place to which we are rounding and mark the digit to its right with an arrow.

$$2④, \overset{\downarrow}{3}\,7\,4$$

2. Change the arrow-marked digit and all digits to its right to zero.

$$2④, \overset{\downarrow}{0}\,0\,0$$

3. Leave the circled digit unchanged or increase it 1 unit as determined by the following rules:
   (a) If the arrow-marked digit is less than 5, do not change the circled digit.
   (b) If the arrow-marked digit is 5 or greater, increase the circled digit by 1 unit.

Rule (a) applies in this problem, so our answer is

24,000

example 2.3   Round 471,326,502 to the nearest ten thousand.

solution   First circle the ten-thousands' digit and mark the digit to its right with an arrow.

$$4\,7\,1,\,3②\overset{\downarrow}{6},\,5\,0\,2$$

Next change the arrow-marked digit and all digits to its right to zero.

$$4\,7\,1,\,3②\overset{\downarrow}{0},\,0\,0\,0$$

Since the arrow-marked digit was greater than 5, we increase the circled digit from 2 to 3 and get

471,330,000

example 2.4 Round 83,752,914,625 to the nearest ten thousand.

solution First we circle the ten-thousands' digit and mark the digit to its right with an arrow.

$$\downarrow$$
$$8\ 3,\ 7\ 5\ 2,\ 9\ \textcircled{1}\ 4,\ 6\ 2\ 5$$

Now we change the arrow-marked digit and all digits to its right to zero.

$$\downarrow$$
$$8\ 3,\ 7\ 5\ 2,\ 9\ \textcircled{1}\ 0,\ 0\ 0\ 0$$

Since the arrow-marked digit was less than 5, we leave the circled digit unchanged and get

83,752,910,000

practice **a.** Round 914,471,752 to the nearest ten thousand.

b. Round 83,625,502 to the nearest thousand.

c. Use a number ray to arrange these numbers in order: 3, 11, 14, 19

d. Arrange the following numbers in order from least to greatest:
736, 367, 376, 673

problem set 2

1. Use a number ray to arrange these numbers in order: 2, 19, 6, 11, 5

Arrange the following numbers in order from least to greatest:

2. 514, 154, 145, 451   **3.** 942, 249, 924, 294, 429

4. Round 4,185,270 to the nearest hundred.

5. Round 83,721,525 to the nearest thousand.

6. Round 415,237,842 to the nearest hundred thousand.

7. A number has nine digits. All the digits are 7 except the ten-thousands' digit, which is 2, and the tens' digit, which is 5. Write the number.

8. A number has seven digits. All the digits are 3 except the hundred-thousands' digit, which is 6; the thousands' digit, which is 4; and the hundreds' digit, which is 7. Write the number.

9. Use digits to write one hundred seven million, forty-seven thousand, twenty.

10. Use digits to write ninety-three billion, four hundred sixty-two million, forty-seven.

Use words to write each number:

11. 731284006   **12.** 903721625

13. 9003001256   **14.** 7234000052

Write in standard form:

15. $(7 \times 10,000) + (6 \times 100) + (5 \times 10) + (4 \times 1)$

16. $(3 \times 100,000) + (9 \times 1000) + (7 \times 100) + (6 \times 10) + (3 \times 1)$

17. $(9 \times 1000) + (6 \times 100) + (9 \times 1)$

Write in expanded form:

18. 109,326   **19.** 68,312   **20.** 903,162

Add. Do not use a calculator.

| 21. | 9317 4526 + 9015 | 22. | 7316 4582 + 9143 | 23. | 88,871 40,012 + 90,375 |

| 24. | 78,524 91,325 70,026 + 91,358 | 25. | 42,715 90,826 41,222 + 39,057 |

26. 37,251 + 81,432 + 90,256 + 21,312

27. 14 + 32 + 16 + 21 + 932 + 21

28. 1 + 2 + 21 + 12 + 122 + 1222

29. 33 + 333 + 313 + 1313   **30.** 4 + 314 + 134 + 13,245

LESSON 3 *Subtraction · Patterns*

3.A
subtraction

When we subtract, we find the difference of two numbers. The difference between 21 and 47 is 26.

$$
\begin{array}{r}
47 \\
- 21 \\
\hline
26
\end{array} \leftarrow \text{ difference}
$$

Often we need to "regroup" the top number in order to subtract. When we regroup, we "borrow" 1 from the column to the left. To subtract

$$
\begin{array}{r}
82 \\
- 26
\end{array}
$$

we must change the form of 82. We do this by regrouping. "Silly counting" can help us understand regrouping.

79	is	seventy-nine
80	is	seventy-ten
81	is	seventy-eleven
82	is	seventy-twelve

When we borrow, we really write 82 as "seventy-twelve."

$$
\begin{array}{r}
82 \\
- 26
\end{array} \rightarrow
\begin{array}{r}
^{7}\llap{}^{1} \\
\cancel{8}2 \\
- 26 \\
\hline
56
\end{array}
$$

example 3.1   Subtract:
$$
\begin{array}{r}
432 \\
- 257
\end{array}
$$

solution   This time we must borrow twice.

$$\begin{array}{r} \overset{1}{\underset{}{3\ 2\ 1}} \\ \cancel{4}\cancel{3}2 \\ -\ 2\ 5\ 7 \\ \hline 1\ 7\ 5 \end{array}$$

3.B
patterns

If one operation will undo another operation, we say that the operations are **inverse operations.** If we add 465 and 357, we find that the sum is 822.

$$\begin{array}{r} 465 \\ +\ 357 \\ \hline 822 \end{array} \leftarrow \text{ largest number}$$

These numbers form an addition pattern. **In an addition pattern the largest number is the bottom number** (unless the middle number is zero).

We can undo what we have done if we subtract 357 from 822 or if we subtract 465 from 822.

$$\begin{array}{r} 822 \\ -\ 357 \\ \hline 465 \end{array} \xleftarrow{\ \text{largest number}\ \rightarrow} \begin{array}{r} 822 \\ -\ 465 \\ \hline 357 \end{array}$$

$$\leftarrow \quad \text{difference} \quad \rightarrow$$

In both of these subtraction patterns we note that the largest number is the top number. The bottom number in both patterns is the difference of the other two numbers. We can use these patterns to find missing digits in addition and subtraction problems.

example 3.2   Find the missing digits:
$$\begin{array}{r} 472 \\ +\ AKY \\ \hline 628 \end{array}$$

solution   The missing number is the difference between 628 and 472. To find this number, we subtract.

$$\begin{array}{r} \overset{5\,1}{\cancel{6}2\,8} \\ -\ 4\,7\,2 \\ \hline 1\,5\,6 \end{array}$$

Now we check by adding.

$$\begin{array}{r} 472 \\ +\ 156 \\ \hline 628 \end{array} \quad \text{check}$$

The missing digits form the number **156.**

example 3.3   Find the missing digits:
$$\begin{array}{r} MNP \\ +\ 257 \\ \hline 493 \end{array}$$

solution   What number added to 257 equals 493? To find the number, we will subtract 257 from 493.

$$\begin{array}{r} \overset{8\,1}{4\cancel{9}3} \\ -\ 2\,5\,7 \\ \hline 2\,3\,6 \end{array}$$

Thus *MNP* is 236. Now we check.

$$\begin{array}{r} 236 \\ +\ 257 \\ \hline 493 \end{array} \quad \text{check}$$

This was an addition pattern, but we had to subtract to find the missing number.

example 3.4 Find the missing digits:

$$\begin{array}{r} 526 \\ -\ XYZ \\ \hline 329 \end{array}$$

solution The bottom number in a subtraction problem is the difference. To find *XYZ*, we subtract 329 from 526.

$$\begin{array}{r} ^{1}_{4\ 1\ 1} \\ \cancel{5}\cancel{2}6 \\ -\ 3\ 2\ 9 \\ \hline 1\ 9\ 7 \end{array}$$

The value of *XYZ* is 197. Thus our subtraction pattern is

$$\begin{array}{r} 526 \\ -\ 197 \\ \hline 329 \end{array}$$

This is correct because the top number in every subtraction pattern must equal the sum of the two bottom numbers.

example 3.5 Find the missing digits:

$$\begin{array}{r} XMP \\ -\ 423 \\ \hline 287 \end{array}$$

solution This one is easy. The top number in a subtraction pattern always equals the sum of the bottom two numbers. So we add the bottom numbers to find *XMP*.

$$\begin{array}{r} 423 \\ +\ 287 \\ \hline 710 \end{array}$$

The missing digits are 7, 1, and 0. Our pattern looks like this.

$$\begin{array}{r} 710 \\ -\ 423 \\ \hline 287 \end{array}$$

practice Find the missing digits:

a.
$$\begin{array}{r} 563 \\ +\ AKC \\ \hline 912 \end{array}$$

b.
$$\begin{array}{r} MXP \\ -\ 364 \\ \hline 376 \end{array}$$

c.
$$\begin{array}{r} A9K \\ +\ 123 \\ \hline 522 \end{array}$$

d.
$$\begin{array}{r} 741 \\ -\ F6P \\ \hline 372 \end{array}$$

problem set 3

1. A number has six digits. Every digit is a 2 except the thousands' digit, which is 5, and the units' digit, which is 3. What is the number?

2. A number has five digits. Every digit is a 7 except the thousands' digit, which is 0. What is the number?

3. A number has seven digits. Every digit is a 4 except the hundred-thousands' digit, which is 1. What is the number?

4. A number has nine digits. Every digit is a 3 except for the ten-millions' digit, which is 5, and the hundred-thousands' digit, which is 2. What is the number?

5. Use digits to write fourteen million, seven hundred five thousand, fifty-two.

6. Use digits to write five hundred billion, four hundred sixty-five thousand, one hundred eighty-two.

7. Write 64,030 in expanded form.

8. Write 79,003 in expanded form.

9. Write 123,419 in expanded form.

Add or subtract as indicated. Do not use a calculator.

10.	551	11.	853	12.	936	13.	839
	− 174		− 284		+ 474		+ 472

Find the missing digits. Do not use a calculator.

14.	*XYZ*	15.	800	16.	735	17.	925
	− 245		− *MPQ*		+ *ABC*		+ *FGH*
	276		436		1211		1111

Write in standard form:

18. $(8 \times 10,000) + (5 \times 1000) + (3 \times 100) + (2 \times 10) + (5 \times 1)$

19. $(6 \times 1000) + (6 \times 100) + (6 \times 1)$

20. $(3 \times 100,000) + (2 \times 10,000) + (9 \times 100) + (7 \times 1)$

Use words to write each number:

21. 707070705

22. 5803125702

Add. Do not use a calculator.

23. 295 + 486 + 588 + 714

24. 205 + 937 + 483 + 286

25. 913 + 405 + 709 + 203

26. 41,325 + 80,926 + 71,452 + 52,061

27.	90,125	28.	1125	29.	63,124	30.	895
	40,061		986		9,876		573
	30,627		139		11,314		698
	+ 95,132		+ 2364		+ 6,573		+ 2164

LESSON 4 *Multiplication · Division · The pattern*

4.A

multiplication   Multiplication is a shorthand notation we use to denote repeated addition. If we wish to add 7 twelve times, we could write

$$7 + 7 + 7 + 7 + 7 + 7 + 7 + 7 + 7 + 7 + 7 + 7 = 84$$

or we could write

$$7 \times 12 = 84$$

If we wish to add 12 seven times, we could write

$$12 + 12 + 12 + 12 + 12 + 12 + 12 = 84$$

or we could write

$$12 \times 7 = 84$$

Because the sum of twelve 7s is the same as the sum of seven 12s, we could write for either addition problem

$$7 \times 12 \qquad or \qquad 12 \times 7$$

Here we used the cross to indicate multiplication. In algebra we often use a center dot instead of the cross. The dot also means to multiply. We can write the multiplications above as

$$7 \cdot 12 \qquad or \qquad 12 \cdot 7$$

When we multiply two numbers, either number can be written first.

In the multiplication of two numbers, one number used to be called the **multiplicand** and the other number used to be called the **multiplier**. The answer was called the **product**.

$$
\begin{array}{rl}
11 & \text{multiplicand} \\
\times\ 7 & \text{multiplier} \\
\hline
77 & \text{product}
\end{array}
$$

We still call the answer the product, but we don't use the words *multiplicand* and *multiplier* much any more. Instead, we call both of the numbers **factors**.

$$
\begin{array}{rl}
11 & \text{factor} \\
\times\ 7 & \text{factor} \\
\hline
77 & \text{product}
\end{array}
$$

example 4.1 Multiply: 421×335

solution Either number may be placed on top. We will put 335 on top.

$$
\begin{array}{r}
335 \\
\times\ 421 \\
\hline
335 \\
670 \\
1340 \\
\hline
141,035
\end{array}
$$

4.B
division

We remember that addition and subtraction are inverse operations because one of these operations will "undo" the other operation. **Multiplication and division are also inverse operations.** To demonstrate, we begin with the number 5,

$$5$$

and then multiply by 4 to get 20.

$$5 \times 4 = 20$$

Now if we divide 20 by 4, we will undo the multiplication by 4 and be back at 5 again.

$$20 \div 4 = 5$$

Here we show three different ways to designate the division of 20 by 4.

(a) $20 \div 4$ (b) $4\overline{)20}$ (c) $\dfrac{20}{4}$

All three notations indicate that 20 is to be divided by 4. The notations (a) and (b) are often used in arithmetic books but are not used much in algebra books. In algebra most authors prefer to use the fractional form (c). In this book we will use all three notations. The number we divide by is the **divisor,** and the number it goes into is the **dividend.** The answer to a division problem is called a **quotient.**

$$\begin{array}{c}\text{dividend} \rightarrow \\ \text{divisor} \rightarrow \end{array} \dfrac{20}{4} = 5 \leftarrow \text{quotient}$$

It is sometimes helpful to think of division as a process of separating the dividend into a number of equal groups. For instance, if we wish to divide 12 by 4, we may write

$$\dfrac{12}{4}$$

The question we are asking is, "Into how many groups of 4 can we divide 12 objects?" We can display the solution visually by using 12 dots and arranging them in groups of 4.

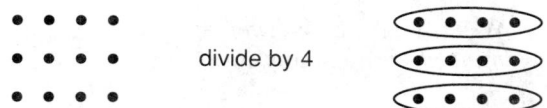

We see that 12 dots can be divided into three groups of 4, so we may say that

$$\dfrac{12}{4} = 3$$

In the same manner, if we write

$$\dfrac{16}{5}$$

we are asking "Into how many groups of 5 can 16 be divided?" Again we use dots to permit a visual solution.

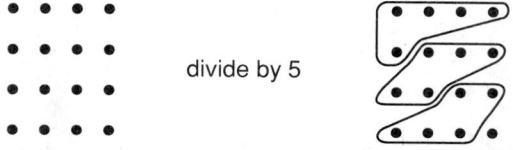

We find that 16 dots can be divided into three groups of 5 with 1 dot left over, and thus we say that the quotient is 3 with a remainder of 1, or

$$\dfrac{16}{5} = \mathbf{3\ R1}$$

If we try to divide 17 objects into groups of 3, we indicate our purpose by writing

$$\dfrac{17}{3}$$

and again we find that our division does not come out with a remainder of zero.

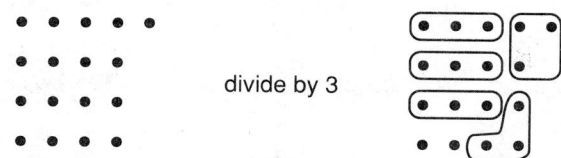

divide by 3

We get five groups of 3 dots and have 2 dots left over. Thus we say that 17 divided by 3 equals 5 with a remainder of 2.

$$17 \div 3 = \textbf{5 R2}$$

We can see from the diagrams why some people use the words **goes into** instead of the word **divides.** They would say that 3 goes into 17 five times with a remainder of 2. The phrase *goes into* is more meaningful, but the word *divides* is preferred by some people.

4.C
two-digit divisors

The word **algorithm** means a way to do something. Many mathematicians like to use this word. When they speak of a division algorithm, they are using unusual words to say a way to do division. For the present, we will restrict our divisors to two digits.

example 4.2 Divide: $\dfrac{251}{40}$

solution We will use the common division algorithm.

$$
\begin{array}{r}
6 \\
40\overline{)251} \\
240 \\
\hline
11\quad R
\end{array}
$$

Thus $\dfrac{251}{40} = \textbf{6 R11}$

example 4.3 Divide: $2183 \div 47$

solution We use the same algorithm.

$$
\begin{array}{r}
46 \\
47\overline{)2183} \\
188 \\
\hline
303 \\
282 \\
\hline
21\quad R
\end{array}
$$

Thus $\dfrac{2183}{47} = \textbf{46 R21}$

This book is designed to permit the reader to automate the upper-level skills of arithmetic while the concepts of algebra are being introduced. The addition, subtraction, multiplication, and division problems in the practice and problem sets are designed to provide paper and pencil practice with the four basic operations. Problems that should be worked with a calculator will be designated.

practice Multiply or divide as indicated. Do not use a calculator.

a. 300×125 **b.** $778 \cdot 563$ **c.** $19 \cdot 20 \cdot 21$

d. $\dfrac{261}{30}$ **e.** $12 \overline{\smash{\big)}\ 4908}$ **f.** $5623 \div 47$

problem set 4

1. A number has four digits. All the digits are 3 except for the thousands' digit, which is 7. What is the number?

2. A number has seven digits. All the digits are 3 except for the hundreds' digit and the thousands' digit, both of which are 9. What is the number?

3. Use digits to write forty-seven million, fourteen.

4. Use digits to write fourteen billion, forty-two thousand, seven hundred fifty-five.

5. Round 716,487,250 to the nearest ten million.

6. Round 716,487,250 to the nearest ten thousand.

7. Write 7650 in expanded form.

8. Write $(5 \times 1000) + (6 \times 10) + (7 \times 1)$ in standard form.

Divide:

9. $\dfrac{2511}{9}$ **10.** $2800 \div 50$ **11.** $45 \overline{\smash{\big)}\ 50{,}217}$

12. $\dfrac{9114}{7}$ **13.** $4165 \div 40$ **14.** $21 \overline{\smash{\big)}\ 30{,}215}$

Multiply:

15. $\begin{array}{r} 285 \\ \times\ 321 \\ \hline \end{array}$ **16.** $\begin{array}{r} 506 \\ \times\ 75 \\ \hline \end{array}$ **17.** $\begin{array}{r} 512 \\ \times\ 320 \\ \hline \end{array}$

18. $25 \times 40 \times 100$ **19.** 500×420 **20.** $6 \times 12 \times 24$

Find the missing digits:

21. $\begin{array}{r} 943 \\ -\ XYZ \\ \hline 274 \end{array}$ **22.** $\begin{array}{r} 605 \\ +\ MPQ \\ \hline 927 \end{array}$ **23.** $\begin{array}{r} X18K \\ -\ 2257 \\ \hline 925 \end{array}$

Use words to write each number:

24. 75400700215   **25.** 39002   **26.** 5000021

Add:

27. $\begin{array}{r} 408{,}627 \\ 915{,}634 \\ 589{,}062 \\ +\ 113{,}093 \\ \hline \end{array}$ **28.** $\begin{array}{r} 957{,}125 \\ 826{,}015 \\ 902{,}121 \\ +\ 313{,}947 \\ \hline \end{array}$

29. $73 + 816 + 92 + 47 + 321 + 5432$

30. $92 + 184 + 3182 + 915 + 21$

LESSON 5   *Addition and subtraction word problems*

The key to working problems that have an addition pattern or a subtraction pattern is recognizing that the problem has a particular pattern.

ADDITION PATTERN   SUBTRACTION PATTERN

$$
\begin{array}{r}
4 \\
+5 \\
\hline
\text{largest number} \rightarrow \quad 9
\end{array}
\qquad
\begin{array}{r}
9 \quad \leftarrow \text{largest number} \\
-4 \\
\hline
5
\end{array}
$$

Both patterns have three numbers. Word problems that have one of these patterns will give us two of the numbers. We put the given numbers in the pattern and add or subtract as required to find the third number in the pattern.

If a problem makes us think **some and then some more,** the pattern is an addition pattern. If a problem makes us think **some went away,** the pattern is a subtraction pattern. The words **how much greater** or **how much less** also indicate a subtraction pattern.

example 5.1   The Duke measured it and got 5260. Then the Count measured it and got 4720. How much greater was the Duke's measurement?

solution   The words *how much greater* tell us that the pattern is a subtraction pattern. We remember that the bottom number in a subtraction pattern is the difference. The largest number goes on top.

$$
\begin{array}{rl}
5260 & \text{Duke} \\
-\ 4720 & \text{Count} \\
\hline
540 & \text{difference}
\end{array}
$$

The Duke's measurement was 540 greater than the Count's measurement.

example 5.2   Four hundred seventeen birds were in the trees at 8 o'clock. By 9 o'clock the number had increased to nine hundred forty-two. How many birds came between 8 and 9 o'clock?

solution   This problem has the thought some and then some more. This means that the problem has an addition pattern. We know the top number and the bottom number.

$$
\begin{array}{rl}
417 & \text{at 8 o'clock} \\
+\ \ N & \text{birds came} \\
\hline
942 & \text{at 9 o'clock}
\end{array}
$$

This pattern is an addition pattern, but we must subtract to find the missing number.

$$
\begin{array}{rl}
942 & \\
-\ 417 & \\
\hline
525 &
\end{array}
\quad \rightarrow \quad
\begin{array}{rl}
417 & \\
+\ 525 & \text{birds came} \\
\hline
942 & \text{check}
\end{array}
$$

We find that **525 birds** came between 8 and 9 o'clock.

example 5.3   At sunrise many ducks were on the pond. Then four hundred twenty-five ducks flew away. Six hundred forty-two ducks remained on the pond. How many ducks were on the pond at sunrise?

solution   A problem that has the thought that some went away has a subtraction pattern. This problem says some flew away, so the pattern is a subtraction pattern.

$$
\begin{array}{r}
N \\
- \ 425 \\
\hline
642
\end{array}
\quad
\begin{array}{l}
\text{ducks on the pond at sunrise} \\
\text{flew away} \\
\text{ducks remaining}
\end{array}
$$

This is an easy pattern to solve because the two bottom numbers in a subtraction pattern can be added to find the top number in the pattern.

$$
\begin{array}{r}
425 \\
+ \ 642 \\
\hline
1067
\end{array}
\quad \rightarrow \quad
\begin{array}{r}
1067 \\
- \ 425 \\
\hline
642
\end{array}
\quad
\begin{array}{l}
\text{at sunrise} \\
\\
\text{check}
\end{array}
$$

There were **1067 ducks** on the pond at sunrise.

example 5.4   George had five hundred forty-two marbles. Then his friend gave him some marbles. Now he has nine hundred sixty-five marbles. How many marbles did his friend give him?

solution   This problem has the thought some and then some more. This means the pattern is an addition pattern.

$$
\begin{array}{r}
542 \\
+ \ \text{some} \\
\hline
965
\end{array}
\quad
\begin{array}{l}
\text{George} \\
\text{friend gave him} \\
\text{total}
\end{array}
$$

This is an addition pattern, but we must subtract to find the missing number. Then we check.

$$
\begin{array}{r}
965 \\
- \ 542 \\
\hline
423
\end{array}
\quad \longrightarrow \quad
\begin{array}{r}
542 \\
+ \ 423 \\
\hline
965
\end{array}
\quad
\begin{array}{l}
\text{George} \\
\text{friend gave him} \\
\text{check}
\end{array}
$$

George's friend gave him **423 marbles.**

practice   **a.**   Hundreds of knights came to the tournament. Then four hundred twenty knights went home. Seven hundred fifty-six knights remained. How many knights came to the tournament?

b.   Rita had nine thousand thirty-five items in her basement. Then she found some more items. Now she has ten thousand eighty-one items. How many items did she find?

problem set 5

1.   When the Duchess estimated it, she estimated 6190. Then the Countess estimated it and got 5320. How much greater was the Duchess' estimate?

2.   At 9 o'clock there were five hundred thirty dancers in the meadow. By 11 o'clock the number had increased to seven hundred seventy-eight. How many dancers came to the meadow between 9 and 11 o'clock?

3.   The estimated world population in the year 1650 was 550,000,000. The estimated world population in 1750 was 725,000,000. By how many people did the world's population increase between 1650 and 1750?

4.   A number has eight digits. All of them are 8 except the ten-thousands' digit, which is 3; the units' digit, which is 7; and the millions' digit, which is 6. What is the number?

5.   A number has four digits. All the digits are 3 except the thousands' digit, which is 7. What is the number?

6. A number has seven digits. All are 3 except for the hundreds' digit and the thousands' digit, both of which are 9. What is the number?

Divide:

7. $\dfrac{9300}{7}$ 8. $4165 \div 23$ 9. $21\,\overline{\smash{)}\,30{,}215}$

Multiply:

10.   285   11.   506   12.   512
     $\times\ 321$ $\times\ 275$ $\times\ 632$

Find the missing digits:

13.   *A6K*   14.   743   15.   *FHG*
     $-\ 526$ $-\ NGS$ $-\ 123$
     $\overline{437}$ $\overline{188}$ $\overline{289}$

16.   *K7S*   17.   643   18.   *ANX*
     $+\ 325$ $+\ ANX$ $+\ 727$
     $\overline{1001}$ $\overline{1112}$ $\overline{915}$

19. Write in standard form: $(7 \times 10{,}000) + (4 \times 100) + (3 \times 10)$

20. Write 2109 in expanded form.

21. Use digits to write forty-seven million, fourteen.

22. Use digits to write fourteen billion, forty-two thousand, seven hundred fifty-five.

23. Write these numbers in order from least to greatest: 916, 691, 619, 961, 196

Use words to write each number:

24. 5000021   25. 75400700215   26. 39002

27. Round 716,487,250 to the nearest ten million.

28. Round 716,487,250 to the nearest ten thousand.

Add:

29. $408{,}627 + 915{,}634 + 589{,}062$ 30.   10,326
                                                    9,012
                                                   91,526
                                               $+\ 47{,}319$

LESSON 6 *Decimal numbers*

6.A
decimal numbers

We have noted that whole numbers have a decimal point just after the last digit in the number. Sometimes the decimal point is written down. Many times it is not written but is understood to be there. Thus the two notations

615   615.

both designate the number six hundred fifteen. In the number on the left we did not write the decimal point, but in the number on the right we did write it.

In some numbers the decimal point is not at the end of the number. We often call these numbers **decimal numbers.** Some people call these numbers **decimal fractions** because they can be written as whole numbers divided by 10, 100, 1000, 10,000, or some other multiple of 10, as shown here.

$$61.23 = \frac{6123}{100}$$

The first place to the right of the decimal point in a decimal fraction is the tenths' place, or $\frac{1}{10}$. The next place is the hundredths' place, or $\frac{1}{100}$. The next place is the thousandths' place, or $\frac{1}{1000}$. And so on. The value of a digit is the digit times the place value. Thus, the 6 in

$$0.0006724$$

has a value of 6 times $\frac{1}{10,000}$, or 6 ten-thousandths, because it is in the ten-thousandths' place.

etc.	millions' place	hundred-thousands' place	ten-thousands' place	thousands' place	hundreds' place	tens' place	ones' place	decimal point	tenths' place	hundredths' place	thousandths' place	ten-thousandths' place	hundred-thousandths' place	millionths' place	etc.
	1,000,000	100,000	10,000	1000	100	10	1	.	$\frac{1}{10}$	$\frac{1}{100}$	$\frac{1}{1000}$	$\frac{1}{10,000}$	$\frac{1}{100,000}$	$\frac{1}{1,000,000}$	

6.B

reading decimal numbers

Although it is not necessary, we can use commas to the right of the decimal point to help us read decimal fractions. We begin on the right end of the number and move left, placing a comma after every three digits.

example 6.1   Place the commas in the following decimal numbers:

    (a) 0.0000416   (b) 0.003102   (c) 0.10705014

solution   **In each case we begin on the right end and move left.**

    (a) 0.0,000,416   (b) 0.003,102   (c) 0.10,705,014

If the number has digits on both sides of the decimal point, we use the above procedure to the right of the decimal point and begin again at the decimal point.

example 6.2   Place the commas in the following:

    (a) 4165283.61805   (b) 7324.0062582

solution   We begin on the right end and move left. Then we begin again at the decimal point and move to the left again.

    (a) 4,165,283.61,805   (b) 7,324.0,062,582

To read a decimal number, we begin on the left end and read according to the following procedure.

1. The digits to the left of the decimal point are read in the same way as they are read in whole numbers.
2. **The decimal point is read as *and*.**
3. **The digits to the right of the decimal point are read as if they formed a whole number, and this reading is followed by naming the *place* of the last digit in the number.**

Commas are placed between the words in the same locations that commas appear in the numbers.

We will demonstrate by reading several decimal fractions.

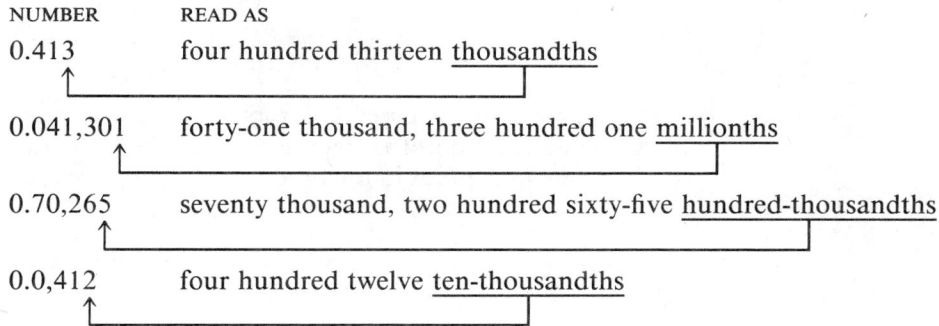

NUMBER	READ AS
0.413	four hundred thirteen <u>thousandths</u>
0.041,301	forty-one thousand, three hundred one <u>millionths</u>
0.70,265	seventy thousand, two hundred sixty-five <u>hundred-thousandths</u>
0.0,412	four hundred twelve <u>ten-thousandths</u>

In the following example we will read two decimal numbers that have nonzero digits on both sides of the decimal point.

example 6.3 Read the numbers: (a) 4165.0162   (b) 7108000.21578

solution First we insert the commas and then we read the numbers.

(a) 4,165.0,162   **Four thousand, one hundred sixty-five *and* one hundred sixty-two ten-thousandths**

(b) 7,108,000.21,578   **Seven million, one hundred eight thousand *and* twenty-one thousand, five hundred seventy-eight hundred-thousandths**

6.C
adding and subtracting decimal numbers

We add and subtract decimal numbers just as we do whole numbers. When we add and subtract decimal numbers, we must remember to write the numbers so that the decimal points are aligned one above the other.

example 6.4 Simplify: (a) 6.231 + 0.044   (b) 6.231 − 0.044

solution To begin, we write the numbers down with the decimal points one above the other. Then we add or subtract as indicated.

(a)   6.231
     + 0.044
     ───────
      6.275

(b)   6.2̸3̸1
     − 0.044
     ───────
      6.187

6.D
multiplying decimal numbers

We do not align the decimal points when we multiply decimal numbers. We multiply decimal numbers just as we multiply whole numbers. Then we add the number of decimal places in both numbers to find the position of the decimal point in the product.

example 6.5   Multiply: 4.12×63.2

solution   We multiply just as if the decimal points were not present.

$$
\begin{array}{r}
4.12 \\
\times\ 63.2 \\
\hline
824 \\
1236 \\
2472 \\
\hline
260384
\end{array}
$$

In the top number, the decimal point is two places from the right end. In the next number, the decimal point is one place from the right end. Two plus one equals three, so the decimal point in the product is placed three places from the right end.

260.384

practice   Use digits to write each number:

    a.   Five thousand and seven hundred forty-two ten-thousandths

    b.   Six hundred forty-two and seventy-five thousandths

Use words to write each number:

    c.   7000.065   **d.**   42.000617

Evaluate (find the value of):

    e. $44.0162 - 0.1420$ **f.** 5.22×0.064

problem set 6

  1.   Thirty-seven million, nine hundred eighteen thousand, five hundred is how much greater than nineteen million, ninety-nine thousand, nine?

  2.   The estimated population of the world in 1750 was 725,000,000. During the next 100 years the population increased by 475,000,000. What was the world's population in 1850?

  3.   A number has nine digits. All the digits are 7 except the millions' digit, which is 3; the ten-thousands' digit, which is 5; and the tens' digit, which is 9. What is the number?

  4.   A number has seven digits. All the digits are 6 except the hundred-thousands' digit, which is 2, and the thousands' digit, which is 4. What is the number?

  5.   A whole number has six digits. All digits are 3 except the ten-thousands' digit, which is 7, and the thousands' digit, which is 2. What is the number?

Multiply:

  6. 0.0732×1.63 **7.** 4.16×0.305 **8.** 41.06×0.0005

Subtract. Add to check.

  9. $14.03 - 0.0132$ **10.** $941.2 - 14.23$

Divide:

11. $\dfrac{3624}{23}$ **12.** $1275 \div 17$ **13.** $51\overline{)41{,}362}$

14. $27\overline{)2198}$ **15.** $2546 \div 41$ **16.** $\dfrac{92{,}438}{51}$

17. Round 5,143,782 to the nearest thousand.

18. Round 90,521,765 to the nearest ten thousand.

Use words to write each number:

19. 4165.0162   **20.** 504327.001510512

21. Use digits to write sixty-three thousand and two hundred fourteen ten-thousandths.

22. Use digits and a decimal point to write twenty-nine millionths.

Find the missing digits:

23. $\begin{array}{r} 625 \\ + \ ZAX \\ \hline 913 \end{array}$ **24.** $\begin{array}{r} 921 \\ - \ YAT \\ \hline 199 \end{array}$ **25.** $\begin{array}{r} XNT \\ - \ 763 \\ \hline 189 \end{array}$

Add:

26. $0.005 + 21.62 + 9.035 + 5165.2$ **27.** $70.02 + 0.0013 + 9.062 + 0.142$

28. Write 3917 in expanded form.

29. Write in standard form:
$$(9 \times 10{,}000) + (4 \times 1000) + (5 \times 100) + (7 \times 10) + (9 \times 1)$$

30. Multiply: $\begin{array}{r} 703 \\ \times 579 \\ \hline \end{array}$

LESSON 7   *Multiplying and dividing by powers of 10*

7.A
powers of 10

When we use 10 as a factor 2 times, the product is 100.

$$10 \times 10 = 100 \qquad \text{second power of 10}$$

When we use 10 as a factor 3 times, the product is 1000.

$$10 \times 10 \times 10 = 1000 \qquad \text{third power of 10}$$

When we use 10 as a factor 4 times, the product is 10,000.

$$10 \times 10 \times 10 \times 10 = 10{,}000 \qquad \text{fourth power of 10}$$

From this, we can see that the number of zeros in each product equals the number of times 10 is used as a factor. The number is called a **power of 10.** Thus, we see that the number

$$100{,}000{,}000$$

has eight 0s and must be the eighth power of 10. This is the product we get if 10 is used as a factor 8 times.

$$10 \times 10 \times 10 \times 10 \times 10 \times 10 \times 10 \times 10 = 100{,}000{,}000 \qquad \text{eighth power of 10}$$

When we multiply a number by a power of 10, all we do is move the decimal point to the right. The number of places the decimal point is moved equals the number of zeros in the power of 10.

example 7.1   Multiply: $47{,}162.314 \times 100$

solution   When we multiply by a power of 10, the digits will not change. The only change is in the position of the decimal point.

$$\begin{array}{r} 47{,}162.314 \\ \times 100 \\ \hline \mathbf{4{,}716{,}231.400} \end{array}$$

The decimal point moved two places to the right.

example 7.2   Multiply: $0.031652 \times 10{,}000$

solution   We see that 10,000 has four zeros. Thus, if we multiply a number by 10,000, all we will do is move the decimal point four places **to the right.** This time we will not show the multiplication but will just write the answer.

316.52

7.B
dividing by powers of 10

When we divide a number by a power of 10, we move the decimal point to the left. The number of places we move the decimal point equals the number of zeros in the power of 10.

example 7.3   Divide: $41.32 \div 1000$

solution   We will do the division this time.

$$\begin{array}{r} 0.04132 \\ 1000\,\overline{)\,41.32} \\ \underline{40\ 00} \\ 1\ 320 \\ \underline{1\ 000} \\ 3200 \\ \underline{3000} \\ 2000 \\ \underline{2000} \end{array}$$

All that happened was that the decimal point was moved three places **to the left.**

example 7.4   Divide: $48.512 \div 10{,}000$

solution   This time we will just write the answer. The number 10,000 has four zeros, so dividing by 10,000 will move the decimal point four places **to the left.**

0.0048512

practice   Do each multiplication or division problem mentally:

a. $4162 \cdot 100$ **b.** $4162 \div 100$

c. $73.426 \cdot 10{,}000$ d. $73.416 \div 10{,}000$

problem set 7

1. Hercules performed 1569 heroic exploits while Theseus performed only 1237. How many heroic exploits did they perform in all?

2. Saturday's football game was attended by 35,264 Hoosier fans and 17,927 Boilermaker fans. How many more Hoosier fans than Boilermaker fans attended the game?

3. A number has eight digits. All of them are 8 except the ten-thousands' digit, which is 3, and the units' digit, which is 7. What is the number?

4. A number has 12 digits. All of them are 7 except the ten-billions' digit, which is 4; the hundred-thousands' digit, which is 6; and the hundred's digit, which is 2. What is the number?

Simplify mentally:

5. 31.621×1000

6. $311.83615 \times 10{,}000$

7. $54.26 \div 1000$

8. $\dfrac{34{,}826}{10{,}000}$

Subtract. Add to check.

9. $1.416 - 0.0168$

10. $23.41 - 2.666$

11. $38.04 - 1.687$

Multiply:

12. 0.00413×0.312

13. 914.23×0.0132

14. 1.413×216

Divide:

15. $9016 \div 23$

16. $41 \overline{)74{,}316}$

17. $\dfrac{90{,}327}{43}$

18. $42{,}153 \div 19$

19. Round 91,648,573 to the nearest hundred thousand.

20. Round 84,165,812 to the nearest thousand.

21. Write these numbers in order from least to greatest: 642, 246, 264, 624, 426

22. Use digits and a decimal point to write four thousand, seven and nine thousand seven hundred forty-two hundred-thousandths.

23. Use digits and a decimal point to write seven hundred two and nine hundred forty-two hundred-thousandths.

Use words to write the following numbers.

24. 14372.015264

25. 9056213.00057328

Find the missing digits:

26.
$$\begin{array}{r} XYZ \\ -\ 493 \\ \hline 409 \end{array}$$

27.
$$\begin{array}{r} SFX \\ +\ 473 \\ \hline 1151 \end{array}$$

28.
$$\begin{array}{r} 126 \\ +\ AZKM \\ \hline 1152 \end{array}$$

Add:

29. $3.164 + 75.236 + 4328.914 + 508.21$

30. $3.0624 + 783.91 + 9053.216$

LESSON 8 *Dividing, ordering, and rounding decimal numbers*

8.A
dividing decimal numbers

To divide by a decimal number, we first move the decimal point in the divisor to the right as necessary to make the divisor a whole number. Then the decimal point in the dividend is moved to the right the same number of places. The decimal point in the answer is placed just above the decimal point in the dividend.

example 8.1   Divide: $0.004415 \div 0.032$

solution   First we record the numbers.

$$0.032 \overline{)0.004415}$$

Next we move the decimal point in the divisor to the right so that the divisor is a whole number. We use a caret (∧) to mark the old location of the decimal point.

$$\wedge032. \overline{)0.004415}$$

Now we must move the decimal point in the dividend the same number of places and write the decimal point for the answer just above it.

$$\wedge032 \overline{)\wedge004.415}$$

This result is untidy so we recopy it, omitting the carets and the extra 0s. Then we divide.

$$
\begin{array}{r}
0.137 \\
32 \overline{)4.415} \\
3\,2 \\
\hline
1\,21 \\
96 \\
\hline
255 \\
224 \\
\hline
31
\end{array}
$$

This division did not come out even. We decide to round the answer to two decimal places, and we get

0.14

8.B
ordering decimal numbers

The number ray can also be used to help us understand the way decimal numbers are arranged in order. If we want to order the numbers 4.36, 4.22, 4.65, and 4.56, we could graph the numbers on the number ray. We will show the portion of the number ray between 4 and 5.

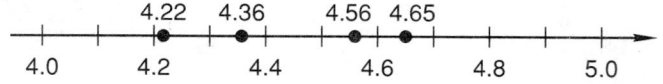

We note that the numbers are arranged in order from left to right. Whenever we graph numbers on a number ray, the graph will show the numbers arranged **in order** from left to right.

example 8.2   Order the numbers 0.426, 0.0732, 0.732, and 0.0426 from least to greatest.

solution   The digit to the left of the decimal point in all the numbers is zero. Next we consider the digits to the right of the decimal point. There are two numbers that begin with a zero to the right of the decimal point. They are

$$0.0732 \qquad \text{and} \qquad 0.0426$$

When we consider the second digits, we see that the first number is greater, so we reverse the positions of the numbers.

$$0.0426 \qquad \text{and} \qquad 0.0732$$

The next greatest digits are 4 and 7. This information lets us determine the final ordering of these numbers. We get

$$0.0426, \qquad 0.0732, \qquad 0.426, \qquad 0.732$$

8.C
rounding decimal numbers

The number ray can help us understand the process of rounding decimal numbers. If we graph 0.0423,

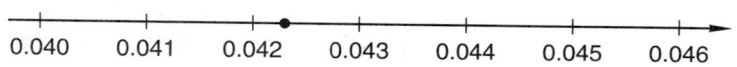

we can see that this number is closer to 0.042 than it is to 0.043. If we round 0.0423 to the nearest thousandth, we get

0.042

Since 0.0423 is closer to 0.04 than it is to 0.05, we can also round 0.0423 to the nearest hundredth, or

0.04

We find that the circle and arrow also can be used to explain how we round decimal numbers. The procedure is the same as the procedure we use for whole numbers.

example 8.3   Round 212.0165725 to the nearest ten-thousandth.

solution   First we circle the digit in the ten-thousandths' place. Then we mark the digit to its right with an arrow.

$$2\ 1\ 2.\ 0\ 1\ 6\ \circled{5}\overset{\downarrow}{7}\ 2\ 5$$

Next we change the arrow-marked digit and the digits to its right to zero.

$$2\ 1\ 2.\ 0\ 1\ 6\ \circled{5}0\ 0\ 0$$

The arrow-marked digit is greater than 5, so we increase the circled digit by 1.

$$2\ 1\ 2.\ 0\ 1\ 6\ \circled{6}0\ 0\ 0$$

and since terminal zeros (those on the end of the number) to the right of the decimal point have no value, we can omit them.

212.0166

example 8.4   Round 4057.2138362 to two decimal places.

solution   We begin by circling the second digit to the right of the decimal point and marking the next digit to the right with an arrow.

$$\overset{\downarrow}{4\ 0\ 5\ 7\ .\ 2\ \textcircled{1}\ 3\ 8\ 3\ 6\ 2}$$

Now we change the arrow-marked digit and the digits to its right to 0.

$$4\ 0\ 5\ 7\ .\ 2\ \textcircled{1}\ 0\ 0\ 0\ 0\ 0$$

Since the arrow-marked digit was less than 5, we do not change the circled digit. Also we discard the terminal zeros and get

4057.21

practice   **a.**   Round 416.042737 to the nearest hundred-thousandth.

b.   Round 4375.23562 to the nearest thousandth.

c.   Round 2837.065248 to the nearest ten-thousandth.

d.   Graph these numbers on a number ray: 0.423, 0.445, 0.453, 0.460

**problem set
8**

1.   The first measurement was fourteen million, six hundred forty-two units. The second measurement was thirty-two million, fifteen thousand, thirty-two units. How much greater was the second measurement?

2.   Plato delivered 1743 orations, whereas Aristotle delivered only 1234. How many orations did they deliver in all?

Simplify mentally:

3. $\dfrac{41,362.68}{100}$

4. 305.2165×100

5. $9315.21 \div 1000$

6. $32.1652 \times 10,000$

Multiply:

7. 0.00526×3.14

8. 2.315×413

9. 0.00312×0.642

10. 313.65×0.0147

Subtract. Add to check.

11. $392.163 - 4.077$

12. $3.2421 - 1.363$

Find the missing digits:

13. $\begin{array}{r} AZ.K \\ + \ 13.4 \\ \hline 28.9 \end{array}$

14. $\begin{array}{r} 604.13 \\ + \ FNS.ZK \\ \hline 1112.01 \end{array}$

15. $\begin{array}{r} AX.M \\ - \ 34.5 \\ \hline 37.8 \end{array}$

16. $\begin{array}{r} 116.04 \\ - \ X.FS \\ \hline 107.06 \end{array}$

Divide and round to two decimal places:

17. $\dfrac{14.045}{0.014}$

18. $0.0020 \div 0.013$

19. $\dfrac{321.4}{0.071}$

20. $3.22 \div 0.0022$

21.   Round 42.12345678 to the nearest millionth.

22.   Round 31.6372052 to two decimal places.

23. Use digits and a decimal point to write the number forty-seven billion, sixty-seven thousand and four hundred seventeen hundred-thousandths.

24. Use digits and a decimal point to write the number one thousand three and four thousand seven hundred forty-two hundred-thousandths.

Use words to write each number:

25. 0.00006184

26. 4000062.0130023

27. Order these decimal numbers from least to greatest:
0.0426   0.0164   0.0461   0.0614

Add:

28. 7852.165 + 7186.132 + 9185.624

29. 42.16 + 0.0032 + 3.165 + 305.321

30. 0.0016 + 32.1005 + 9.0312 + 0.00324

LESSON 9   *Points, lines, and rays · Angles · Perimeter*

9.A
points, lines, and rays

Here we show a series of dots, each one smaller than the one to its left.

If we continue drawing the dots with each one smaller than the one to its left, we would finally have a dot so small that it could not be seen without magnification. **This dot would still be larger than a mathematical point because a mathematical point is so small that it has no size at all.**

A curve is an endless connection of mathematical points. A line is a straight curve. Any two points on a line can be used to name a line. Because a line is made of mathematical points, a line has no width. When we draw a line with a pencil, the line we draw is a graph of the mathematical line and marks the location of the mathematical line. We sometimes put arrowheads on the ends of lines to emphasize that the lines continue without end in both directions.

Any two points on a line can be used to name a line. It is customary to use an overbar with two arrowheads to indicate a line. Any of the following notations designate the line shown above.

$$\overleftrightarrow{AB} \quad \overleftrightarrow{BA} \quad \overleftrightarrow{AM} \quad \overleftrightarrow{MA} \quad \overleftrightarrow{MB} \quad \overleftrightarrow{BM}$$

These notations are read as "line *AB*," "line *BA*," "line *AM*," "line *MA*," "line *MB*," and "line *BM*." Of course, all these name the same line.

We use the overbar with no arrowheads to name a line segment between the

two points. We would use either

$$\overline{AB} \quad \text{or} \quad \overline{BA}$$

to name the line segment between points A and B.

A **ray** is sometimes called a **half line.** A ray begins at one point and continues without end through the other point. A single-arrowhead overbar is often used to name a ray. We could call the ray from the preceding figure either

$$\overrightarrow{PQ} \quad \text{or} \quad \overrightarrow{PX}$$

9.B
angles

An **angle** is formed by the intersection of two rays. Some people define an angle to be the opening between the two rays. Others say that the angle is the rays themselves. The **vertex** of an angle is the point of intersection of the two rays.

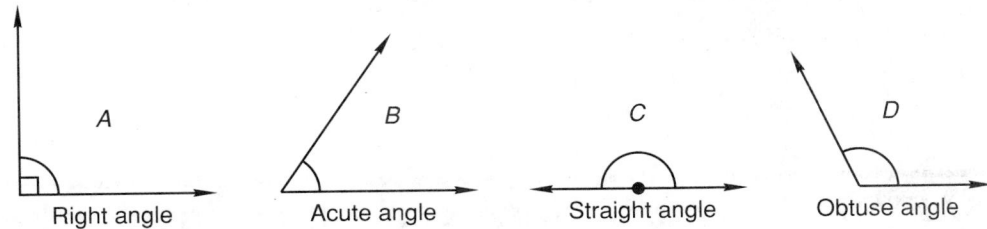

Angle A is formed by two perpendicular rays. We say the angle A is a **right angle.** The little square drawn in this angle tells us that the angle is a right angle, or a 90° angle. Angles smaller than right angles are called **acute angles.** If the two rays point in opposite directions, we say that the angle formed is a **straight angle.** A straight angle is a 180° angle. Angle C is a straight angle. An angle such as angle D that is greater than a right angle but less than a straight angle is an **obtuse angle.**

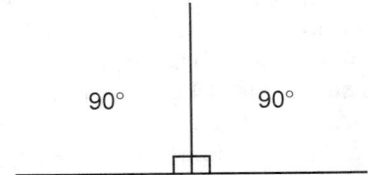

In this figure we note that a straight angle can be formed by two 90° angles. From this we see that a straight angle is a 180° angle.

9.C
perimeter

The word **perimeter** comes from the Greek prefix *peri-*, which means "around" and the Greek word *metron,* which means "measure." Thus, perimeter means the measure around or the distance around.

example 9.1

Find the perimeter of this figure. The dimensions are in feet. All angles are right angles.

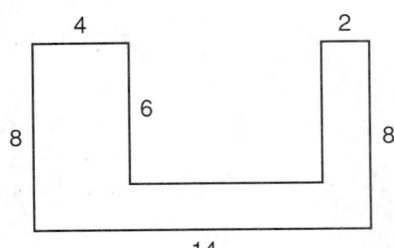

solution Several lengths are not given. Since all angles are right angles, we can determine the missing lengths. The dip on top goes down 6 feet so it must go up 6 feet. Finally, since it is 14 feet across the bottom, the missing length on top must be 8 feet.

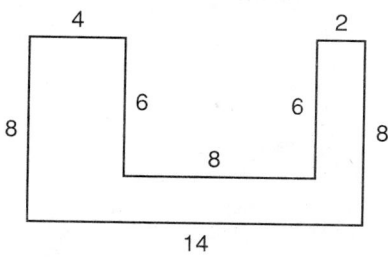

The perimeter is the distance around, or the sum of these lengths.

$$8 + 4 + 6 + 8 + 6 + 2 + 8 + 14 = \textbf{56 feet}$$

practice Find the perimeter of this figure. All angles are right angles. Dimensions are in centimeters.

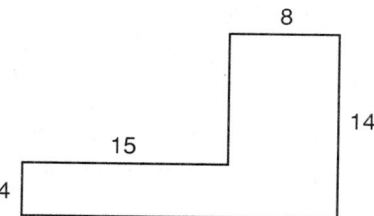

problem set 9

1. Thousands of people attended the music conference. If 35,000 people attended the conference and 19,763 were not musicians, how many musicians attended the conference?

2. Bosch prepared 23,215 canvasses, whereas Picasso prepared only 16,219. How many more canvasses did Bosch prepare?

Simplify mentally:

3. $\dfrac{3164.215}{100}$

4. 3164.215×100

5. $\dfrac{417,365.20}{1000}$

6. $2.1532 \times 10,000$

Multiply:

7. 0.0316×2.4

8. 2.862×0.013

9. 0.08421×0.22

10. 8.123×3.13

Subtract. Add to check.

11. $3.065 - 1.423$

12. $43.24 - 0.000613$

13. $3.065 - 0.2121$

Find the perimeter of each figure. Dimensions are in inches. All angles are right angles.

14.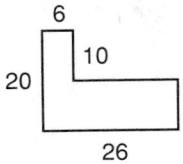

15.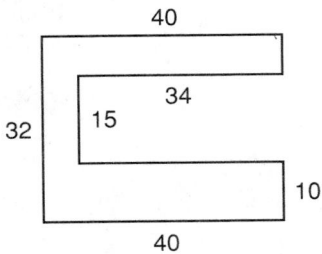

Divide. Round each answer to two decimal places.

16. $0.002215 \div 0.042$ **17.** $\dfrac{16.032}{0.024}$ **18.** $0.0040 \div 0.024$

19. $\dfrac{416.5}{0.073}$ **20.** $3.44 \div 0.0066$

21. Round 61.373737842 to the nearest ten-millionth.

22. Round 433.6851472 to five decimal places.

23. Use digits and a decimal point to write the number seven hundred forty-two million, five hundred thirty-seven and ten thousand nine hundred forty-eight millionths.

24. Use digits and a decimal point to write the number one thousand seven hundred forty-eight ten-millionths.

Use words to write each number:

25. 0.00128647   **26.** 27000316.08156   **27.** 51786.00785

Add:

28. $904.682 + 513.976 + 214.685$ **29.** $4293.015 + 2172.062 + 5091.799$

30. Find the missing digits:
$$
\begin{array}{r}
204.63 \\
-\ AZK.NS \\
\hline
39.67
\end{array}
$$

LESSON 10 *Divisibility*

The number 2 will divide into the number 30 and will have no remainder.

$$\frac{30}{2} = 15$$

Thus we say that 2 is a divisor of 30 and also that 30 is divisible by 2. Other divisors of 30 are 3, 5, 6, 10, and 15.

$$\frac{30}{3} = 10 \qquad \frac{30}{5} = 6 \qquad \frac{30}{6} = 5 \qquad \frac{30}{10} = 3 \qquad \frac{30}{15} = 2$$

To be called a divisor, a number must be a whole number. If we list all the divisors of 30, we get

1, 2, 3, 5, 6, 10, 15, and 30

It is often helpful to know if a whole number is divisible by 2, 3, 5, or 10. There are rules that we can use to find out. There are also rules to test for divisibility by 4 and 8 and a few other numbers. These rules are not used as often as the rules for divisibility by 2, 3, 5, and 10. Thus we will concentrate on the rules for these four numbers.

The rules are:

1. A whole number is divisible by 2 if its last digit is either 0, 2, 4, 6 or 8.
2. A whole number is divisible by 10 if its last digit is 0.
3. A whole number is divisible by 5 if its last digit is either 5 or 0.
4. A whole number is divisible by 3 if the sum of its digits is divisible by 3.

example 10.1 Which of these numbers is divisible by 3?
(a) 99   (b) 1239   (c) 1561

solution We will add the digits in each number and divide the sum by 3.

(a) 99	$9 + 9 = 18$	$\dfrac{18}{3} = 6$
(b) 1239	$1 + 2 + 3 + 9 = 15$	$\dfrac{15}{3} = 5$
(c) 1561	$1 + 5 + 6 + 1 = 13$	$\dfrac{13}{3} = $ not divisible

The numbers (a) **99** and (b) **1239 are divisible by 3.** The number 1561 is not divisible by 3 because 13 is not divisible by 3.

example 10.2 Which of these numbers is divisible:
(a) by 2?   (b) by 5?   (c) by 10?

4   32   75   99   4165   4020

solution (a)   Any even number is divisible by 2 so 2 is a divisor of **4, 32,** and **4020** because the last digit of each of these numbers is an even digit.

(b)   The numbers that are divisible by 5 have either a 5 or 0 as their last digit. These numbers are

75,   4165,   and   4020

(c)   To be divisible by 10, the last digit must be 0. Thus **4020** is the only number that is divisible by 10.

practice **a.** Which of these numbers is divisible by 3?
(1) 47285   (2) 45285   (3) 305961

b. Which of these numbers is divisible by 5?
(1) 235725   (2) 84603   (3) 72840

c. Which of these numbers is divisible by 10?
(1) 235725   (2) 84603   (3) 72840

problem set 10 **1.** In all there were two million, three hundred thousand, six hundred nineteen people in the city. If one million, two hundred nineteen thousand, three hundred twelve of them were adolescents, how many were not adolescents?

2. Aristotle was called the peripatetic philosopher because he walked as he talked. He walked 1627 meters the first week and 2941 meters the second week. How much farther did he walk the second week?

3. Use digits to write the number seven hundred sixty-two million, four hundred forty-two and twelve thousand, seven hundred ninety-two hundred-thousandths.

4. Use digits to write the number fourteen thousand, seven hundred two millionths.

Find the perimeter of each figure. Dimensions are in meters. All angles are right angles.

5.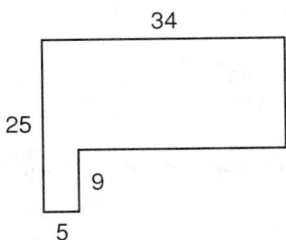

6.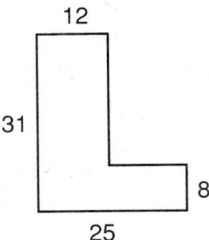

Multiply:

7. 0.0352×2.24 8. 305×2.42

9. 3.062×410 10. 3.06×4.18

Subtract. Add to check.

11. $4.016 - 3.217$ 12. $23.21 - 0.0034$

Find the missing digits:

13.   2.049   14.   *FXY.ZK*
     $-$ *N.AZK* $+ 963.09$
     ‾‾‾‾‾‾‾   ‾‾‾‾‾‾‾‾
       0.684   1750.24

Divide. Round each answer to two places.

15. $0.0030 \div 0.031$ 16. $0.003326 \div 0.021$ 17. $\dfrac{18.034}{0.047}$

18. $2.77 \div 0.0055$ 19. $\dfrac{428.3}{0.067}$

20. Use the rules for divisibility to tell which of the following numbers is divisible:
    (a) by 2   (b) by 3   (c) by 5   (d) by 10

          235   300   4888   9132   72,654

21. Round 4283.52162 to the nearest hundred.

22. Round 478.64385 to three decimal places.

Simplify mentally:

23. $47,123 \div 1000$ 24. 40.265×1000 25. $\dfrac{0.00143}{100}$

Use words to write each number:

26. 472.058003   27. 5000162.0008

28. Order from least to greatest: 0.417, 0.341, 0.471, 0.714, 0.704

Add:

29. 723.528 + 804.526 + 613.912 + 844.504

30. 0.00032 + 416.52 + 3.006 + 215.006

LESSON 11 *Word problems about equal groups*

Some word problems about equal groups can be solved by multiplying. Some word problems about equal groups can be solved by dividing. **The key to working these problems is to recognize that the problems are about equal groups.** To illustrate the pattern, we will show three stacks of pancakes. Each stack has 4 pancakes, so there are 12 pancakes in all.

If we use addition for this problem, we write

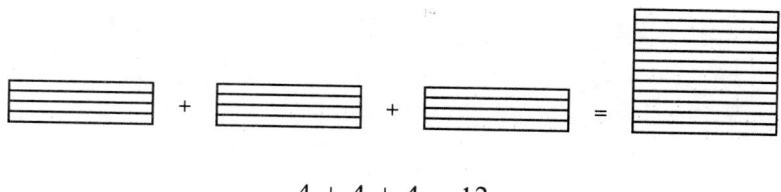

$$4 + 4 + 4 = 12$$

We can also use multiplication and write

$$\boxed{4} \times \boxed{3} = \boxed{12}$$

This is the multiplication/division pattern. The first number is the number in each equal group. The second number is the number of groups. The third number is the total.

$$\boxed{\text{Number in each group}} \times \boxed{\text{number of groups}} = \boxed{\text{total}}$$

If we know the first two numbers, we multiply to find the total. If we know the total and one of the first two numbers, we divide to find the other number.

example 11.1 Forty-two students can ride on each bus. If there are 12 buses, how many students can ride to the game?

solution **If we recognize that the problem is about equal groups, the problem is easy to solve. All we have to do is use the pattern.** Forty-two students are in each equal group. There are 12 equal groups. We have

$$\boxed{42} \times \boxed{12} = \boxed{\text{total}}$$

Number in one Number of
equal group   equal groups

We multiply 42 by 12 to find the total.

$$
\begin{array}{r}
42 \\
\times 12 \\
\hline
84 \\
42 \\
\hline
504 \quad \text{total}
\end{array}
$$

There are **504 students** who can ride to the game.

example 11.2 Forty-two students can ride on each bus. There are 1264 students who want to go to the game. How many buses are needed to take all the students to the game?

solution Aha! This is an equal groups problem. All we have to do is put the numbers into the pattern. There are 42 in each group and there are 1264 total students.

$$\boxed{42} \times \boxed{N} = 1264$$

We divide to find N, which is the number of equal groups.

$$
\begin{array}{r}
30 \quad \text{buses} \\
42 \overline{)1264} \\
\underline{126} \\
4
\end{array}
$$

We can fill 30 buses. However, we need **31 buses** because we can't leave the 4 extra students behind.

example 11.3 Five hundred sixty students will be evenly divided between 42 buses. How many students will ride on each bus?

solution We see that this is an equal groups problem. All we have to do is to use the pattern.

$$\boxed{N} \times \boxed{42} = 560$$

To find the number of students on each bus, we divide 560 by 42.

$$
\begin{array}{r}
13 \\
42 \overline{)560} \\
\underline{42} \\
140 \\
\underline{126} \\
14
\end{array}
$$

We cannot put an equal number on each bus. If we put 13 on each bus, there will be 14 left over. So we put one extra student on 14 buses.

14 buses will have 14 students

28 buses will have 13 students

practice **a.** Forty kids stood patiently in each line. There were 35 lines of kids. How many kids were standing patiently in all the lines?

b. Forty kids could crawl into one space. If there were 1600 kids, how many spaces would it take to hold them all?

c. There were 148 kids that had to be crowded evenly into 8 spaces. How many kids would be in each space?

problem set 11

1. Fifty players could ride each bus. If there were 15 buses, how many players could ride to the tournament?

2. Six hundred sixty students will be evenly distributed between 15 classes. How many students will be in each class?

3. On the first jump the frog jumped nineteen and eight hundred sixty-three ten-thousandths inches. On the next jump the frog jumped twenty-four and four thousand, five hundred six ten-thousandths inches. How far did the frog jump in all?

4. Of the forty-three thousand, two hundred nineteen who attended, twenty-six thousand, three hundred fourteen were multilingual. How many were not multilingual?

Find the perimeter of each figure. Dimensions are in <u>inches</u>. All angles are right angles.

5.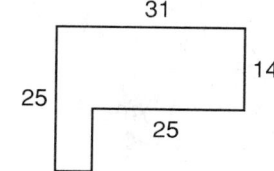

6.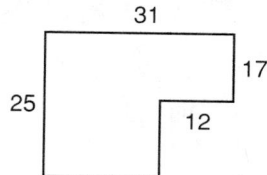

Simplify mentally:

7. $132,116 \div 10,000$ 　　8. 136.131×1000 　　9. $\dfrac{123.631}{1000}$

Multiply:

10. 162×2.25 　　　　　　　　11. 1.811×20.1

12. 6.61×3.16 　　　　　　　　13. 30.5×1.62

Subtract. Add to check.

14. $6.316 - 3.124$ 　　15. $61.81 - 0.0012$ 　　16. $129.631 - 2.48$

17. $2.11 - 1.031$

18. Use the rules for divisibility to tell which of the following numbers is divisible:
 (a) by 2　　(b) by 3　　(c) by 5　　(d) by 10
 　　　　6132　　6325　　9130　　6111　　6130

Divide. Round each answer to two decimal places.

19. $\dfrac{411.23}{61}$ 　　　　　　　　20. $0.0016 \div 0.011$

21. $\dfrac{11.031}{3.1}$ 　　　　　　　　22. $\dfrac{0.123}{0.013}$

23. Round 223.092870 to the nearest hundredth.

24. Round 1,621.32161 to the nearest hundred.

25. Use digits to write the number one trillion, six hundred twenty-five billion, two hundred fifty thousand, twenty-five and one hundred twenty-three thousandths.

Use words to write each number:

26. 123.9621

27. 223092870

Add:

28. 1135.62 + 32.61 + 311.82 + 99.01

29. 1132.8 + 6251.6 + 312.1 + 11.3

30. Find the missing digits:

$$\begin{array}{r} AZ.KN \\ -\ 22.49 \\ \hline 68.73 \end{array}$$

LESSON 12 *Prime numbers and composite numbers ·*
Products of primes

12.A
prime numbers and composite numbers

The number 6 can be composed by multiplying the two whole numbers 3 and 2.

$$3 \times 2 = 6$$

We say that 6 is a **composite number** because it can be composed by multiplying two other whole numbers. The number 21 is also a composite number. We can compose 21 by multiplying 3 and 7.

$$3 \times 7 = 21$$

The number 11 can be composed in only one way. That way is to multiply 11 by 1.

$$11 \times 1 = 11$$

There are many other numbers whose only factors are the numbers themselves and the number 1. Some are

$$5 \times 1 = 5 \qquad 23 \times 1 = 23 \qquad 31 \times 1 = 31$$

There is no other way to compose 5, 23, or 31 by multiplying. We call these numbers **prime numbers.**

> PRIME NUMBERS
>
> A prime number is a whole number greater than 1 whose only whole number divisors are 1 and the number itself.

example 12.1 Write the whole numbers 1 through 40 and circle the prime numbers. Do not circle 1, because 1 is not a prime number.

solution We begin with 2 and circle the prime numbers. Note that there are no even numbers circled except 2 because every other even number has 2 as a factor.

1 (2) (3) 4 (5) 6 (7) 8 9 10 (11) 12 (13) 14 15 16 (17) 18 (19) 20
21 22 (23) 24 25 26 27 28 (29) 30 (31) 32 33 34 35 36 (37) 38
39 40

12.B products of primes

Sometimes it is necessary to write a composite number as a product of prime numbers. We can write 12 as a product of prime numbers as

$$2 \times 2 \times 3$$

Here we found the prime factors of 12 by inspection. However, if we wish to write large numbers such as 84 or 1260 as products of prime numbers, it is nice to have a procedure to follow. A procedure often used is to divide the given number by the prime numbers 2, 3, 5, etc., until the prime factors are found. To find the prime factors of 84, we begin by dividing by 2.

$$\frac{84}{2} = 42$$

Now 42 can be divided by 2 so we divide again.

$$\frac{42}{2} = 21$$

and 21 can be divided by 3

$$\frac{21}{3} = 7$$

If we string these steps together, we get

$$\frac{84}{2} \longrightarrow \frac{42}{2} \longrightarrow \frac{21}{3} \longrightarrow 7$$

So we can write 84 as a product of prime numbers as

$$2 \times 2 \times 3 \times 7$$

Many people use the following division format to find the prime factors of a number.

$$\begin{array}{r|l} 2 & 84 \\ \hline 2 & 42 \\ \hline 3 & 21 \\ \hline & 7 \end{array} \qquad \text{so } 84 = 2 \times 2 \times 3 \times 7$$

We did the same divisions as before, but this time we used a more compact format.

example 12.2 Write 1260 as a product of prime numbers.

solution We will work the problem twice to demonstrate both formats.

$$\frac{1260}{2} \longrightarrow \frac{630}{2} \longrightarrow \frac{315}{5} \longrightarrow \frac{63}{3} \longrightarrow \frac{21}{3} \longrightarrow 7$$

Now we use the other format.

```
2 | 1260
2 |  630
5 |  315
3 |   63
3 |   21
         7
```

Both formats yield the same answer.

$$1260 = 2 \times 2 \times 3 \times 3 \times 5 \times 7$$

practice Write each number as a product of prime numbers:

a. 120   **b.** 640   **c.** 2520

problem set 12

1. The circumference of the great wheel was twelve thousand and forty-one thousandths inches. The circumference of the lesser wheel was only one thousand twenty-one and two hundred-thousandths inches. How much larger was the great wheel?

2. When the smoke cleared, the judges found that on the first try Roger had covered fourteen million, seven hundred sixty-two and seventy-five ten-thousandths units. The second try was only eight hundred forty-two thousand, fifteen and seven thousandths units. What was the sum of both tries?

3. Five pounds of beans will fit in each container. How many containers are needed for 740 pounds of beans?

4. Six hundred two fanatic fans wish to travel to the game on 25 buses. If the fans are divided as evenly as possible, how many fanatic fans will travel on each bus?

5. Use the rules for divisibility to determine which of the following numbers is divisible:
   (a) by 2   (b) by 3   (c) by 5   (d) by 10

   625   302   9172   3132   62,120

Find the perimeter of each figure. Dimensions are in feet. All angles are right angles.

6.   7.

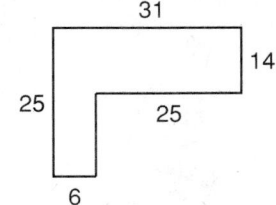   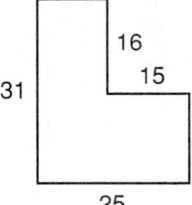

Simplify mentally:

8. 91,865 ÷ 100   9. 36.8211 × 1000

10. What is the sum of the prime numbers that are greater than 3 and less than 28?

11. What is the sum of the eight smallest prime numbers?

Multiply:

12. 3.16 × 71.3   13. 913 × 6.19   14. 0.0316 × 72.1

15. Write each number as a product of primes:
    (a) 95   (b) 720   (c) 2862

Subtract. Add to check.

16. $162.133 - 0.0123$ **17.** $1.329 - 0.999$

18. Find the missing digits:
$$\begin{array}{r} \$AX.KN \\ - \$41.29 \\ \hline \$\ 9.93 \end{array}$$

Divide and round each answer to two decimal places:

19. $18.621 \div 6.1$ **20.** $\dfrac{32.631}{0.03}$

21. $\dfrac{3012.3}{12}$ **22.** $\dfrac{621}{3.1}$

23. Round 4692.83215 to the nearest hundred.

24. Round 4113.62185 to two decimal places.

25. Use digits to write the number nine hundred sixty-one billion, three hundred thirteen million, twenty-five.

Use words to write the following numbers:

26. 0.001621   **27.** 16.0562   **28.** 6231562.01

Add:

29. $931.62 + 621.73 + 631.81 + 713.13$

30. $0.0031 + 612.13 + 0.721 + 16.11$

LESSON 13 *Three-digit divisors · Multiplication word problems*

13.A

three-digit divisors

In division problems whose divisors contain three or more digits, we use the same procedure that we use in problems in which the divisors have only two digits.

example 13.1 Divide: $0.41623 \div 0.0215$

solution The decimal point in 0.0215 must be moved four places to the right to make a whole number. Thus, the decimal point in 0.41623 is also moved four places to the right.

$$\begin{array}{r} 19.359 \\ 215\,\overline{)\,4162.300} \\ \underline{215} \\ 2012 \\ \underline{1935} \\ 773 \\ \underline{645} \\ 1280 \\ \underline{1075} \\ 2050 \\ \underline{1935} \end{array}$$

We round the answer to two decimal places and get **19.36.**

13.B
multiplication word problems

Some word problems require that numbers be multiplied to find the answer. These word problems sometimes contain the word **product.** Many of them contain the word **times** used in a phrase such as "5 times as many."

example 13.2   The second game score was 25 times the score of the first game. If Harriet had scored 14,025 points in the first game, how many points did she score in the second game?

solution   The word **times** tells us to multiply.

$$25 \times 14{,}025 = \textbf{350,625}$$

example 13.3   Roger scored 26,142 points in the first game. He scored 7 times this many points in the second game. How many points did he score in all?

solution   The word **times** tells us to multiply, and the words **in all** tell us to add.

$$26{,}142 \times 7 = \begin{array}{ll} 26{,}142 & \text{first game} \\ \underline{182{,}994} & \text{second game} \\ 209{,}136 & \text{in all} \end{array}$$

He scored **209,136** points in all.

practice   Inflation caused the price of everything to increase. The price of an item after inflation was 4 times the price before inflation. If an item cost $5.67 before inflation, what did it cost after inflation?

problem set 13

1. The pinball wizard set a new record on a pinball machine. Her new record score was 26 times the previous high score of 79,864 points. What was her new record score?

2. If 100 pounds of pinto beans fit in one bag, how many bags are needed to hold 26,000 pounds of pinto beans?

3. The first one measured fourteen million, seven hundred forty-two thousand and seventeen hundred-thousandths. The second one measured only eight hundred thousand and forty-two millionths. By how much was the first one larger?

4. Find the sum of the prime numbers that are between 12 and 42.

5. Use the rules for divisibility to determine which of the following numbers is divisible:
   (a) by 2   (b) by 5   (c) by 3   (d) by 10

   | 1020 | 125 | 130 | 1332 | 185 | 132 |

Find the perimeter of each figure. Dimensions are in centimeters. All angles are right angles.

6.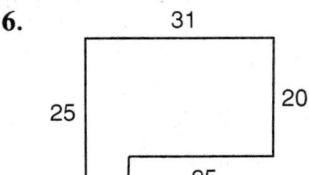

7.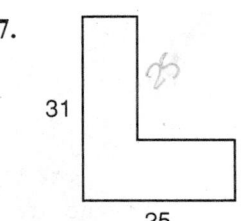

Simplify mentally:

8. 31,621 ÷ 1000   **9.** 311.836152 × 10,000

Multiply:

10. 0.0121 × 62.1   **11.** 621 × 8.11   **12.** 2.28 × 22.4

13. Write each number as a product of primes:
(a) 360   (b) 720   (c) 1440

Subtract. Add to check:

14. 16.162 − 12.373   **15.** 1.6132 − 0.1316   **16.** 17.132 − 1.693

Find the missing digits:

17.
$$\begin{array}{r} \$643.28 \\ - \ \$NFK.AZ \\ \hline \$257.29 \end{array}$$

18.
$$\begin{array}{r} 1019.05 \\ + \ SFAN.ZK \\ \hline 2364.41 \end{array}$$

Divide and round the answers to two decimal places:

19. $\dfrac{0.41623}{0.0215}$ **20.** 19.312 ÷ 0.061

21. $\dfrac{311.12}{1.2}$ **22.** $\dfrac{621}{0.31}$

23. Round 1231.62567 to three decimal places.

24. Round 6,469,693,230 to the nearest ten million.

25. Use digits to write the number three hundred twenty-one million, six hundred seventeen thousand, two hundred twelve and two hundred thirty-one thousandths.

Use words to write each number:

26. 161.016   **27.** 613.162   **28.** 111111112

29. Add: 621.81 + 31.62 + 62.11 + 12.61

30. Order from least to greatest: 1.119, 1.191, 1.911, 1.091, 0.901

LESSON 14 *Fractions · Expanding and reducing fractions*

14.A
fractions

When we write two numbers vertically and draw a line between them, we have written a fraction. Thus 3 over 4, written as follows,

$$\begin{array}{l} \text{Numerator} \quad \longrightarrow \\ \text{Denominator} \quad \longrightarrow \end{array} \ \dfrac{3}{4} \ \longleftarrow \ \text{Fraction line}$$

is a fraction. We call the top number the **numerator** of the fraction. We call the bottom number the **denominator** of the fraction. We call the line between the numbers a **fraction line.**

Fractions are used to designate parts of a whole. The bottom of the fraction tells us how many parts there are in all. The top of the fraction tells us how many of these parts we are considering. Here we show what we mean when we write $\frac{1}{4}$, $\frac{2}{4}$, $\frac{3}{4}$, and $\frac{4}{4}$.

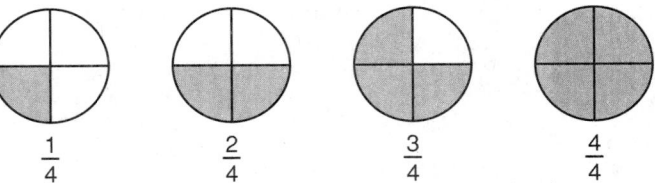

If two fractions have the same value, we say that the fractions are **equivalent fractions.** The two fractions

$$\frac{1}{2} \quad \text{and} \quad \frac{2}{4}$$

are equivalent fractions, for they have the same value. We can show this by drawing a picture that represents each of these fractions. First, we draw two circles. We divide the first circle into two equal parts, or halves. Then we divide the second circle into four equal parts, or fourths.

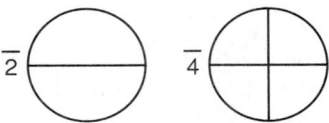

On the left we will shade one of the halves to represent $\frac{1}{2}$, and on the right we will shade two of the fourths to represent $\frac{2}{4}$.

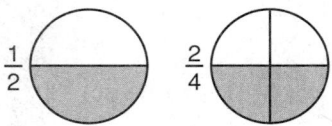

We see that we have shaded equal parts of each whole. We say that $\frac{1}{2}$ and $\frac{2}{4}$ are equivalent fractions because they represent equal parts of the whole and thus have the same value. **Thus, one-half and two-fourths are different names for the same number.**

14.B
expanding and reducing fractions

Any number divided by itself has a value of 1. All of the following have a value of 1.

$$\frac{3}{3} \qquad \frac{1.743}{1.743} \qquad \frac{\frac{1}{2}}{\frac{1}{2}} \qquad \frac{\frac{7}{3}}{\frac{7}{3}} \qquad \frac{5}{5} \qquad \frac{1.42}{1.42}$$

We can change the name of any fraction by multiplying or dividing both the numerator and the denominator by the same number. We will call this rule the **denominator-numerator rule for fractions.**

DENOMINATOR-NUMERATOR RULE FOR FRACTIONS

1. The bottom and top of a fraction can be multiplied by the same number (except zero) without changing the value of the fraction.
2. The bottom and top of a fraction can be divided by the same number (except zero) without changing the value of the fraction.

When we change the name of a fraction by making the denominator larger, we say that we have **expanded** the fraction. When we change the name of a fraction by making the denominator smaller, we say we have **reduced** the fraction. When we have reduced a fraction so that no prime number will divide both the numerator and denominator, we say that we have reduced the fraction to **lowest terms.**

example 14.1 Expand the fraction $\frac{3}{4}$ so that the denominator is 24.

solution If we multiply 4 by 6, we get 24. The denominator-numerator rule says we must also multiply 3 by 6.

$$\frac{3}{4} \cdot \frac{6}{6} = \frac{18}{24}$$

The number 18 over 24 is another way to write the number 3 over 4. Both numerals have the same value, so they represent the same number.

example 14.2 Write each number with a denominator of 20:

(a) 5   (b) $\frac{3}{5}$

solution (a) Five can be written as 5 over 1. We will multiply above and below by 20.

$$\frac{5}{1} \cdot \frac{20}{20} = \frac{100}{20}$$

(b) We will multiply 3 over 5 by 4 over 4.

$$\frac{3}{5} \cdot \frac{4}{4} = \frac{12}{20}$$

example 14.3 Reduce $\frac{30}{45}$ to lowest terms.

solution As the first step, we divide the top and the bottom of the fraction by 5.

$$\frac{\overset{6}{\cancel{30}}}{\underset{9}{\cancel{45}}}$$

Next we divide the top and bottom of the new fraction by 3.

$$\frac{\overset{2}{\cancel{\overset{6}{\cancel{30}}}}}{\underset{3}{\cancel{\underset{9}{\cancel{45}}}}} = \frac{2}{3}$$

This fraction is an equivalent fraction to the original fraction. The fraction is now in lowest terms because 2 and 3 do not have a common factor.

practice   Write each number with a denominator of 18:

a. 3   **b.** $\frac{1}{6}$ **c.** $\frac{4}{9}$

Reduce each fraction to lowest terms:

d. $\frac{18}{24}$ **e.** $\frac{36}{72}$ **f.** $\frac{15}{25}$

problem set
14

1. Ramirez had 5282 taco shells. If each box would hold 12 taco shells, how many completely full boxes could he ship?

2. The fairy queen found that 14 dryads could sit comfortably on one toadstool. If 780 dryads were coming to the forest convocation, how many toadstools would she have to provide so that they could all sit comfortably?

3. The first microbe had a measure of one hundred sixty-two ten-millionths. The measure of the second microbe was four hundred twenty-three hundred-millionths. By how much was the first microbe larger?

4. In the flang-flung contest, Jesse Lee flang his fourteen million, six hundred forty-two units. When Jethroe tried, he flung his thirty-two million, fifteen thousand, thirty-two units. How much farther did Jethroe flung than Jesse Lee flang?

5. Use the rules for divisibility to determine which of these numbers is divisible:
   (a) by 10   (b) by 5   (c) by 2   (d) by 3

   120   135   122   1332   1620

Find the perimeter of each figure. Dimensions are in kilometers. All angles are right angles.

6.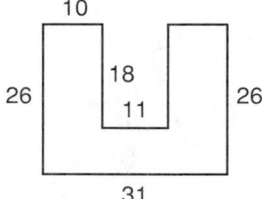

7.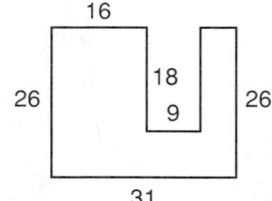

8. Reduce each fraction to lowest terms:

   (a) $\frac{45}{60}$ (b) $\frac{30}{75}$ (c) $\frac{80}{220}$ (d) $\frac{16}{80}$

9. Write each fraction with a denominator of 20:

   (a) $\frac{1}{2}$ (b) $\frac{3}{15}$ (c) 7   (d) $\frac{4}{16}$

Simplify mentally:

10. $12.361 \div 1000$ **11.** 11.36121×1000

Multiply:

12. 16.21×11.3 **13.** 113×1.28 **14.** 2.14×11.6

15. Write each of the following as a product of primes:
  (a) 1800   (b) 900   (c) 450

Subtract. Add to check.

16. 131.61 − 11.87

17. 181.811 − 6.329

Find the missing digits:

18.
$$\begin{array}{r} \$MNP.ZX \\ + \ \$463.12 \\ \hline \$657.06 \end{array}$$

19.
$$\begin{array}{r} \$653.28 \\ - \ \$SPX.YN \\ \hline \$468.29 \end{array}$$

Divide and round each answer to two decimal places:

20. $\dfrac{612.13}{0.603}$

21. 913.62 ÷ 0.025

22. 6111.12 ÷ 7.5

23. $\dfrac{611.21}{1.2}$

24. Round 1612.316289 to four decimal places.

25. Round 20,056,049,013 to the nearest hundred million.

Use words to write the following numbers:

26. 1231.161   **27.** 11123.121   **28.** 1612.12

Add:

29. 1093.06 + 113.1016 + 915.09

30. 43,116.013 + 647.112 + 2641.094 + 9158.0109

LESSON 15 *Fractions to decimals*

15.A
fractions and decimals

Both fractions and decimal numbers can be used to represent the parts of a whole.

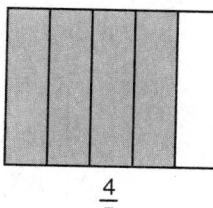

$\dfrac{4}{5}$

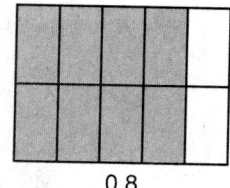

0.8

On the left we have divided the whole figure into 5 parts and shaded 4 of them to represent $\frac{4}{5}$. On the right we have divided the same figure into 10 parts and shaded 8 of them to represent 0.8, which is eight-tenths. We see that the shaded areas are equal. This is a picture that shows us that

$$\frac{4}{5} \quad \text{equals} \quad 0.8$$

Every fraction can be written as a decimal number. To write a fraction of whole numbers as a decimal number, we divide the bottom number into the top number.

The answer will always be a terminating decimal number or will be a decimal number that repeats in a definite pattern. Sometimes it is easy to see whether the digits terminate or repeat. Other times there are a large number of repeating digits in the pattern, and the pattern is not easy to see. When this happens, we will just round to a convenient number of digits. The important thing is to remember that if the number does not terminate the repeating pattern is always there even if we don't see it. We show several examples here.

(a) $\frac{1}{8} = 0.125$

(b) $\frac{3}{7} = 0.4285714285714285714\cdots$

(c) $\frac{21}{99} = 0.2121212121\cdots$

(d) $\frac{5}{6} = 0.8333333333\cdots$

(e) $\frac{56{,}156}{99{,}000} = 0.5672323232323\cdots$

We can use a bar over the repeating digits to designate a repeating pattern. Here we use bars as necessary to write the repeating answers again.

(b) $0.\overline{428571}$

(c) $0.\overline{21}$

(d) $0.8\overline{3}$

(e) $0.567\overline{23}$

15.B
rounding repeaters

To round repeating decimal numbers, it is helpful to write out the repeating pattern several times.

example 15.1   Round $42.\overline{617}$ to the nearest millionth.

solution   We begin by writing the repeating digits until the millionths' place is passed.

$$42.617617617617\cdots$$

Then we circle the digit in the millionths' place and mark the digit to its right with an arrow.

$$42.61761\,⑦\,617617\cdots$$

Then we change the arrow-marked digit and the digits to its right to zero.

$$42.61761\,⑦\,0000\cdots$$

Since the arrow-marked digit was greater than 5, we increase the circled digit by 1 and get

$$42.6176180000\cdots$$

Terminal zeros to the right of the decimal point have no value. Thus we can omit these zeros and write the answer as

42.617618

example 15.2 Round $718.0\overline{73}$ to eight decimal places.

solution First we write quite a few of the repeating digits, circle the digit in the eighth decimal place, and mark the next digit with an arrow.

$$718.0737373\textcircled{7}\overset{\downarrow}{3}73$$

Since the arrow-marked digit was less than 5, we do not change the circled digit when we round. Thus, the answer is

718.07373737

15.C
fractions to decimals

To write a fraction as a decimal number, we perform the indicated division.

example 15.3 Write each fraction as a decimal number: (a) $\dfrac{1}{8}$ (b) $\dfrac{1}{30}$

solution (a) We divide 1 by 8.
      (b) We divide 1 by 30.

```
      0.125
   8 ) 1.000
       8
       20
       16
       40
       40
```

```
        0.0333
   30 ) 1.0000
        90
        100
         90
         100
          90
          10
```

solution In (a) the number terminated. In (b) the 3s repeated. Thus our answers are

(a) **0.125**   (b) **$0.0\overline{3}$**

example 15.4 Write each fraction as a decimal: (a) $\dfrac{2}{5}$ (b) $\dfrac{5}{7}$

solution We will perform the indicated divisions.

```
           0.4
   (a)  5 ) 2.0
           2 0
```

```
            0.7142857
   (b)  7 ) 5.0000000
            4 9
             10
              7
             30
             28
             20
             14
             60
             56
             40
             35
             50
             49
```

The answer to (a) is 0.4 and the answer to (b) is a six-digit repeater, but we decide we want a shorter answer so we round to two digits.

(a) **0.4**   (b) **0.71**

practice **a.** Round $4.\overline{613}$ to the nearest millionth.

b. Round $1.4\overline{16}$ to five decimal places.

Write each fraction as a decimal number:

c. $\frac{1}{3}$

d. $\frac{1}{12}$

problem set 15

1. Their approach was inexorable, and there were fourteen million, seven thousand, nine hundred twenty in the first wave. In the second wave, there were only nine hundred thousand, sixty-seven. How many were there in all?

2. Harriet ran 14 times as far after she got her second wind. If she ran three thousand and seven hundred eighty-seven millionths feet on her first wind, how far did she run in all?

3. Ward 4 reported 8 times as many votes as Ward 7 reported. Ward 6 reported 12 times as many votes as Ward 7 reported. If Ward 7 reported nine thousand forty-three votes, how many votes did the three wards report in all?

4. The ladybugs grouped themselves into bunches of 20. If there were nine thousand forty-two ladybugs, how many bunches did they form?

5. Which of the following numbers is divisible (a) by 3 and (b) by 2?

   212   2133   312   610   630

6. Find the sum of the prime numbers that are between 30 and 42.

7. Find the perimeter of this figure. Dimensions are in feet. All angles are right angles.

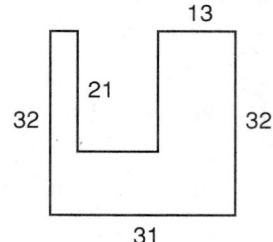

Write each fraction as a decimal number. Round to two decimal places.

8. $\frac{5}{9}$

9. $\frac{1}{20}$

10. Write $\frac{3}{14}$ with a denominator of 42.

11. Reduce each fraction to lowest terms:

   (a) $\frac{70}{60}$ (b) $\frac{16}{24}$ (c) $\frac{24}{36}$

Simplify mentally:

12. $169,211 \div 10,000$

13. 123.61311×100

Multiply:

14. 89.21×62.1

15. 2.16×32.8

16. Write each number as a product of primes:
   (a)   3600   (b)   450   (c)   4500

Subtract. Add to check.

17. $1131.13 - 131.98$ **18.** $192.68 - 6.321$

Find the missing digits:

19. $NAZ.XY$ **20.** 16.0395
          $-\ 364.82$ $+\ \ 0.ABZX$
          $\overline{59.69}$ $\overline{17.0094}$

Divide and round each answer to two places:

21. $\dfrac{629.1}{2.31}$ **22.** $117.2 \div 0.012$

23. $7.81 \div 3.11$ **24.** $\dfrac{2310}{13}$

25.   Round $4017.\overline{336}$ to the nearest millionth.

26.   Round $946.0\overline{54}$ to eight decimal places.

27.   Use words to write the number 1876211.32.

Add:

28. $1921.6 + 1872.7 + 1321.3 + 62.1$

29. $613.1 + 7214.6 + 11.2 + 3.1$

30.   Order from least to greatest: 0.0119, 0.091, 0.0191, 0.9

LESSON 16 *Decimals to fractions*

Both terminating and repeating decimal numbers can be written as fractions of whole numbers. The method of writing repeaters as fractions is rather complicated and will be taught in a subsequent course. We will discuss terminating decimals now.

Remember from Lesson 14 that a particular number can be multiplied and divided by another number without changing the value of the original number. This is because a number divided by itself equals 1.

$$\frac{47}{47} = 1 \qquad \frac{5}{5} = 1 \qquad \frac{100}{100} = 1 \qquad \frac{1000}{1000} = 1$$

We use this fact and the rule for multiplying by powers of 10 to write terminating decimal numbers as fractions. We multiply and divide by the power of 10 necessary to make the top number a whole number.

example 16.1   Write 0.041 as a fraction.

solution   If we multiply 0.041 by 1000, we will make it a whole number.

$$0.041 \times 1000 = 41$$

But we must also divide by 1000 so we do not change the value of 0.041.

$$0.041 \times \frac{1000}{1000} = \frac{41}{1000}$$

example 16.2   Write 43.21657 as a fraction.

solution   This time the decimal point must be moved five places, so we multiply and divide by 100,000, which has five 0s.

$$43.21657 \times \frac{100,000}{100,000} = \frac{4,321,657}{100,000}$$

practice   Write each decimal number as a fraction:

a. 4.76325   **b.** 757.623

problem set 16

1. Only fourteen could crawl into a single space. If three thousand eight hundred eight had to be sheltered, how many spaces were necessary?

2. Gene could muster up a total of fourteen and seven hundred forty-two ten-thousandths. Mary could muster up 7 times that much. How much could the two of them muster up all together?

3. When the shakedown was finished, Roberto found that he had shaken down ninety-one thousand, forty-two. Raoul was chagrined because this exceeded his total by twelve thousand, fifteen. What was Raoul's total?

4. Each box would hold 142 apples. If the crew filled 432 boxes one shift and had 5 apples left over, how many apples were there in all?

5. Which of the following is divisible (a) by 5, and (b) by 10?

   650   625   15   20   30

6. What are the prime numbers that are greater than 40 but less than 64?

7. Find the perimeter of this figure. Dimensions are in yards. All angles are right angles.

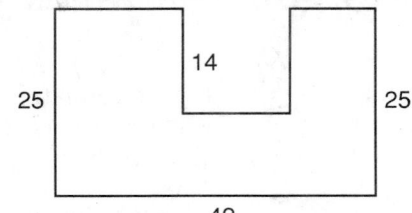

Write each fraction as a decimal number. Round to two decimal places.

8. $\frac{11}{16}$ **9.** $\frac{1}{17}$ **10.** $\frac{4}{7}$

Reduce each fraction to lowest terms:

11. $\frac{36}{42}$ **12.** $\frac{72}{120}$

13. Write $\frac{9}{27}$ with a denominator of 243.

Multiply:

14. 13.61 × 71.3   **15.** 61.3 × 11.2

16.   Simplify mentally: 12,389.32 ÷ 100

Subtract. Add to check.

17.   169,211.36 − 1892.98   **18.**   181,131.62 − 1.9876

Divide and round each answer to two places:

19. $\dfrac{613.1}{3.17}$ **20.** $\dfrac{123.8}{9.98}$ **21.**   181.3 ÷ 1.2

Write each number as a product of primes:

22.   288   **23.**   1080   **24.**   10,800

25.   Round 87,621.32178939 to the nearest millionth.

26.   Round 437.00$\overline{621}$ to the nearest hundred-millionth.

27.   Use words to write the number 172.312.

28.   Add: 1361.31 + 21.14 + 112.17 + 1.18

29.   Write the decimal number 0.85 as a fraction reduced to lowest terms.

30.   Find the missing digits: 14,392.091
                                         + *AZ,KXM.YNP*
                                         ─────────────
                                            107,072.060

LESSON 17   *Rectangular area*

The left-hand diagram represents a table top that is 4 feet long and 3 feet wide. On the right we show how this top can be divided into 12 squares. All the sides of the squares are 1 foot (1 ft) long. We say that the area of each square is 1 square foot, or 1 ft².

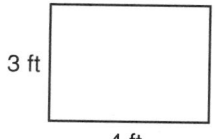   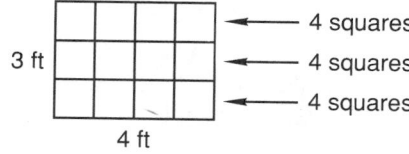

There are 4 squares in each row and 3 rows, so there are 12 squares.

$$4 \text{ squares} \times 3 = 12 \text{ squares}$$

Thus, we say that the total area is 12 square feet, or 12 ft². If there were 6 rows of 4 squares each, then there would be a total of 24 squares.

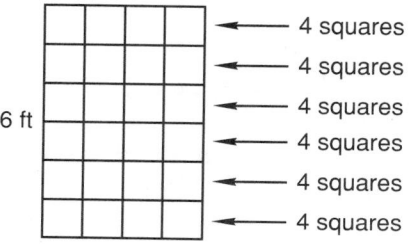

4 squares × 6 = 24 squares

The length of the first row tells us the number of squares in this row, and the width tells us the number of rows. Thus, the number of squares is the length times the width.

Number of squares = length × width

Each square could be covered by a floor tile that is 1 foot on a side. It is helpful to think "floor tiles" when we hear the word *area*. Floor tiles can be touched and felt and understood, while area is more abstract.

example 17.1   How many 1-inch-square floor tiles would it take to cover this figure?

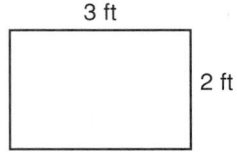

solution   The dimensions are in feet, so we first change feet to inches (in.).

3 ft = 3 × 12 in. = 36 in.   2 ft = 2 × 12 in. = 24 in.

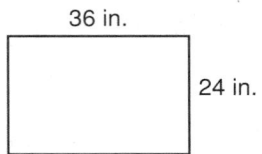

Each row will have 36 tiles, and there will be 24 rows. Now we multiply.

24 × 36 tiles = **864 tiles**

example 17.2   Find the area of this figure in square centimeters (cm²). Dimensions are in centimeters (cm). All angles are right angles.

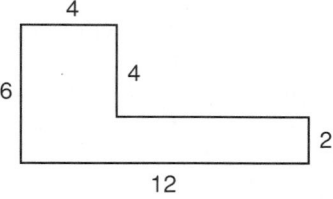

solution   We will divide the given area into rectangles and find the areas of the rectangles. Then we will add the areas. Two ways are shown.

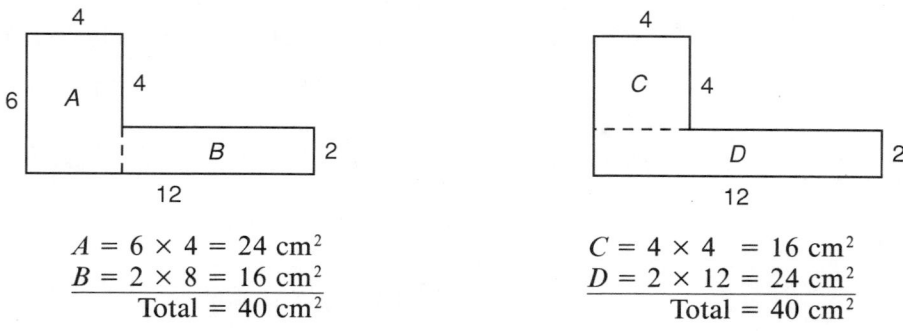

$A = 6 × 4 = 24$ cm² $C = 4 × 4 = 16$ cm²
$B = 2 × 8 = 16$ cm² $D = 2 × 12 = 24$ cm²
Total = 40 cm²   Total = 40 cm²

The area of the figure is **40 cm²**. Thus 40 tiles, each 1-centimeter square, would be required to cover the figure.

practice Find the area of each figure. Dimensions are in feet. All angles are right angles.

a.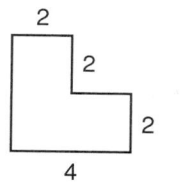

b.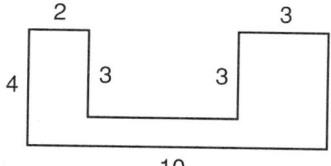

c.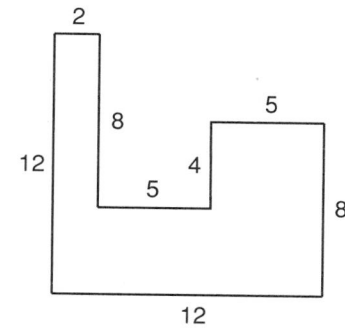

problem set 17

1. The third avatar of Happiness was a smile. Happiness is said to have 11 avatars in all. If she used 1 avatar a day, how many times would she appear as a smile in 10,802 days?

2. There were four thousand eight hundred forty-two ants in each anthill. If there were three hundred thirty anthills, how many ants were there in all?

3. In the next valley the anthills each contained one hundred eight thousand fifteen ants. How many more ants lived in one of these anthills than lived in an anthill from Problem 2?

4. Which of the following numbers is divisible (a) by 3 and (b) by 5?

   135   1050   335   4145   1010

5. Find the sum of the first 13 prime numbers.

6. Find the perimeter of this figure. Dimensions are in meters. All angles are right angles.

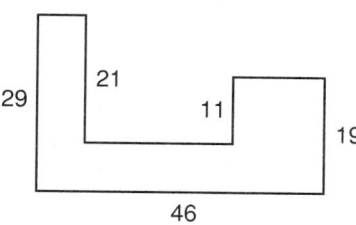

7. Find the area of this figure. Dimensions are in centimeters. All angles are right angles.

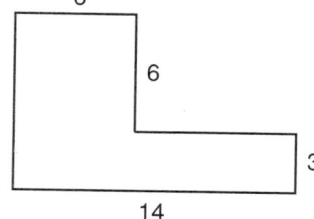

8. How many 1-inch-square floor tiles would cover this figure? Dimensions are in inches. All angles are right angles.

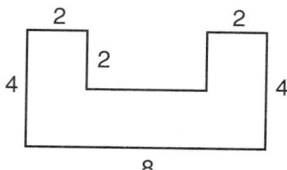

Write each fraction as a decimal. Round to two decimal places.

9. $\dfrac{3}{17}$ **10.** $\dfrac{7}{11}$ **11.** $\dfrac{12}{13}$

Write each number as a product of primes:

12. 540 **13.** 30 **14.** 210 **15.** 1260

Reduce each fraction to lowest terms:

16. $\dfrac{21}{49}$ **17.** $\dfrac{36}{48}$

18. Write $\dfrac{3}{7}$ with a denominator of 210.

Multiply:

19. 17.31×31.11 **20.** 62.13×7.81

21. Simplify mentally: 625.3876×100

22. Subtract (add to check): $185.3617 - 17.3169$

23. Find the missing digits:

$$\begin{array}{r} 179.3622 \\ -\ ZKY.XNPR \\ \hline 57.4643 \end{array}$$

Divide and round to two places:

24. $\dfrac{231.2}{3.89}$ **25.** $190.1 \div 2.1$

26. Round 3.141,592,654 to the nearest ten-millionth.

27. Round $2.0\overline{70}$ to seven decimal places.

28. Use words to write the number 62987621.1.

29. Order from least to greatest: 0.3, 0.091, 0.0091, 0.090109

30. Add: $3.14159 + 2.621 + 82.71$

LESSON 18 *Products of primes in cancellation*

If we look at the expression

$$\frac{2 \times 3 \times 7 \times 5}{2 \times 3 \times 7 \times 7}$$

We see that both the top and bottom contain $2 \times 3 \times 7$, which equals 42.

$$\frac{\overbrace{2 \times 3 \times 7} \times 5}{\underbrace{2 \times 3 \times 7} \times 7} = \frac{42 \times 5}{42 \times 7}$$

Because 42 over 42 equals 1, the original expression can be written in lowest terms as $\frac{5}{7}$. We can use this thought to reduce some fractions to lowest terms. We write both the top and the bottom as products of prime factors and cancel the factors that appear both above and below.

example 18.1 Use products of primes to reduce $\frac{360}{900}$ to lowest terms.

solution We begin by writing both numbers as products of primes.

$$\frac{360}{900} = \frac{2 \times 2 \times 2 \times 3 \times 3 \times 5}{2 \times 2 \times 3 \times 3 \times 5 \times 5}$$

We note the combinations of factors that equal 1.

$$\frac{2}{2} \times \frac{2}{2} \times \frac{2}{3} \times \frac{3}{3} \times \frac{3}{5} \times \frac{5}{5}$$

Now $\frac{2}{2}$ and $\frac{2}{2}$ and $\frac{3}{3}$ and $\frac{3}{3}$ and $\frac{5}{5}$ all have a value of 1. All that is left is a 2 on top and a 5 on the bottom. Thus, this fraction reduces to

$$\frac{2}{5}$$

example 18.2 Use products of primes to reduce $\frac{42}{78}$ to lowest terms.

solution We write both numbers as products of primes and cancel the common factors.

$$\frac{42}{78} = \frac{2 \times 3 \times 7}{2 \times 3 \times 13} = \frac{7}{13}$$

practice Use products of primes to reduce each fraction to lowest terms.

   a. $\frac{84}{90}$ **b.** $\frac{105}{287}$

problem set 18

1. Paul picked 5482 peaches. Peter picked four times as many as Paul picked. Penny picked twice as many as Peter picked. How many peaches did they pick altogether?

2. Nineteen million and forty-seven hundred-thousandths was Ahab's first guess. His second guess was seventeen thousand, five and two hundred thirty-seven ten-thousandths greater than his first guess. What was his second guess?

3. Even with a pry bar, Judy could not force more than sixty-three into each container. If she had seven hundred thousand, thirty-three containers, how many did she need to fill all the containers?

4. When she came to work the next day, she was delighted because four hundred sixteen thousand, nine hundred forty-three containers had been delivered. How many things (Problem 3) would fill them all?

5. What prime numbers are less than 50 but greater than 25?

6. Find the perimeter of this figure. Dimensions are in centimeters. All angles are right angles.

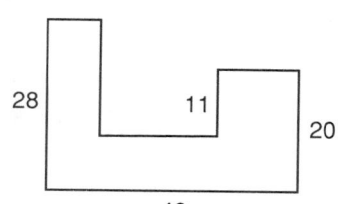

Find the area of each figure. Dimensions are in centimeters. All angles are right angles.

7.
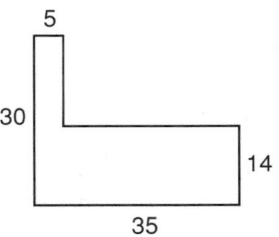

8.
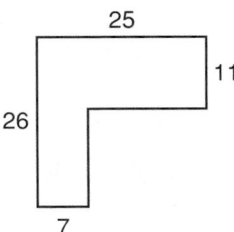

Write each number as a product of prime factors and simplify:

9. $\dfrac{480}{1000}$ **10.** $\dfrac{66}{104}$

11. $\dfrac{210}{294}$ **12.** $\dfrac{738}{882}$

Write each fraction as a decimal number. Round to two decimal places.

13. $\dfrac{16}{17}$ **14.** $\dfrac{6}{7}$ **15.** $\dfrac{11}{13}$

Expand each fraction so that the denominator is 90.

16. $\dfrac{3}{10}$ **17.** $\dfrac{14}{15}$

Multiply:

18. 62.13×11.31 **19.** 78.11×113

20. Simplify mentally: $621.1378 \div 10,000$

Subtract. Add to check.

21. $183.1782 - 1.8999$ **22.** $6211.31 - 689.13$

Divide and round to two places:

23. $\dfrac{6.89}{2.17}$ **24.** $\dfrac{62.15}{2.1}$

25. Round 9.8696044 to five decimal places.

26. Use words to write the number 1 123 689 113.

Add:

27. $6251.13 + 231.61 + 81.62 + 998.11$

28. $6.91 + 8.23 + 0.1135 + 113$

29. Write the decimal number 0.095 as a fraction and simplify.

30. Write $\dfrac{7}{12}$ with a denominator of 108.

LESSON 19 *Multiplying fractions · Dividing fractions*

19.A
symbols for multiplication

Thus far, we have used a cross or a center dot to designate multiplication. To write 4 times 3, we have written

$$4 \times 3 \quad \text{or} \quad 4 \cdot 3$$

In algebra, the letter x is used often instead of using a particular number. The symbol for the letter x is similar to the cross $\times$ used to indicate multiplication. Therefore, in algebra we do not use the cross and designate multiplication in other ways (but we will still use the cross when it is convenient). Sometimes we use a dot to designate multiplication, and sometimes we use parentheses. All the following indicate that 4 and 3 are to be multiplied.

$$4 \cdot 3 \qquad 4(3) \qquad (4)3 \qquad (4)(3)$$

None of these forms is necessarily a preferred form, and all of them designate the same thing.

19.B
multiplication of fractions

To multiply fractions, we multiply the tops to get the new top and multiply the bottoms to get the new bottom.

example 19.1 Simplify: $\dfrac{4}{9} \cdot \dfrac{5}{7}$

solution We get the new top by multiplying 4 and 5 and the new bottom by multiplying 9 and 7.

$$\frac{4}{9} \cdot \frac{5}{7} = \frac{20}{63}$$

If the tops and bottoms contain equal factors, it is often helpful to simplify before multiplying, as shown in the next example.

example 19.2 Simplify: $\dfrac{4}{9} \cdot \dfrac{18}{30}$

solution We reduce the fractions before we multiply.

$$\frac{\overset{2}{\cancel{4}}}{\cancel{9}} \cdot \frac{\overset{2}{\cancel{18}}}{\underset{15}{\cancel{30}}} = \frac{4}{15}$$

19.C
division of fractions

We divide fractions by inverting the divisor and multiplying. Both of these expressions indicate that $\frac{5}{7}$ is to be divided by $\frac{4}{5}$.

$$\text{(a)} \quad \frac{\frac{5}{7}}{\frac{4}{5}} \qquad \text{(b)} \quad \frac{5}{7} \div \frac{4}{5}$$

Thus, in both expressions, the divisor is $\frac{4}{5}$. We simplify both expressions in the same way—by inverting the divisor and multiplying.

$$\frac{5}{7} \cdot \frac{5}{4} = \frac{25}{28}$$

example 19.3   Simplify: $\frac{21}{6} \div \frac{27}{4}$

solution   First we invert the divisor. Then we reduce and multiply.

$$\frac{21}{6} \cdot \frac{4}{27} = \frac{\overset{7}{\cancel{21}}}{\underset{3}{\cancel{6}}} \cdot \frac{\overset{2}{\cancel{4}}}{\underset{9}{\cancel{27}}} = \frac{14}{27}$$

practice   Simplify:

a. $\frac{4}{9} \cdot \frac{3}{12}$

b. $\dfrac{\frac{3}{8}}{\frac{4}{6}}$

c. $\frac{4}{5} \div \frac{10}{5}$

problem set 19

1. The prize was for the closest guess. The exact answer was 14,016.2163. Charles guessed thirteen thousand, forty-one and six ten-thousandths. Mary guessed fourteen thousand, nine hundred ninety-one and two hundred thirty thousandths. By how much did each of them miss the exact answer? Whose guess was closer?

2. They crawled in until 51,416 had arrived. Then 4 times this number crawled in on the next day. How many of them crawled in in all?

3. Five thousand, seven hundred fifteen stood and roared their approval. Nineteen thousand seven remained seated and were discontented. How many more were discontented than approved?

4. The fire department had rules that occupancy by more than 315 in a room was unlawful. If 14,321 attended, how many rooms would it take to hold them lawfully?

5. What is the smallest prime number that is at least 14 greater than 21?

Simplify. If possible, cancel before multiplying.

6. $\frac{5}{8} \cdot \frac{4}{7}$

7. $\frac{5}{9} \cdot \frac{15}{20}$

8. $\frac{3}{8} \div \frac{2}{3}$

9. $\frac{9}{4} \div \frac{14}{3}$

10. Find the perimeter of this figure. Dimensions are in meters. All angles are right angles.

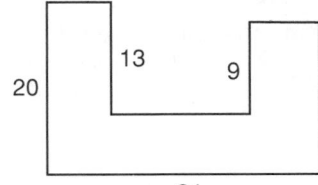

Find the area of each figure. Dimensions are in yards. All angles are right angles.

11.

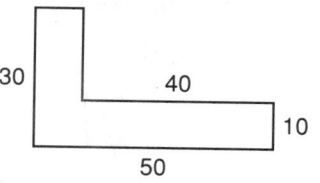

12.

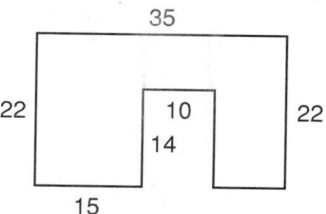

Write each number as a product of prime factors and simplify:

13. $\dfrac{360}{540}$ **14.** $\dfrac{196}{360}$ **15.** $\dfrac{256}{720}$

Write each fraction as a decimal number. Round to two decimal places.

16. $\dfrac{17}{23}$ **17.** $\dfrac{5}{17}$

Write each number as a product of prime numbers:

18. 4200   **19.** 10,080

Multiply:

20. 11.23×62.14 **21.** 1.28×62.1

Subtract. Add to check.

22. $1921.761 - 18.32$ **23.** $612.81 - 19.362$

Divide and round to two places:

24. $\dfrac{7.32}{0.12}$ **25.** $\dfrac{61.31}{3.1}$

26. Round $28.30\overline{57}$ to the nearest ten-millionth.

27. Use words to write the number 9699690.

Add:

28. $31.613 + 1.812 + 61.234 + 111.621$

29. $118.32 + 116.15 + 311.13$

30. Simplify mentally: $62,158.3211 \times 100$

LESSON 20 *Multiples*

Multiples, factors, and **products** are words that go together. We know that 3 times 2 equals 6.

$$3 \cdot 2 = 6$$

In this example, the factors are 3 and 2, and the product is 6. Because we can multiply 3 by 2 and get 6, we say that 6 is a multiple of 3 and that 6 is also a multiple of 2. The numbers 3, 6, 9, 12, 15, and 18 are also multiples of 3 because they can be composed by multiplying 3 by itself or by another whole number.

$$3 \cdot 1 = 3$$
$$3 \cdot 2 = 6$$
$$3 \cdot 3 = 9$$
$$3 \cdot 4 = 12$$
$$3 \cdot 5 = 15$$
$$3 \cdot 6 = 18$$

Similarly, the numbers 2, 4, 6, 8, 10, 12, and 14 are all multiples of 2 because they can be composed by multiplying 2 by itself or by another whole number, as shown here.

$$2 \cdot 1 = 2$$
$$2 \cdot 2 = 4$$
$$2 \cdot 3 = 6$$
$$2 \cdot 4 = 8$$
$$2 \cdot 5 = 10$$
$$2 \cdot 6 = 12$$
$$2 \cdot 7 = 14$$

The words *multiple, factor,* and *product* are not difficult, but they are new and thus are different. Problems that use these words will help us remember what they mean.

example 20.1   The number 8 is a multiple of which numbers?

solution   We can compose 8 by multiplying 2 and 4 or by multiplying 1 and 8.

$$2 \cdot 4 = 8 \qquad 8 \cdot 1 = 8$$

So 8 is a multiple of **1, 2, 4,** and **8.** Note that we say that 8 is a multiple of itself and also of 1.

example 20.2   Find all multiples of 9 that are less than 50.

solution   A multiple of 9 is the product of 9 and a whole number greater than zero.

$$9 \cdot 1 = 9$$
$$9 \cdot 2 = 18$$
$$9 \cdot 3 = 27$$
$$9 \cdot 4 = 36$$
$$9 \cdot 5 = 45$$

Thus the multiples of 9 that are less than 50 are **9, 18, 27, 36,** and **45.**

practice   **a.**   Find all multiples of 7 that are less than 56.

b.   Find all multiples of 6 that are greater than 29 and are also less than 50.

problem set 20

1. Seven thousand, forty-two wore red hats. Ninety-three thousand, nine hundred seventy-five wore blue hats. If one hundred thirty-seven thousand, eight hundred forty-two attended, how many did not wear either red hats or blue hats?

2. When the first box was opened, the chapeaux inside totaled one thousand, nine hundred three. If 47,300 students were expected to attend, how many boxes of chapeaux would be required so that every student could have one?

3. When the first frost came, nineteen thousand, two shriveled and died. If 5 times this number survived the frost, how many had there been before the frost came?

4. Eloise drove 636.57 miles on the first day. Elvira drove 3 times as far on the second day. On the third day Noni drove twice as far as Eloise drove. What was the total distance driven by the three women?

5. Find all multiples of 7 that are less than 70.

Simplify. If possible, cancel before multiplying.

6. $\dfrac{42}{15} \cdot \dfrac{3}{7}$

7. $\dfrac{5}{18} \cdot \dfrac{90}{120}$

8. $\dfrac{5}{8} \div \dfrac{7}{4}$

9. $\dfrac{7}{18} \div \dfrac{14}{9}$

10. Find the perimeter of this figure. Dimensions are in feet. All angles are right angles.

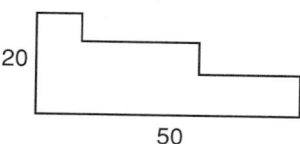

Find the area of each figure. Dimensions are in meters. All angles are right angles.

11.

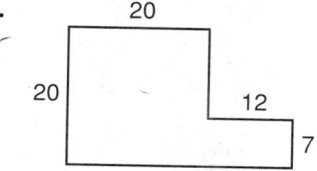

12.

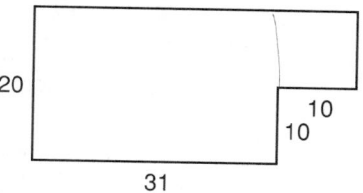

Write each number as a product of prime factors and simplify:

13. $\dfrac{540}{720}$

14. $\dfrac{144}{360}$

15. $\dfrac{432}{720}$

Write each fraction as a decimal number. Round to two decimal places.

16. $\dfrac{21}{23}$

17. $\dfrac{6}{13}$

18. Find the number which is divisible by both 2 and 3:
    (a) 1215　　(b) 572　　(c) 1111　　(d) 9924

19. Order from least to greatest: 0.04, 0.0049, 0.1, 0.0096

Multiply:

20. 11.92 × 12.32

21. 1792 × 61.2

Subtract. Add to check.

22. 62.0325 − 7.9813

23. 17.3252 − 8.9211

Divide and round to two places:

24. $\dfrac{7.91}{3.14}$ **25.** $\dfrac{9.86}{3.14}$

26. Round 194,591,014.62 to the nearest million.

27. Use words to write the number 111546435.

28. Simplify mentally: 8361.2361 ÷ 100

Add:

29. 16.17 + 181.325 + 0.00131   **30.** 14.159 + 2.71828 + 2236.21

LESSON 21 *Average*

The average of a group of numbers is the sum of the numbers divided by the number of numbers. The average of the four numbers 1, 2, 4, and 5 is 3.

$$\text{Average} = \frac{1 + 2 + 4 + 5}{4} = \frac{12}{4} = 3$$

If we have 4 piles of pancakes as shown here,

the average is the number in each group if we put the same number in each group.

If we know the average of a group of numbers, we know the sum of the numbers. The sum is the average times the number of numbers.

example 21.1   Find the average of (a) 1765, (b) 93, (c) 742, and (d) 21,050.

solution   We add the four numbers and divide the sum by 4.

$$\text{Average} = \frac{1765 + 93 + 742 + 21{,}050}{4} = \frac{23{,}650}{4} = \mathbf{5912.5}$$

example 21.2   Find the average of 8, 17, 14, 10, 18, 6, 13, and 12.

solution   We add the numbers and divide by 8.

$$\text{Average} = \frac{8 + 17 + 14 + 10 + 18 + 6 + 13 + 12}{8} = \frac{98}{8} = \mathbf{12.25}$$

example 21.3   The average of four numbers is 10. Three of the four numbers are 2, 14, and 7. What is the fourth number?

solution If the average of four numbers is 10, the sum of the numbers is 4 times 10, or 40. The sum of 2, 14, and 7 is 23.

$$2 + 14 + 7 = 23$$

The other number must be **17** because 17 plus 23 equals 40.

$$\underbrace{2 + 14 + 7}_{23} + 17 = 40$$

practice **a.** Find the average of 74.2, 87.3, and 106.5.

b. The average of four numbers is 100.1. Three of the numbers are 24.2, 13.5, and 48.7. What is the fourth number?

c. The first four items cost $5.21, $7.42, $8.64, and $9.23. What was the average cost per item?

problem set 21

1. Jed guessed ten billion, four hundred seventy-five thousand, fifteen and nine hundred twenty-three ten-thousandths. Ned guessed eight billion, four hundred seventy-three million, forty-two and seventy-five thousandths. How much greater was Jed's guess than Ned's guess?

2. On the first day the counts were 4012.06 and 0.00418. On the next day the counts were 732.05 and 9.016. What was the total of the counts for both days?

3. Only 243 could squeeze through at one time. If 416,202 came, how many times would it take all of them to squeeze through?

4. Which of the following numbers are divisible (a) by 3, and (b) by 5?

   11,682   10,517   2193   4200

5. List the prime numbers that are greater than 30 and are less than 50.

6. Find the average of 1242, 87, 521, and 169,810.

7. The average cost of five items is $316.05. The first four items cost $48.20, $40.60, $63.75, and $70.15. Find the cost of the fifth item.

Simplify. If possible, reduce before multiplying.

8. $\dfrac{36}{28} \cdot \dfrac{7}{6}$

9. $\dfrac{21}{32} \cdot \dfrac{4}{7}$

10. $\dfrac{5}{7} \div \dfrac{15}{21}$

11. $\dfrac{9}{25} \div \dfrac{27}{15}$

12. Find the perimeter of the following figure. Dimensions are in meters. All angles are right angles.

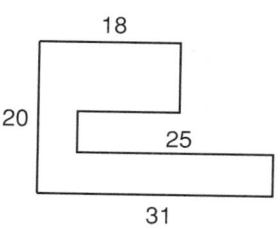

13. Find the area of the following figure. Dimensions are in centimeters. All angles are right angles.

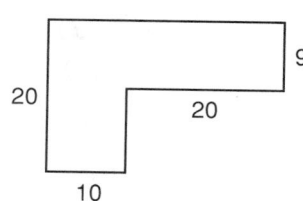

Write each number as a product of prime factors and simplify:

14. $\dfrac{270}{360}$ **15.** $\dfrac{35}{42}$

16. Find all the multiples of 3 that are less than 90.

Write each fraction as a decimal number. Round to two decimal places.

17. $\dfrac{11}{17}$ **18.** $\dfrac{5}{13}$

Write each number as a product of primes:

19. 2520   **20.** 8400

Multiply:

21. 1.83×61.3 **22.** 66.12×1.7

23. Subtract (add to check): $123.789 - 1.899$

24. Find the missing digits:
$$\begin{array}{r} 12.8760 \\ -\ N.XYMZ \\ \hline 3.0931 \end{array}$$

Divide and round to two places:

25. $\dfrac{6.25}{0.87}$ **26.** $\dfrac{71.3}{3.7}$

27. Round $34.7\overline{18}$ to the nearest ten-thousandth.

28. Use words to write the number 2718.2818.

29. Simplify mentally: $9.8762 \div 1000$

30. Add: $7.36 + 99.825 + 1070.0327$

LESSON 22 *Multiple fractional factors*

There is no change in the procedure when a problem has three or more fractions to be multiplied. First we cancel common factors and then we multiply.

example 22.1   Simplify: $\dfrac{56}{27} \cdot \dfrac{14}{21} \cdot \dfrac{15}{22}$

solution   We begin by writing each number as a product of prime factors. Then we cancel common factors and multiply as the last step.

$$\frac{7 \cdot 2 \cdot 2 \cdot 2}{3 \cdot 3 \cdot \cancel{3}} \cdot \frac{\cancel{2} \cdot \cancel{7}}{3 \cdot \cancel{7}} \cdot \frac{\cancel{3} \cdot 5}{\cancel{2} \cdot 11} = \frac{\mathbf{280}}{\mathbf{297}}$$

example 22.2   Simplify: $\dfrac{26}{30} \cdot \dfrac{21}{39} \div \dfrac{4}{15}$

solution First we must invert $\frac{4}{15}$ and change the division symbol to a dot for multiplication.

$$\frac{26}{30} \cdot \frac{21}{39} \cdot \frac{15}{4}$$

Now we simplify and then multiply.

$$\frac{\cancel{2} \cdot \cancel{13}}{2 \cdot \cancel{3} \cdot \cancel{5}} \cdot \frac{\cancel{3} \cdot 7}{\cancel{3} \cdot \cancel{13}} \cdot \frac{\cancel{3} \cdot \cancel{5}}{\cancel{2} \cdot 2} = \frac{7}{4}$$

practice Use products of prime factors to help simplify.

a. $\dfrac{12}{105} \times \dfrac{14}{30} \times \dfrac{15}{14}$

b. $\dfrac{4}{15} \times \dfrac{3}{16} \times \dfrac{6}{18}$

c. $\dfrac{6}{21} \times \dfrac{14}{30} \div \dfrac{6}{15}$

d. $\dfrac{10}{35} \times \dfrac{14}{28} \div \dfrac{12}{45}$

problem set 22

1. Ninety million, four thousand and sixty-two millionths is a big number. By how much is it greater than eighty-six million, four hundred twenty-seven thousand and fourteen ten-thousandths?

2. In the first wave there were 742,000. The second and third waves contained 96,016 and 1,001,892, respectively. How many were there in all?

3. The average time for the three downhill runs was 49.8 seconds. If the first run took only 45.2 seconds and the second run slowed to 51.8 seconds, what was the time for the third run?

4. Harry cornered 146 in the backyard. Jennet cornered 5 times that number in the side yard. How many did they corner all together?

5. Find the average of 1862, 1430, and 276.

6. (a) What are the prime numbers between 80 and 90?
    (b) List the multiples of 3 that are greater than 80 and less than 90.

Simplify. Cancel where possible as the first step.

7. $\dfrac{63}{54} \cdot \dfrac{21}{35} \cdot \dfrac{15}{22}$

8. $\dfrac{26}{90} \cdot \dfrac{24}{25} \div \dfrac{4}{15}$

9. $\dfrac{36}{24} \cdot \dfrac{8}{6} \div \dfrac{3}{6}$

10. Find the perimeter of the following figure. Dimensions are in inches. All angles are right angles.

11. Find the area of the following figure. Dimensions are in feet. All angles are right angles.

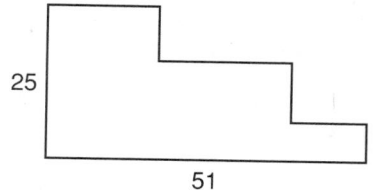

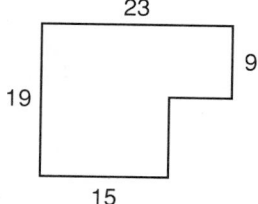

Write each number as a product of primes:

12. 14,000

13. 5040

Write each fraction as a decimal number. Round to two decimal places.

14. $\dfrac{13}{17}$

15. $\dfrac{3}{11}$

16. $\dfrac{3}{7}$

Write each number as a product of prime factors and simplify:

17. $\dfrac{540}{1440}$ **18.** $\dfrac{128}{360}$ **19.** $\dfrac{720}{840}$ **20.** $\dfrac{240}{256}$

Multiply:

21. 1.91×31.7 **22.** 581×0.163

Subtract. Add to check.

23. $6854.32 - 1.871$ **24.** $725.113 - 26.91$

Divide and round to two places:

25. $\dfrac{11.7}{3.1}$ **26.** $\dfrac{1132.1}{0.7}$

27. Round 0.301029996 to the nearest hundredth.

28. Use words to write the number 3678922117.32.

29. Simplify mentally: 123.7136×1000

30. Add: $11.762 + 0.8171 + 1162.18$

LESSON 23 *English units · Unit multipliers*

23.A
English units

When we attach words to numbers such as

4 feet   13 inches   27 yards   42 miles

we call each combination a **denominate number,** and we call the words **units.** We say that the units of the denominate number

4 feet

are feet. In the denominate number

32 miles per hour

we say that the units are miles per hour. The basic units of length in the English system are the inch, foot, yard, and mile. They are related by the following equivalences.

12 inches = 1 foot   3 feet = 1 yard   5280 feet = 1 mile

23.B
unit multipliers

If we multiply a number by a fraction that equals 1, we do not change the value of the number. We do change the name of the fraction. If we multiply 5 by 3 over 3, we get 15 over 3.

$$5 \times \frac{3}{3} = \frac{15}{3}$$

Fifteen divided by 3 has a value of 5 and is another way to write 5. Each of these fractions

$$\frac{12 \text{ in.}}{1 \text{ ft}} \qquad \frac{1 \text{ yd}}{3 \text{ ft}} \qquad \frac{5280 \text{ ft}}{1 \text{ mi}}$$

has a value of 1 because the denominators and the numerators have equal values. We can use these fractions as multipliers to change the names of denominate numbers. We call these fractions **unit multipliers** because **unity** is a word that means 1.

example 23.1   Use a unit multiplier to convert 72 inches to feet.

solution   We have a choice of two unit multipliers.

$$\text{(a)} \quad \frac{12 \text{ in.}}{1 \text{ ft}} \qquad \text{and} \qquad \text{(b)} \quad \frac{1 \text{ ft}}{12 \text{ in.}}$$

We want to change the name of 72 inches, so we choose (b) because it has inches on the bottom. We will multiply by 1 foot over 12 inches and cancel the inches.

$$72 \text{ inches} \times \frac{1 \text{ foot}}{12 \text{ inches}} = \frac{72}{12} \text{ feet} = \textbf{6 feet}$$

We note that we can cancel units just as if they were numbers.

example 23.2   Use a unit multiplier to convert 430 feet to yards (yd).

solution   The two unit multipliers that we can use are

$$\text{(a)} \quad \frac{3 \text{ ft}}{1 \text{ yd}} \qquad \text{and} \qquad \text{(b)} \quad \frac{1 \text{ yd}}{3 \text{ ft}}$$

Beginners sometimes don't know which one to use. If we decide to use (a), we get

$$430 \text{ ft} \times \frac{3 \text{ ft}}{1 \text{ yd}} = 430 \cdot 3 \frac{(\text{ft})(\text{ft})}{\text{yd}}$$

This is not incorrect but is not what we want. Let's try (b).

$$430 \text{ ft} \times \frac{1 \text{ yd}}{3 \text{ ft}} = \frac{430}{3} \text{ yd} = \textbf{143}\frac{\textbf{1}}{\textbf{3}} \textbf{ yd}$$

Notice that this time we can cancel the unit *ft* above with the unit *ft* below.

example 23.3   Use a unit multiplier to convert 4.6 miles to feet.

solution   There are two unit multipliers that we can use.

$$\text{(a)} \quad \frac{5280 \text{ ft}}{1 \text{ mi}} \qquad \text{and} \qquad \text{(b)} \quad \frac{1 \text{ mi}}{5280 \text{ ft}}$$

We decide to use (a) because it has *miles* on the bottom.

$$4.6 \text{ mi} \times \frac{5280 \text{ ft}}{1 \text{ mi}} = \textbf{(4.6)(5280) ft}$$

We will not do the multiplication. This problem is a unit conversion problem. If a numerical answer is desired, a calculator can be used to do the multiplication. The answer (4.6)(5280) ft is satisfactory. Other problems designed for practice in arithmetic appear in the homework problem sets.

practice   Use unit multipliers to convert:

a.   412 feet to yards   **b.**   17,200 feet to inches

c.   412 feet to inches

problem set   **1.**   Hortense ran nineteen thousand, seven and four hundred twenty-three thou-
23   sandths centimeters in the allotted time. Then in the allotted time Jim ran
                  eight thousand, forty-two and seven hundred sixty-five ten-thousandths centi-
                  meters. What was the sum of the distances they ran?

2.   At first there were only 2,046,021 microbes. Then Muzowbe added 3 times this
         many. Then how many microbes were there in all?

3.   The first factory could produce 243 cars in an allotted work period. If 19,197
         cars were required, how many work periods had to be allotted?

4.   From this list Kunta selected the numbers that were divisible by 3: 40,225;
         93,663; 72,205; 20,163. What was the sum of the numbers he selected?

5.   (a)   List the prime numbers between 60 and 70.
         (b)   Write the multiple of 9 between 60 and 70.

6.   The average of four numbers is 638.4. The first three numbers are 216, 159.8,
         and 301.25. Find the fourth number.

Use one unit multiplier to convert:

7.   120 inches to feet   **8.**   999 feet to yards   **9.**   42 feet to inches

Simplify. If possible, cancel as the first step.

10. $\dfrac{42}{12} \cdot \dfrac{36}{14} \div \dfrac{18}{2}$ **11.** $\dfrac{12}{16} \cdot \dfrac{4}{3} \div \dfrac{5}{6}$ **12.** $\dfrac{16}{24} \cdot \dfrac{12}{4} \cdot \dfrac{1}{2}$

13.   Find the perimeter of the follow-   **14.**   Find the area of the following
          ing figure. Dimensions are in   figure. Dimensions are in meters.
          yards. All angles are right angles.   All angles are right angles.

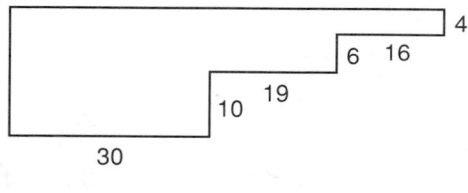   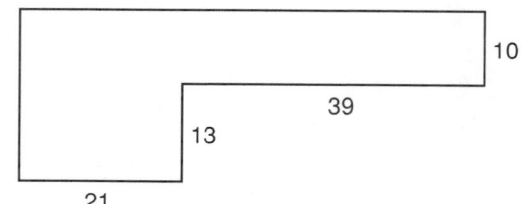

Write each number as a product of primes:

15.   2880   **16.**   6750

Write each fraction as a decimal number. Round to two decimal places.

17. $\dfrac{13}{19}$ **18.** $\dfrac{17}{19}$

Write each number as a product of prime factors and simplify:

19. $\dfrac{180}{256}$ **20.** $\dfrac{240}{600}$ **21.** $\dfrac{540}{2160}$

Multiply:

22.   71.2 × 0.173   **23.**   61.7 × 11.2

Subtract. Add to check.

24. $1812.721 - 729.871$ **25.** $61,131.2 - 71.823$

Divide and round to two places.

26. $\dfrac{713.7}{2.5}$ **27.** $\dfrac{7181.3}{0.3}$

28. Round 728,651.32 to the nearest thousand.

29. Use words to write the number 67211361.72.

30. Add: $71.161 + 0.8121 + 177.62$

LESSON 24 *Metric length conversions*

The advantage to the metric system is its simplicity. All you have to do to convert from one metric unit of length to another metric unit of length is to move the decimal point. The basic unit of length in the metric system is the meter. A meter is just a little longer than a yard. If we divide a meter into 100 equal parts, we call each of the parts a centimeter. One thousand meters is one kilometer, which is about six-tenths of a mile.

$$100 \text{ centimeters} = 1 \text{ meter (m)} \qquad 1000 \text{ meters} = 1 \text{ kilometer (km)}$$

The four unit multipliers that we get from these two equivalences are

$$\frac{100 \text{ cm}}{1 \text{ m}} \quad \text{and} \quad \frac{1 \text{ m}}{100 \text{ cm}} \qquad \frac{1000 \text{ m}}{1 \text{ km}} \quad \text{and} \quad \frac{1 \text{ km}}{1000 \text{ m}}$$

There are other units of length such as the millimeter, the decimeter, and the micrometer. These units are special-purpose units and can be researched when their use is required. We will concentrate on the three basic metric units of length.

example 24.1   Convert 528 centimeters to meters.

solution   We have our choice of using

$$\text{(a)} \quad \frac{100 \text{ cm}}{1 \text{ m}} \quad \text{or} \quad \text{(b)} \quad \frac{1 \text{ m}}{100 \text{ cm}}$$

We will use (b) because *cm* is on the bottom and will cancel the *cm* on the top.

$$528 \text{ cm} \times \frac{1 \text{ m}}{100 \text{ cm}} = \frac{528}{100} \text{ m} = \textbf{5.28 m}$$

Note that 528 cm equals 5.28 m. All we had to do was to move the decimal point!

example 24.2   Convert 486 kilometers to meters.

solution   We will use the unit multiplier that has *km* on the bottom.

$$486 \text{ km} \times \frac{1000 \text{ m}}{1 \text{ km}} = \textbf{486,000 m}$$

Again the digits are the same. The only difference is the position of the decimal point.

practice Use unit multipliers to convert:

   a. 58 meters to centimeters   **b.** 100 meters to kilometers

   c. 48,000 centimeters to meters

problem set
24

1. At the turn of the century, fourteen thousand, seven things were on the proscribed list. Thirty years later only seven thousand, nine hundred forty-two remained on this list. How many had been removed?

2. Mary measured it first. She got one hundred forty-seven and nine hundred twenty-three millionths inches. Then Jim tried. He got one hundred forty-six and three thousand, one hundred forty-two ten-thousandths inches. By how much was Mary's measurement greater?

3. Each slot could hold only 47 balls. If there were 1982 balls, how many slots were necessary to hold them all?

4. Which of the following numbers are divisible (a) by 2, and (b) by 10?

       40,613   90,528   71,423   4020

5. (a) List the prime numbers between 20 and 30.
   (b) List the multiples of 4 between 20 and 30.

6. Find the average of 98, 142, 76, 81, and 6.

Use unit multipliers to convert:

   7. 859 centimeters to meters   **8.** 6400 kilometers to meters

   9. 204 inches to feet   **10.** 47 yards to feet

Simplify. If possible, cancel as the first step.

   11. $\dfrac{16}{24} \cdot \dfrac{12}{4} \cdot \dfrac{1}{3}$ **12.** $\dfrac{18}{24} \cdot \dfrac{8}{9} \cdot \dfrac{3}{2}$ **13.** $\dfrac{14}{21} \cdot \dfrac{7}{2} \div \dfrac{14}{6}$

14. Find the perimeter of the following figure. Dimensions are in feet. All angles are right angles.

15. Find the area of the following figure. Dimensions are in inches. All angles are right angles.

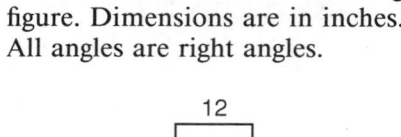

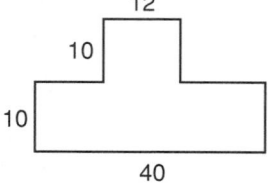

Write each number as a product of primes:

16. 480   **17.** 2400

Write each fraction as a decimal. Round to two decimal places.

   18. $\dfrac{3}{14}$ **19.** $\dfrac{4}{17}$ **20.** $\dfrac{81}{135}$

Write each number as a product of prime factors and simplify:

21. $\frac{24}{72}$ **22.** $\frac{18}{24}$

Multiply:

23. 61.7×121 **24.** $712(1.23)$

25. Subtract (add to check): $712.61 - 699.781$

26. Divide and round to two places: $\frac{7.162}{0.7}$

27. Write each fraction with a denominator of 42:

(a) $\frac{3}{14}$ (b) $\frac{4}{7}$ (c) $\frac{33}{126}$

28. Round $218.0\overline{97}$ to the nearest millionth.

29. Round $3,817,321.127$ to the nearest hundred.

30. Use words to write the number 1317621.13.

LESSON 25 *Area as a difference*

We have been finding the areas of figures by adding rectangular areas. Sometimes we can find the total area by finding the difference of rectangular areas.

example 25.1 Use the difference of areas to find the area of this figure. Dimensions are in meters. All angles are right angles.

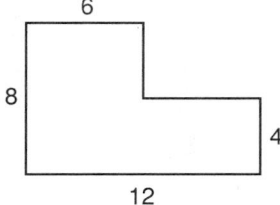

solution We add dashed lines to complete the rectangle.

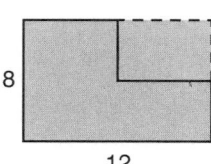   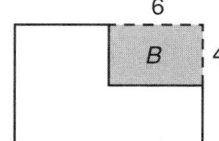

Then we can find the area of the whole rectangle shown on the left and subtract from it the area of the small rectangle *B* on the right.

$$8 \text{ m} \times 12 \text{ m} = 96 \text{ m}^2 \qquad 6 \text{ m} \times 4 \text{ m} = 24 \text{ m}^2$$

Thus the area of the figure is

$$96 \text{ m}^2 - 24 \text{ m}^2 = \textbf{72 m}^2$$

example 25.2 Use the difference of areas to find the area of this figure. Dimensions are in feet. All angles are right angles.

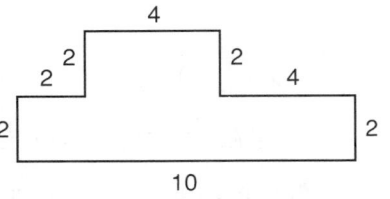

solution We use dashed lines to complete the big rectangle. Then we find the area of the big rectangle, and from this we subtract the areas of the two small rectangles we have labeled A and B.

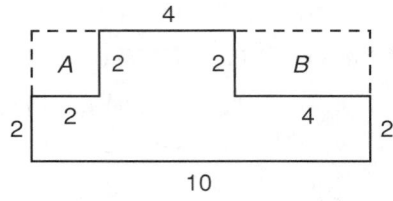

$$\begin{aligned}
\text{Big rectangle:} \quad & 4 \text{ ft} \times 10 \text{ ft} = 40 \text{ ft}^2 \\
A: \quad & 2 \text{ ft} \times 2 \text{ ft} \;\; = \;\; 4 \text{ ft}^2 \\
B: \quad & 2 \text{ ft} \times 4 \text{ ft} \;\; = \;\; 8 \text{ ft}^2
\end{aligned}$$

$$\text{Area} = 40 \text{ ft}^2 - 4 \text{ ft}^2 - 8 \text{ ft}^2 = \mathbf{28 \text{ ft}^2}$$

practice Use the difference of areas to find the area of each figure. All angles are right angles.

    a. Dimensions are in centimeters.   **b.** Dimensions are in feet.

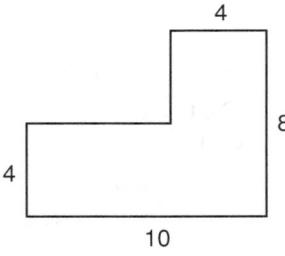
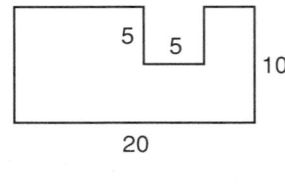

problem set 25

1. In the first month ten thousand, forty-two students used the library. In the second month 4 times this number used the library. In the third month, only four thousand, seventy-five used the library. How many students used the library in all?

2. The flowers proliferated until they numbered 17,842. Then the first frost killed 9085. How many flowers were still alive?

3. The first night 482 came. Five times this number came on the second night. On the third night only 392 came. How many came in all?

4. Which of the following numbers is divisible by 3?

315,234   906,185   21,387   4072

5. (a) List the prime numbers between 40 and 50.
(b) List the multiples of 3 between 40 and 50.

6. The average of six numbers is 48.6. The first five numbers are 6.2, 5.1, 9.8, 8.8, and 10. Find the sixth number.

Use unit multipliers to convert:

7. 46.31 meters to centimeters

8. 48 miles to feet

9. 416 meters to kilometers

Simplify. If possible, cancel as the first step.

10. $\dfrac{42}{24} \cdot \dfrac{6}{7} \cdot \dfrac{2}{3}$

11. $\dfrac{72}{48} \cdot \dfrac{8}{9} \cdot \dfrac{3}{4}$

12. $\dfrac{2}{3} \cdot \dfrac{27}{64} \div \dfrac{3}{16}$

13. Find the perimeter of the following figure. Dimensions are in feet. All angles are right angles.

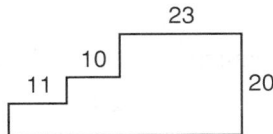

14. Use the difference of areas to find the area of this figure. Dimensions are in meters. All angles are right angles.

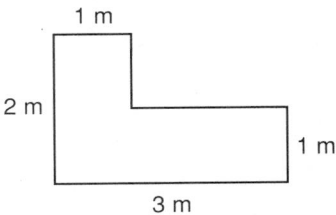

Write each number as a product of primes:

15. 288

16. 630

Write each fraction as a decimal. Round to two decimal places.

17. $\dfrac{3}{19}$

18. $\dfrac{8}{23}$

Write each fraction as a product of prime factors and simplify:

19. $\dfrac{39}{72}$

20. $\dfrac{210}{420}$

21. $\dfrac{125}{360}$

Multiply:

22. 1123.1×3.1

23. 0.187×92.8

24. Subtract (add to check): $718.87 - 9.876$

25. Find the missing digits:

$$\begin{array}{r} MN.AK \\ -\ 1.38 \\ \hline 9.93 \end{array}$$

Divide and round to two places:

26. $\dfrac{625.8}{1.7}$

27. $\dfrac{1361.4}{0.4}$

28. Round 18,976,511.31$\overline{052}$ to the nearest hundred thousand.

29. Use words to write the number 111321654.7

30. Add: $713.85 + 0.3162 + 3.14$

LESSON 26 *Mode, median, and mean · Average in word problems*

26.A
mode, median,
and mean

In statistics courses, there are three words that are used to describe the distribution of a group of numbers. In this list of numbers

2, 2, 2, 2, 2, 7, 8, 8, 9, 10, and 784

the number 2 appears more than any other number. In statistics we say that the number that appears the most often is the **mode** of a group of numbers. In this list the number 7 is the middle number. In statistics we call the middle number the **median** of a group of numbers. If we add the numbers and divide this sum by the number of numbers, we find what is called in statistics the **mean** of a group of numbers.

$$\text{Mean} = \frac{2 + 2 + 2 + 2 + 2 + 7 + 8 + 8 + 9 + 10 + 784}{11} = 76$$

Beginners often have difficulty in remembering that *mean* and *average* are two words with the same meaning. We will use the word *mean* often in the problem sets so that this word becomes a familiar word. We will consider the mode and the median again in future lessons.

26.B
average in
word problems

Many word problems require that we find the average, or the mean, of a group of numbers.

example 26.1 When the boys measured the distance 3 times, they got three different answers. If the measurements were 4.017 meters, 4.212 meters, and 3.996 meters, what was the mean of the three measurements?

solution We add the numbers and then divide the sum by 3.

$$\begin{array}{r} 4.017 \\ 4.212 \\ 3.996 \\ \hline 12.225 \end{array} \qquad \frac{12.225}{3} = \mathbf{4.075}$$

example 26.2 When Athena sprang in full armor from the head of Zeus, two other gods guessed the weight of her panoply. One guess was 41.802 kilograms and the other guess was 37.408 kilograms. What was the mean of the two guesses?

solution We add the two numbers and divide the sum by 2.

$$\begin{array}{r} 41.802 \\ 37.408 \\ \hline 79.210 \end{array} \qquad \frac{79.210}{2} = \mathbf{39.605 \ kg}$$

example 26.3 The mean weight of the three pigs was 320 pounds. The first pig weighed 220 pounds and the second pig weighed 240 pounds. What did the third pig weigh?

solution The average weight was 320 pounds, so the sum of the weight was 3 times 320.

$$\text{Sum} = 3 \times 320 = 960 \text{ pounds}$$

The first two pigs weighed 220 pounds and 240 pounds for a total of 460 pounds.

$$
\begin{array}{rl}
960 & \text{total} \\
-\ 460 & \text{weight of first two pigs} \\
\hline
500 &
\end{array}
$$

The third pig weighed **500 pounds.**

$$220 + 240 + 500 = 960 \quad \textbf{check}$$

practice The height of the first of the three buildings was 420 ft. The height of the second building was 380 ft. What was the height of the third building if the mean height of the buildings was 430 ft?

problem set 26

1. The mean weight of the four trucks was 3700 pounds. The first truck weighed 3250 pounds, the second truck weighed 3890 pounds, and the third truck weighed 3640 pounds. What did the fourth truck weigh?

2. The altitude to be achieved by the club's first rocketry experiment was estimated to be 214.063 meters. Later, club members estimated that the rocket would reach an altitude of at least 435.09 meters. What was the mean of their estimations?

3. Find the average of the first eight prime numbers.

4. Which of the following numbers is divisible by 3? What is the average of the numbers that are divisible by 3?

$$41{,}625 \quad\quad 9081 \quad\quad 20{,}733 \quad\quad 10{,}662$$

5. There were 325 pellets in every shotgun shell. The shipment contained 495,625 pellets. This would be enough pellets to make how many shotgun shells?

6. The first toss was 60,152.035 units. The second toss was 82,006.012 units. By how many units was the second toss greater?

Use unit multipliers to convert:

7. 136.15 centimeters to meters

8. 12 miles to feet

9. 1899 meters to kilometers

Simplify.

10. $\dfrac{21}{36} \cdot \dfrac{12}{14} \cdot \dfrac{2}{3}$

11. $\dfrac{36}{42} \cdot \dfrac{21}{9} \cdot \dfrac{1}{2}$

12. $\dfrac{28}{36} \cdot \dfrac{8}{21} \div \dfrac{16}{27}$

13. Find the area. Dimensions are in meters. All angles are right angles.

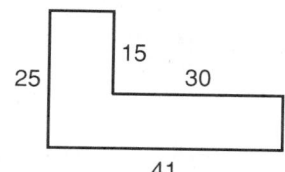

14. Use the difference of areas to find the area of this figure. Dimensions are in feet. All angles are right angles.

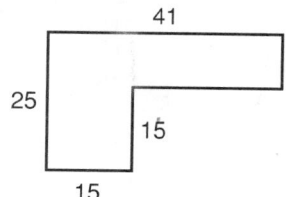

Write each number as a product of primes:

15. 6300

16. 1050

Write each fraction as a decimal. Round to two decimal places.

17. $\frac{7}{13}$

18. $\frac{11}{17}$

Write each fraction as a product of prime factors and simplify:

19. $\frac{24}{28}$

20. $\frac{240}{300}$

21. $\frac{144}{256}$

Multiply:

22. 132.1×6.1

23. 17.21×0.11

Subtract. Add to check.

24. $1361.78 - 31.921$

25. $789.891 - 77.892$

Divide and round to two places.

26. $\frac{7621.1}{2.1}$

27. $\frac{3.623}{0.02}$

28. Round 781,136.36293 to three decimal places.

29. Use words to write the number 6211357.5.

30. Add: $613.7211 + 0.0178 + 17.369$

LESSON 27 *Areas of triangles*

The **altitude,** or **height,** of a triangle is the perpendicular distance from the base of the triangle or extension of the base to the opposite corner. Any one of the three sides can be designated as the base.

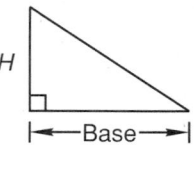
(a)

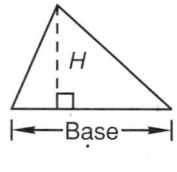
(b)

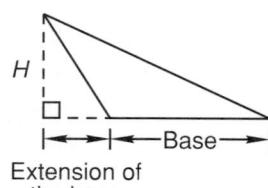
Extension of the base
(c)

The altitude can (a) be one of the sides of the triangle, (b) fall inside the triangle, or (c) fall outside the triangle. When the altitude falls outside the triangle, we have to extend the base so that the altitude can be drawn. This extension of the base is not part of the length of the base.

The formula for the area of any triangle is the product of the base and altitude divided by 2.

$$\text{Area} = \frac{\text{base} \times \text{height}}{2}$$

example 27.1   Find the areas of these triangles. Dimensions are in inches.

(a) 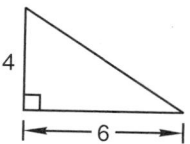   (b) 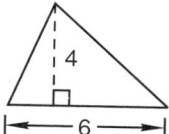   (c)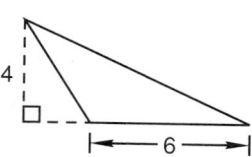

solution   The base of each triangle is 6 in and the height of each triangle is 4 in. Thus, all the triangles have the same area.

$$\text{Area} = \frac{\text{base} \times \text{height}}{2} = \frac{6 \text{ in.} \times 4 \text{ in.}}{2} = \textbf{12 in.}^2$$

practice   Find the area of each triangle. Dimensions are in feet.

a. 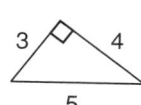   b. 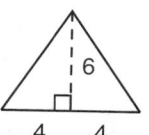   c.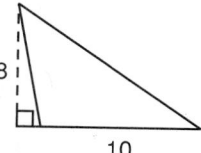

problem set 27

1. The first shift produced 14,782 units. The second shift produced 18,962 units. The third shift produced only 2085 units. By how many units did the production of the second shift exceed the sum of the production of the first and third shifts?

2. Seven thousand, four hundred eight hundred-thousandths was Mary's guess. Tom guessed six thousand, five hundred eighty-two ten-thousandths. Whose guess was larger and by how much?

3. Nine hundred eighty-four was increased by one hundred forty-seven thousand, seventeen. Then this sum was doubled. What was the final result?

4. Find the average of the prime numbers that are greater than 20 but less than 42.

5. The first number was 743. The second number was 486. The third number was twice the sum of the first two. What was the mean of the three numbers?

6. Which of the following numbers is divisible by 5?

   41,320   90,183   76,142   9405   50,172

Use one of the unit multipliers to convert:

7. 81 yards to feet

8. 81 feet to yards

9. 4899 meters to centimeters

Simplify:

10. $\dfrac{20}{24} \cdot \dfrac{6}{15} \cdot \dfrac{2}{3}$

11. $\dfrac{16}{20} \cdot \dfrac{5}{8} \div \dfrac{1}{4}$

12. $\dfrac{18}{20} \cdot \dfrac{5}{3} \div \dfrac{6}{9}$

13. Find the perimeter of this figure. Dimensions are in feet. All angles are right angles.

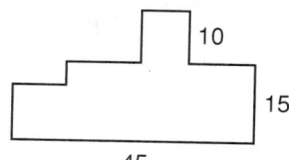

Find the area of each figure. Dimensions are in feet.

14.
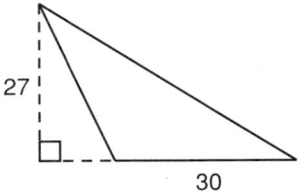

15.
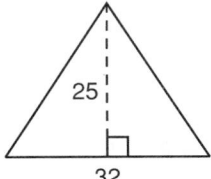

Write each number as a product of primes:

16. 324   **17.** 2916

Write each fraction as a decimal. Round to two decimal places.

18. $\frac{7}{17}$ **19.** $\frac{7}{13}$ **20.** $\frac{6}{11}$

Write each number as a product of prime factors and simplify:

21. $\frac{72}{120}$ **22.** $\frac{280}{360}$

23. Multiply: 612.1×2.8

24. Subtract (add to check): $18{,}521.62 - 3.8972$

25. Simplify mentally: $850{,}311.092 \div 100{,}000$

26. Order from least to greatest: 112.091, 112.9, 1129.0, 112.00956

Divide and round each answer to two decimal places.

27. $\frac{611.32}{0.04}$ **28.** $\frac{0.0062}{0.07}$

29. Round $781.10\overline{563}$ to the nearest hundred-thousandth.

30. Use words to write the number 7811.7821.

LESSON 28   *Improper fractions and mixed numbers*

28.A

improper fractions and mixed numbers

If the numerator of the fraction is less than the denominator of the fraction, the fraction represents a number less than 1 and is called a **proper fraction.** The fraction $\frac{2}{3}$ has a numerator that is less than the denominator and is therefore a proper fraction.

$\frac{2}{3}$ designates a number less than 1, as we see in the figure

If the numerator of a fraction is greater than or equal to the denominator of the fraction, the fraction represents a number greater than or equal to 1 and the fraction is called an **improper fraction.** Thus the number

$$\frac{10}{3}$$

can be called an improper fraction. We see that the denominator is 3, so the basic part is $\frac{1}{3}$ of a whole.

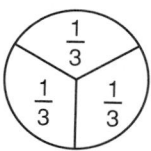

Each part of this circle is $\frac{1}{3}$ of the whole, and the fraction $\frac{10}{3}$ tells us that we are considering 10 of these parts, each of which is $\frac{1}{3}$ of a whole.

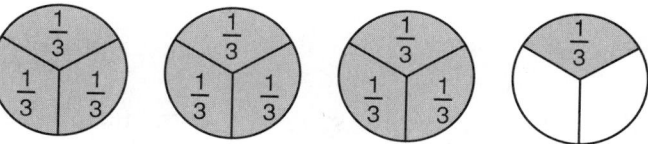

Each of the first three circles has 3 shaded parts equal to $\frac{1}{3}$ of a whole. The last circle has 1 shaded part for a total of 10 shaded parts. From this we can see that the number represented by $\frac{10}{3}$ is the same number as the number represented by $3\frac{1}{3}$.

$$\frac{10}{3} \quad \text{equals} \quad 3\frac{1}{3}$$

The number $3\frac{1}{3}$ has two parts. One part is the number 3, which is a whole number. The other part is the fraction $\frac{1}{3}$. We call a number that has **both a whole part and a fractional part a mixed number.** We note that $3\frac{1}{3}$ really means

$$\text{three plus one-third} \quad \text{or} \quad 3 + \frac{1}{3}$$

Because we use mixed numbers so often, we omit the plus sign to make the number easier to write. But we must remember that

$$3\frac{1}{3} \quad \text{means} \quad 3 + \frac{1}{3}$$

28.B
improper fraction to mixed number

To change an improper fraction to a mixed number, we divide the denominator into the numerator. The whole part of the answer is the whole part of the mixed number. The remainder will be the numerator of the fraction of the mixed number.
To convert the improper fraction $\frac{10}{3}$ to a mixed number, we divide 10 by 3.

$$3\overline{)10} = 3 \text{ R}1$$

This tells us that $\frac{10}{3}$ represents 3 wholes, with a remainder of 1. So

$$\frac{10}{3} = 3\frac{1}{3}$$

example 28.1 Convert the following improper fractions to mixed numbers or to whole numbers.

(a) $\frac{14}{5}$ (b) $\frac{21}{8}$

solution   In each case we divide the denominator into the numerator to find the number of wholes represented by the fraction. The remainder will be the numerator of the fraction of the mixed number.

(a) $\dfrac{14}{5}$ $\longrightarrow$ $5\overline{\smash{)}14}$ so $\dfrac{14}{5} = 2$ R4   Thus $\dfrac{14}{5}$ can be written as $2\dfrac{4}{5}$.

$$\begin{array}{r} 2 \\ 5\overline{)14} \\ 10 \\ \hline 4 \end{array}$$

(b) $\dfrac{21}{8}$ $\longrightarrow$ $8\overline{\smash{)}21}$ so $\dfrac{21}{8} = 2$ R5   Thus $\dfrac{21}{8}$ can be written as $2\dfrac{5}{8}$.

$$\begin{array}{r} 2 \\ 8\overline{)21} \\ 16 \\ \hline 5 \end{array}$$

28.C
mixed number to improper fraction

In the preceding section we found that $\frac{10}{3}$ and $3\frac{1}{3}$ were equal numbers,

$$\frac{10}{3} = 3\frac{1}{3}$$

and we found that the mixed number that is equal to $\frac{10}{3}$ can be found by dividing 10 by 3.

$$\begin{array}{r} 3 \\ 3\overline{)10} \\ 9 \\ \hline 1 \end{array} \qquad \text{so} \qquad \frac{10}{3} \text{ equals } 3\frac{1}{3}$$

To go the other way and find the improper fraction represented by the mixed number $3\frac{1}{3}$, we will use our picture of $3\frac{1}{3}$ again.

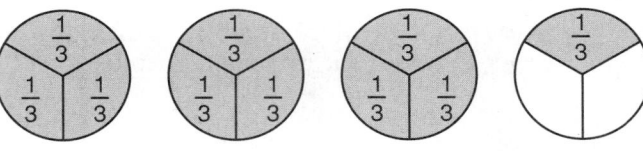

We see here that the number $3\frac{1}{3}$ is composed of 10 parts, each with a size of $\frac{1}{3}$. Nine of these one-thirds come from the whole number 3, and one of them comes from the fraction $\frac{1}{3}$. We can say, therefore, that the mixed number $3\frac{1}{3}$ is the same as the improper fraction $\frac{10}{3}$.

For another example, we will write $2\frac{1}{8}$ as an improper fraction. To do this we must find out how many $\frac{1}{8}$'s there are in 2 and then add one more $\frac{1}{8}$.

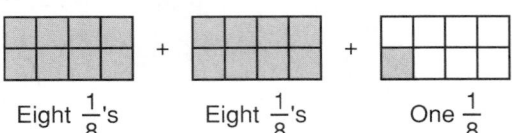

Eight $\frac{1}{8}$'s   Eight $\frac{1}{8}$'s   One $\frac{1}{8}$

Eight $\frac{1}{8}$'s plus eight $\frac{1}{8}$'s plus one $\frac{1}{8}$ equals seventeen $\frac{1}{8}$'s, or $\dfrac{17}{8}$

Thus, the procedure for finding the improper fraction represented by a mixed number is to (1) find the number of fractional parts in the whole number and then (2) add to this the number of fractional parts designated by the fraction. We can do this in two steps by

1. Multiplying the denominator of the fraction by the whole number.
2. Adding to this product the numerator of the fraction and recording the sum over the denominator of the fraction.

Thus to write $5\frac{3}{11}$ as an improper fraction, we write

$$\frac{5 \times 11 + 3}{11} = \frac{58}{11}$$

Some people find it helpful to remember the procedure by thinking of it as a circular process.

$$5\frac{3}{11} = 5\overbrace{\underset{\times}{+}}^{3}\underset{11} = \frac{11 \times 5 + 3}{11} = \frac{55 + 3}{11} = \frac{58}{11}$$

We begin at the bottom with 11, multiply by 5, then add 3, and record over 11.

example 28.2 Write $7\frac{4}{9}$ as an improper fraction.

$$7\overbrace{\underset{\times}{+}}^{4}\underset{9} = \frac{9 \times 7 + 4}{9} = \frac{63 + 4}{9} = \frac{67}{9}$$

We multiply 9 by 7 to get 63 and then add 4 to get the numerator of 67; then we record this over the denominator of 9.

practice Write each mixed number as an improper fraction:

a. $5\frac{2}{3}$ **b.** $6\frac{7}{8}$

Write each improper fraction as a mixed number:

c. $\frac{14}{3}$ **d.** $\frac{23}{5}$

problem set 28

1. Four hundred seventeen ten-thousandths is how much greater than four hundred seventeen hundred-thousandths?

2. The first flock contained 5283 birds. The second flock contained 5 times as many birds. The third flock had twice as many birds as did the second flock. How many birds were there in all?

3. Seven times 4,820,718 is a very large number. Write this number in words.

4. Find the average of the prime numbers that are greater than 10 and less than 25.

Convert each fraction to a mixed number:

5. $\frac{17}{9}$ 6. $\frac{24}{7}$

7. Draw a diagram that shows why $4\frac{1}{4}$ equals $\frac{17}{4}$.

Convert each mixed number to an improper fraction:

8. $6\frac{2}{9}$ 9. $8\frac{4}{7}$ 10. $5\frac{2}{8}$

Use unit multipliers to convert:

11. 192.72 centimeters to meters 12. 17 miles to feet

Simplify:

13. $\dfrac{28}{36} \cdot \dfrac{24}{21} \cdot \dfrac{3}{16}$ **14.** $\dfrac{18}{24} \cdot \dfrac{36}{28} \cdot \dfrac{14}{27}$ **15.** $\dfrac{16}{18} \cdot \dfrac{16}{12} \div \dfrac{8}{9}$

16. Find the area of the following figure (total area = area of rectangle + area of triangle). Dimensions are in feet. All angles that look like right angles are right angles.

17. Express the perimeter in centimeters. Dimensions are in meters. All angles are right angles.

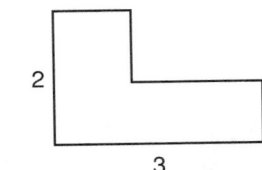

18. Write 1440 as a product of prime factors.

Write each fraction as a decimal. Round to two decimal places.

19. $\dfrac{16}{23}$ **20.** $\dfrac{7}{19}$

Write each number as a product of prime factors and simplify:

21. $\dfrac{36}{42}$ **22.** $\dfrac{280}{320}$ **23.** $\dfrac{125}{175}$

24. List the multiples of 3 that are greater than 15 and are less than 30.

25. Multiply: 61.72×1.89

26. Subtract (add to check): $172.325 - 61.89$

27. Divide and round to two places: $\dfrac{7811.3}{0.03}$

28. Round 6,789,211.82 to the nearest million.

29. Use words to write 78256113.7.

30. Simplify mentally: $62,562.13 \div 100$

LESSON 29 *Multiplying fractions and whole numbers ·*
Fractional part of a number

29.A
multiplying fractions and whole numbers

When we multiply two fractions, we multiply the tops to get the new top. We multiply the bottoms to get the new bottom.

$$\frac{1}{4} \cdot \frac{3}{5} = \frac{1 \cdot 3}{4 \cdot 5} = \frac{3}{20}$$

We remember that every whole number can be written as a fraction by writing a 1 below the number. Thus both of these

$$\frac{4}{1} \qquad 4$$

are ways to write the number 4. To multiply 4 by $\frac{2}{9}$, we can either write the 1 under the 4, as we do on the left, or omit the 1, as we do on the right.

$$\frac{2}{9} \cdot \frac{4}{1} = \frac{8}{9} \qquad \text{or} \qquad \frac{2}{9} \cdot 4 = \frac{8}{9}$$

29.B
top, bottom, and middle

A fraction has a top. A fraction has a bottom. A fraction does not have a middle. Some people are confused by the notation used to indicate the multiplication of a fraction and a whole number, such as

$$4 \cdot \frac{3}{5}$$

because they cannot remember whether to multiply the 4 by 3 or by 5. The way it is written, it appears that the 4 is somehow in the middle. Of course,

$$4 \cdot \frac{3}{5} \qquad \text{means} \qquad \frac{4}{1} \cdot \frac{3}{5}$$

For this reason, students often find it helpful to write the 1 below the whole number, as we have done here.

29.C
fractional part of a number

When we multiply a whole number by a fraction, the answer is a part of the whole number. In (a) we show the number 15 and see that if we divide it into 5 equal parts (b) each part is 3. Each of these parts is $\frac{1}{5}$ of 15.

$$\frac{1}{5} \times 15 = 3 \qquad \frac{2}{5} \times 15 = 6 \qquad \frac{4}{5} \times 15 = 12$$

In (c) we show $\frac{2}{5} \times 15$ means 2 of the parts, and the answer is 6. In (d) we show that $\frac{4}{5} \times 15$ means 4 of the parts, and the answer is 12.

We use the word **of** to designate multiplication. Thus,

$$\frac{2}{5} \text{ of } 15 \qquad \text{means} \qquad \frac{2}{5} \times 15$$

$$\frac{4}{5} \text{ of } 15 \qquad \text{means} \qquad \frac{4}{5} \times 15$$

example 29.1 Find three-sevenths of 42.

solution The word **of** means to multiply. So we multiply 42 by $\frac{3}{7}$.

$$\frac{3}{7} \times 42 = \frac{126}{7} = \mathbf{18}$$

example 29.2 What number is $\frac{5}{7}$ of 12?

solution   The word **of** means to multiply.

$$\frac{5}{7} \cdot 12 = \frac{60}{7} = 8\frac{4}{7}$$

practice   **a.**   Find $\frac{4}{3}$ of 30.   **b.**   What number is $\frac{3}{4}$ of 15?

c.   What number is $\frac{5}{7}$ of 3?

problem set   **1.**   The first try flew nineteen thousand and seventy-five millionths feet. The
29   second try flew twenty-one thousand and one thousand, three millionths feet.
                        What was the sum of the two tries?

2.   First came fourteen thousand nine hundred eighty-two. Then 10 times this
         number came. How many came in all?

3.   What is the mean of the prime numbers greater than 30 but less than 45?

4.   One truck could carry 840 units. If 19,420 units had to be transported, how
         many trucks were required?

5.   In the preceding problem, if only 18 trucks could be located, how many units
         had to be left on the loading dock?

6.   Find $\frac{5}{6}$ of 48.   **7.**   Find $\frac{6}{7}$ of 35.

Convert each fraction to a mixed number:

8. $\frac{15}{7}$ **9.** $\frac{21}{5}$

10.   Draw a diagram that shows why $2\frac{1}{4}$ equals $\frac{9}{4}$.

Convert each mixed number to an improper fraction:

11. $7\frac{3}{8}$ **12.** $6\frac{2}{3}$ **13.** $5\frac{7}{11}$

Use one unit multiplier to convert:

14.   199.62 meters to centimeters   **15.**   18 miles to feet

Simplify. If possible, cancel as the first step.

16. $\frac{42}{48} \cdot \frac{32}{14} \cdot \frac{3}{4}$ **17.** $\frac{18}{20} \cdot \frac{16}{8} \div \frac{27}{2}$

18.   Convert centimeter measures to
          meters and then find the area of
          this figure in square meters. All
          angles that look like right angles
          are right angles.

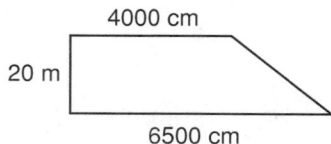

Write each number as a product of primes:

19.   2100   **20.**   5250

Write each fraction as a decimal. Round to two decimal places.

21. $\frac{16}{23}$ **22.** $\frac{8}{13}$

Write each number as a product of prime factors and simplify:

23. $\dfrac{183}{270}$ **24.** $\dfrac{360}{420}$

Multiply:

25. 31.25×0.0012 **26.** 17.01×0.12

27. Subtract (add to check): $61.892 - 9.299$

28. Divide and round the answer to two decimal places: $\dfrac{6218.3}{0.007}$

29. Round 2831.8211261 to the hundred-thousandths' place.

30. Use words to write 13.82567.

LESSON 30 *Graphs*

Graphs are used to present numerical information in picture form. The two most common forms of graphs are bar graphs and broken-line graphs. The two graphs shown here present the same information.

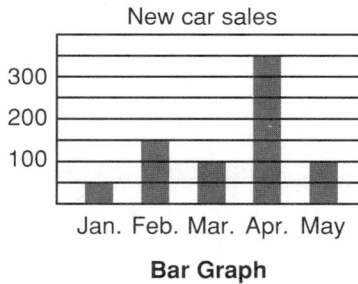

Bar Graph

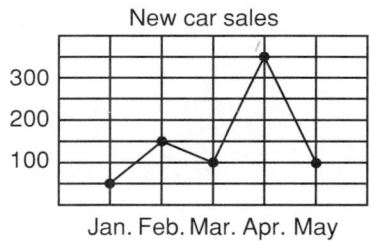

Broken-Line Graph

Both graphs show sales of new cars each month from January through May. Some people believe that the bar graph is easier to interpret, whereas others prefer broken-line graphs. Note that exact information cannot be obtained from either kind of graph.

When we draw graphs, the quantities plotted vertically and horizontally should be clearly labeled. Also, care should be taken when choosing the scale. Whenever possible, the major divisions on the scales should be divisible by 2 or by 10. Scales in which the major divisions are numbers such as 3 or 7 are often difficult to read.

example 30.1 In the graphs above, how many more cars were sold in April than were sold in January?

solution We cannot read the graphs exactly, but it appears that 350 cars were sold in April, whereas only 50 were sold in January.

$$350 - 50 = \textbf{300 more cars sold in April}$$

example 30.2   The highest temperatures recorded during the week were as follows: Monday, 86°; Tuesday, 80°; Wednesday, 75°; Thursday, 82°; Friday, 88°; Saturday, 74°; and Sunday, 84°. Make a bar graph and a broken-line graph that present this information.

solution   It is not necessary to show the full vertical scale. We will use a vertical scale that goes between 70° and 100°.

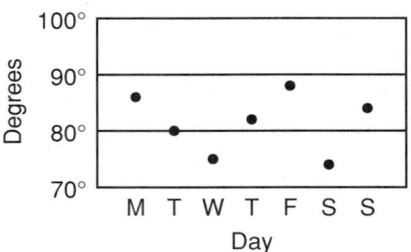

In the figure on the left below we draw bars of proper heights to make a bar graph, and in the figure on the right, we connect the points with straight lines to make a broken-line graph.

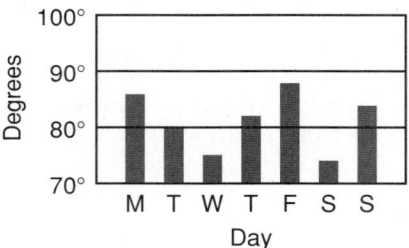 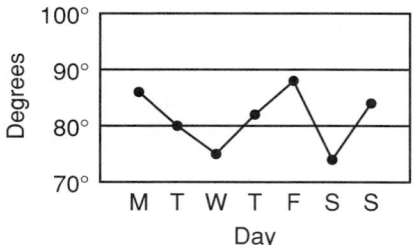

practice   In this graph, what was the average number of type B cars sold in the 3-year period?

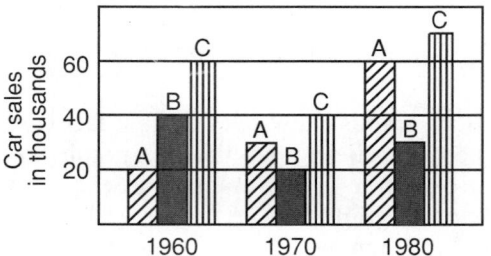

problem set 30

1. How many fewer cars were sold in February than were sold in April?

2. The highest temperature on Monday was 80°. The highest temperature on Wednesday was 60°. The highest temperature on Friday was 70°. Make a bar graph that displays this information.

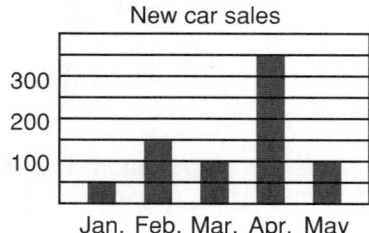
New car sales

3. Hamilcar saw an average of 12 elephants daily for the 3-day period. He saw 8 elephants on the first day and 7 elephants on the second day. How many elephants did he see on the third day?

4. When Hannibal opened the boxes, he found that each box contained 53 spear points. If 5062 spears needed new points, how many boxes did he have to open?

5. The first batch contained 5142 red marbles and 7821 green marbles. The second batch had twice as many red marbles and 3 times as many green marbles. How many total red marbles and green marbles were there in both batches?

6. Find $\frac{3}{8}$ of 42.

7. Find $\frac{6}{7}$ of 49.

Convert each fraction to a mixed number:

8. $\frac{17}{8}$

9. $\frac{22}{5}$

10. Draw a diagram that shows that $3\frac{1}{5}$ equals $\frac{16}{5}$.

Convert each mixed number to an improper fraction:

11. $7\frac{2}{3}$

12. $7\frac{6}{7}$

13. $4\frac{5}{8}$

Use one unit multiplier to convert:

14. 721 yards to feet

15. 19,262 centimeters to meters

Simplify. If possible, cancel as the first step.

16. $\frac{28}{32} \cdot \frac{24}{21} \cdot \frac{3}{4}$

17. $\frac{16}{18} \cdot \frac{20}{24} \div \frac{10}{9}$

18. Convert meter measures to centimeters and then express the area of this figure in square centimeters.

19. Convert centimeters measures to meters and then express the perimeter of this figure in meters. All angles are right angles.

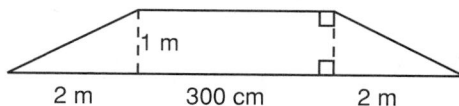

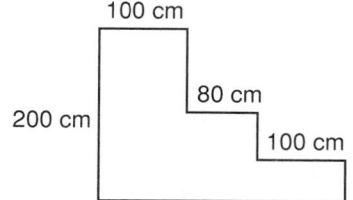

Write each number as a product of prime factors and simplify:

20. $\frac{180}{200}$

21. $\frac{256}{720}$

Write each fraction as a decimal. Round to two decimal places.

22. $\frac{16}{21}$

23. $\frac{9}{17}$

Write each number as a product of primes:

24. 2450

25. 1650

Multiply:

26. 612.7×0.0017

27. 1821.1×0.025

28.   Subtract (add to check): $71.328 - 19.6214$

29.   Divide and round the answer to two decimal places: $\dfrac{689.2}{0.005}$

30.   Round $0.02\overline{735}$ to the nearest ten-millionth.

LESSON 31   *Least common multiple*

The smallest whole number into which several other whole numbers will divide evenly is called the **least common multiple (LCM)** of the several whole numbers. The least common multiple of the numbers

2, 3, and 5

is 30 because 30 is the smallest number into which the three numbers will divide and have a remainder of zero. We can find the LCM of some numbers by making mental calculations, but it is nice to have a special procedure to use if some of the numbers are large numbers. The procedure is as follows:

1. Write each number as a product of prime factors.
2. Compute the LCM by using every factor of the given numbers as a factor of the LCM. Use each factor the greatest number of times it is a factor in any of the numbers.

To demonstrate this procedure, we will find the LCM of

18, 81, and 500

First we write each number as a product of prime factors.

$$18 = 2 \cdot 3 \cdot 3 \qquad 81 = 3 \cdot 3 \cdot 3 \cdot 3 \qquad 500 = 2 \cdot 2 \cdot 5 \cdot 5 \cdot 5$$

Now we find the LCM by using the procedure in step 2. The number 2 is a factor of both 18 and 500. It is a factor twice in 500 so it will appear twice in the LCM.

$$2 \cdot 2$$

The number 3 is a factor of both 18 and 81. It appears 4 times in 81 so it will appear 4 times in the LCM.

$$2 \cdot 2 \cdot 3 \cdot 3 \cdot 3 \cdot 3$$

The number 5 is the other factor. It appears 3 times in 500 so it will appear 3 times in the LCM.

$$\text{LCM} = 2 \cdot 2 \cdot 3 \cdot 3 \cdot 3 \cdot 3 \cdot 5 \cdot 5 \cdot 5 = \mathbf{40,500}$$

example 31.1   Find the LCM of 24, 55, and 80.

solution   First we write the numbers as products of prime factors.

$$24 = 2 \cdot 2 \cdot 2 \cdot 3 \qquad 55 = 5 \cdot 11 \qquad 80 = 2 \cdot 2 \cdot 2 \cdot 2 \cdot 5$$

To compute the LCM, we use each factor the greatest number of times it is a factor of any of the numbers.

$$LCM = 3 \cdot 5 \cdot 11 \cdot 2 \cdot 2 \cdot 2 \cdot 2 = \mathbf{2640}$$

example 31.2 Below we show three numbers expressed as products of prime factors.
(a) What are the numbers?
(b) What is the LCM of the numbers?

  (1) $2 \cdot 2 \cdot 3 \cdot 7 \cdot 7$ (2) $2 \cdot 2 \cdot 2 \cdot 2 \cdot 5 \cdot 7$ (3) $5 \cdot 5 \cdot 5 \cdot 5 \cdot 2$

solution (a) We multiply to find the numbers.

  (1) $2 \cdot 2 \cdot 3 \cdot 7 \cdot 7 = \mathbf{588}$ (2) $2 \cdot 2 \cdot 2 \cdot 2 \cdot 5 \cdot 7 = \mathbf{560}$

  (3) $5 \cdot 5 \cdot 5 \cdot 5 \cdot 2 = \mathbf{1250}$

(b) $LCM = 2 \cdot 2 \cdot 2 \cdot 2 \cdot 3 \cdot 5 \cdot 5 \cdot 5 \cdot 5 \cdot 7 \cdot 7 = \mathbf{1,470,000}$

Thus, we find that the smallest number that is evenly divisible by 588, 560, and 1250 is 1,470,000.

practice Use products of prime numbers to find the LCM of each group of numbers.
a. 35, 60, and 90   **b.** 17, 33, and 45

problem set 31

1. Mildred bought 20 show tickets for $2.50 each. If she gave the agent a $100 bill, how much change did she receive?

2. How many more type B cars were sold in 1980 than were sold in 1970?

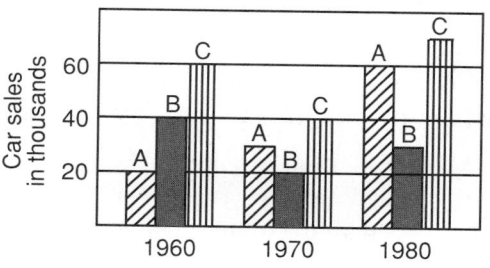

3. Try as he might, Waltersheid found that he could do only 175 in an hour. If he had to do 2975 in all, how many hours would it take him?

4. The average cost of 5 kits was $9.50. The first kit cost $11.25, the second kit cost $8.75, the third kit cost $9.35, and the fourth kit cost $8.90. Find the cost of the fifth kit.

5. In Saudi Arabia it takes 20 kursh to equal 1 riyal.
(a) Write the two unit multipliers implied by this relationship.
(b) Use one of the unit multipliers to convert 16,000 kursh to riyals.

6. What number is $\frac{5}{6}$ of 960?   **7.** What number is $\frac{11}{5}$ of 55?

Convert each fraction to a mixed number:

8. $\frac{41}{3}$ **9.** $\frac{52}{16}$

10. Draw a diagram that shows why $4\frac{1}{3}$ equals $\frac{13}{3}$.

Find the least common multiple of:

11. 14, 94, and 300 **12.** 27, 66, and 90 **13.** 40, 85, and 100

14. Write $15\frac{2}{3}$ as an improper fraction.

15. Use one unit multiplier to convert 14,780,000 centimeters to meters.

Simplify:

16. $\frac{15}{18} \cdot \frac{6}{30} \div \frac{3}{2}$ **17.** $\frac{160}{180} \cdot \frac{10}{12} \div \frac{20}{18}$

18. Convert meter measures to centimeters and then find the area of this figure in square centimeters.

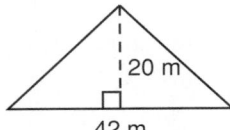

20 m
42 m

19. Convert meter measures to centimeters and then find the area of this figure in square centimeters. All angles are right angles.

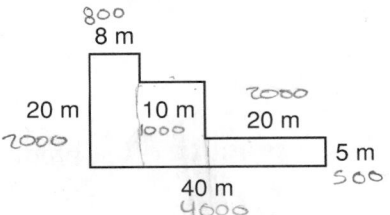

8 m
20 m   10 m   20 m
                     5 m
40 m

20. Write 460 as a product of prime factors.

21. Simplify $\frac{210}{240}$. Begin by writing both numerator and denominator as products of prime factors.

Write each fraction as a decimal number. Round to two decimal places.

22. $\frac{42}{17}$ **23.** $\frac{19}{23}$

Multiply:

24. 4.016 × 0.0027   **25.** 30.60 × 0.0409   **26.** 3.0162 × 4.008

27. Subtract (add to check): 71.428 − 14.649

28. Divide and round the answer to two decimal places: $\frac{412.36}{0.0061}$

29. Round 415.62852 to the nearest thousandth.

30. Use words to write 42000.00162.

LESSON 32 *Adding fractions*

32.A
adding fractions

The denominator (bottom) of a fraction tells us into how many parts the whole is divided. The numerator (top) of a fraction indicates how many of these parts we are considering. Since the fractions $\frac{1}{5}$ and $\frac{3}{5}$ both have 5 in the denominator, we know that each of the wholes has been divided into 5 parts. In one of the fractions we are considering 1 of the parts, and in the other we are considering 3 of the parts.

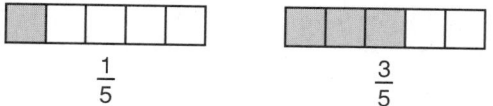

If we add the fractions $\frac{1}{5}$ and $\frac{3}{5}$, we get $\frac{4}{5}$,

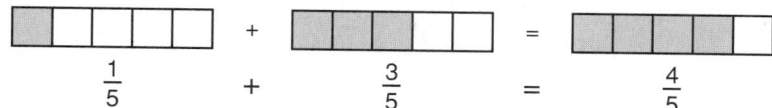

for we have one $\frac{1}{5}$ from the first and three $\frac{1}{5}$'s from the second for a total of four $\frac{1}{5}$'s. This gives us a picture of why we add the numerators when we add fractions whose denominators are equal.

FRACTIONS WITH EQUAL DENOMINATORS

To add or subtract fractions **whose denominators are equal,** the numerators are added or subtracted as indicated by the + and − signs, and the result is recorded over a single denominator.

32.B
unequal denominators

When we wish to add fractions whose denominators are not equal

$$\frac{1}{4} + \frac{1}{2}$$

we must first transform one or both fractions as necessary so that the fractions will have equal denominators. If we multiply the top and bottom of the second fraction above by 2,

$$\frac{1}{4} + \frac{1(2)}{2(2)} = \frac{1}{4} + \frac{2}{4}$$

we can change it to $\frac{2}{4}$. Now both bottoms are equal, and we can add.

$$\frac{1}{4} + \frac{2}{4} = \frac{3}{4}$$

In this problem the original denominators were 2 and 4. The new denominator for both fractions is 4 because 4 is the smallest number that is evenly divisible by both 2 and 4. Thus 4 is the LCM of 2 and 4. When we use the least common multiple of several numbers as the new denominator, we call it the **least common denominator,** which we abbreviate as **LCD.**

We remember that the rule we used to change $\frac{1}{2}$ to $\frac{2}{4}$ is the denominator-numerator rule for fractions. We could call this rule the **bottom-top rule for fractions.**

BOTTOM-TOP RULE FOR FRACTIONS

Both the bottom and the top of a fraction can be multiplied or divided by the same number (except zero) without changing the value of the fraction.

In later courses, this rule is called the fundamental theorem of fractions. We will call it the bottom-top rule because this name is descriptive and is easy to remember. We will use four steps to add fractions with unequal denominators.

1. Find the least common denominator of the fractions.
2. Write all fractions with this new denominator.
3. Find the new numerators.
4. Add the fractions.

example 32.1   Simplify: $\dfrac{5}{11} - \dfrac{1}{5}$

solution   Both of these denominators are prime numbers so the least common denominator is 5×11 or 55. We write the fractions with this new denominator.

$$\overline{55} - \overline{55}$$

In the first fraction, we have multiplied the old bottom of 11 by 5 to get 55, so we must multiply the old top by 5. We get

$$\dfrac{25}{55} - \overline{55}$$

The old bottom of the second fraction was 5. Now it is 55. We have multiplied the old bottom by 11. Thus, we must multiply the old top by 11. Now we have

$$\dfrac{25}{55} - \dfrac{11}{55} = \mathbf{\dfrac{14}{55}}$$

example 32.2   Simplify: $\dfrac{1}{4} + \dfrac{5}{9} + \dfrac{2}{3}$

solution   First we find the LCD of the fractions.

$$2 \cdot 2 \quad \text{and} \quad 3 \cdot 3 \quad \longrightarrow \quad 2 \cdot 2 \cdot 3 \cdot 3 = 36$$

Now we use 36 as the new denominator for each fraction.

$$\overline{36} + \overline{36} + \overline{36}$$

In the leftmost fraction, we have multiplied the old bottom by 9 to get 36, so we also multiply the old top by 9.

$$\dfrac{9}{36} + \overline{36} + \overline{36}$$

In the center fraction, we have multiplied the old bottom by 4 to get 36, so we must also multiply the old top by 4.

$$\dfrac{9}{36} + \dfrac{20}{36} + \overline{36}$$

In the rightmost fraction, we have multiplied the old bottom by 12 to get 36, so we must also multiply the old top by 12. Then we add.

$$\dfrac{9}{36} + \dfrac{20}{36} + \dfrac{24}{36} = \dfrac{53}{36} = 1\mathbf{\dfrac{17}{36}}$$

practice Simplify:

a. $\dfrac{3}{4} - \dfrac{2}{5}$

b. $\dfrac{5}{3} + \dfrac{1}{2} - \dfrac{1}{4}$

c. $\dfrac{5}{8} - \dfrac{1}{2} + \dfrac{1}{4}$

d. $\dfrac{2}{3} + \dfrac{3}{4} - \dfrac{1}{5}$

problem set 32

1. The mean cost of four items was $36.96. The first item cost $28.50, the second item cost $41.25, and the third item cost $50. Find the cost of the fourth item.

2. The attendance at the first six home games was 10,400; 8,000; 14,600; 7,000; 12,000; 15,700. Make a broken-line graph that presents this information.

3. In the first cache, Hazel uncovered 1481 dull ones. The next cache held 1300 shiny ones, and the last cache had 300 that were both shiny and dull. What was the average number per cache?

4. The time for the first increment was seven hundred thousand and one thousand, four hundred forty hundred-thousandths seconds. The time for the second increment was only six hundred forty-two thousand, fourteen and seventy-five ten-thousandths seconds. By how much was the time for the first increment greater?

5. Four thousand, fifty-seven screaming football fans could be seated in a single section of the stadium. If the stadium had 17 sections, how many screaming football fans would it hold?

6. What number is $\dfrac{4}{13}$ of 39?

7. What number is $\dfrac{11}{5}$ of 500?

Convert each fraction to a mixed number:

8. $\dfrac{93}{13}$

9. $\dfrac{41}{7}$

10. Draw a diagram that shows why $3\dfrac{4}{5}$ equals $\dfrac{19}{5}$.

Find the least common multiple of:

11. 100, 40, and 90

12. 200, 120, and 180

13. 12, 20, and 72

Simplify:

14. $\dfrac{2}{3} + \dfrac{5}{8} + \dfrac{1}{2}$

15. $\dfrac{6}{7} - \dfrac{4}{5}$

16. $\dfrac{3}{5} + \dfrac{4}{7} - \dfrac{1}{3}$

17. $\dfrac{4}{5} \cdot \dfrac{20}{40} \div \dfrac{5}{10}$

18. $\dfrac{2}{8} \cdot \dfrac{16}{24} \div \dfrac{15}{8}$

19. 0.0023×1.047

20. $3.024 \div 0.042$

21. $483.124 - 1.632$

22. Twenty shillings equals 1 pound.
    (a) Write the two unit multipliers implied by this comparison.
    (b) Use one of these unit multipliers to convert 1000 pounds to shillings.

23. Find the area of this figure. Dimensions are in feet.

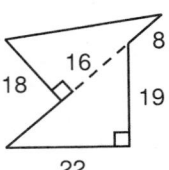

24. Find the perimeter of this figure. Dimensions are in centimeters. All angles are right angles.

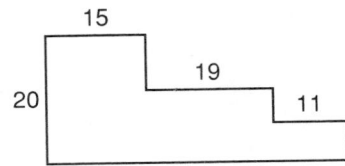

25. Simplify this fraction. Begin by writing both the numerator and the denominator as a product of prime factors:

$$\frac{260}{305}$$

26. (a) List the prime numbers between 50 and 60.
(b) List the multiples of 4 between 50 and 60.

Write each fraction as a decimal. Round to two decimal places.

27. $\frac{52}{63}$

28. $\frac{14}{19}$

29. Round 42,062,918,004.01372 to the nearest thousand.

30. Use words to write 42062918004.001372.

LESSON 33 *Order of operations*

There seems to be two ways that the following expression can be simplified.

$$4 + 3 \times 5$$

It looks as though we could add first and then multiply, or multiply first and then add.

(a) ADD FIRST	(b) MULTIPLY FIRST
$4 + 3 \times 5$	$4 + 3 \times 5$
7×5	$4 + 15$
35	**19**

Thus, there seem to be two possible answers to this problem. Since two answers to the same problem would cause considerable confusion, mathematicians have had to agree on one way or the other. They have agreed to always

<u>**Multiply before adding or subtracting**</u>

Thus, the value of the expression above is 19 and is not 35.

example 33.1 Simplify: $41 - 3 \cdot 5 + 4 \cdot 6$

solution Two multiplications are indicated. We do these first and then move from left to right, performing the additions and subtractions in the order we encounter them.

$$41 - 15 + 24 \qquad \text{multiplied}$$
$$26 + 24 \qquad \text{subtracted 15 from 41}$$
$$\mathbf{50} \qquad \text{added}$$

example 33.2 Simplify: $4 + 5 \cdot 3 - 2 \cdot 4 + 6 \cdot 3$

solution We do the multiplications first and get

$$4 + 15 - 8 + 18$$

We finish by doing the additions and subtractions from left to right.

$$19 - 8 + 18 \qquad \text{added 4 and 15}$$
$$11 + 18 \qquad \text{subtracted 8 from 19}$$
$$\mathbf{29} \qquad \text{added 11 and 18}$$

practice Simplify:

a. $4 + 3 \cdot 6 - 2 \cdot 3$ **b.** $2 + 3 \cdot 6 - 4 \cdot 3$

problem set 33

1. For every 37 tickets she sold, Martine was given a free ticket for one of the children. If 175 children wanted to go, how many tickets did Martine have to sell?

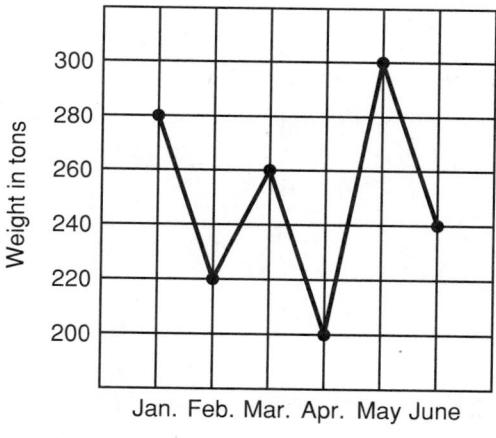

Tons per month

2. By how many pounds was the weight in April less than the weight in January. There are 2000 pounds in 1 ton.

3. Ralph's average score for the 5 tests was 82. On four of the tests his scores were 85, 73, 92, and 66. What was his score on the other test?

4. Nineteen million, four hundred eighty-four thousand and seventy-five thousandths is a big number. Twenty-two million, thirteen and nine hundred eighty-four ten-thousandths is even bigger. What is the difference between these numbers?

5. Red ones cost $5 each, and Don bought 7 red ones. Blue ones were $3.40 each, and he bought 9 of these. Green ones were only $1.30 each, so he bought 20 of these. How much money did Don spend?

6. What number is $\frac{5}{16}$ of 128?   7. What number is $\frac{14}{3}$ of 30?

Write each fraction as a mixed number:

8. $\frac{93}{12}$ 9. $\frac{40}{7}$

Find the least common multiple of:

10. 24, 36, and 40   11. 8, 108, and 180

Simplify:

12. $\frac{3}{7} + \frac{2}{5} - \frac{3}{10}$ 13. $\frac{5}{8} + \frac{3}{5} - \frac{1}{4}$

14. $32 - 2 \cdot 5 + 3 \cdot 6$ 15. $7 + 5 \cdot 3 - 2 \cdot 4 + 5 \cdot 3$

16. $3 + 3 \cdot 5 - 4 \cdot 2$ 17. $\frac{4}{5} \cdot \frac{25}{20} \div \frac{5}{10}$

18. $\frac{25}{36} \cdot \frac{12}{5} \div \frac{5}{6}$ 19. 0.016×0.0023

20. $\frac{400.7}{0.0016}$ 21. $\frac{7}{11} - \frac{1}{3}$

22. There are 16 ounces in 1 pint of liquid.
   (a) Write the two unit multipliers implied by this relationship.
   (b) Use one of these unit multipliers to convert 640 ounces to pints.

23. Find the area of this figure. Dimensions are in feet.   24. Find the perimeter of this figure. Dimensions are in yards. All angles are right angles.

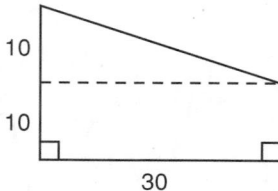

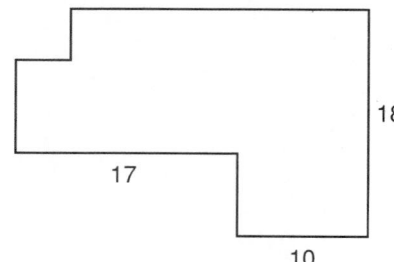

25. Write 4160 as a product of prime factors.

Write each fraction as a decimal number:

26. $\frac{13}{5}$ 27. $\frac{17}{4}$

28. Write $\frac{7}{18}$ as a decimal number. Round to two places.

29. Round 99,535,620 to the nearest ten thousand.

30. Use words to write 0.0021587.

LESSON 34 *Variables and evaluation*

In algebra we often use letters as **variables** to stand for or to take the places of numbers. The letters themselves have no value. The value of the expression

$$x + 4$$

depends on the number we use for x. If we replace x with 11, the value of the expression is 15.

$$11 + 4 = 15$$

When two letters are written together, such as

$$xy$$

the notation means that x and y are to be multiplied.

example 34.1   Evaluate: $xy + x$ if $x = 2$ and $y = 4$

solution   The word **evaluate** means "find the value of." We replace x with 2 and y with 4.

$$xy + x = 2 \cdot 4 + 2$$

We remember to multiply before we add. Thus, we get

$$8 + 2 = \mathbf{10}$$

example 34.2   Evaluate: $xmy - xy$ if $x = 2$, $m = 5$, and $y = 4$

solution   We replace x with 2, replace m with 5, and replace y with 4.

$$xmy - xy = 2 \cdot 5 \cdot 4 - 2 \cdot 4$$

We multiply before we subtract and get

$$40 - 8 = \mathbf{32}$$

example 34.3   Evaluate: $mx + 4m$ if $x = \dfrac{2}{3}$ and $m = \dfrac{9}{11}$

solution   First we replace m and x with $\dfrac{9}{11}$ and $\dfrac{2}{3}$, respectively.

$$mx + 4m = \frac{9}{11} \cdot \frac{2}{3} + 4 \cdot \frac{9}{11}$$

Now we multiply and then add.

$$\frac{6}{11} + \frac{36}{11} = \frac{42}{11} = \mathbf{3\frac{9}{11}}$$

practice   Evaluate:

a. $xy + yx$ if $x = 2$ and $y = 4$

b. $mpx + mx$ if $x = 3$, $p = 4$, and $m = 6$

problem set 34

1. Silicon chips were packed 420 to the box. If 130,420 chips were to be shipped, how many boxes would be needed?

2. In the catalog, apple trees were $15.95 each, and Sam bought 3 apple trees. He bought 7 nectarine trees at $17.75 each and 4 peach trees for $11.95 each. How much money did he spend for fruit trees?

3. What was the average of the car sales for the five months listed on this broken-line graph?

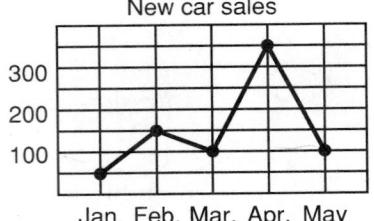
New car sales

4. The weights of the first four loads were 14,000 pounds, 12,000 pounds, 18,200 pounds, and 16,280 pounds. What was the mean weight of the four loads?

5. If 32 books could fit on a shelf, how many shelves would be required for 544 books?

6. What number is $\frac{4}{9}$ of 72?

7. What number is $\frac{5}{17}$ of 136?

Write each fraction as a mixed number:

8. $\frac{37}{3}$

9. $\frac{421}{5}$

Find the least common multiple of:

10. 12, 210, and 600

11. 60, 84, and 120

Simplify:

12. $\frac{3}{4} + \frac{5}{8} + \frac{2}{3} - \frac{1}{6}$

13. $\frac{5}{8} + \frac{1}{16} + \frac{1}{2} - \frac{1}{4}$

14. $3 + 2 \cdot 6 - 4 \cdot 3$

15. $5 \cdot 2 - 3 \cdot 2 + 4 \cdot 3$

16. $\frac{16}{25} \times \frac{15}{8} \div \frac{3}{2}$

17. $\frac{30.03}{0.0021}$

18. $\frac{4}{3} - \frac{7}{10}$

19. $1742.05 - 11.06$

20. 0.0016×4000.085

21. 40 gallons = 1 barrel
    (a) Write the two unit multipliers implied by this relationship.
    (b) Use one of these unit multipliers to convert 2500 barrels to gallons.

Evaluate:

22. $xy + 2m$ if $x = 2$, $y = 4$, and $m = 3$

23. $xym + xy$ if $x = 3$, $m = 4$, and $y = 6$.

24. $mx + 4m$ if $x = \frac{2}{3}$ and $m = 5$

25. Find the area of this figure. Dimensions are in feet.

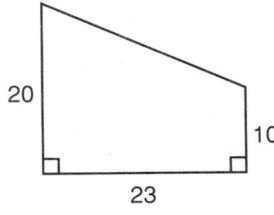

20

10

23

26. Find the perimeter of this figure. Dimensions are in meters. All angles are right angles.

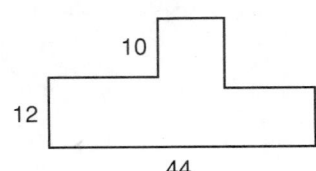

10

12

44

Write each fraction as a decimal:

27. $\dfrac{3}{5}$ **28.** $\dfrac{1.3}{5}$

29. Round 41,062,315 to the nearest ten thousand.

30. Use words to write 41002.00164.

LESSON 35 *Multiple unit multipliers*

Unit multipliers are used to help prevent mistakes. We can use unit multipliers almost without thinking and be sure the answer we get is correct. We will often use an extra step in unit conversions if the extra step makes the process more automatic. To convert from inches to yards, we could remember that 1 yard equals 36 inches. This conversion would require only one unit multiplier but would require a little extra thought. A more automatic approach would be to go from inches to feet and from feet to yards.

example 35.1 Convert 360 inches to yards.

solution We will use one unit multiplier to convert from inches to feet. Then we will use another to convert from feet to yards.

$$360 \text{ in.} \cdot \frac{1 \text{ ft}}{12 \text{ in.}} \cdot \frac{1 \text{ yd}}{3 \text{ ft}} = \frac{360}{36} \text{ yd} = \textbf{10 yd}$$

example 35.2 Use two unit multipliers to convert 1.4 kilometers to centimeters.

$$1.4 \text{ km} \cdot \frac{1000 \text{ m}}{1 \text{ km}} \cdot \frac{100 \text{ cm}}{1 \text{ m}} = \textbf{140,000 cm}$$

We might have remembered that 1 kilometer equals 100,000 centimeters, but attempting shortcuts like this often leads to errors.

example 35.3 Use two unit multipliers to convert 24 miles to inches.

solution We will go from miles to feet to inches.

$$24 \text{ mi} \cdot \frac{5280 \text{ ft}}{1 \text{ mi}} \cdot \frac{12 \text{ in.}}{1 \text{ ft}} = \textbf{24(5280)(12) in.}$$

problem set 35

1. The first measurement was four hundred seventeen ten-thousandths. The second measurement was forty-five thousandths. Which measurement was larger and by how much?

2. The two hundred tatterdemalions had an average of 45 cents each. The ragamuffins numbered 450, and they had an average of 55 cents each. What was the total value of the money of both groups?

3. How much more does a 25-year employee make in 2 years than a 10-year employee makes in 2 years?

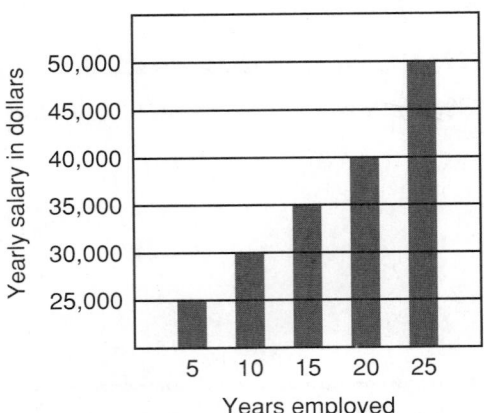

4. The legal limit for the number of cars in a parade was 450 cars. If 9000 drivers wanted to be in a parade, how many parades were required to accommodate all of them?

5. The fat ones weighed 417 lb, 832 lb, and 619 lb. The leans weighed only 148 lb, 212 lb, and 184 lb. What was the average weight of the six fats and leans?

6. What number is $\frac{4}{7}$ of 28?   7. What number is $\frac{5}{12}$ of 48?

Write each fraction as a mixed number:

8. $\frac{214}{5}$ 9. $\frac{47}{2}$

10. Find the least common multiple of 50, 60, and 72.

Simplify:

11. $\frac{3}{4} + \frac{5}{8} + \frac{3}{16}$ 12. $\frac{13}{15} - \frac{1}{5}$

13. $3 \cdot 12 - 4 \cdot 2 + 3 \cdot 5$ 14. $2 \cdot 5 \cdot 2 - 3 \cdot 5 + 2 - 5$

15. 4.014×0.027 16. $\frac{3.052}{0.07}$

17. $513.002 - 0.0009$ 18. $\frac{4}{6} \times \frac{9}{14} \div \frac{2}{5}$

19. $\frac{3}{7} \times \frac{21}{6} \div \frac{2}{4}$

Evaluate:

20. $xy - y$ if $x = 5$ and $y = 4$

21. $m - xy$ if $m = 10$, $x = 2$, and $y = 3$

22. $xym - m$ if $x = \frac{1}{2}$, $y = 4$, and $m = 2$

Use two unit multipliers to convert:

23. 400 inches to yards

24. 3.6 kilometers to centimeters

25. 32 miles to inches

26. Express the area of this figure in square centimeters. Begin by changing 0.4 m to centimeters.

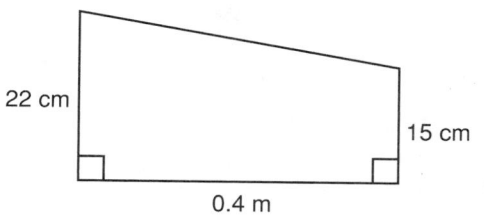

27. Express the perimeter of this figure in centimeters. Begin by changing 0.1 m to centimeters. All angles are right angles.

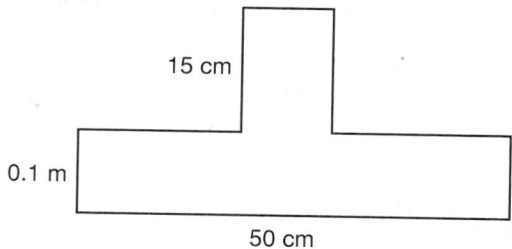

28. Write $\frac{7}{8}$ as a decimal number.

29. Round $5.\overline{5}$ to the nearest hundred-thousandth.

30. Simplify mentally: $6193.06091 \times 1,000,000$

LESSON 36 *Adding mixed numbers · Rate*

36.A
adding mixed numbers

There are <u>two ways to add mixed numbers</u>. When the fractions are simple and the whole parts are small, it is sometimes convenient to begin by changing the mixed numbers to improper fractions. Then we find a common denominator and add the fractions.

example 36.1   Add: $2\frac{1}{4} + 3\frac{1}{8} + 7\frac{1}{2}$

solution   First we rewrite each number as an improper fraction.

$$\frac{9}{4} + \frac{25}{8} + \frac{15}{2}$$

Next we change the fractions to equivalent fractions whose denominators are 8 and add.

$$\frac{18}{8} + \frac{25}{8} + \frac{60}{8} = \frac{103}{8} = 12\frac{7}{8}$$

36.B
adding like parts

When the whole numbers are large numbers, it is convenient to add the whole numbers and the fractions separately.

example 36.2   Add: $528\frac{1}{3} + 7142\frac{3}{4}$

solution   We rewrite the problem in a vertical format. Then we rewrite the fractions with a common denominator and add.

$$528\frac{1}{3} = \ \ 528\frac{4}{12}$$

$$7142\frac{3}{4} = 7142\frac{9}{12}$$

$$7670\frac{13}{12} = \textbf{7671}\frac{\textbf{1}}{\textbf{12}} \quad \text{added and simplified}$$

example 36.3   Add: $421\frac{3}{5} + 274\frac{1}{20}$

solution   Again we will use the vertical format.

$$421\frac{3}{5} = 421\frac{12}{20}$$

$$274\frac{1}{20} = 274\frac{1}{20}$$

$$\textbf{695}\frac{\textbf{13}}{\textbf{20}} \quad \text{added}$$

36.C
rate   A **rate** is a ratio. A **ratio** is a comparison of two numbers. A ratio can be written as a fraction. If two apples can be purchased for 16 cents, we can write either

$$\frac{2 \text{ apples}}{16 \text{ cents}} \quad \text{or} \quad \frac{16 \text{ cents}}{2 \text{ apples}}$$

Often we use the word *per* instead of writing the units as a fraction and write

$$\frac{2 \text{ apples}}{16 \text{ cents}} = \frac{1}{8} \text{ apple per cent} \quad \text{or} \quad \frac{16 \text{ cents}}{2 \text{ apples}} = 8 \text{ cents per apple}$$

example 36.4   Waldo bought 5 toys for 40 dollars.
(a)   What was the rate (ratio) in dollars per toy?
(b)   What was the rate (ratio) in toys per dollar?

solution   (a)   The denominate number that comes first goes on top. So for dollars per toy, we write

$$\frac{40 \text{ dollars}}{5 \text{ toys}} = \frac{8 \text{ dollars}}{1 \text{ toy}} = \textbf{8 dollars per toy}$$

(b)   For the rate in toys per dollar, we write the toys on top.

$$\frac{5 \text{ toys}}{40 \text{ dollars}} = \frac{1 \text{ toy}}{8 \text{ dollars}} = \frac{\textbf{1}}{\textbf{8}} \textbf{ toy per dollar}$$

practice   Add:

a. $2\frac{1}{5} + 3\frac{1}{10}$

b. $472\frac{2}{5} + 312\frac{3}{4}$

c.   Roger could travel 48 yards in 6 seconds. What was his rate in yards per second and what was his rate in seconds per yard?

d.   Eight apples can be purchased for 2 dollars. What is the rate in apples per dollar and what is the rate in dollars per apple?

problem set 36

1. The mean weight of the seven linebackers was 236 pounds. The first six linebackers weighed 215, 305, 265, 196, 221, and 236 lb, respectively. What was the weight of the seventh linebacker?

2. The values of the crops sold each year were as follows: 1940, $700,000; 1945, $800,000; 1950, $1,400,000; 1955, $2,000,000; 1960, $3,000,000. Draw a bar graph that presents this information.

3. The children bought 7 notebooks for $5.40 each, 200 pencils at 30 cents each, and 40 reams of paper at $22.50 a ream. How much did they spend in all?

4. Four hundred sixty quarts could be packaged in one shift. If 10,120 quarts were needed, how many shifts would be required?

5. Marco Polo bought 40 skins for 8 lira. What was the rate in skins per lira and in liras per skin?

6. What number is $\frac{3}{11}$ of 33?

7. What number is $\frac{4}{5}$ of 200?

Write each fraction as a mixed number:

8. $\frac{21}{4}$

9. $\frac{86}{11}$

10. Find the least common multiple of 40, 50, and 70.

Simplify:

11. $\frac{3}{5} - \frac{1}{15}$

12. $\frac{1}{10} + \frac{3}{5} - \frac{1}{20}$

13. $3 + 5 - 2 \cdot 4 + 3 \cdot 5$

14. $4 + 3(2) + 5 \cdot 4$

15. $2\frac{1}{4} + 3\frac{1}{8}$

16. $429\frac{1}{5} + 8162\frac{4}{15}$

17. $534\frac{3}{8} + 371\frac{1}{40}$

18. 51.67×0.081

19. $\frac{716.2}{0.008}$

20. $713.891 - 0.8917$

21. $\frac{3}{8} \times \frac{24}{9} \div \frac{3}{7}$

Evaluate:

22. $zy - z$ if $z = 3$ and $y = 4$

23. $xyz + yz$ if $x = \frac{1}{3}$, $y = 9$, and $z = 2$

Use two unit multipliers to convert:

24. 540 inches to yards

25. 187,625.8 cm to kilometers

26. Find the area of this figure. Dimensions are in inches.

27. Find the perimeter of this figure. Dimensions are in centimeters. All angles are right angles.

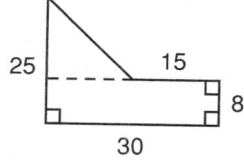

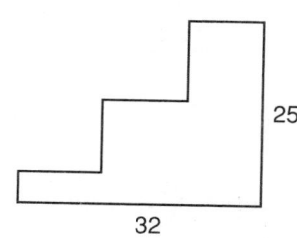

28. (a) List the prime numbers between 20 and 36.
     (b) List the multiples of 7 between 20 and 36.

29. Round 658,976,581 to the nearest hundred thousand.

30. Simplify mentally: $230,976,092 \div 100,000$

LESSON 37   *Mixed number subtraction*

37.A
mixed number subtraction

When the whole number parts are small and the fractions are simple, both numbers can be written as improper fractions and then subtracted.

example 37.1   Subtract: $4\frac{1}{4} - 1\frac{1}{8}$

solution   We first write both numbers as improper fractions.

$$\frac{17}{4} - \frac{9}{8}$$

Next we rewrite both numbers with equal denominators and subtract.

$$\frac{34}{8} - \frac{9}{8} = \frac{25}{8} = 3\frac{1}{8}$$

37.B
subtracting like parts

When the numbers are large, it is convenient to subtract the parts individually. A vertical format is often used.

example 37.2   Subtract: $416\frac{3}{4} - 21\frac{1}{16}$

solution   We use the vertical format and rewrite the fractions with equal denominators. Then we subtract.

$$416\frac{3}{4} = \quad 416\frac{12}{16}$$
$$21\frac{1}{16} = - \ 21\frac{1}{16}$$
$$\overline{\qquad\qquad 395\frac{11}{16}}$$

Sometimes it is necessary to borrow before we can subtract, as we see in the next example.

example 37.3   Subtract: $461\frac{1}{3} - 82\frac{13}{15}$

solution We use the vertical format and equal denominators.

$$461\frac{1}{3} = \quad 461\frac{5}{15}$$

$$- \; 82\frac{13}{15} = - \; 82\frac{13}{15}$$

We can't subtract $\frac{13}{15}$ from $\frac{5}{15}$. Thus, we borrow $\frac{15}{15}$ from 461.

$$461\frac{5}{15} = 460 + \frac{15}{15} + \frac{5}{15} = 460\frac{20}{15}$$

Now we can subtract.

$$460\frac{20}{15}$$

$$- \; 82\frac{13}{15}$$

$$378\frac{7}{15}$$

practice Subtract:

a. $7\frac{1}{5} - 4\frac{2}{3}$

b. $316\frac{1}{5} - 42\frac{7}{8}$

problem set 37

1. Winifred paid $180 for the first 10 items. Write the two rates (ratios) implied by this statement.

2. The mean daily rate of production for 5 days at the bottling plant was 1 million bottles of Big Pop per day. The totals for the first 4 days were 998,163 bottles, 899,989 bottles, 1,200,316 bottles, and 987,900 bottles. How many bottles of Big Pop were produced on the fifth day?

3. By how much did the income in 1950 exceed the income in 1910?

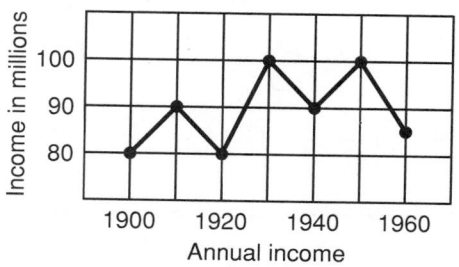

Annual income

4. If 415 came in each unit and if 29,050 were required, how many units had to be obtained?

5. What number is $\frac{5}{16}$ of 32?

6. What number is $\frac{4}{7}$ of 280?

Write each fraction as a mixed number:

7. $\frac{31}{5}$

8. $\frac{93}{7}$

9. Find the least common multiple of 8, 36, and 70.

Simplify:

10. $\dfrac{3}{5} - \dfrac{2}{10}$ **11.** $\dfrac{3}{8} + \dfrac{1}{2} - \dfrac{1}{4}$ **12.** $4 + 3 \cdot 2 - 5$

13. $3\dfrac{1}{8} + 2\dfrac{1}{4} + 5\dfrac{1}{2}$ **14.** $428\dfrac{1}{11} + 22\dfrac{1}{44}$ **15.** $3\dfrac{2}{5} + 748\dfrac{2}{10}$

16. $4\dfrac{4}{10} - 1\dfrac{1}{5}$ **17.** $548\dfrac{6}{8} - 31\dfrac{1}{16}$ **18.** $991\dfrac{1}{3} - 791\dfrac{17}{18}$

19. 71.82×8.01 **20.** $\dfrac{936.7}{0.04}$ **21.** $691.872 - 17.816$

22. $\dfrac{21}{8} \times \dfrac{4}{14} \div \dfrac{9}{2}$

Evaluate:

23. $xy + yz - z$ if $x = 1$, $y = 7$, and $z = 2$

24. $xyz - xy$ if $x = 2$, $y = 3$, and $z = 3$

25. Use two unit multipliers to convert 10 miles to inches.

26. Find the area of this figure. Dimensions are in yards.

27. Find the perimeter of this figure. Dimensions are in inches. All angles are right angles.

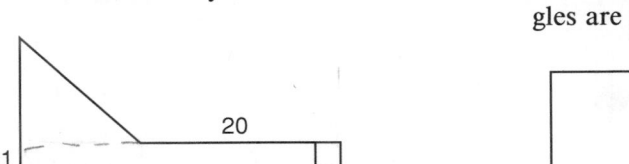

28. Write $\dfrac{11}{13}$ as a decimal number. Round to two decimal places.

29. Round $1.05\overline{2}$ to six decimal places.

30. Use words to write 71326811.23.

LESSON 38   *Rate word problems*

Rate word problems are equal group problems. The rate is the number in an equal group. Suppose 20 apples cost 10 dollars. The two rates (ratios) are

(a) $\dfrac{20 \text{ apples}}{10 \text{ dollars}}$ (b) $\dfrac{10 \text{ dollars}}{20 \text{ apples}}$

If we multiply (a) by 8 dollars, we get apples because the dollars will cancel.

$$\frac{20 \text{ apples}}{10 \text{ dollars}} \times 8 \text{ dollars} = 16 \text{ apples}$$

If we multiply (b) by 8 apples, we get dollars because the apples will cancel.

$$\frac{10 \text{ dollars}}{20 \text{ apples}} \times 8 \text{ apples} = 4 \text{ dollars}$$

When the problem asks "how many apples," we will use the rate that has apples on top. When the problem asks "how many dollars" or "what will be the cost," we will use the rate that has dollars on top. When we solve rate problems, we can always multiply to find the answer. This is because we have our choice of two rates.

example 38.1   Twenty apples cost 5 dollars. How many apples can we buy for 80 dollars?

solution   The pattern is

$$\boxed{\text{rate}} \times \boxed{\text{number}} = \text{total}$$

The two rates (ratios) for this problem

$$\text{(a)} \quad \frac{20 \text{ apples}}{5 \text{ dollars}} \qquad \text{(b)} \quad \frac{5 \text{ dollars}}{20 \text{ apples}}$$

We want to know the number of apples, so we use rate (a) because it has apples on top.

$$\frac{20 \text{ apples}}{5 \text{ dollars}} \times 80 \text{ dollars} = \textbf{320 apples}$$

example 38.2   Twenty apples cost 5 dollars. What will be the cost of 40 apples?

solution   The pattern is

$$\boxed{\text{rate}} \times \boxed{\text{number}} = \text{total}$$

The two rates (ratios) are

$$\text{(a)} \quad \frac{20 \text{ apples}}{5 \text{ dollars}} \qquad \text{(b)} \quad \frac{5 \text{ dollars}}{20 \text{ apples}}$$

We want to know the cost. We want to know the number of dollars. We will use (b) because it has dollars on top.

$$\frac{5 \text{ dollars}}{20 \text{ apples}} \times 40 \text{ apples} = \textbf{10 dollars}$$

practice   **a.**   Chestnuts cost 40 cents for 2 ounces. Write the two rates (ratios) implied by this statement. How many ounces of chestnuts can Joan buy for $1.20?

b.   Chestnuts cost 40 cents for 2 ounces. Write the two rates (ratios) implied by this statement. What would be the cost of 50 ounces of chestnuts?

problem set 38   **1.**   Rubella was paid $28 for working 7 hours. Write the two rates (ratios) implied by this statement. How much was she paid for a 40-hour week?

2. The first number was eight million, four hundred thirty-two thousand, four hundred sixteen and one hundred thirty-seven ten-thousandths. The second number was three hundred seven thousand and two thousand, two hundred-millionths. What was the sum of the numbers?

3. The three weights were 4016 tons, 7132 tons, and 9831 tons. What was the average of the three weights?

4. If 942 tries were permitted and John averaged 462 points on each try, how many points did John get if he used all his tries?

5. What number is $\frac{5}{17}$ of 34?   6. What number is $\frac{1}{8}$ of 320?

Write each fraction as a mixed number:

7. $\frac{428}{17}$ 8. $\frac{521}{3}$

9. Find the least common multiple of 50, 47, and 120.

Simplify.

10. $\frac{15}{17} - \frac{2}{34}$ 11. $\frac{3}{8} + 1\frac{1}{5} - \frac{1}{10}$ 12. $4 + 13 - 5 \cdot 2 + 7$

13. $2 \cdot 8 - 3 \cdot 1 + 2 \cdot 4$ 14. $3\frac{1}{5} + 2\frac{1}{8}$ 15. $376\frac{4}{5} + 142\frac{3}{10}$

16. $42\frac{7}{8} - 15\frac{3}{4}$ 17. $513\frac{11}{20} - 21\frac{4}{5}$ 18. 813.2×0.032

19. $\frac{13.62}{0.05}$ 20. $1317.2 - 18.861$ 21. $\frac{14}{16} \cdot \frac{24}{21} \div \frac{2}{3}$

22. 5970 is divisible by which of the following numbers?
   (a) 2   (b) 3   (c) 5   (d) 10

23. Simplify mentally: $11,615,948 \times 10,000$

Evaluate:

24. $xz - yz$ if $x = 7$, $y = 1$, and $z = 3$

25. $xyz + yz - y$ if $x = 6$, $y = 3$, and $z = \frac{1}{3}$

26. Use unit multipliers to convert 144 inches to yards.

27. Express the area of this figure in square centimeters.   28. Find the perimeter of this figure. Dimensions are in yards. All angles are right angles.

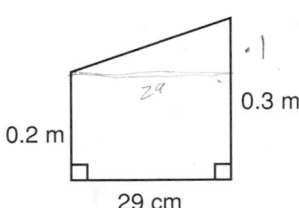

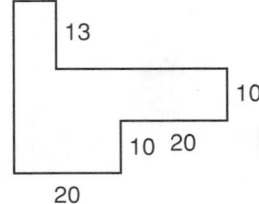

29. Write $\frac{17}{23}$ as a decimal number. Round to two decimal places.

30. Round 681,972,531 to the nearest hundred thousand.

LESSON 39   *Equations; answers and solutions*

If we connect two meaningful arrangements of numbers or of numbers and letters with an equals sign, we have written an **equation**. Some equations are **true equations,** and some equations are **false equations.**

(a) $4 + 2 = 6$ true   (b) $4 + 2 = 3$ false

Equation (a) is a true equation because 4 plus 2 is equal to 6. Equation (b) is a false equation because 4 plus 2 does not equal 3. Some equations that contain variables are called **conditional equations.**

(c) $x + 4 = 7$ conditional

Equation (c) is a conditional equation, and it is neither a true equation nor a false equation. Its truth or falsity depends on the number we use to replace x. If we replace x with 5, the equation becomes a false equation.

$5 + 4 = 7$ false

But if we replace x with 3, the equation becomes a true equation.

$3 + 4 = 7$ true

If replacing x with a number turns a conditional equation into a true equation, we say that the number is the **solution** of the equation or is an **answer** to the equation. We also say that the number **satisfies the equation.** Many times we can look at a conditional equation and tell by inspection what number will satisfy the equation.

example 39.1   Solve: $x + 4 = 10$

solution   *Solve* means to find the number that will make this equation a true equation. Since $6 + 4 = 10$, the solution is **6.**

example 39.2   Solve: $x - 4 = 10$

solution   The equation tells us that 4 subtracted from some number equals 10. Then the number must be 14 because

$14 - 4 = 10$

So the solution is **14.**

practice   Solve:

   a. $x + 6 = 12$ **b.** $x - 6 = 12$ **c.** $x - 14 = 6$

problem set 39

1. The squirrel could store 48 nuts in a hiding place. If there were 768 nuts to hide, how many hiding places did the squirrel need?

2. The squirrel could store 48 nuts per hour. Write the two rates (ratios) implied by this statement. How many nuts could the squirrel store in 20 hours?

3. When the engineers measured the distance 4 times, they found four different measurements. If the measurements were 5.6 meters, 5.4 meters, 5.3 meters, and 5.7 meters, what was the mean of the four measurements?

4. The average of the first five numbers drawn was 2499. The first four numbers were 4165, 320, 7142, and 64. What was the fifth number?

5. What number is $\frac{3}{5}$ of 40?   **6.** What number is $\frac{9}{11}$ of 99?

Write each fraction as a mixed number:

7. $\frac{316}{13}$ **8.** $\frac{428}{5}$

9. Find the least common multiple of 8, 12, and 72.

Simplify:

10. $\frac{14}{15} - \frac{1}{5}$ **11.** $4\frac{2}{5} + 3\frac{5}{10}$ **12.** $3 + 2 + 2 \cdot 5 - 3 \cdot 2$

13. $3 - 2 \cdot 1 + 4 \cdot 7$ **14.** $3\frac{2}{7} + 5\frac{20}{21}$ **15.** $420\frac{3}{5} + 262\frac{1}{40}$

16. $43\frac{5}{7} - 12\frac{2}{14}$ **17.** $210\frac{5}{8} - 17\frac{3}{40}$ **18.** 611.2×0.0061

19. $\frac{0.8136}{0.008}$ **20.** $2836.31 - 19.691$ **21.** $\frac{18}{20} \cdot \frac{24}{16} \div \frac{9}{10}$

22. Simplify mentally: $0.0003165 \times 1,000,000$

Solve:

23. $x + 9 = 10$ **24.** $x - 9 = 10$

Evaluate:

25. $x + xy + xyz$ if $x = 1$, $y = 2$, and $z = 3$

26. $xy + x - y$ if $x = 5$ and $y = 1$

27. Use unit multipliers to convert 625.611 centimeters to kilometers.

28. Find the area of this figure. Dimensions are in feet.

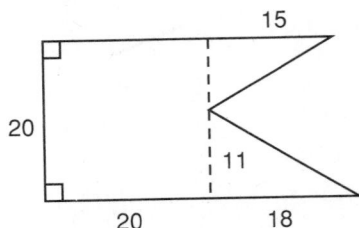

29. Write $\frac{20}{23}$ as a decimal number. Round to two places.

30. Round 16,998,711 to the nearest thousand.

LESSON 40 *Equivalent equations · Addition-subtraction rule for equations*

The five equations shown here are different equations.

$$\text{(a)} \quad x = 6 \qquad \text{(b)} \quad x + 3 = 9 \qquad \text{(c)} \quad x - 3 = 3$$

$$\text{(d)} \quad 3x = 18 \qquad \text{(e)} \quad \frac{x}{3} = 2$$

Yet the number 6 is a solution to all five equations, for if we replace x with 6, each of the equations becomes a true equation.

$$\text{(a')} \quad 6 = 6 \qquad \text{(b')} \quad 6 + 3 = 9 \qquad \text{(c')} \quad 6 - 3 = 3$$
$$6 = 6 \qquad\qquad\quad 9 = 9 \qquad\qquad\quad 3 = 3$$

$$\text{(d')} \quad 3(6) = 18 \qquad \text{(e')} \quad \frac{6}{3} = 2$$
$$18 = 18 \qquad\qquad\quad 2 = 2$$

Equation (a) was the easiest because it told us that the answer was 6. We didn't even have to think. We say that all five equations are **equivalent equations** because equivalent equations are equations that have the same solutions (answers).

Some equations are very easy to solve. We can mentally calculate the numbers that are solutions to the simple equations (b) through (e) shown here.

$$\text{(b)} \quad x + 3 = 9 \qquad \text{(c)} \quad x - 3 = 3 \qquad \text{(d)} \quad 3x = 18 \qquad \text{(e)} \quad \frac{x}{3} = 2$$

However, many equations are not easy to solve mentally. It is helpful to have rules that can be used to solve difficult equations. The first rule is the addition-subtraction rule.

ADDITION-SUBTRACTION RULE

1. The same number can be added to both sides of an equation without changing the solution to the equation.
2. The same number can be subtracted from both sides of an equation without changing the solution to the equation.

We will begin practicing the use of this rule by using it to solve rather simple equations.

example 40.1 Solve: $x - 3 = 7$

solution We can undo the subtraction of 3 by adding 3. So we add 3 to both sides of the equation.

$$x - 3 + 3 = 7 + 3 \qquad \text{added 3 to both sides}$$

$$x = 10 \qquad\qquad \text{simplified}$$

example 40.2   Solve: $x + 3 = 7$

solution   This time we must undo the addition of 3 by subtracting 3. We subtract 3 from both sides of the equation.

$$x + 3 - 3 = 7 - 3 \qquad \text{subtracted 3 from both sides}$$

$$\mathbf{x = 4} \qquad \text{simplified}$$

example 40.3   Solve: $x + \dfrac{3}{4} = \dfrac{9}{11}$

solution

We can undo adding $\frac{3}{4}$ by subtracting $\frac{3}{4}$. We subtract $\frac{3}{4}$ from both sides of the equation.

$$x + \frac{3}{4} - \frac{3}{4} = \frac{9}{11} - \frac{3}{4} \qquad \text{subtracted } \tfrac{3}{4} \text{ from both sides}$$

$$x = \frac{9}{11} - \frac{3}{4} \qquad \text{simplified}$$

$$x = \frac{36}{44} - \frac{33}{44} \qquad \text{common denominator}$$

$$\mathbf{x = \frac{3}{44}} \qquad \text{subtracted}$$

example 40.4   Solve: $x - \dfrac{3}{5} = \dfrac{2}{10}$

solution   We can undo subtracting $\frac{3}{5}$ by adding $\frac{3}{5}$. We add $\frac{3}{5}$ to both sides of the equation.

$$x - \frac{3}{5} + \frac{3}{5} = \frac{2}{10} + \frac{3}{5} \qquad \text{added } \tfrac{3}{5} \text{ to both sides}$$

$$x = \frac{2}{10} + \frac{3}{5} \qquad \text{simplified}$$

$$x = \frac{2}{10} + \frac{6}{10} \qquad \text{common denominator}$$

$$\mathbf{x = \frac{8}{10} = \frac{4}{5}} \qquad \text{simplified}$$

practice   Solve:

    a. $a + 4 = 6$ **b.** $x - 2 = 4$

    c. $x + \dfrac{1}{2} = \dfrac{3}{4}$ **d.** $x - \dfrac{1}{3} = \dfrac{11}{12}$

problem set 40

1.   Sandra counted them as they slid around the corner. The first hour she counted 415, and the second hour she counted 478. If 526 were counted in the third hour, what was the average for the 3 hours?

2.   Watonga picked up 420 in the first 6 minutes. Write the two rates (ratios) implied by this statement. How many could Watonga pick up in 48 hours?

3. Only 420 could crawl into 1 space. If there were 12 spaces, how many could crawl in?

4. Nineteen thousand, four hundred forty came to see the spectacle. If 32 could be seated in a bus, how many buses were required to haul them all away?

5. What number is $\frac{3}{8}$ of 40?   **6.** What number is $\frac{4}{5}$ of 80?

Write each fraction as a mixed number:

7. $\frac{41}{3}$ **8.** $\frac{93}{21}$

9. Find the least common multiple of 27, 28, and 30.

Simplify:

10. $\frac{5}{16} - \frac{1}{8}$ **11.** $2\frac{1}{5} + 3\frac{1}{3} - \frac{2}{10}$ **12.** $14 - 2 \cdot 3 + 4 \cdot 5$

13. $2 \cdot 5 - 2 \cdot 2 + 3$ **14.** $7\frac{1}{8} + 3\frac{2}{5}$ **15.** $674\frac{2}{5} - 13\frac{7}{10}$

16. $2\frac{1}{4} + 3\frac{1}{8} + 4\frac{5}{12}$ **17.** $461\frac{3}{4} - 65\frac{7}{8}$ **18.** 117.1×2.01

19. $\frac{171.6}{0.006}$ **20.** $6132.81 - 621.981$ **21.** $\frac{6}{21} \cdot \frac{24}{3} \div \frac{8}{14}$

22. Use two unit multipliers to convert 1 mile to inches.

Solve:

23. $x - 8 = 9$ **24.** $x + \frac{3}{8} = \frac{9}{14}$ **25.** $x - \frac{3}{7} = \frac{5}{14}$

26. Simplify mentally: $160,394,000 \div 1,000,000$

27. Evaluate: $xyz + xy + yz - z$ if $x = \frac{1}{3}$, $y = 12$, and $z = 2$

28. Find the area of this figure. Dimensions are in inches.

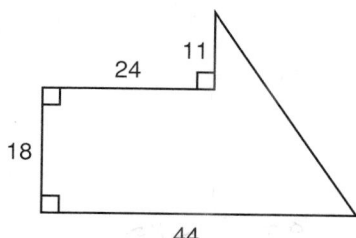

29. (a) List the prime numbers between 37 and 46.
(b) List the multiples of 6 between 37 and 46.

30. Round 8,265,891,131 to the nearest ten thousand.

LESSON 41 Reciprocals · Multiplication rule

41.A
reciprocals

When we write a fraction upside down, we say that we have written the reciprocal of the fraction. Thus,

$$\frac{2}{5} \quad \text{is the reciprocal of} \quad \frac{5}{2}$$

$$\frac{5}{2} \quad \text{is the reciprocal of} \quad \frac{2}{5}$$

$$\frac{3}{7} \quad \text{is the reciprocal of} \quad \frac{7}{3}$$

$$\frac{7}{3} \quad \text{is the reciprocal of} \quad \frac{3}{7}$$

Since whole numbers can be written with a denominator of 1, all nonzero whole numbers have a reciprocal.

$$\frac{1}{4} \quad \text{is the reciprocal of} \quad 4$$

$$4 \quad \text{is the reciprocal of} \quad \frac{1}{4}$$

$$\frac{1}{16} \quad \text{is the reciprocal of} \quad 16$$

$$16 \quad \text{is the reciprocal of} \quad \frac{1}{16}$$

The number 0 can be written with a denominator of 1

$$\frac{0}{1}$$

and this still means zero. When we turn it upside down, we get

$$\frac{1}{0}$$

We say that this expression has no meaning because we cannot divide by 0. **For this reason we say that the number 0 does not have a reciprocal. The number 0 is the only number that does not have a reciprocal.**

41.B
products of reciprocals

The reciprocal of 4 is $\frac{1}{4}$. If we multiply 4 by its reciprocal, the answer is 1.

$$\frac{4}{1} \cdot \frac{1}{4} = \frac{4}{4} = \mathbf{1}$$

The reciprocal of $\frac{1}{3}$ is 3. If we multiply $\frac{1}{3}$ by its reciprocal, the answer is 1.

$$\frac{1}{3} \cdot \frac{3}{1} = \frac{3}{3} = \mathbf{1}$$

The reciprocal of $\frac{2}{5}$ is $\frac{5}{2}$. If we multiply $\frac{2}{5}$ by its reciprocal, the answer is 1.

$$\frac{2}{5} \cdot \frac{5}{2} = \frac{10}{10} = 1$$

> **The product of any number and its reciprocal is the number 1.**

We can use this property of reciprocals to help us solve equations.

41.C
multiplication rule

If a letter is multiplied by a number, we say that the number is the **coefficient** of the letter. In the expression

$$4x$$

we say that 4 is the coefficient of x. The multiplication rule for equations can help us solve equations in which the variable has a coefficient.

> **MULTIPLICATION RULE FOR EQUATIONS**
>
> Both sides of an equation can be multiplied by the same number (except zero) without changing the solution to the equation.

To use this rule, we multiply both sides of the equation by the reciprocal of the coefficient of x.

example 41.1   Solve: $4x = 12$

solution   To "get rid of" the 4, we will multiply by $\frac{1}{4}$. We must remember to multiply both sides of the equation by $\frac{1}{4}$.

$$4x = 12 \qquad \text{equation}$$

$$\left(\frac{1}{\cancel{4}} \cdot \cancel{4}\right)x = \cancel{12}^{\,3}\left(\frac{1}{\cancel{4}}\right) \qquad \text{multiplied by } \frac{1}{4}$$

$$x = 3 \qquad \text{simplified}$$

Now we check by using 3 for x in the original equation.

$$4x = 12 \qquad \text{equation}$$
$$4(3) = 12 \qquad \text{substituted}$$
$$12 = 12 \qquad \text{check}$$

example 41.2   Solve: $\frac{2}{3}x = \frac{1}{5}$

solution   To get rid of the $\frac{2}{3}$, we will multiply **both sides** of the equation by $\frac{3}{2}$.

$$\frac{2}{3}x = \frac{1}{5} \qquad \text{equation}$$

$$\frac{3}{2} \cdot \frac{2}{3}x = \frac{1}{5} \cdot \frac{3}{2} \qquad \text{multiplied by } \frac{3}{2}$$

$$x = \frac{3}{10} \qquad \text{solution}$$

Now we use $\frac{3}{10}$ in the original equation and check.

$$\frac{2}{3}x = \frac{1}{5} \qquad \text{equation}$$

$$\frac{2}{3} \cdot \frac{3}{10} = \frac{1}{5} \qquad \text{substituted } \frac{3}{10} \text{ for } x$$

$$\frac{1}{5} = \frac{1}{5} \qquad \text{check}$$

example 41.3 Solve: $\frac{x}{3} = 6$

solution The coefficient of x is $\frac{1}{3}$, so we multiply both sides of the equation by 3.

$$\frac{x}{3} = 6 \qquad \text{equation}$$

$$3 \cdot \frac{x}{3} = 6 \cdot 3 \qquad \text{multiplied by 3}$$

$$x = \mathbf{18} \qquad \text{solution}$$

Now we check by using 18 for x in the original equation.

$$\frac{x}{3} = 6 \qquad \text{equation}$$

$$\frac{18}{3} = 6 \qquad \text{substituted}$$

$$6 = 6 \qquad \text{check}$$

practice Solve each equation by multiplying both sides of the equation by the reciprocal of the coefficient of x. Check each answer.

a. $\frac{5}{11}x = 7$ b. $\frac{5}{3}x = 4$ c. $\frac{2}{3}x = \frac{1}{4}$

d. $3x = 15$ e. $\frac{x}{3} = 5$ f. $4x = \frac{1}{3}$

problem set 41

1. The average rainfall for 5 days was 3 inches per day. It rained 1 inch on the first day, 4.5 inches on the second day, 2.9 inches on the third day, and 4 inches on the fourth day. How many inches did it rain on the fifth day?

2. Ronald paid $600 and got 30 hanging plants. Write the two rates (ratios) implied by this statement. How much would he have had to pay for 70 hanging plants?

3. If 47 could be crammed into each compartment and if 2820 were waiting patiently in line, how many compartments would it take for all of them?

4. What was the average rainfall for the first 5 months of the year as indicated by this graph?

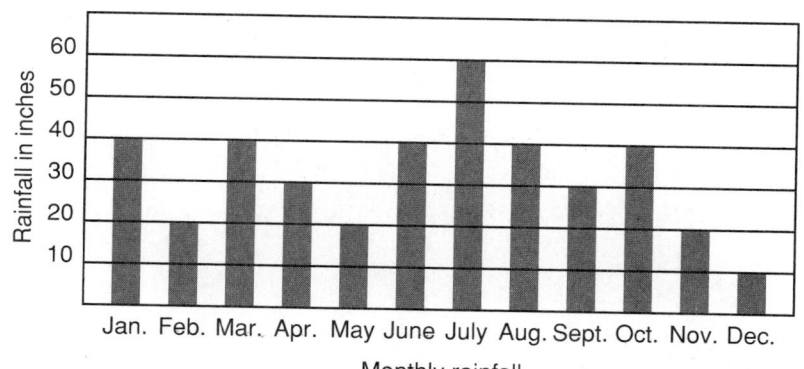

Monthly rainfall

5. What number is $\frac{3}{7}$ of 21?

6. What number is $\frac{6}{13}$ of 39?

Write each fraction as a mixed number:

7. $\frac{82}{5}$

8. $\frac{121}{15}$

9. Find the least common multiple of 35, 40, and 120.

Solve:

10. $x + \frac{3}{4} = \frac{7}{8}$

11. $x - \frac{1}{2} = \frac{5}{6}$

12. $6x = 18$

13. $\frac{x}{4} = 15$

14. $5x = 20$

15. $\frac{x}{7} = 5$

Simplify:

16. $\frac{7}{15} - \frac{1}{5}$

17. $3 \cdot 8 - 2 \cdot 6 + 1 \cdot 7$

18. $36\frac{3}{4} - 21\frac{7}{8}$

19. $\frac{171.6}{0.6}$

20. 112.4×0.071

21. $6781.8 - 179.89$

22. $\frac{16}{18} \cdot \frac{24}{36} \div \frac{8}{9}$

23. Simplify mentally: $7,296,000,000,000 \div 1,000,000,000$

Evaluate:

24. $x + zx - y$ if $x = 10$, $y = 3$, and $z = 2$

25. $xy + xz + yz$ if $x = 2$, $y = 4$, and $z = 6$

26. Use two unit multipliers to convert 5280 inches to yards.

27. Find the area of this figure. Dimensions are in meters.

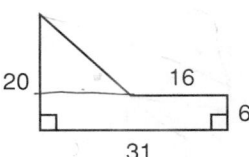

28. Find the perimeter of this figure. Dimensions are in centimeters.

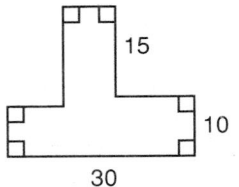

29. Write $\frac{19}{24}$ as a decimal number. Round to two decimal places.

30. Round $4.0\overline{325}$ to the nearest millionth.

LESSON 42 *Overall average*

The average of the averages is seldom the overall average. Suppose we have 10 numbers and the first 3 numbers are the numbers 3, 4, and 5. The average of the first 3 numbers is 4.

$$\frac{3 + 4 + 5}{3} = \frac{12}{3} = 4$$

Suppose the last 7 numbers are 7, 8, 8, 8, 8, 8, and 9. The average of these numbers is 8.

$$\frac{7 + 8 + 8 + 8 + 8 + 8 + 9}{7} = \frac{56}{7} = 8$$

The average of the first 3 numbers is 4. The average of the other 7 numbers is 8. **We cannot find the overall average by finding the average of the average.**

Overall average is not $\frac{4 + 8}{2} = \frac{12}{2} = 6$ **NO**

We must compute the overall average by dividing the sum of all the numbers by the number of numbers. If we do, we get

Overall average $= \dfrac{3 + 4 + 5 + 7 + 8 + 8 + 8 + 8 + 8 + 9}{10} = \dfrac{68}{10} = 6.8$

The overall average must be some number between 4 and 8, but it is not 6, which is halfway between 4 and 8. The overall average of 6.8 is closer to 8 than it is to 4 because there are more numbers in the group whose average is 8.

example 42.1   The average of the first 2 numbers was 6. The average of the next 8 numbers was 20. What was the overall average?

solution   We compute the overall average the same way we compute any average. **The average is always the sum of the numbers divided by the number of numbers.** The average of the first 2 numbers was 6, so their sum was 6 × 2, or 12. The average of the next 8 numbers was 20, so their sum was 8 × 20, or 160. The total number of numbers was 10.

$$\text{Average} = \frac{(2 \times 6) + (8 \times 20)}{10} = \frac{172}{10} = \textbf{17.2}$$

practice   The average price of the first four items was $40. The average price of the next six items was $80. What was the overall average price?

problem set 42

1. The first shift lasted 4 hours and produced 800 units per hour. The second shift was a 6-hour shift that produced 600 units per hour. The third shift was a 4-hour shift that produced 400 units per hour. What was the average number of units per hour for the entire period?

2. Ashanti averaged 89 points per test. Write the two rates (ratios) implied by this statement. How many points could she score if she took 22 tests?

3. Seventy people had been employed for 20 years. Thirty people had been employed for 10 years. What was their overall average salary?

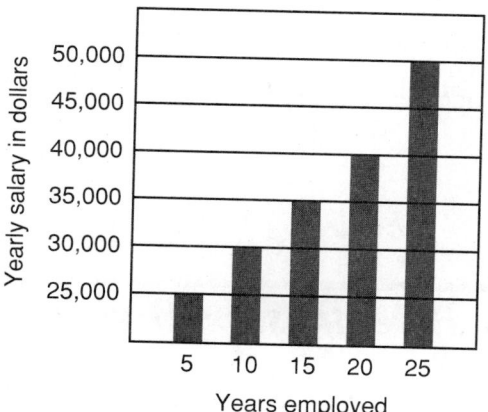

4. Four quarts equals 1 gallon.
   (a) Write the two unit multipliers implied by this statement.
   (b) Use one of the unit multipliers to convert 1,000,000 gallons to quarts.

5. What number is $\frac{5}{8}$ of 64?   6. What number is $\frac{6}{14}$ of 126?

7. Find the least common multiple of 28, 56, and 350.

Write each fraction as a mixed number:

8. $\frac{94}{7}$ 9. $\frac{289}{12}$

Solve:

10. $x + \frac{5}{8} = \frac{11}{16}$ 11. $x - \frac{1}{4} = \frac{5}{12}$ 12. $14x = 56$

13. $\frac{x}{4} = 9$ 14. $9x = 81$ 15. $\frac{x}{13} = 6$

Simplify:

16. $\dfrac{11}{12} - \dfrac{5}{6}$

17. $4 \cdot 7 - 3 \cdot 4 + 2 \cdot 9$

18. $19\dfrac{1}{8} - 8\dfrac{3}{4}$

19. $\dfrac{195.8}{1.1}$

20. 163.09×0.063

21. $9876.5 - 643.99$

22. $\dfrac{14}{16} \cdot \dfrac{6}{32} \div \dfrac{3}{4}$

23. Simplify mentally: $73,962 \div 100,000$

Evaluate:

24. $mn + zy - y$ if $m = 4$, $n = 3$, $z = 2$, and $y = {}^-1$

25. $ax + bx - ab$ if $a = 2$, $b = 3$, and $x = 2$

26. Use two unit multipliers to convert 628 kilometers to centimeters.

27. Find the area of this figure. Dimensions are in feet. Corners that look square are square.

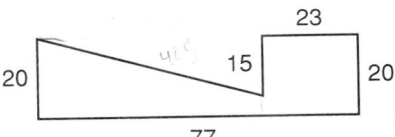

28. Write $\dfrac{5}{8}$ as a decimal number.

29. Round 189,762,581.32 to the nearest million.

30. Use words to write 6213625326.12.

LESSON 43 *Parentheses · Division in order of operations*

43.A
parentheses

We have been using crosses and dots to indicate multiplication. We can also use parentheses to indicate multiplication. To do this we put parentheses around one number or around both numbers. All of the following tell us to multiply 4 by 5.

    (a) 4×5 (b) $4 \cdot 5$ (c) $(4)(5)$ (d) $4(5)$ (e) $(4)5$
                                        ↑   ↑   ↑

Multiplication is indicated when there is no plus or minus sign between parentheses, as in (c). Multiplication is also indicated when there is no plus or minus sign between a number and a parenthesis, as in (d) and (e).

    Parentheses can be used to group numbers together. We always simplify within the parentheses first. Then we multiply, and finally we add and subtract.

example 43.1 Simplify: $4 + (3 \cdot 5)2 + 2(15 - 3)$

solution We begin by simplifying within the two sets of parentheses.

$$4 + (15)2 + 2(12)$$

Now we perform all the multiplication and get

$$4 + 30 + 24$$

We finish by adding these numbers and get

58

example 43.2 Simplify: $8(20 - 3) - (17 - 11 + 3)2 + 5$

solution We begin by simplifying inside the two sets of parentheses.

$$8(17) - (9)2 + 5$$

Now we multiply.

$$136 - 18 + 5$$

We finish by adding and subtracting from left to right.

$$118 + 5 \quad \text{subtracted}$$

$$\mathbf{123} \quad \text{added}$$

43.B
division in order of operations

The rules for the order of operations when we have addition, subtraction, multiplication, and division are as follows:

1. Simplify within the parentheses.
2. Proceed from left to right, doing the multiplications and divisions in the order they are encountered.
3. Then proceed from left to right, doing the additions and subtractions in the order they are encountered.

example 43.3 Simplify: $5 + 12 \div 2 + 3 \cdot 4 - 2(5 - 2)$

solution First we simplify within the parentheses and get

$$5 + 12 \div 2 + 3 \cdot 4 - 2(3)$$

We must always multiply and divide before we add or subtract. Thus we skip the first operation, which is addition. The next operation is division. We divide 12 by 2 and get 6. Now we have

$$5 + 6 + 3 \cdot 4 - 2(3)$$

The next multiplication or division is $3 \cdot 4$. We multiply and get

$$5 + 6 + 12 - 2(3)$$

Now we do the last multiplication and get

$$5 + 6 + 12 - 6$$

We finish by adding and subtracting from left to right.

$$5 + 6 + 12 - 6 = \mathbf{17}$$

43.C
conversion of units of area

Two unit multipliers are required to convert units of area, as shown in the next two examples.

example 43.4 Use two unit multipliers to convert 4 square feet to square inches.

solution   First we write 4 square feet as

$$4 \text{ ft}^2$$

Now since "ft²" means "ft" times "ft," we can write this as

$$4 \text{ ft} \cdot \text{ft}$$

Two unit multipliers are required because one unit multiplier is required to change each "ft" to "in."

$$4 \cancel{\text{ft}} \cdot \cancel{\text{ft}} \times \frac{12 \text{ in.}}{1 \cancel{\text{ft}}} \times \frac{12 \text{ in.}}{1 \cancel{\text{ft}}} = 4(12)(12) \text{ in.}^2 = \textbf{576 in.}^2$$

example 43.5   Use two unit multipliers to convert 4 square meters to square centimeters.

solution   We use two unit multipliers so we can cancel "m" twice.

$$4 \cancel{\text{m}}^2 \times \frac{100 \text{ cm}}{\cancel{\text{m}}} \times \frac{100 \text{ cm}}{\cancel{\text{m}}} = \textbf{40,000 cm}^2$$

example 43.6   Use two unit multipliers to convert 400 square inches to square feet.

solution   Since square inches means inches times inches, we use two unit multipliers so we can cancel "in." twice.

$$400 \cancel{\text{in.}} \cdot \cancel{\text{in.}} \times \frac{1 \text{ ft}}{12 \cancel{\text{in.}}} \times \frac{1 \text{ ft}}{12 \cancel{\text{in.}}} = \frac{\textbf{400}}{\textbf{(12)(12)}} \text{ ft}^2$$

If a decimal answer is required, a calculator can be used to do the arithmetic.

practice   Simplify:

a. $4 + 10 \div 2 - 3(6 - 4)$ **b.** $5 + 10 \cdot 6 - 2(3 \cdot 2) - 2$

Use two unit multipliers to convert:

c.   500 square inches to square feet

d.   2000 square centimeters to square meters

problem set 43

1. Nineteen thousand, one hundred forty-two millionths was the first guess. The second guess was eight thousand, five hundred forty-one hundred-thousandths. Which guess was greater and by how much?

2. They were happy because 80 of them could be seated comfortably in each vehicle. If there were 1420 vehicles in the parking lot, how many could be seated comfortably?

3. Pavo bought 20 bunches at $40 a bunch. Write the two rates (ratio) implied by this statement. How much would he have to pay for 100 bunches?

4. Fran did 42 a minute for the first hour. For the next 30 minutes she did 20 a minute, and for the last 50 minutes she did only 10 every minute. How many did she do in all?

5. The first go contained 40. The second go contained twice that many. The third go contained four times the number the second go contained. By how many did the third go exceed the sum of the first go and the second go?

6. What number is $\frac{11}{7}$ of 70?   7. What number is $\frac{13}{5}$ of 20?

8. Write $12\frac{5}{7}$ as an improper fraction.   **9.** Write $\frac{271}{15}$ as a mixed number.

10. Find the least common multiple of 12, 22, and 40.

Solve:

11. $\frac{2}{3}x = 3$

12. $\frac{3}{7}x = \frac{2}{9}$

13. $\frac{x}{5} = 60$

14. $5x = 60$

15. $x + \frac{1}{8} = \frac{1}{2}$

16. $x - \frac{1}{4} = \frac{1}{2}$

Simplify:

17. $61\frac{11}{15} - 15\frac{3}{5}$

18. $\frac{5}{8} + 2\frac{1}{4} - \frac{1}{10}$

19. $6 \cdot 8 - 4 \cdot 2 + 6$

20. 611.3×0.016

21. $2362.8 - 189.87$

22. $\frac{612.5}{0.07}$

23. $\frac{12}{16} \cdot \frac{12}{21} \div \frac{6}{14}$

24. $7 + (6 \cdot 3)2 + 3(11 - 2)$

25. $6(17 - 5) - (15 - 13 + 4)4 + 6$

Evaluate:

26. $xy + yz - z$ if $x = \frac{1}{5}$, $y = 20$, and $z = 3$

27. $xyz - x$ if $x = 6$, $y = 16$, and $z = \frac{1}{12}$

28. Convert 5 square feet to square inches.

29. Express the perimeter of this figure in centimeters. All angles are right angles.

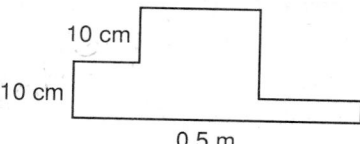

30. Use words to write 6256187.87621.

LESSON 44   *Multiplying and dividing mixed numbers*

44.A
multiplying mixed numbers

The easiest way to multiply mixed numbers is to convert them to improper fractions and then multiply the improper fractions.

example 44.1   Multiply: $4\frac{1}{3} \cdot \frac{6}{5}$

solution   We first write $4\frac{1}{3}$ as an improper fraction.

$$\frac{13}{3} \cdot \frac{6}{5}$$

We always cancel if we can. Then we multiply.

$$\frac{13}{\cancel{3}} \cdot \frac{\cancel{6}^{\,2}}{5} = \frac{26}{5}$$

This answer can be written as $5\frac{1}{5}$ if desired.

example 44.2   Multiply: $2\frac{1}{4} \times 3\frac{1}{3} \times 5\frac{1}{12}$

solution   First we convert each mixed number to an improper fraction.

$$\frac{9}{4} \times \frac{10}{3} \times \frac{61}{12}$$

Now we cancel where possible and then multiply.

$$\frac{\cancel{9}^{\,3}}{4} \times \frac{\cancel{10}^{\,5}}{\cancel{3}} \times \frac{61}{\cancel{12}_{\,2}} = \frac{305}{8}$$

In arithmetic, we usually convert all improper fractions to mixed numbers. In algebra and other advanced courses, the improper fraction form is preferred by many. We will leave this answer as an improper fraction. In future lessons and in the answers in the back of the book, sometimes we will use the improper fraction form and sometimes we will use the mixed number form. **Either form is correct.**

44.B
dividing mixed numbers

To divide mixed numbers, we first write the mixed numbers as improper fractions. Then we invert the divisor and multiply.

example 44.3   Simplify: $\dfrac{2\frac{1}{8}}{3\frac{2}{3}}$

solution   First we write both numbers as improper fractions.

$$\frac{\frac{17}{8}}{\frac{11}{3}}$$

Now we invert and multiply.

$$\frac{17}{8} \cdot \frac{3}{11} = \frac{51}{88}$$

example 44.4   Simplify: $4\frac{1}{8} \div 2\frac{1}{5} \cdot 3\frac{1}{4} \div 5\frac{1}{2}$

solution   First we write each number as an improper fraction.

$$\frac{33}{8} \div \frac{11}{5} \cdot \frac{13}{4} \div \frac{11}{2}$$

Next we invert the divisors and change every division sign to a multiplication sign. Then we cancel and multiply:

$$\frac{\overset{3}{\cancel{33}}}{\underset{4}{\cancel{8}}} \cdot \frac{5}{\cancel{11}} \cdot \frac{13}{4} \cdot \frac{\cancel{2}}{11} = \frac{195}{176}$$

practice Simplify:

a. $4\frac{1}{4} \times \frac{8}{9}$

b. $2\frac{1}{4} \times 1\frac{1}{2} \times 3\frac{1}{3}$

c. $\dfrac{3\frac{1}{4}}{2\frac{1}{5}}$

d. $\frac{14}{3} \div \frac{7}{6} \cdot \frac{21}{9} \div \frac{5}{3}$

problem set 44

1. Virago bought 78 pots for $312. Write the two rates (ratios) implied by this statement. What would she have had to pay for 400 pots?

2. What was the average number of tons per month for the first 4 months of the year?

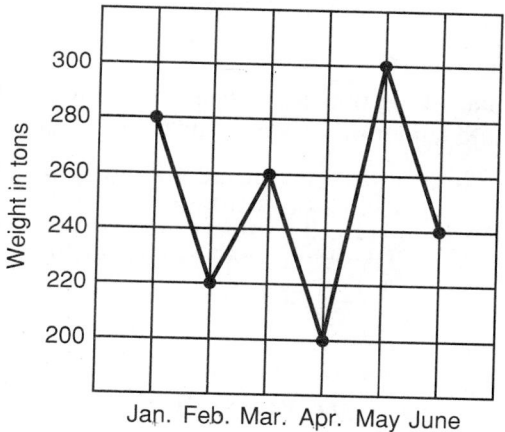

Tons per month

3. During the first hour, twenty-two million, forty were born. During the second hour, only fourteen million, eight hundred sixty-five thousand, nine hundred thirty-two were born. How many more were born during the first hour?

4. The average number of pills in each of the 10 small containers was 1100. The average number of pills in each of the 20 big containers was 1400. What was the average number of pills in all the containers?

5. What number is $\frac{2}{9}$ of 180?

6. What number is $\frac{1}{16}$ of 320?

7. Write $7\frac{5}{16}$ as an improper fraction.

8. Find the least common multiple of 14, 21, and 49.

Solve:

9. $\frac{7}{8}x = 4$

10. $4x = 80$

11. $\frac{x}{7} = 84$

12. $x - \dfrac{3}{7} = \dfrac{9}{14}$ **13.** $x + \dfrac{3}{11} = \dfrac{9}{22}$

Simplify:

14. $3 + (2 \cdot 5)3 + 4(2 \cdot 8)$ **15.** $5(3 - 1) + 2(6 - 5) - 2$

16. $5\dfrac{1}{2} \times \dfrac{12}{7}$ **17.** $3\dfrac{1}{5} \times 1\dfrac{3}{8} \times 1\dfrac{1}{11}$ **18.** $\dfrac{1\frac{2}{5}}{2\frac{1}{7}}$

19. $\dfrac{7}{6} \div \dfrac{9}{3} \cdot \dfrac{27}{4} \div \dfrac{3}{2}$ **20.** $\dfrac{5}{6} + 1\dfrac{5}{12} - \dfrac{3}{4}$ **21.** $17\dfrac{7}{12} - 12\dfrac{3}{4}$

22. $117.89 - 112.341$ **23.** 14.02×0.0015 **24.** $\dfrac{7812}{0.003}$

25. Evaluate: $xz + yz - xy$ if $x = 4$, $y = \dfrac{1}{2}$, and $z = 8$

26. Use two unit multipliers to convert 3 miles to inches.

27. Find the area of this figure. Dimensions are in feet. Corners that look square are square.

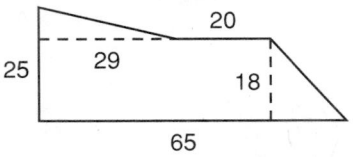

28. If we use 10 as a factor twice, we get 100 as the product. How many times must we use 10 as a factor to get 1,000,000 as the product?

29. (a) List the prime numbers that are greater than 37 but less than 52.
(b) List the multiples of 5 that are greater than 37 but less than 52.

30. Use two unit multipliers to convert 5 square meters to square centimeters.

LESSON 45 *Division rule for equations*

If we divide 20 by 4, we get the same answer that we get if we multiply 20 by $\frac{1}{4}$.

$$\frac{20}{4} = 5 \qquad \frac{1}{4} \cdot 20 = 5$$

Thus dividing by a number and multiplying by the reciprocal of the number produce the same result. We know that we can solve equations by multiplying both sides of the equation by the reciprocal of the coefficient of x. We can also solve equations by dividing both sides of the equation by the coefficient of x.

> **DIVISION RULE FOR EQUATIONS**
>
> Both sides of an equation can be divided by the same number (except zero) without changing the solution of the equation.

example 45.1 Solve by dividing: $4x = 3$

solution Of course we could solve this equation by multiplying both sides by $\frac{1}{4}$.

$$\left(\frac{1}{4}\right)4x = 3\left(\frac{1}{4}\right)$$

$$x = \frac{3}{4} \qquad \text{solved}$$

But we were told to divide. The answer will be exactly the same.

$$\frac{4x}{4} = \frac{3}{4} \qquad \text{divided both sides by 4}$$

$$x = \frac{3}{4} \qquad \text{simplified}$$

Now we check our solution.

$$4x = 3 \qquad \text{equation}$$

$$4\left(\frac{3}{4}\right) = 3 \qquad \text{substituted}$$

$$3 = 3 \qquad \text{check}$$

example 45.2 Solve by dividing: $N \times 0.04 = 28$

solution The coefficient of N is 0.04. So we divide both sides by 0.04.

$$\frac{N \times 0.04}{0.04} = \frac{28}{0.04}$$

$$N = 700 \qquad \text{solved}$$

Now we check our solution.

$$N \times 0.04 = 28 \qquad \text{equation}$$

$$(700) \times 0.04 = 28 \qquad \text{substituted 700 for } N$$

$$28 = 28 \qquad \text{check}$$

example 45.3 Solve by dividing: $\frac{1}{3}x = \frac{4}{5}$

solution We can solve this equation by dividing both sides by $\frac{1}{3}$.

$$\frac{\frac{1}{3}x}{\frac{1}{3}} = \frac{\frac{4}{5}}{\frac{1}{3}}$$

To simplify on the right-hand side, we invert the denominator and multiply.

$$x = \frac{4}{5} \times \frac{3}{1} = \frac{12}{5}$$

It would have been easier to solve this equation by multiplying by 3, as we will show.

$$\frac{1}{3}x = \frac{4}{5} \qquad \text{equation}$$

$$\left(3 \cdot \frac{1}{3}\right)x = \frac{4}{5} \cdot 3 \qquad \text{multiplied by 3}$$

$$x = \frac{12}{5} \qquad \text{solved}$$

Now we check.

$$\frac{1}{3}x = \frac{4}{5} \qquad \text{equation}$$

$$\frac{1}{\cancel{3}}\left(\frac{\overset{4}{\cancel{12}}}{5}\right) = \frac{4}{5} \qquad \text{substituted } \frac{12}{5} \text{ for } x$$

$$\frac{4}{5} = \frac{4}{5} \qquad \text{check}$$

practice   Solve by multiplying or by dividing:

a. $0.04x = 28$ **b.** $\frac{1}{3}x = 5$ **c.** $\frac{3}{5}x = \frac{2}{3}$

**problem set
45**

1. The average number of teachers in each of the 40 small towns was 116. The average number of teachers in each of the 60 large towns was 216. What was the average number of teachers in all the towns?

2. Shylock went to court 12 times a month for the first year. For the next 10 years he went to court twice a month. For the last year he went to court 3 times a month. How many times did he go to court in all?

3. The grocer sold avocados for 79 cents each. Write the two rates (ratios) implied by this statement. How many avocados would the grocer sell for $23.70?

4. Antigone's estimate of 16,319.06 cm exceeded the actual measurement by three hundred nine and twelve-thousandths. What was the actual measurement?

5. What number is $\frac{7}{8}$ of 128?   6. What number is $\frac{1}{26}$ of 520?

7. Find the least common multiple of 16, 24, and 32.

Solve:

8. $\frac{3}{8}x = 1$ 9. $5x = 90$ 10. $\frac{x}{4} = 35$

11. $x - \frac{2}{5} = \frac{1}{100}$ 12. $x + \frac{3}{13} = \frac{9}{26}$

Simplify:

13. $5 + (3 \cdot 2)4 + 3(2 \cdot 5)$ 14. $4(6 - 2) + 3(5 - 1) - 6$

15. $4\frac{1}{3} \times \frac{7}{5}$ 16. $2\frac{1}{4} \times 1\frac{2}{5} \times 1\frac{3}{14}$ 17. $\dfrac{2\frac{1}{3}}{3\frac{2}{7}}$

18. $\dfrac{6}{5} \div \dfrac{3}{2} \cdot \dfrac{15}{8} \div \dfrac{9}{8}$

19. $\dfrac{7}{8} + 1\dfrac{3}{16} - \dfrac{1}{2}$

20. $14\dfrac{3}{16} - 1\dfrac{1}{2}$

21. $114.97 - 108.391$

22. 23.04×0.00012

23. $\dfrac{9663}{0.0006}$

24. Evaluate: $km + zm - kz$ if $k = 3$, $m = 6$, and $z = \dfrac{1}{3}$

25. Use two unit multipliers to convert 17 kilometers to centimeters.

26. Find the area of this figure. Dimensions are in meters. Angles that look like right angles are right angles.

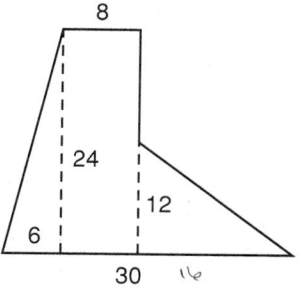

27. The number 10 is used as a factor how many times to get a product of 100,000,000?

28. (a) List the prime numbers that are greater than 41 but less than 67.
    (b) List the multiples of 3 that are greater than 41 but less than 67.

29. Use two unit multipliers to convert 120 square miles to square feet.

30. How much more did the teacher spend for foil and paper than she spent for scissors?

Art Supplies

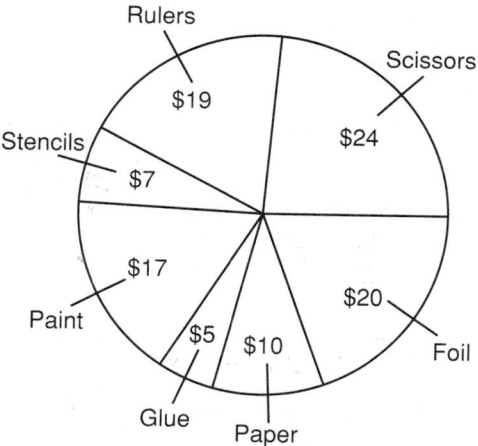

LESSON 46 *Exponents and roots*

46.A
exponents

Often it is necessary to indicate that a number is to be multiplied by itself a given number of times. If we wish to indicate that 7 is to be used as a factor 6 times, we could write

$$7 \cdot 7 \cdot 7 \cdot 7 \cdot 7 \cdot 7$$

We can also designate repeated multiplication of this kind by using **exponential notation.** Exponential notation lets us express the same thought in a more concise form. To indicate the same multiplication by using exponential notation, we would write

$$\text{base} \longrightarrow 7^6 \longleftarrow \text{exponent or power}$$

The lower number is called the **base,** and the upper number is called the **exponent.** The exponent tells how many times the base is used as a factor. The whole expression is called a **power.** Sometimes we call the exponent the power.

$$2^3 = 8$$

We read this by saying "two to the third power equals eight." We also can say that 8 is the third power of 2.

$$3^4 = 3 \cdot 3 \cdot 3 \cdot 3 \qquad \text{The base is 3 and the exponent is 4.}$$
$$4^3 = 4 \cdot 4 \cdot 4 \qquad \text{The base is 4 and the exponent is 3.}$$
$$\left(\frac{2}{3}\right)^3 = \frac{2}{3} \cdot \frac{2}{3} \cdot \frac{2}{3} \qquad \text{The base is } \frac{2}{3} \text{ and the exponent is 3.}$$

example 46.1   Simplify: $4^2 + 3^4 + 2^3$

solution   We write each power in expanded form and get

$$4 \cdot 4 + 3 \cdot 3 \cdot 3 \cdot 3 + 2 \cdot 2 \cdot 2$$

Next we do the multiplications and then we add.

$$16 + 81 + 8 = \mathbf{105}$$

46.B
roots of numbers

The inverse operation of raising to a power is called **taking the root.** If we use 2 as a factor 4 times, the answer is 16.

$$2^4 = 16$$

If we wish to undo this, we can ask, "What number used as a factor 4 times equals 16?" by writing

$$\sqrt[4]{16}$$

The answer is 2 because

$$2 \cdot 2 \cdot 2 \cdot 2 = 16$$

so we write

$$\sqrt[4]{16} = 2$$

We read this by saying "the fourth root of sixteen equals two." We say that 16 is the **radicand,** 4 is the **index,** and 2 is the **root.** We call the symbol

$$\sqrt{}$$

the **radical sign.** If the index is not written, it is understood to be 2. We call the whole expression a **radical expression** or just a **radical.** For the present, we will restrict our attention to radicals that represent whole numbers. In later lessons, we will investigate radicals that represent decimal numbers.

example 46.2 Simplify: (a) $\sqrt[4]{81}$ (b) $\sqrt[3]{27}$ (c) $\sqrt{16}$ (d) $\sqrt[3]{8}$

solution (a) Since $3 \cdot 3 \cdot 3 \cdot 3 = 81$ $\sqrt[4]{81} = \mathbf{3}$

(b) Since $3 \cdot 3 \cdot 3 = 27$ $\sqrt[3]{27} = \mathbf{3}$

(c) Since $4 \cdot 4 = 16$ $\sqrt{16} = \mathbf{4}$

(d) Since $2 \cdot 2 \cdot 2 = 8$ $\sqrt[3]{8} = \mathbf{2}$

When we simplify expressions that contain powers and/or radicals, we begin by simplifying the powers and radicals. Then we simplify within symbols of inclusion. Lastly, we remember to do all multiplications and division before we add or subtract.

example 46.3 Simplify: (a) 1^5 (b) $\sqrt[3]{1}$

solution (a) If we use 1 as a factor 5 times, the answer is 1.

$$1 \cdot 1 \cdot 1 \cdot 1 \cdot 1 = 1$$

We see that **1 raised to any power equals 1.**

(b) When we write

$$\sqrt[3]{1}$$

we are asking "what number used as a factor 3 times equals 1?" Of course, the answer is 1.

$$\sqrt[3]{1} = 1$$

The fourth root of 1 is 1. The fifth root of 1 is 1. **Any root of 1 is 1.**

example 46.4 Simplify: $4(3 - 2 + 8) + 2^2 - \sqrt[3]{27} + 12 \div 2$

solution First we simplify the powers and radicals.

$$4(3 - 2 + 8) + 4 - 3 + 12 \div 2$$

Then we simplify within the parentheses.

$$4(9) + 4 - 3 + 12 \div 2$$

Now we multiply and divide where indicated.

$$36 + 4 - 3 + 6$$

We finish by adding and subtracting from left to right.

$$40 - 3 + 6 \qquad \text{added 36 and 4}$$

$$37 + 6 \qquad \text{subtracted 3}$$

$$\mathbf{43} \qquad \text{added 6}$$

practice   Simplify:

    a. 2^6 **b.** 1^3 **c.** $\sqrt[4]{625}$

    d. $\sqrt[3]{64}$ **e.** $\sqrt{81}$

    f. $5(4 - 1 + 7) + 5^2 - \sqrt[3]{64}$

problem set
46

1. Four large ones cost $40. Write the two rates (ratios) implied by this statement. How much would the customer have to pay for 120 large ones?

2. The order was for 4 bunches of asparagus at 50 cents a bunch, 9 bushels of okra at $5.50 a bushel, and 4 pecks of beans at $7.50 a peck. What was the total cost of the order?

3. The average elapsed time for the first 10 time trials was 26 minutes. The average elapsed time for the next 90 time trials was 30 minutes. Find the overall average. (Express the result as a decimal number.)

4. William has 42 boxes and had 5040 items to put in the boxes. If he divided the items evenly, how many would go in each box?

5. What number is $\frac{7}{8}$ of 400?   **6.** What number is $\frac{11}{6}$ of 120?

7. Write $3\frac{2}{5}$ as an improper fraction.

8. Find the least common multiple of 18, 42, and 50.

Solve:

9. $\frac{5}{3}x = 20$ **10.** $4x = 2$

11. $\frac{x}{4} = 7$ **12.** $x - 7 = 2$

Simplify:

13. $3^3 + 4^4 + 2^5$ **14.** $\sqrt[5]{243}$ **15.** $\sqrt[3]{125}$

16. $5(9 - 7 + 4) + 3^2 - \sqrt[3]{27} + 3 \cdot 2$ **17.** $\frac{3}{4} + 7\frac{11}{12} - \frac{5}{6}$

18. $321\frac{7}{12} - 123\frac{3}{4}$ **19.** $(6)(3) + 4(2 + 12)$ **20.** 111.8×0.007

21. $7816.7 - 982.67$ **22.** $\frac{179.32}{0.004}$ **23.** $7\frac{1}{8} \div 2\frac{1}{4} \times 3\frac{1}{6}$

24. $2\frac{1}{4} \times 6\frac{3}{4} \div 3\frac{1}{8}$ **25.** $\dfrac{7\frac{2}{3}}{6\frac{5}{6}}$

26. Evaluate: $xyz + y + x + z$ if $x = 3$, $y = 4$, and $z = 5$

27. Express the perimeter of this figure in meters. Dimensions are in centimeters. All angles are right angles.

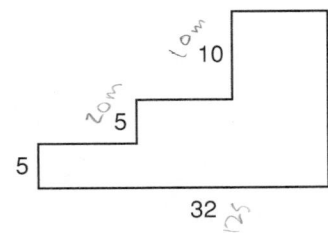

28. Use words to write 25.000131.

29. Factor 1,000,000,000 in multiples of 10.

30. Use two unit multipliers to convert 5 square yards to square feet.

LESSON 47 *Volume*

We use the word **area** to describe the size of a surface. When we tell how large an area is, we describe how many squares of a certain size can be drawn on the surface. The area of a surface also tells us the number of floor tiles it will take to cover the surface. Area has only length and width.

We use the word **volume** to describe a space or a solid that has depth as well as length and width. **Volume is not flat, for volume describes how many cubes of a certain size a thing will hold.** We can use sugar cubes to help us think about volume. A cube is a six-sided figure each of whose faces is square.

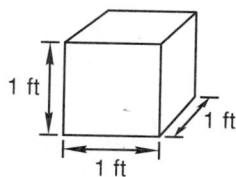

If each of the edges is 1 foot long, we say that the cube has a volume of 1 cubic foot. If each of the edges is 1 centimeter long, the volume is 1 cubic centimeter. If each of the edges is 1 mile long, the volume is 1 cubic mile, etc. We use exponents to help us abbreviate the units for volume.

$$1 \text{ cubic foot} = 1 \text{ ft}^3 \qquad 1 \text{ cubic centimeter} = 1 \text{ cm}^3$$

$$1 \text{ cubic mile} = 1 \text{ mi}^3$$

If we have a rectangular area that measures 4 feet by 2 feet, it has an area of 8 square feet.

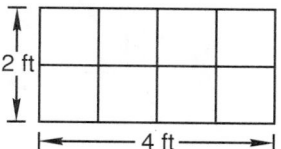

One sugar cube with a volume of 1 cubic foot can be set on each of the squares shown. If we place 1 cube on each square, we will use 8 cubes. If we stack the cubes 2 deep, we will use 16 cubes.

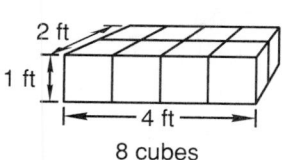

8 cubes

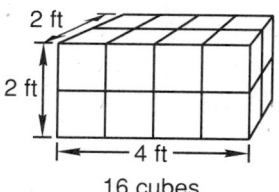

16 cubes

The figure on the left at the bottom of the preceding page has a volume of 8 ft³, and the figure on the right has a volume of 16 ft³.

If we stack the cubes 3 deep, we would have 3 layers of 8 cubes; and if we stack them 4 deep, we would have 4 layers of 8 cubes.

$$3 \text{ layers of 8 cubes} = 24 \text{ cubes}$$

$$4 \text{ layers of 8 cubes} = 32 \text{ cubes}$$

We call a geometric figure that occupies space a **geometric solid.** If the sides of the solid go straight up from the base, the sides make a right angle where they contact the base. We call these geometric solids **right geometric solids** or just **right solids.** From the discussion about sugar cubes above, we see that the volume of a right solid equals the area of the base times the height.

Volume of a right solid = area of the base × height

example 47.1   The rectangle shown is the base of a right solid whose sides are 4 centimeters high. Dimensions are in centimeters. Find the volume of the solid.

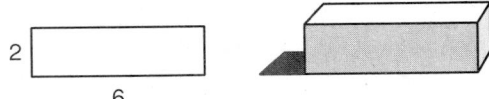

solution   The base measures 2 cm × 6 cm so the area of the base is 12 cm².

$$\text{Area of base} = 6 \text{ cm} \times 2 \text{ cm} = 12 \text{ cm}^2$$

This means that 12 one-centimeter sugar cubes will cover the base to a depth of 1 centimeter. The height of the solid is 4 centimeters, so we will stack the cubes 4 deep.

$$\text{Volume} = \text{area of base} \times \text{height}$$

$$= (12 \text{ cm}^2)(4 \text{ cm}) = \textbf{48 cm}^3$$

This tells us that 48 one-centimeter sugar cubes will completely fill the space occupied by this right solid.

example 47.2   The two-dimensional drawing shows the base of a right solid that is 6 feet high. Find the volume of the solid. Dimensions are in feet.

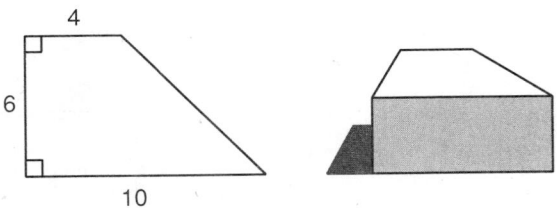

solution   First we find the area of the base.

$$\text{Area} = A + B$$

$$= (6 \times 4) + \frac{(6 \times 6)}{2}$$

$$= 24 + 18 = 42 \text{ ft}^2$$

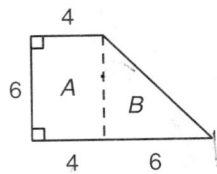

Now we know that the base can be covered to a depth of 1 foot with 42 one-cubic-foot sugar cubes. Of course, we would have to crush the sugar cubes to fit into the corners. If we stack the sugar cubes 6 feet deep, we get a volume of 252 cubic feet.

$$\text{Volume} = (42 \text{ ft}^2) \times (6 \text{ ft}) = \mathbf{252 \text{ ft}^3}$$

It will take 252 one-foot sugar cubes to completely fill the space occupied by this right solid.

practice

a. Find the number of 1-cm cubes that can be placed in a rectangular box measuring 4 cm by 6 cm by 10 cm.

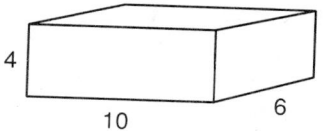

b. On the left we show the base of a right solid with a height of 12 meters. Dimensions are in meters. Find the volume of the solid.

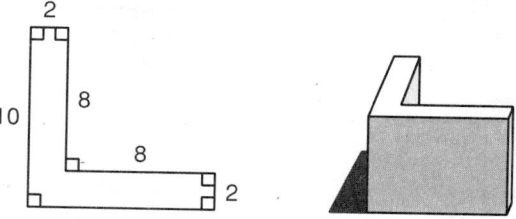

c. On the left we show the base of a right solid whose height is 2 meters. Dimensions are in meters. Find the volume of the solid.

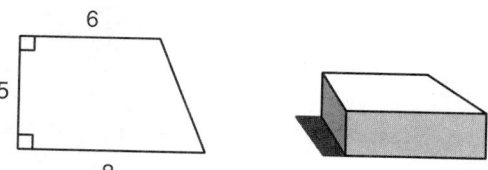

problem set 47

1. There were 3 times as many castles as there were kings. There were 7 times as many princesses as kings. If there were 12 kings, how many kings, castles, and princesses were there in all?

2. Joe tossed 4 into the pot. Their average weight was 6 lb. Then he tossed in 5 more whose average weight was 7 lb. Then he threw in a big one. The overall average weight of all of them was 7 lb. What did the big one weigh?

3. The number of red frogs exceeded the number of blue frogs by 80. The number of green frogs was 20 less than the number of blue frogs. If there were 120 blue frogs, what was the sum of the reds, the blues, and the greens?

4. Spinach Pop could be produced at 640 bottles per hour. Write the two rates (ratios) implied by this statement. How many hours would it take to produce 9600 bottles of Spinach Pop?

5. On the left we show the base of a right solid whose height is 6 inches. Dimensions are in inches, and all angles are right angles. Find the volume of the solid.

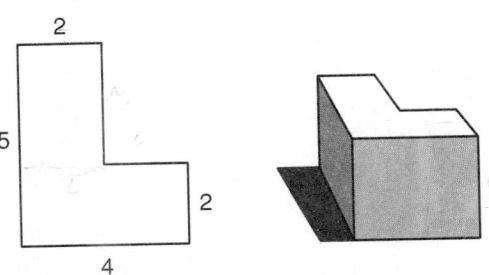

6. On the left we show the base of a right solid whose height is 15 meters. Dimensions are in meters. Find the volume of the solid.

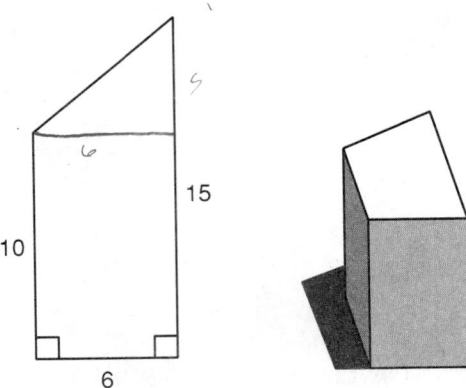

7. What number is $\frac{7}{9}$ of 1800?

8. Write $7\frac{5}{6}$ as an improper fraction.

9. Find the least common multiple of 16, 24, and 36.

Solve:

10. $\frac{4}{3}x = 160$

11. $x + 6 = 12$

12. $x - 5 = 13$

Simplify:

13. $2^2 + 3^3 + 4^2$

14. $\sqrt[4]{16}$

15. $7(2 + 6 - 4) + 2^3 - \sqrt[3]{8}$

16. $\frac{4}{5} + 6\frac{3}{10} - \frac{6}{15}$

17. $615\frac{3}{8} - 138\frac{3}{4}$

18. $7(8 - 3) + 4 + (6)(2)$

19. 134.8×0.013

20. $725.89 - 62.871$

21. $\dfrac{381.42}{0.006}$

22. $3 + (6 \cdot 12)5 - 7$

23. $7(6 - 4) + 2(4 + 2) - 7 \cdot 2$

24. $3\frac{3}{5} \times 3\frac{1}{2} \div 6\frac{3}{4}$

25. $\dfrac{3\frac{4}{5}}{2\frac{7}{8}}$

26. Evaluate: $xyz + yz - y$ if $x = \dfrac{1}{6}$, $y = 3$, and $z = 2$

27. Find the area of this figure. Dimensions are in feet.

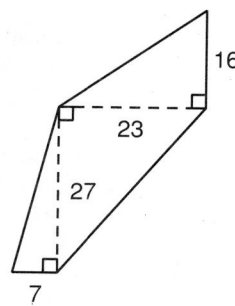

28. Use two unit multipliers to convert 6200 centimeters to kilometers.

29. Round 621.72727 to three decimal places.

30. Express the perimeter of this figure in feet. Dimensions are in inches. All angles are right angles.

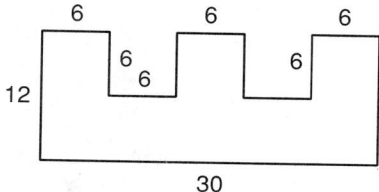

LESSON 48   *Order of operations with fractions*

Thus far, we have restricted order-of-operations problems to those containing whole numbers. The procedure is the same for fractions and mixed numbers. Unless otherwise indicated, we always multiply and divide before adding or subtracting.

example 48.1   Simplify: $\dfrac{3}{20} + \dfrac{3}{2} \div \dfrac{5}{2}$

solution   We skip the addition because we must divide first. We change the division to a multiplication and then multiply.

$$\frac{3}{20} + \frac{3}{2} \cdot \frac{2}{5} = \frac{3}{20} + \frac{3}{5}$$

Now we find a common denominator and add.

$$\frac{3}{20} + \frac{3}{5} = \frac{3}{20} + \frac{12}{20} = \frac{15}{20} = \frac{3}{4}$$

example 48.2   Simplify: $\frac{5}{7} - \frac{1}{10} \cdot \frac{4}{5}$

solution   We skip the subtraction and do the multiplication first. We get

$$\frac{5}{7} - \frac{2}{25}$$

Now we subtract.

$$\frac{5}{7} - \frac{2}{25} = \frac{125}{175} - \frac{14}{175} = \mathbf{\frac{111}{175}}$$

practice   Simplify:

a. $\frac{9}{7} - \frac{3}{7} \cdot \frac{5}{2}$

b. $4\frac{1}{5} + 2\frac{1}{3} \cdot 5$

problem set 48

1.   The pompon girls could put 44 pompons in one box. If there were 1760 pompons that had to be packed, how many boxes did the pompon girls need?

2.   When the boxes came, a count revealed an overshipment. They had shipped 62 boxes. Now how many pompons were required to fill all the boxes?

3.   Ronk bought 14 games for $70. Write the two rates (ratios) for this statement. How many games could he buy for $200?

4.   Sarah bought 7 games for $35. Write the two rates (ratios) for this statement. What would she have to pay for 200 games?

5.   The times for the race came down with practice. The average time for the 5 racers in the first race was 57 seconds. The average time for the second race was 55 seconds. For the third race the average time was 52 seconds. What was the average time for all three races?

6.   What number is $\frac{6}{7}$ of 217?

7.   Write $\frac{316}{25}$ as a mixed number.

8.   Find the least common multiple of 8, 21, and 24.

Solve:

9. $\frac{6}{7}x = 18$

10. $x + \frac{5}{13} = \frac{18}{26}$

11. $7x = 315$

Simplify:

12. $4 + (3 \cdot 5)6 + 4(2 \cdot 3)$

13. $6(3 - 1) + 2(3 - 1) - 4$

14. $\frac{4}{9} + \frac{5}{4} \cdot \frac{1}{6}$

15. $4\frac{2}{5} \cdot \frac{3}{7} - \frac{4}{7}$

16. $4^2 + 3^3 - 2^3$

17. $\sqrt[3]{8} + \sqrt[3]{27}$

18. $\frac{7}{8} + 5\frac{5}{16} - \frac{3}{4}$

19. $117\frac{9}{10} - 12\frac{3}{5}$

20. 181.3×0.012

21. $1.825 - 0.981$

22. $\frac{262.15}{0.003}$

23. $4\frac{3}{4} \div 2\frac{1}{3} \times 3\frac{1}{3} \div 1\frac{1}{4}$

24. $6\frac{3}{8} \times 2\frac{4}{5} \div 2\frac{1}{2}$

25. $\dfrac{8\frac{6}{7}}{3\frac{5}{14}}$

26. Evaluate: $xy + xyz - z$ if $x = 24$, $y = \frac{1}{4}$, and $z = 2$

27. Find the perimeter of this figure in centimeters. Dimensions are in meters. All angles are right angles.

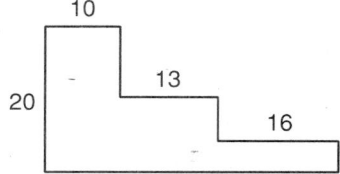

28. On the left we show the base of a right solid whose height is 17 centimeters. Dimensions are in centimeters, and all angles are right angles. Find the volume of the solid.

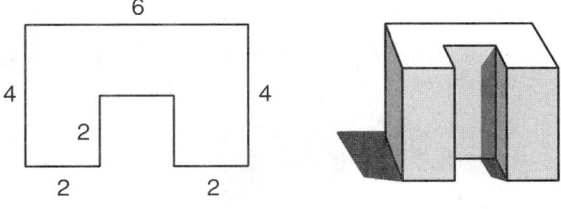

29. Use two unit multipliers to convert 360 yards to inches.

30. Round $621{,}721.8\overline{17}$ to four decimal places.

LESSON 49   *Evaluation of exponential expressions and radicals*

Exponential expressions designate a base and tell us how many times the base is to be used as a factor. The expression 5^3 means

$$5 \cdot 5 \cdot 5$$

When the base is a letter, the exponent tells us how many times the letter is to be used as a factor. Thus the expression x^3 means

$$x \cdot x \cdot x$$

When we evaluate an exponential expression that contains letters, we replace the letters with the proper numbers.

example 49.1   Evaluate m^2 if $m = 9$

solution We replace *m* with 9 and get

$$9^2$$

and 9^2 means 9 times 9, so we get

$$9^2 = \mathbf{81}$$

example 49.2 Evaluate 4^x if $x = 3$.

solution We replace *x* with 3 and get

$$4^3$$

This means to use 4 as a factor 3 times.

$$4^3 = 4 \cdot 4 \cdot 4 = \mathbf{64}$$

example 49.3 Evaluate $\sqrt[n]{64}$ if $n = 3$.

solution We replace *n* with 3 and get

$$\sqrt[3]{64}$$

The cube root of 64 is 4 because $4 \cdot 4 \cdot 4$ equals 64.

$$\sqrt[3]{64} = \mathbf{4}$$

example 49.4 Evaluate $\sqrt[3]{n}$ if $n = 27$.

solution We replace *n* with 27 and get

$$\sqrt[3]{27}$$

The third root of 27 is 3 because $3 \cdot 3 \cdot 3 = 27$.

$$\sqrt[3]{27} = \mathbf{3}$$

practice Evaluate:

a. 4^x if $x = 3$ **b.** x^3 if $x = 5$

c. $\sqrt[n]{27}$ if $n = 3$ **d.** $\sqrt[n]{81}$ if $n = 2$

problem set 49

1. Sarah sold whortleberries after school. She sold 40 pecks for $640. Write the two rates (ratios) implied by this statement. How much money would she get for 100 pecks?

2. Irwin stretched but could only reach seventy-one and one thousand, four hundred three ten-millionths inches. He lost because Eisel could reach seventy-two and one thousand, six hundred, forty-two hundred-thousandths inches. How much farther could Eisel reach?

3. Knapp hid them in discrete bunches of 48. Some were under the front porch. In all he hid 1305 discrete bunches. How many did he hide in all?

4. Forty bottles could be capped in 3 hours. Write the two rates (ratios) that this statement gives. How long would it take to cap 360 bottles?

5. Find the number of 1-foot cubes this right prism will hold. Dimensions are in feet. First find the area of one triangular end. Remember that the area of a triangle is one-half the base times the height.

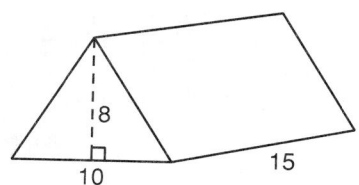

6. What number is $\frac{3}{8}$ of 256?

7. Write $\frac{213}{8}$ as a mixed number.

8. Find the least common multiple of 16, 24, and 30.

Solve:

9. $\frac{7}{8}x = 14$

10. $x - \frac{1}{8} = \frac{3}{16}$

11. $x + \frac{5}{8} = \frac{14}{5}$

Simplify:

12. $6 \cdot 3 + (3 \cdot 2)5 - 4(2 \cdot 2)$

13. $4(6 - 2 + 1) + 3^2 - \sqrt[3]{8}$

14. $\frac{7}{20} + \frac{4}{5} \cdot \frac{2}{3}$

15. $3\frac{1}{6} \cdot \frac{1}{8} - \frac{4}{12}$

16. $3^2 + 4^2 - 5^2$

17. $\frac{3}{4} + 7\frac{1}{16} - \frac{5}{8}$

18. $113\frac{4}{7} - 32\frac{1}{14}$

19. 31.62×0.08

20. $89.265 - 6.898$

21. $\frac{261.82}{0.004}$

22. $3\frac{2}{3} \times 2\frac{3}{4} \div 3\frac{1}{2}$

23. $\dfrac{6\frac{3}{4}}{7\frac{11}{12}}$

Evaluate:

24. $xyz - x + xy$ if $x = 12$, $y = \frac{1}{4}$, and $z = 6$

25. 3^x if $x = 5$

26. $\sqrt[p]{8}$ if $p = 3$

27. Find the area of this figure. Dimensions are in meters.

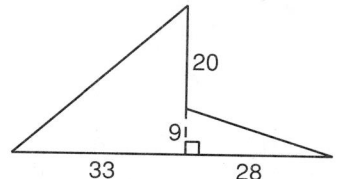

28. Use two unit multipliers to convert 250,001 centimeters to kilometers.

29. Write 7164.003186 in words.

30. Divide 364.285 by 100,000 mentally.

LESSON 50 *Fractional part of a number ·*
Fractional equations

50.A

fractional part of a number

We remember that a fraction designates a part of a whole. On the left we show a rectangle that represents the number 1 or the whole thing. On the right, we divide the whole thing into 5 parts and shade 2 of the parts. This is what we mean by the fraction $\frac{2}{5}$.

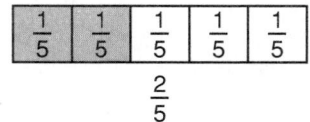

If the whole is a number greater than 1, then $\frac{2}{5}$ of the number means 2 of the 5 equal parts of the number. For example, if the number is 150, each of the 5 equal parts is 30.

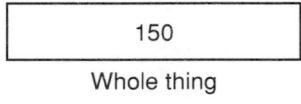   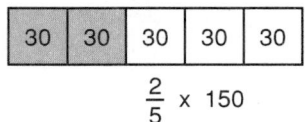

Then 2 of these parts equals $30 + 30 = 60$.

$$\frac{2}{5} \times 150 = 60$$

50.B

fractional equations

The general equation for the fractional part of a number is very important. The equation above is read as

> Two-fifths of 150 is 60

This statement contains a fraction and two numbers. One of the numbers associates with the word **of** and the other associates with the word **is.** We can write the general form of this equation as

$$F \times of = is$$

The letter F stands for **fraction** and the words *of* and *is* designate the two numbers.

example 50.1   What fraction of 75 is 25?

solution   First we write the general equation.

$$F \times of = is$$

Now we replace F with WF for "what fraction." The problem says "of 75," so we replace *of* with 75. The problem says "is 25," so we replace *is* with 25.

$$WF \times 75 = 25$$

We can undo multiplication by 75 by dividing by 75. Thus, we divide both sides by 75.

$$\frac{WF \times 75}{75} = \frac{25}{75} \longrightarrow WF = \frac{1}{3}$$

Thus we find that $\frac{1}{3}$ of 75 is 25.

example 50.2   Five-sevenths of what number is $\frac{5}{2}$?

solution   We will use the equation

$$F \times of = is$$

We replace F with $\frac{5}{7}$, replace *of* with *WN*, and replace *is* with $\frac{5}{2}$.

$$\frac{5}{7} \times WN = \frac{5}{2}$$

Then we solve by multiplying both sides by $\frac{7}{5}$.

$$\frac{7}{5} \cdot \frac{5}{7} \times WN = \frac{5}{2} \cdot \frac{7}{5} \quad \longrightarrow \quad WN = \frac{7}{2}$$

Thus we find that $\frac{5}{7}$ of $\frac{7}{2}$ is $\frac{5}{2}$.

example 50.3   Five-sevenths of $\frac{5}{2}$ is what number?

solution   Again we use the equation

$$F \times of = is$$

We replace F with $\frac{5}{7}$, *of* with $\frac{5}{2}$, and *is* with *WN*.

$$\frac{5}{7} \cdot \frac{5}{2} = WN$$

Then we multiply.

$$\frac{25}{14} = WN$$

Thus we find that five-sevenths of $\frac{5}{2}$ is $\frac{25}{14}$.

practice   **a.**   What fraction of 20 is 140?

b.   Seven-fifths of what number is $\frac{3}{8}$?

c.   Five-thirteenths of 26 is what number?

problem set   **1.**   Big Daddy put all his money in gilt-edged stock certificates. If he invested
50 $64,000 and paid $40 for each certificate, how many certificates did Big
                       Daddy buy?

2.   Marsha guessed forty-one thousand, fifteen millionths. Then Kevin the Pipe
         Man guessed twenty-seven ten-thousandths. What was the sum of their
         guesses?

3. Forty good ones cost $12. Write the two rates (ratios) implied by this statement. How many good ones could be purchased for $3.40?

4. The sum of the weights of the first four contestants was 760 pounds. The first three contestants weighed 210 pounds, 185 pounds, and 220 pounds. What was the average weight of the four contestants?

5. Five-sevenths of what number is $\frac{7}{3}$?

6. Three-fifths of what number is $\frac{3}{7}$?

7. Find the number of 1-ft cubes that a rectangular box which measures 3 ft × 4 ft × 6 ft will hold.

8. What number is $\frac{4}{7}$ of 210?   9. Write $6\frac{27}{31}$ as an improper fraction.

10. Find the least common multiple of 18, 20, and 24.

Solve:

11. $\frac{8}{9}x = 16$ 12. $x - 3\frac{1}{4} = \frac{1}{2}$

Simplify:

13. $7 \cdot 2 + (3 \cdot 2)4 - 3(2 \cdot 1)$ 14. $6(3 - 2 + 1) + 3^2 - \sqrt[3]{8}$

15. $\frac{3}{20} + \frac{4}{5} \cdot \frac{3}{4}$ 16. $4\frac{2}{3} \cdot \frac{3}{4} - \frac{7}{12}$ 17. $137\frac{5}{7} - 16\frac{1}{14}$

18. $\frac{4}{5} + 3\frac{1}{10} - \frac{14}{25}$ 19. $3^2 - \sqrt{9} + \sqrt[3]{8}$ 20. 21.32×0.06

21. $213.81 - 11.713$ 22. $\frac{6111.2}{0.005}$ 23. $\frac{6\frac{4}{5}}{5\frac{3}{4}}$

Evaluate:

24. $xyz + yz - xy$ if $x = 20$, $y = \frac{1}{5}$, and $z = 5$

25. y^2 if $y = 3$ 26. $\sqrt[p]{27}$ if $p = 3$

27. Find the volume of this rectangular solid. Dimensions are in feet.

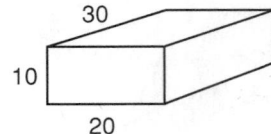

28. Use two unit multipliers to convert 3,059,000 miles to inches.

29. Use words to write 700.0000563.

30. Simplify mentally: $100 \times 10,000$

LESSON 51   *Surface area*

The surface area of a solid is the total area of all the exposed surfaces of the solid. A rectangular box has six surfaces. It has a top and a bottom, a front and a back, and two sides. The surface area is the sum of these six areas.

example 51.1   Find the surface area of this rectangular box. All dimensions are in meters.

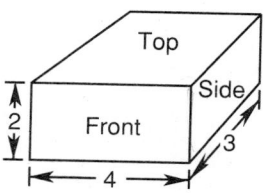

solution   Since the box is rectangular, the areas of the top and the bottom are equal. The area of the front equals the area of the back, and the areas of the two sides are equal.

$$
\begin{aligned}
\text{Area of front} &= 4 \text{ m} \times 2 \text{ m} = 8 \text{ m}^2 \\
\text{Area of back} &= 4 \text{ m} \times 2 \text{ m} = 8 \text{ m}^2 \\
\text{Area of top} &= 4 \text{ m} \times 3 \text{ m} = 12 \text{ m}^2 \\
\text{Area of bottom} &= 4 \text{ m} \times 3 \text{ m} = 12 \text{ m}^2 \\
\text{Area of side} &= 3 \text{ m} \times 2 \text{ m} = 6 \text{ m}^2 \\
\text{Area of side} &= 3 \text{ m} \times 2 \text{ m} = \underline{6 \text{ m}^2} \\
\text{Surface area} &= \quad\text{total}\quad = \mathbf{52 \text{ m}^2}
\end{aligned}
$$

example 51.2   Find the surface area of this right prism. All dimensions are in centimeters.

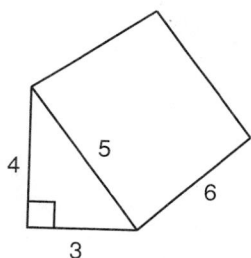

solution   The prism has two ends that are triangles. It has three faces that are rectangles.

$$
\text{Area of one end} = \frac{4 \text{ cm} \times 3 \text{ cm}}{2} = 6 \text{ cm}^2
$$

$$
\text{Area of one end} = \frac{4 \text{ cm} \times 3 \text{ cm}}{2} = 6 \text{ cm}^2
$$

$$
\begin{aligned}
\text{Area of bottom} &= 3 \text{ cm} \times 6 \text{ cm} = 18 \text{ cm}^2 \\
\text{Area of back} &= 4 \text{ cm} \times 6 \text{ cm} = 24 \text{ cm}^2 \\
\text{Area of front} &= 5 \text{ cm} \times 6 \text{ cm} = \underline{30 \text{ cm}^2} \\
\text{Surface area} &= \quad\text{total}\quad = \mathbf{84 \text{ cm}^2}
\end{aligned}
$$

practice   **a.**   Find the surface area of this rectangular solid. The dimensions are in feet.

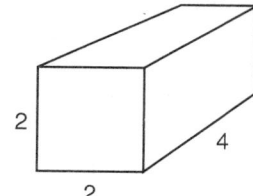

b.   Find the surface area of this right prism. The dimensions are in centimeters.

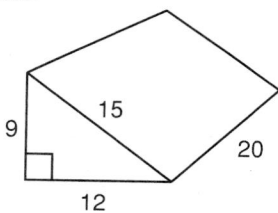

problem set 51

1. On one hand there were one hundred forty-two thousand, seven hundred sixty-three. On the other hand there were two hundred twenty-eight thousand, fourteen. How many more were there on the other hand?

2. The whole batch cost $28,000 and contained 140 items. Write two rates (ratios) implied by this statement. What would be the price for 200 items?

3. What was the average number of tons for the 4-month period of March, April, May, and June?

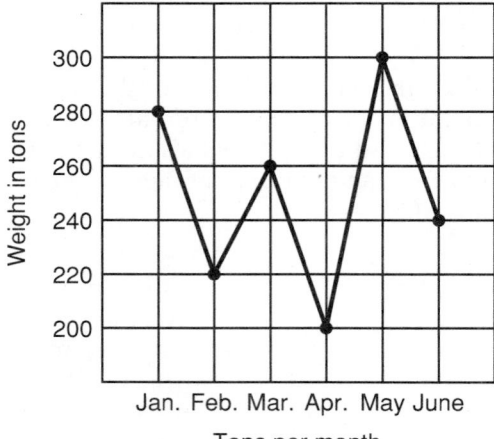

Tons per month

4. The average time for the four races was 10.2 seconds. If the times for the first three races were 10.1, 10.6, and 10.3 seconds, what was the time of the fourth race?

5. Find the surface area of this rectangular solid. All dimensions are in meters.

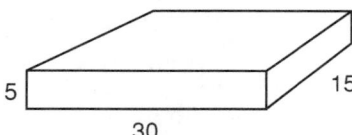

6. Find the number of 1-centimeter cubes this right prism will hold. Dimensions are in centimeters. First find the area of one end. Remember that the area of a triangle is one-half the base times the height.

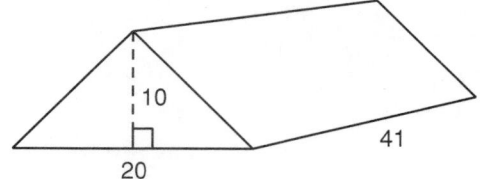

7. What number is $\frac{5}{8}$ of 168?

8. Four-fifths of what number is $\frac{7}{2}$?

9. Find the least common multiple of 18, 24, and 36.

10. Five-sixths of what number is $4\frac{1}{3}$?

11. Write $\frac{168}{7}$ as a mixed number.

Solve:

12. $\frac{8}{9}x = 16$

13. $x - \frac{3}{5} = 2\frac{3}{10}$

Simplify:

14. $6 \cdot 2 + (6 \cdot 3)7 - 4(3 \cdot 1)$

15. $5(6 - 3 + 1) + 6^2 - \sqrt{100}$

16. $\dfrac{11}{14} + \dfrac{3}{7} \cdot \dfrac{3}{4}$

17. $6\dfrac{2}{3} \cdot \dfrac{1}{4} - \dfrac{3}{4}$

18. $3^2 + \sqrt[3]{27} - \sqrt[3]{8}$

19. $\dfrac{3}{5} + 3\dfrac{4}{15} - \dfrac{7}{30}$

20. $117\dfrac{3}{8} - 14\dfrac{7}{16}$

21. 16.82×0.016

22. $118.321 - 81.34$

23. $\dfrac{16.25}{0.03}$

24. $4\dfrac{2}{3} \times 2\dfrac{3}{4} \div 1\dfrac{5}{12}$

25. $\dfrac{7\frac{2}{3}}{2\frac{5}{6}}$

Evaluate:

26. $x + xy + xyz$ if $x = 1$, $y = 10$, and $z = \dfrac{1}{5}$

27. p^3 if $p = 2$

28. Use two unit multipliers to convert 6 miles to inches.

29. Write 41000.0000002 in words.

30. Use two unit multipliers to convert 12 square kilometers to square meters.

LESSON 52 *Brackets*

We call the parentheses symbols of **inclusion.** In this expression

$$4 + (5 \cdot 3)$$

the number included inside the parentheses is 5 times 3, or 15.

$$4 + (15) = 19$$

We also use brackets as symbols of inclusion.

(parentheses)   [brackets]

When an expression contains more than one symbol of inclusion, we begin by simplifying within the innermost symbols of inclusion.

example 52.1   Simplify: $24 - 2[(5 - 2)(14 - 12) + 3]$

solution   We begin by simplifying within the parentheses.

$$24 - 2[(3)(2) + 3]$$

Now we simplify within the brackets. We remember to multiply before we add.

$$24 - 2[9]$$

Again we multiply first and then subtract.

$$24 - 18 = \mathbf{6}$$

example 52.2   Simplify: $33 - 2[3(3 + 12) - (5 \cdot 2)3]$

solution   We begin by simplifying within the parentheses.

$$33 - 2[3(15) - (10)3]$$

Now we simplify within the brackets.

$$33 - 2[45 - 30]$$
$$33 - 2[15]$$

Next we multiply and then we subtract.

$$33 - 30 = \mathbf{3}$$

practice   Simplify:

a. $55 - 3[(4 - 1)(10 - 6) + 2]$ **b.** $2[3(6 - 3)(4 - 2) + 2] - 2$

**problem set
52**

1. On the first day the attendance was forty-seven thousand, three hundred sixty-four. On the second day the attendance was fifty-three thousand, seven. How many more attended on the second day?

2. Each bin held 14 uniforms. Write the two rates (ratios) implied by this statement. If there were 140 bins in all, how many uniforms did they hold?

3. Fifteen could be purchased for only $315. Write the two rates (ratios) implied by this statement. How much would 140 cost?

4. The bulls weighed 2153 lb, 1491 lb, and 1840 lb. What was the average weight of the bulls?

5. Find the volume of the right tri-angular prism. Dimensions are in feet. First find the area of the triangle and then multiply by the length of the long side.

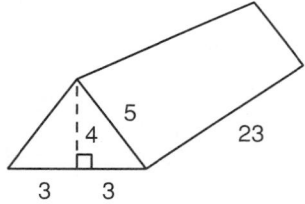

6. Find the surface area of the prism given in Problem 5.

7. Find the least common multiple of 14, 21, and 27.

8. Five-eighths of what number is 100?

9. What fraction of 64 is 56?

10. Seven-elevenths of 88 is what number?

11. Write $\dfrac{131}{15}$ as a mixed number.

Solve:

12. $\dfrac{4}{7}x = 112$ **13.** $x + \dfrac{7}{12} = 3\dfrac{5}{12}$

Simplify:

14. $28 - 2[(6 - 2)(10 - 9) + 3]$

15. $15 + 3[(6 - 4)(7 - 4) - 2]$

16. $4[(3 - 1)(3 + 2) - 1] + 25$

17. $7\frac{3}{4} \cdot \frac{2}{3} - \frac{5}{12}$

18. $\frac{13}{14} + \frac{2}{7} \cdot \frac{3}{4}$

19. $\frac{4}{5} + 2\frac{7}{10} - \frac{3}{20}$

20. $2^3 + 3^3 - \sqrt[3]{8}$

21. $36\frac{6}{7} - 14\frac{3}{14}$

22. 171.3×0.012

23. $62.891 - 18.812$

24. $\frac{17.025}{0.003}$

25. $4\frac{1}{8} \div 2\frac{1}{4} \times 3\frac{1}{2} \div 1\frac{1}{16}$

26. $\dfrac{8\frac{1}{2}}{4\frac{1}{7}}$

Evaluate:

27. $xyz + yz - x$ if $x = 16$, $y = \frac{1}{8}$, and $z = 24$

28. $p^3 + p + q$ if $p = 2$ and $q = 3$

29. Use two unit multipliers to convert 2162.18 centimeters to kilometers.

30. Write 16821621.00321 in words.

LESSON 53 *Scientific notation*

53.A
scientific notation

When we multiply 0.0534 by 1000, we move the decimal point three places to the right.

$$0.0534 \times 1000 = 53.4$$

Because 10^3 equals 1000, when we multiply 0.0534 by 10^3, we also move the decimal point three places to the right.

$$0.0534 \times 10^3 = 53.4$$

When we multiply a number by a power of 10, we move the decimal point to the right that many places.

We can use this fact to write numbers in a special way that makes very large numbers easy to handle. We call this special way **scientific notation.** When we write a number in scientific notation, we use two steps.

1. Place the decimal point just to the right of the first digit that is not zero.
2. Multiply this number by 10^b to tell where the decimal point really is (b is a counting number).

example 53.1   Write 7,024,000 in scientific notation.

solution   First we mark the old position of the decimal point with a caret. Then we put the new decimal point just to the right of the 7.

$$7.024000_\wedge$$

Now we count the places between the decimal point and the caret.

$$7.024000_\wedge$$

There are six places between the decimal point and the caret. So we multiply by 10^6 and get

$$\mathbf{7.024 \times 10^6}$$

The 10^6 tells us that the decimal point **is really** six places to the right of where it is written.

example 53.2   Write 476.24 in scientific notation.

solution   We use a caret. We put the decimal point just behind the 4. Then we count the places.

$$4.76_\wedge 24$$

There are two places between the decimal point and the caret, so we write

$$\mathbf{4.7624 \times 10^2}$$

The 10^2 tells us that the decimal point **is really** two places to the right of where it is written.

53.B
negative exponents

We use 10 with a **negative exponent** to show that the decimal point should be moved **to the left**. The notation

$$4.76 \times 10^{-3}$$

tells us that the decimal point **is really** three places to the left of where we have written it. Thus

$$4.76 \times 10^{-3} \quad \text{means} \quad 0.00476$$

example 53.3   Write 0.0652 in scientific notation.

solution   As usual we begin by using a caret to mark the old position of the decimal point. We write the new decimal point just to the right of the 6.

$$_\wedge 06.52$$

We see that the true position of the decimal point is two places **to the left** of where we have written it. So we write

$$6.52 \times 10^{-2}$$

The 10^{-2} tells us that the true position of the decimal point **is really** two places to the left of where it is written.

example 53.4   Write these numbers in standard form: (a) 4.6×10^{-4} (b) 4.6×10^4

solution   (a)   The negative exponent tells us to move the decimal point to the left.

$$4.6 \times 10^{-4} = \textbf{0.00046}$$

(b)   The positive exponent tells us to move the decimal point to the right.

$$4.6 \times 10^{4} = \textbf{46,000}$$

practice   Write each number in scientific notation:

a.   476,000   b.   0.000476

c.   305,600   d.   0.000003056

Write each number in standard form:

e. 4.06×10^{5} f. 4.72×10^{-6}

problem set 53

1.   At the sock hop there were 420 pairs of red socks, 375 pairs of black socks, and 835 pairs that were mismatched. By how many pairs did the mismatched socks outnumber the sum of the reds and blacks?

2.   The original diameter of the microbe was one thousand, four hundred seventy-five millionths inches. Later the diameter increased to one hundred three ten-thousandths inches. By how much did the diameter increase?

3.   Jimmy bought 1900 oscillators for $38,000. Write the two rates (ratios) implied by this statement. How much would he have to pay for 5000 oscillators?

4.   In the first five games the team scored 86 points, 92 points, 80 points, 70 points, and 104 points. What was their average score?

5.   Write in scientific notation:
(a)   47,000,000   (b)   0.00000047

6.   Write in standard form:
(a) 6.3×10^{9} (b) 6.3×10^{-9}

7.   Find the volume of the rectangular box. Dimensions are in yards.

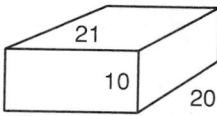

8.   Find the surface area of the rectangular box shown in Problem 7.

9.   Find the least common multiple of 20, 30, and 36.

10.   Three-fourths of what number is 51?

11.   Three-sevenths of 91 is what number?

12.   Write $7\frac{3}{22}$ as an improper fraction.

Solve:

13. $\frac{5}{4}x = 120$ 14. $x + \frac{3}{11} = \frac{51}{22}$

Simplify:

15. $17 - 2[(6-4)(7-3) - 7]$ 16. $3^{2} + 2[(7+1)(7-5) - 12]$

17. $6\frac{2}{3} \cdot \frac{1}{4} - \frac{5}{12}$ **18.** $\frac{5}{6} + 1\frac{7}{12} - \frac{2}{3}$ **19.** $2 + 2^2 + 2^3$

20. $\sqrt[3]{27} - \sqrt[2]{9} + 3$ **21.** $16\frac{3}{5} - 5\frac{1}{10}$ **22.** 181.4×0.0012

23. $1189.26 - 91.872$ **24.** $\dfrac{181.02}{0.006}$ **25.** $7\frac{1}{4} \div 3\frac{1}{3} \times 1\frac{2}{3} \times 1\frac{1}{6}$

26. $\dfrac{3\frac{2}{7}}{2\frac{1}{14}}$

Evaluate:

27. $xyzt + xyz + yzt$ if $x = \frac{1}{6}$, $y = 3$, $z = 4$, and $t = 5$

28. $\sqrt[x]{8} + y$ if $x = 3$ and $y = 8$

29. Use four unit multipliers to convert 625 square yards to square inches.

30. Round 621.5621812 to the nearest thousandth.

LESSON 54   *Decimal part of a number*

The equation for the fractional part of a number is

$$F \times of = is$$

The equation for the decimal part of a number is exactly the same except that we use D for decimal in place of F for fraction.

$$D \times of = is$$

example 54.1   0.4 of what number is 72?

solution   We will use the equation for the decimal part of a number

$$D \times of = is$$

and replace D with 0.4, *of* with *WN*, and *is* with 72.

$$0.4 \times WN = 72$$

We divide by 0.4 to solve.

$$\frac{0.4 \times WN}{0.4} = \frac{72}{0.4}$$

$$WN = 180$$

Thus, 0.4 of 180 is 72.

example 54.2   What decimal part of 240 is 90?

solution   We use the decimal form of the equation

$$D \times of = is$$

and replace *of* with 240 and *is* with 90.

$$WD \times 240 = 90$$

Then we divide both sides by 240.

$$\frac{WD \times 240}{240} = \frac{90}{240} \qquad \text{divided by 240}$$

$$WD = 0.375 \qquad \text{simplified}$$

example 54.3   Four-tenths of 72 is what number?

solution   We use the decimal form of the equation

$$D \times of = is$$

and replace *D* with 0.4 and replace *of* with 72.

$$0.4 \times 72 = WN$$

We multiply to find the answer.

$$0.4 \times 72 = 28.8$$

practice   Use the equation $D \times of = is$ to solve these problems.

a.   Seven-tenths of what number is 0.14?

b.   Seven-hundredths of 86 is what number?

c.   What decimal part of 72 is 30.24?

problem set 54

1. Willikin hid one hundred forty million, fourteen in the boscage of the weald. The next week he hid an additional fifteen million, nine hundred eighty-two thousand. How many did Willikin hide in all?

2. Wilhelmina bought 80 horses at the auction. She paid 320 guineas for them. Write the two rates (ratios) implied. How much would she have had to pay for 320 horses?

3. Richard the Lion-Hearted counted his troops on the outskirts of Accra. He counted seven thousand, nine hundred forty-two. If this was two hundred forty-two more than he counted yesterday, what was yesterday's count?

4. The first number was 186,925. The second number was 905,106. The third number was 2,105,061. How much greater was the third number than the sum of the first two numbers?

5. 0.6 of what number is 72?

6. What decimal part of 360 is 60?

7. Use two unit multipliers to convert 1000 square centimeters to square meters.

8. Find the volume of this right prism. Dimensions are in meters. First find the area of a triangular end.

9. Find the surface area of the prism given in Problem 8.

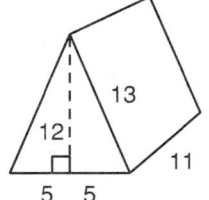

10. Find the least common multiple of 15, 20, and 30.

11. What fraction of 56 is 21?

12. Eight-thirteenths of what number is 16?

13. Write $\dfrac{181}{20}$ as a mixed number.

Solve:

14. $\dfrac{8}{7}x = 104$

15. $x - \dfrac{5}{12} = \dfrac{15}{4}$

Simplify:

16. $24 + 2[6(3 - 1)(7 - 3) - 6]$

17. $2^3 + 2[(8 + 1)(6 - 5) - 3]$

18. $5\dfrac{6}{7} \cdot \dfrac{3}{2} - 1\dfrac{1}{14}$

19. $\dfrac{5}{6} + 1\dfrac{5}{12} - 1\dfrac{1}{3}$

20. $1 + 2^2 + 4^2 - 3^2$

21. $\sqrt[3]{8} + \sqrt[3]{27} - 5$

22. $23\dfrac{4}{5} - 6\dfrac{1}{15}$

23. 181.4×0.0013

24. $\dfrac{182.101}{0.0006}$

25. $2811.62 - 13.981$

26. $8\dfrac{3}{4} \div 1\dfrac{1}{3} \times 1\dfrac{3}{4} \div \dfrac{1}{12}$

27. $\dfrac{2\dfrac{3}{7}}{3\dfrac{3}{14}}$

Evaluate:

28. $xyt - yt + x$ if $x = 6$, $y = \dfrac{1}{3}$, and $t = 12$

29. $\sqrt[x]{125}$ if $x = 3$

30. (a) Write 100,000 in scientific notation.
    (b) Write 0.0072 in scientific notation.
    (c) Write 3.2×10^5 in standard form.
    (d) Write 3.2×10^{-5} in standard form.

LESSON 55   *Fractions and symbols of inclusion*

We have restricted our practice with symbols of inclusion to problems that have whole numbers. Symbols of inclusion can also be used in the simplification of expressions that contain fractions or mixed numbers.

example 55.1   Simplify: $\frac{3}{2}\left(\frac{1}{4} + \frac{7}{16}\right) - \frac{1}{8}$

solution   First we simplify within the parentheses.

$$\frac{3}{2}\left(\frac{4}{16} + \frac{7}{16}\right) - \frac{1}{8} = \frac{3}{2}\left(\frac{11}{16}\right) - \frac{1}{8}$$

Next we multiply

$$\frac{33}{32} - \frac{1}{8}$$

and now we subtract

$$\frac{33}{32} - \frac{4}{32} = \frac{29}{32}$$

example 55.2   Simplify: $\frac{1}{3}\left(3\frac{1}{4} - \frac{1}{8}\right) + \frac{1}{48}$

solution   First we simplify within the parentheses.

$$\frac{1}{3}\left(\frac{26}{8} - \frac{1}{8}\right) + \frac{1}{48} = \frac{1}{3}\left(\frac{25}{8}\right) + \frac{1}{48}$$

Then we multiply. Then we add and simplify the result.

$$\frac{25}{24} + \frac{1}{48} = \frac{50}{48} + \frac{1}{48} = \frac{51}{48} = \frac{17}{16}$$

example 55.3   Simplify: $\frac{2}{5}\left[\left(\frac{2}{3} - \frac{1}{6}\right) + \left(\frac{1}{3} + \frac{5}{6}\right)\right]$

solution   First we simplify within the parentheses.

$$\frac{2}{5}\left[\left(\frac{1}{2}\right) + \left(\frac{7}{6}\right)\right]$$

Now we simplify within the brackets and get

$$\frac{2}{5}\left[\frac{10}{6}\right]$$

and then we multiply.

$$\frac{2}{5}\left[\frac{10}{6}\right] = \frac{2}{3}$$

practice   Simplify:

a. $\frac{3}{4}\left[\left(\frac{4}{5} - \frac{1}{2}\right) + \left(\frac{1}{6} + \frac{1}{3}\right)\right]$ **b.** $\frac{1}{4}\left(2\frac{1}{2} - \frac{1}{7}\right) + \frac{1}{28}$

problem set 55

1. Bill's funds were limited, and he bought only 70 items, for which he paid $3500. Write the two rates (ratios) implied. What would he have had to pay for 720 items?

2. The fishmonger sold 40 codfish at 10 shillings each, 59 salmon at 30 shillings each, and 1 grouper for 300 shillings. What was the average price paid for each fish?

3. The crones and the curmudgeons forced their way in until 900,062 had arrived. This was 202,020 more than had come last time. How many came last time?

4. The forecaster made multiple forecasts every day. For the first 10 days, the forecaster averaged 8.2 correct forecasts per day. For the next 90 days, 5.1 correct forecasts per day was the average. What was the overall average number of correct forecasts per day?

5. 0.3 of what number is 36?

6. What decimal part of 480 is 60?

7. (a) Write 0.000387 in scientific notation.
   (b) Write 8.69×10^{11} in standard form.

8. Find the area of the following figure. Dimensions are in centimeters.

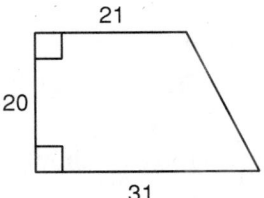

9. What is the volume of a right solid whose base is the figure shown in Problem 8 and whose height is 10 centimeters?

10. Find the least common multiple of 10, 15, and 25.

11. Six-sevenths of 98 is what number?

12. What fraction of 64 is 48?

13. Write $7\frac{21}{23}$ as an improper fraction.

Solve:

14. $\frac{12}{13}x = 60$

15. $x + \frac{3}{7} = \frac{29}{14}$

Simplify:

16. $36 - 3[(6 - 3)(3 - 1) - 4]$

17. $2^2 + 2[2(3 + 1)(3 - 2) - 1]$

18. $4\frac{3}{5} \cdot \frac{2}{3} - \frac{14}{15}$

19. $5\frac{2}{3} - 3\frac{5}{6} + \frac{5}{18}$

20. $3^2 + 2^2 - 3 + \sqrt[3]{8}$

21. 197.3×0.013

22. $6211.89 - 8.987$

23. $\frac{192.03}{0.05}$

24. $\dfrac{1\frac{3}{4}}{2\frac{1}{5}}$

25. $\frac{5}{3}\left(\frac{1}{7} + \frac{3}{8}\right) - \frac{1}{4}$

26. $\frac{1}{4}\left(2\frac{1}{4} - \frac{1}{8}\right) + \frac{3}{16}$

27. $3\frac{2}{3} \div 1\frac{1}{4} \times 2\frac{1}{6} \times 1\frac{1}{3}$

Evaluate:

28. $xyz + xz - z$ if $x = \frac{1}{3}$, $y = 9$, and $z = 12$

29. $\sqrt[3]{y} + x^3$ if $x = 2$ and $y = 8$

30. Round $18.0\overline{618}$ to the nearest millionth.

LESSON 56 *Percent*

56.A

percent   The language spoken by the ancient Romans was Latin. The Latin word for *by* is *per* and the word for *one hundred* is *centum*. We put these words together in English to form the word **percent**. From this we see that percent means *by 100*.

One percent of a number is one-hundredth of the number. To demonstrate we will use the number 256.

If we divide 256 into 100 parts, each part is 2.56. Thus 2.56 is 1 percent of 256 because 1 percent is exactly the same thing as one-hundredth. If we want 5 percent of 256, we must use 5 of these parts.

$$5 \text{ percent} = 5(2.56) = 12.8$$

One hundred sixty percent of 256 would be 160 of these parts.

$$160 \text{ percent} = 160(2.56) = 409.6$$

From this we see that we can find a given percent of a number if we first divide the number into 100 equal parts. Each of these parts is 1 percent of the number.

example 56.1   (a)   What is 1 percent of 7500?
(b)   What is 34 percent of 7500?

solution   (a)   To find 1 percent of 7500, we divide 7500 by 100.

$$1 \text{ percent of } 7500 = \frac{7500}{100} = \mathbf{75}$$

(b)   34 percent is 34 times the value of 1 percent.

$$34 \times 75 = \mathbf{2550}$$

example 56.2   (a)   What is 1 percent of 67?
(b)   What is 140 percent of 67?

solution   (a)   To find 1 percent of 67, we divide by 100.

$$1 \text{ percent of } 67 = \frac{67}{100} = \mathbf{0.67}$$

(b) 140 percent is 140 of these parts.

$$140 \times 0.67 = \textbf{93.8}$$

practice **a.** What is 1 percent of 64?
b. What is 40 percent of 64?
c. What is 1 percent of 5?
d. What is 35 percent of 5?

problem set **1.** The new tires cost $780, and for this price the dealer got 10 tires. Write the two
56 rates (ratios) implied by this statement.
(a) What would the dealer have to pay for 58 tires?
(b) How many tires could the dealer get for $156?

2. Two numbers were proposed. The first number was nine million, forty-seven.
The second number was eight million, seven hundred ninety-three thousand,
two hundred fifteen. By how much was the first number greater?

3. The first four cows through the chute were nice and fat. The first one weighed
1200 lb, the second one weighed 900 lb, and the third one weighed 840 lb.
What did the last cow weigh if the average weight was 998 pounds?

4. By how much did the rainfall in July exceed the total rainfall in May and
December?

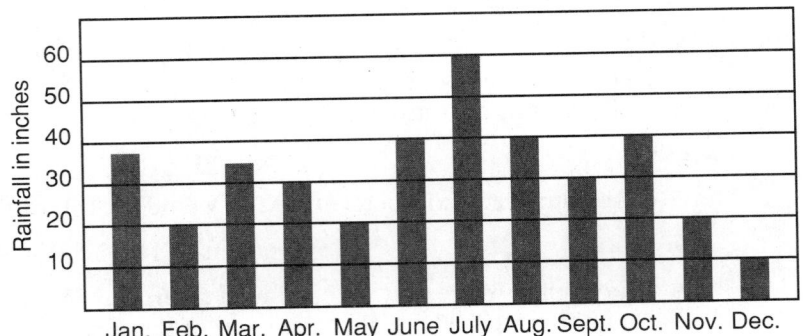

Monthly rainfall

5. (a) What is 1 percent of 6200?
(b) What is 24 percent of 6200?

6. (a) What is 1 percent of 52?
(b) What is 140 percent of 52?

7. Three-fourths of what number is 60?

8. 0.6 of what number is 42?

9. What fraction of 57 is 45?

10. What decimal of 280 is 70?

11. Four-ninths of 99 is what number?

12. What is the volume of a solid whose base is the figure shown on the left and
whose height is 5 centimeters? Dimensions are in centimeters.

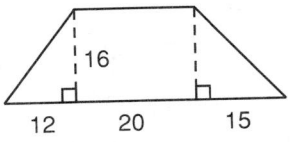
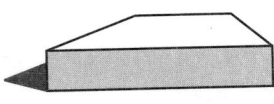

13. Find the least common multiple of 18, 24, and 30.

14. Write $6\frac{16}{17}$ as an improper fraction.

Solve:

15. $\frac{4}{15}x = 8$

16. $x + 1\frac{3}{5} = 3\frac{7}{10}$

Simplify:

17. $72 - 3[(7 - 4)(3 - 1) - 4]$

18. $3^3 + 2^2[2(2 + 1)(2 - 1) - 5]$

19. $4\frac{4}{5} - 3\frac{2}{3} + \frac{7}{15}$

20. $2^3 + 3^2 - \sqrt[3]{27}$

21. 132.7×0.012

22. $18{,}251.3 - 62.982$

23. $\dfrac{135.06}{0.003}$

24. $\dfrac{1\frac{4}{5}}{2\frac{1}{3}}$

25. $\frac{1}{3}\left(\frac{1}{6} + \frac{5}{12}\right) - \frac{1}{36}$

26. $\frac{1}{4}\left(3\frac{1}{3} - \frac{1}{6}\right) + \frac{1}{12}$

27. $2\frac{1}{3} \div 1\frac{1}{6} \times 3\frac{1}{4} \div 1\frac{1}{3}$

Evaluate:

28. $xy + x + xyz - z$ if $x = 24$, $y = \frac{1}{6}$, and $z = 3$

29. $\sqrt[3]{q} + pq$ if $p = 3$ and $q = 8$

30. (a) Write $16{,}000{,}000{,}000$ in scientific notation.
     (b) Write 1.6×10^{-8} in standard form.

LESSON 57 *Ratio and proportion · P^Q and $\sqrt[Q]{P}$*

57.A
ratio and proportion

Remember that a ratio is a comparison of two numbers. We can write a ratio as a fraction. If we write

$$\frac{3}{4}$$

we can say that we have written the fraction three-fourths. We can also say that we have written the ratio of 3 to 4. There are many equivalent forms of this ratio (fraction).

$$\frac{3}{4} \qquad \frac{6}{8} \qquad \frac{15}{20} \qquad \frac{90}{120} \qquad \frac{150}{200} \qquad \frac{450}{600}$$

All these ratios have the same value.

When we write an equation that consists of two ratios connected by an equals sign, such as

$$\frac{3}{4} = \frac{6}{8}$$

we say that we have written a **proportion.** Proportions can be either true proportions or false proportions.

$$\frac{1}{2} = \frac{5}{10} \quad \text{true} \qquad \frac{1}{2} = \frac{7}{10} \quad \text{false}$$

The proportion on the left is a true proportion, and the proportion on the right is a false proportion.

We find it helpful to note that the cross products of a true proportion are equal to each other.

$$\frac{3}{4} \diagup\!\!\!\!\diagdown \frac{6}{8} \qquad \begin{array}{l} 4 \cdot 6 = 24 \\ 3 \cdot 8 = 24 \end{array}$$

Here one cross product is 4 times 6, and the other cross product is 3 times 8. Both cross products equal 24.

When one part of a proportion is a variable, the proportion is a **conditional proportion.** Conditional proportions can also be called **conditional equations.**

$$\frac{20}{15} = \frac{4}{x}$$

This proportion is a conditional proportion and is neither true nor false. There is a value of x that will make this proportion a true proportion. To find this value of x, we begin by setting the cross products equal to each other.

example 57.1   Solve: $\dfrac{20}{15} = \dfrac{4}{x}$

solution   We begin by setting the cross products equal to each other.

$$20x = 4 \cdot 15$$

To solve, we multiply 4 by 15 and then divide both sides by 20.

$$\frac{20x}{20} = \frac{60}{20}$$

$$x = 3$$

example 57.2   Solve: $\dfrac{4}{5} = \dfrac{p}{7}$

solution   First we set the cross products equal to each other.

$$4 \cdot 7 = 5p$$

And then we divide both sides by 5.

$$\frac{4 \cdot 7}{5} = \frac{5p}{5}$$

$$\frac{28}{5} = p$$

We decide to leave the answer in the form of an improper fraction.

57.B

P^Q and $\sqrt[Q]{P}$ We have been evaluating exponential expressions where one of the numbers is represented by a variable. We know how to find the value of

$$\sqrt[3]{P} \quad \text{if } P = 8 \qquad \text{and the value of} \qquad P^3 \quad \text{if } P = 2$$

Sometimes both numbers are represented by variables.

example 57.3 Evaluate x^y if $x = 4$ and $y = 3$.

solution We replace x with 4 and replace y with 3 and get

$$(4)^3 = \mathbf{64}$$

example 57.4 Evaluate $\sqrt[P]{Q}$ if $P = 3$ and $Q = 64$.

solution We replace P with 3 and replace Q with 64 and get

$$\sqrt[3]{64} = \mathbf{4}$$

practice Solve:

a. $\dfrac{4}{p} = \dfrac{7}{2}$ **b.** $\dfrac{s}{5} = \dfrac{4}{3}$

Evaluate:

c. x^y if $x = 4$ and $y = 2$

d. $\sqrt[x]{y}$ if $x = 4$ and $y = 16$

problem set 57

1. Jimmy bought 80 small trees and paid $3520. Write the two ratios implied. What would he have to pay for just 10 trees?

2. The average weight of the first two dogs was 43 pounds. The average weight of the next three dogs was 58 pounds. What was the average weight of all the dogs?

3. The first try resulted in 1436. The second try resulted in 1892. The third try was the big one, as it came to 4400. By how much did the third try exceed the sum of the first two?

4. Ninety-seven was a bad guess. However, 5 times this number increased by six hundred forty-two was the correct number. What was the correct number?

5. (a) What is 1 percent of 3600?
   (b) What is 16 percent of 3600?

6. (a) What is 1 percent of 25?
   (b) What is 150 percent of 25?

7. Use two unit multipliers to change 10,000 square inches to square feet.

8. 0.8 of what number is 48? 9. What fraction of 60 is 48?

10. What decimal of 350 is 70?

11. What is the volume of a right solid whose base is the figure shown on the left and whose height is 2 feet? Dimensions are in feet.

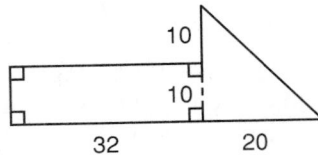

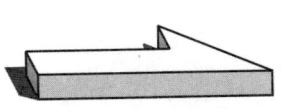

12. Find the surface area of this triangular prism. Dimensions are in meters.

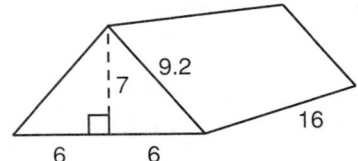

13. Write $\dfrac{192}{7}$ as a mixed number.

Solve:

14. $\dfrac{15}{7} = \dfrac{4}{x}$

15. $\dfrac{5}{6} = \dfrac{p}{4}$

16. $\dfrac{7}{15}x = 3$

17. $x + 3\dfrac{3}{14} = 4\dfrac{5}{28}$

Simplify:

18. $64 - 2[(3 - 1)(5 - 2) + 1]$

19. $2^3 + 2^2[3(2 - 1)(2 + 2) - 7]$

20. $2\dfrac{1}{3} \cdot 1\dfrac{3}{4} - \dfrac{7}{12}$

21. $6\dfrac{2}{5} - 2\dfrac{1}{4} + \dfrac{3}{40}$

22. $2^4 + 3^3 - 2^3 + \sqrt[3]{8}$

23. 1921×0.0016

24. $9218.821 - 61.872$

25. $\dfrac{1721.2}{0.02}$

26. $\dfrac{1}{5}\left(3\dfrac{1}{4} - 2\dfrac{1}{3}\right) + \dfrac{7}{15}$

27. $\dfrac{2\dfrac{3}{4}}{1\dfrac{7}{8}}$

28. $6\dfrac{2}{3} \div 1\dfrac{1}{6} \times 3\dfrac{1}{3} \div 2\dfrac{4}{5}$

29. (a) Write 1.3×10 in standard form.
   (b) Write 0.0392 in scientific notation.

30. Evaluate:
   (a) x^y if $x = 3$ and $y = 3$
   (b) $\sqrt[x]{y}$ if $x = 3$ and $y = 125$

LESSON 58 *Decimals, fractions, and percent · Reference numbers*

58.A
decimals, fractions, and percent

We can designate the same part of a whole by using a fraction or a decimal number or by using percent.

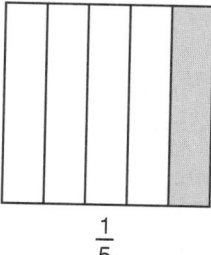
$$\frac{1}{5}$$

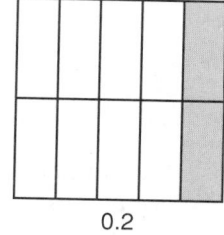
0.2

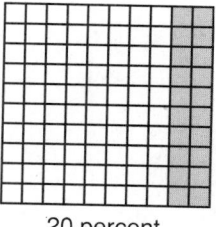
20 percent

In the figure on the left we have shaded 1 of the 5 parts to represent $\frac{1}{5}$. The center figure has 10 parts, and we have shaded 2 of them to represent two-tenths, which is 0.2. The figure on the right has been divided into 100 parts; 20 of these parts have been shaded to represent 20 percent. All three shaded areas are equal. From these diagrams we see that

$$\frac{1}{5} \quad \text{and} \quad 0.2 \quad \text{and} \quad 20 \text{ percent}$$

are three ways of saying the same thing.

example 58.1 What are (a) the decimal equivalent and (b) the percent equivalent of $\frac{51}{100}$?

solution (a) To find the decimal equivalent of $\frac{51}{100}$, we divide.

$$\frac{51}{100} = \mathbf{0.51}$$

(b) To change a decimal to percent, we move the decimal point two places to the right.

$$0.51 = \mathbf{51 \text{ percent}}$$

58.B
reference numbers

Many people understand how to change from one of the three forms to the others, but still they make mistakes. The mistake made most often is moving the decimal point the wrong way. To prevent mistakes, it is helpful to memorize one set of numbers that is correct. Then we use these numbers as reference numbers. In this book we will use the numbers from the preceding example as reference numbers.

example 58.2   Complete the table.

Fraction	Decimal	Percent
(b)	(a)	25

solution   First we write the reference numbers.

Fraction	Decimal	Percent
$\frac{51}{100}$	0.51	51
(b)	(a)	25

(a)   The reference numbers remind us to move the decimal point two places to the left to get the decimal form. Thus

25 percent is equivalent to the decimal number 0.25

(b)   To change 0.25 to a fraction, we write it as twenty-five hundredths and then simplify.

$$\frac{25}{100} = \frac{1}{4}$$

Thus we have

Fraction	Decimal	Percent
$\frac{51}{100}$	0.51	51
$\frac{1}{4}$	**0.25**	25

example 58.3   Complete the table.

Fraction	Decimal	Percent
$\frac{1}{16}$	(a)	(b)

solution   The three blank spaces are provided for the reference numbers. We write them in these spaces.

Fraction	Decimal	Percent
$\frac{51}{100}$	0.51	51
$\frac{1}{16}$	(a)	(b)

(a) To find the decimal form, we divide.

$$
\begin{array}{r}
0.0625 \\
16\overline{)1.0000} \\
\underline{96} \\
40 \\
\underline{32} \\
80 \\
\underline{80}
\end{array}
$$

(b) The reference numbers show that we move the decimal point two places to the right to find the percent form. Thus

0.0625 is equivalent to 6.25 percent

Fraction	Decimal	Percent
$\frac{51}{100}$	0.51	51
$\frac{1}{16}$	**0.0625**	**6.25**

practice   Complete each table. Begin by inserting the reference numbers.

Fraction	Decimal	Percent
$\frac{1}{8}$	**a**	**b**

Fraction	Decimal	Percent
c	0.22	**d**

**problem set
58**

1. The first guess was sixty-nine thousand and seven hundred forty-one hundred-thousandths. The second guess was only forty-two thousand and seventy-five hundred-thousandths. By how much was the first guess greater?

2. While new ideas abounded, there was a dearth of suggestions as to how they were to be used. There were one hundred forty-three thousand new ideas but only nine thousand, six hundred fourteen suggestions for their use. How many more new ideas were there than suggestions for their use?

3. Jill could buy 53 new ones for only $742. Write the two rates (ratios) implied by this statement. How much would Jill have to pay for only 25 new ones?

4. One hundred forty-five thousand came on the first day. On the second day 5 times this many came. How many came in all?

5. Complete the table. Begin by inserting the reference numbers.

6. (a) Write 106,000,000 in scientific notation.
    (b) Write 4.13×10^{-8} in standard form.

7. 0.7 of what number is 42?

Fraction	Decimal	Percent
(a)	(b)	25
$\frac{1}{2}$	(c)	(d)

8. What fraction of 72 is 64?

9. What decimal of 420 is 273?

10. What is the volume of a right solid whose base is the figure shown on the left and whose height is 3 inches? Dimensions are in inches. Angles that look square are square.

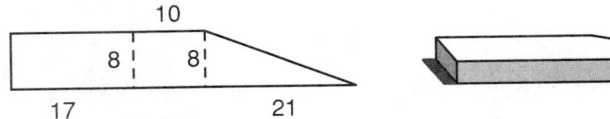

11. Find the least common multiple of 15, 18, and 20.

12. Write $17\frac{1}{12}$ as an improper fraction.

Solve:

13. $\dfrac{16}{5} = \dfrac{3}{x}$

14. $\dfrac{6}{7} = \dfrac{p}{14}$

15. $\dfrac{5}{12}x = 9$

16. $x - 2\dfrac{3}{14} = 1\dfrac{1}{21}$

Simplify:

17. $38 - 2[(6 - 5)(4 - 1) + 1]$

18. $2^3 + 3^2[(7 - 2)(3 - 1)3 - 25]$

19. $1\dfrac{1}{3} \cdot 2\dfrac{3}{4} - \dfrac{7}{12}$

20. $5\dfrac{3}{5} - 1\dfrac{3}{4} + \dfrac{7}{20}$

21. $2^3 + 3^2 - 2^2 + \sqrt{16}$

22. 16.82×0.013

23. $1262.81 - 12.981$

24. $\dfrac{618.21}{0.004}$

25. $\dfrac{1}{5}\left(\dfrac{2}{3} + \dfrac{1}{2}\right) - \dfrac{2}{15}$

26. $\dfrac{1}{4}\left(2\dfrac{3}{4} - 1\dfrac{1}{8}\right) + \dfrac{3}{16}$

27. $\dfrac{1\frac{4}{7}}{2\frac{3}{4}}$

28. $5\dfrac{1}{3} \div 1\dfrac{1}{6} \times 3\dfrac{1}{3} \div \dfrac{1}{6}$

Evaluate:

29. $x^2 + 2xy + y^x$ if $x = 2$ and $y = 3$

30. p^q if $p = 2$ and $q = 3$

LESSON 59 *Equations with mixed numbers*

When equations contain mixed numbers, it is often helpful if we convert the mixed numbers to improper fractions as the first step.

example 59.1   Solve: $2\frac{1}{3}m = 5$

solution   As the first step, we write $2\frac{1}{3}$ as an improper fraction.

$$\frac{7}{3}m = 5$$

Now we solve by multiplying both sides by $\frac{3}{7}$.

$$\frac{3}{7} \cdot \frac{7}{3}m = 5 \cdot \frac{3}{7} \qquad \text{multiplied by } \frac{3}{7}$$

$$\boldsymbol{m = \frac{15}{7}} \qquad \text{simplified}$$

example 59.2   Solve: $3\frac{1}{2}k = 4\frac{1}{5}$

solution   This time there are two mixed numbers. As the first step, we write them both as improper fractions.

$$\frac{7}{2}k = \frac{21}{5}$$

To solve, we multiply both sides by $\frac{2}{7}$.

$$\frac{2}{7} \cdot \frac{7}{2}k = \frac{21}{5} \cdot \frac{2}{7} \qquad \text{multiplied by } \frac{2}{7}$$

$$\boldsymbol{k = \frac{6}{5}} \qquad \text{simplified}$$

practice   Solve:

a. $2\frac{1}{4}p = 8\frac{1}{2}$

b. $3\frac{1}{2}x = 4\frac{1}{5}$

problem set 59

1. The average of the first three numbers was 42. The average of the next seven numbers was only 12. What was the average of all the numbers?

2. Big ones did not come cheap, as 7 of them cost $280,000. Write the two rates (ratios) implied by this statement. How many big ones could be purchased for $120,000?

3. The first one weighed one hundred forty-thousand, twenty-six pounds. The second weighed only one hundred thirty-two thousand, seven hundred eighty-one pounds. The first one weighed how many pounds more than the second one?

4. Eighty-seven could fit into one compartment. How many could fit into sixty compartments?

5. Complete the table. Begin by inserting the reference numbers.

Fraction	Decimal	Percent
(a)	(b)	16
$\frac{1}{8}$	(c)	(d)

6. (a) Write 0.093 in scientific notation.
   (b) Write 1.2×10^6 in standard form.

7. 0.9 of what number is 72?

8. What fraction of 48 is 40?

9. What decimal of 630 is 441?

10. $\frac{1}{4}$ of what number is $\frac{1}{3}$?

11. $\frac{1}{4}$ of $\frac{1}{3}$ is what number?

12. What is the volume of a right solid whose base is the figure shown on the left and whose height is 4 inches? Dimensions are in inches.

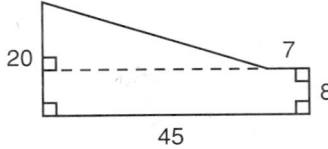 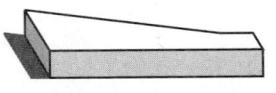

13. Find the least common multiple of 8, 18, and 20.

14. Find the surface area of this rectangular solid. Dimensions are in meters.

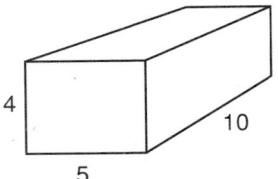

Solve:

15. $\frac{15}{4} = \frac{10}{x}$

16. $\frac{3}{5} = \frac{p}{12}$

17. $1\frac{3}{5}x = 6$

18. $3\frac{1}{4}p = 5$

Simplify:

19. $49 - 2[(5 - 2^2)(4 + 2) - 5]$

20. $\sqrt[3]{8} + 2^3[2^2(2^3 - 5) - 4]$

21. $2\frac{1}{3} \cdot 3\frac{1}{4} - \frac{11}{12}$

22. $14\frac{2}{3} - 12\frac{7}{8} + \frac{11}{48}$

23. 171.6×0.007

24. $1171.61 - 13.321$

25. $\frac{611.51}{0.03}$

26. $\frac{1}{6}\left(\frac{1}{3} + \frac{1}{2}\right) - \frac{5}{36}$

27. $\frac{1}{5}\left(\frac{1}{4} - \frac{1}{8}\right) + 2\frac{7}{8}$

28. $\frac{6\frac{2}{3}}{2\frac{1}{4}}$

29. $6\frac{1}{4} \div 3\frac{2}{3} \times 2\frac{1}{4} \div \frac{1}{8}$

30. Evaluate: $x^y + 3xy^2 + 3x^2y + y^x$ if $x = 1$ and $y = 2$

LESSON 60 *Mixed number problems*

We can write any mixed number as an improper fraction. We can write $2\frac{1}{3}$ as the improper fraction $\frac{7}{3}$.

$$2\frac{1}{3} \text{ equals } \frac{7}{3}$$

When we encounter word problems that contain mixed numbers, a good first step is to rewrite the mixed numbers as improper fractions.

example 60.1 $2\frac{1}{2}$ of what number is $7\frac{1}{3}$?

solution We will use the general equation

$$F \times of = is$$

We replace F with $2\frac{1}{2}$, replace *of* with *WN*, and replace *is* with $7\frac{1}{3}$.

$$2\frac{1}{2} \times WN = 7\frac{1}{3}$$

Now we replace the mixed numbers with their improper fraction equivalents.

$$\frac{5}{2} \times WN = \frac{22}{3}$$

We solve by multiplying both sides by $\frac{2}{5}$.

$$\frac{2}{5} \cdot \frac{5}{2} \times WN = \frac{22}{3} \cdot \frac{2}{5} \qquad \text{multiplied by } \frac{2}{5}$$

$$WN = \frac{44}{15} \qquad \text{simplified}$$

example 60.2 $2\frac{1}{5}$ of $8\frac{1}{8}$ is what number?

solution This time we rewrite the problems using improper fractions.

$$\frac{11}{5} \text{ of } \frac{65}{8} \text{ is what number?}$$

Now we write the fractional part of a number equation.

$$F \times of = is$$

We replace F with $\frac{11}{5}$, replace *of* with $\frac{65}{8}$, and replace *is* with *WN*.

$$\frac{11}{5} \cdot \frac{65}{8} = WN \qquad \text{substituted}$$

Now we cancel and then multiply.

$$\frac{11}{1} \cdot \frac{13}{8} = WN \qquad \text{canceled}$$

$$\frac{143}{8} = WN \qquad \text{multiplied}$$

practice   Solve:

    a. $3\frac{1}{2}$ of $6\frac{1}{4}$ is what number?

    b. $3\frac{1}{4}$ of what number is $6\frac{1}{8}$?

    c.   What fraction of $3\frac{2}{5}$ is $9\frac{7}{8}$?

**problem set
60**

1. Eudemonia bought 560 red ones for 7 pesos. Write the two rates (ratios) implied. How many red ones could she buy with 60 pesos?

2. The Black Prince reined in after traveling 56 miles in 8 hours. If he traveled at the same rate on the next day, how far could he go in 14 hours?

3. Eau de Vapid meandered 40 miles in 5 hours. Then he sauntered 60 miles in 10 hours. By how much did his first rate exceed his second rate?

4. On the next leg of the trip, Eau covered 40 miles in 20 hours. What was his average speed for all three legs of the trip? (*Note:* Averages cannot be averaged; thus, to find the average speed for all three legs of the trip, the total distance must be divided by the total time.)

5. Complete the table. Begin by inserting the reference numbers.

6. (a) Write 0.0701 in scientific notation.
   (b) Write 7.03×10^7 in standard form.

Fraction	Decimal	Percent
(a)	0.24	(b)
$\frac{3}{5}$	(c)	(d)

7. 0.7 of what number is 490?

8. What fraction of $3\frac{1}{9}$ is $\frac{24}{5}$?

9. What decimal part of 720 is 420?

10. $3\frac{1}{3}$ of what number is $4\frac{1}{2}$?

11. $2\frac{1}{4}$ of $2\frac{1}{3}$ is what number?

12. What is the volume of a right solid whose base is the figure shown on the left and whose height is 6 feet? Dimensions are in feet.

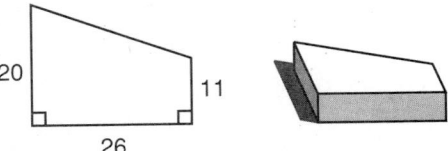

13. Find the least common multiple of 16, 24, and 36.

14. Use two unit multipliers to convert 8000 square kilometers to square meters.

Solve:

15. $\dfrac{16}{5} = \dfrac{8}{x}$

16. $\dfrac{3}{4} = \dfrac{5}{p}$

17. $2\frac{4}{5}x = 5$

18. $4\frac{1}{5}p = 6$

Simplify:

19. $54 - 2[(6 - 2^2)(3 + 1) - 5]$

20. $\sqrt{16} + 2^2[2(3^2 - 2^2) - 5]$

21. $3\frac{1}{3} \cdot 2\frac{1}{4} - \frac{5}{12}$

22. $15\frac{6}{7} - 3\frac{3}{14} + \frac{9}{14}$

23. 621.8×0.018

24. $2612.81 - 14.313$

25. $\frac{1821.5}{0.7}$

26. $\frac{1}{5}\left(\frac{1}{2} + 2\frac{1}{3}\right) - \frac{4}{15}$

27. $\frac{1}{4}\left(\frac{1}{6} + 1\frac{1}{4}\right) - \frac{1}{4}$

28. $\dfrac{3\frac{1}{3}}{1\frac{1}{6}}$

29. $3\frac{1}{2} \times 6\frac{1}{3} \div 2\frac{1}{3} \times 1\frac{1}{3}$

30. Evaluate: $xyz + z^2 + y^2 - y^z$ if $x = \frac{1}{3}$, $y = 6$, and $z = 2$

LESSON 61 *The distance problem*

Jimmy can drive 60 miles in 1 hour. This statement gives us two rates (ratios).

$$\frac{60 \text{ miles}}{1 \text{ hour}} \quad \text{and} \quad \frac{1 \text{ hour}}{60 \text{ miles}}$$

If we want to know how far he could drive in 2 hours, we would use the rate with miles on top.

$$\frac{60 \text{ miles}}{1 \text{ hour}} \times 2 \text{ hours} = 120 \text{ miles}$$

Whenever you see the equation

(a) **Rate × time = distance**

the rate used has time on the bottom. This rate is called **speed.** If we want to know how long it would take to drive 300 miles, we could use the rate with time on top.

$$\frac{1 \text{ hour}}{60 \text{ miles}} \times 300 \text{ miles} = \frac{300 \text{ hours}}{60} = 5 \text{ hours}$$

This equation tells us that

(b) **Rate × distance = time**

If we use the rate with time on top, we can multiply to find the time required. Almost all books teach only rate (a). If we are required to use rate (a), we must divide to find the time required.

example 61.1 Helen could travel 40 miles per hour. How long would it take her to travel 600 miles?

solution First we write the two rates.

(a) $\dfrac{40 \text{ miles}}{1 \text{ hour}}$ (b) $\dfrac{1 \text{ hour}}{40 \text{ miles}}$

If we use rate (a), we use the equation

$$\text{Rate} \times \text{time} = \text{distance}$$

We replace rate with 40 and replace distance with 600. We get

$$40 \times \text{time} = 600$$

To solve we divide both sides of the equation by 40.

$$\frac{\cancel{40} \times \text{time}}{\cancel{40}} = \frac{600}{40}$$

$$\text{Time} = 15 \text{ hours}$$

If we use rate (b), we can find the same answer by multiplying

$$\frac{1 \text{ hour}}{40 \text{ miles}} \times 600 \text{ miles} = \textbf{15 hours}$$

We see that using rate (b) is much easier because we do not have to solve an equation. Using rate (b) also allows us to keep track of the units. But we must learn how to use rate (a) because it involves a general procedure that may be used in other problems.

example 61.2   Harry could travel 350 miles in 7 hours.
(a)   What was his speed?
(b)   Use the formula

$$R \times T = D$$

to find how long it would take him to travel 800 miles.

solution   (a)   Speed is a rate with time on the bottom.

$$\text{Speed} = \frac{350 \text{ miles}}{7 \text{ hours}} = \textbf{50 miles per hour}$$

(b)   Now we write the equation

$$\text{Rate} \times \text{time} = \text{distance}$$

We replace rate with 50 and replace distance with 800.

$$50 \times \text{time} = 800$$

To solve we divide both sides of the equation by 50.

$$\frac{50 \times \text{time}}{50} = \frac{800}{50}$$

$$\text{Time} = \textbf{16 hours}$$

practice   Mildred traveled the first 120 miles in 3 hours. What was her speed? How long would it take her to travel 400 miles at the same speed? Work the problem two ways.

problem set 61   **1.** The knight found that his palfrey could run 4600 feet at 230 feet per second. Write the two rates indicated by this statement. How long would it take the palfrey to run 100,000 feet?

2. Ethelred the Unready was not ready. Yet he covered the 480 miles in 4 hours. Write the two rates (ratios) indicated by this statement.
   (a)   What was his speed?
   (b)   How long would it take him to go 1440 miles?

3. The last leg of the journey was to be 1200 miles, so the tour group got an early start. If their speed was 60 miles per hour, how long would it take to get there?

4. Hadrian's men increased their efforts and built 430 feet of wall in 1 day. At this rate, how long would it take to build 7310 feet of wall?

5. Complete the table. Begin by inserting the reference numbers.

6. (a)   Write 0.639 in scientific notation.
   (b)   Write 7.01×10^4 in standard form.

Fraction	Decimal	Percent
(a)	0.22	(b)
$\frac{21}{25}$	(c)	(d)

7. 0.8 of what number is 96?

8. What fraction of 52 is 30?

9. What decimal of 700 is 581?

10. $3\frac{1}{5}$ of what number is $7\frac{1}{3}$?

11. $2\frac{1}{10}$ of $1\frac{3}{4}$ is what number?

12. What is the volume of a right solid whose base is the figure shown on the left and whose height is 3 feet? Dimensions are in feet. Corners that look square are square.

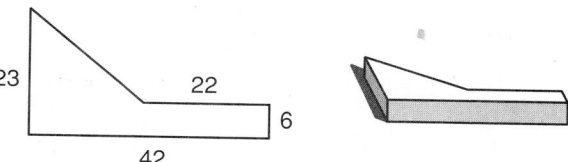

13. Find the surface area of this triangular prism. Dimensions are in inches.

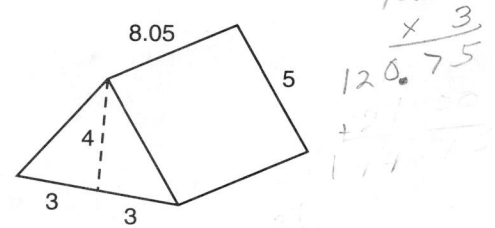

$$\frac{8.05}{\times 5}$$
$$\overline{40.25}$$
$$\frac{\times 3}{120.75}$$

14. Use two unit multipliers to convert 25,000 square centimeters to square meters.

Solve:

15. $\frac{5}{2} = \frac{12}{x}$

16. $\frac{7}{3} = \frac{x}{15}$

17. $\frac{3}{7}x = 7\frac{1}{4}$

18. $x + 6\frac{3}{14} = 12\frac{2}{7}$

Simplify:

19. $96 - 4[(7 - 2^2)(1 + 2) - 2]$

20. $\sqrt{9} + \sqrt{25}[3(3 - 1) - 2]$

21. $2\frac{1}{3} \cdot 3\frac{1}{4} - \frac{5}{6}$

22. $14\frac{4}{5} - 4\frac{3}{4} + \frac{9}{10}$

23. 6111×0.0013

24. $22.8971 - 9.89121$

25. $\dfrac{187.61}{0.005}$

26. $\frac{1}{3}\left(\frac{1}{2} + 2\frac{1}{3}\right) - \frac{7}{18}$

27. $\frac{1}{4}\left(\frac{1}{6} + 3\frac{1}{2}\right) - \frac{4}{5}$

28. $\dfrac{4\frac{1}{3}}{5\frac{5}{6}}$

29. $2\frac{1}{2} \times 3\frac{2}{3} \div 1\frac{5}{6} \times \frac{1}{3}$

30. Evaluate: $x^2 + xy + xyz + \sqrt[z]{x}$ if $x = 9$, $y = \frac{1}{3}$, and $z = 2$

LESSON 62 *Proportions with fractions*

There is no change in the method of solving conditional proportions when they contain fractions or mixed numbers. The first step is to cross multiply. Then we divide or multiply as required to complete the solution.

example 62.1 Solve: $\dfrac{\frac{2}{3}}{x} = \dfrac{\frac{5}{8}}{\frac{1}{5}}$

solution As the first step, we cross multiply.

$$\frac{2}{3} \cdot \frac{1}{5} = \frac{5}{8}x \qquad \text{cross multiplied}$$

$$\frac{2}{15} = \frac{5}{8}x \qquad \text{simplified}$$

We finish by multiplying both sides by $\frac{8}{5}$:

$$\frac{8}{5} \cdot \frac{2}{15} = \frac{8}{5} \cdot \frac{5}{8}x \qquad \text{multiplied by } \frac{8}{5}$$

$$\frac{16}{75} = x \qquad \text{simplified}$$

example 62.2 Solve: $\dfrac{x}{\frac{3}{4}} = \dfrac{\frac{1}{3}}{\frac{2}{5}}$

solution   As the first step, we cross multiply and get

$$\frac{2}{5}x = \frac{3}{4} \cdot \frac{1}{3} \qquad \text{cross multiplied}$$

$$\frac{2}{5}x = \frac{1}{4} \qquad \text{simplified}$$

We finish by multiplying both sides by $\frac{5}{2}$.

$$\frac{5}{2} \cdot \frac{2}{5}x = \frac{1}{4} \cdot \frac{5}{2} \qquad \text{multiplied by } \frac{5}{2}$$

$$x = \frac{5}{8} \qquad \text{simplified}$$

practice   Solve:

a. $\dfrac{\frac{3}{2}}{\frac{1}{5}} = \dfrac{\frac{1}{4}}{x}$

b. $\dfrac{y}{\frac{2}{3}} = \dfrac{\frac{1}{5}}{\frac{1}{6}}$

c. $\dfrac{\frac{1}{2}}{\frac{1}{8}} = \dfrac{2}{x}$

d. $\dfrac{\frac{1}{3}}{x} = \dfrac{\frac{1}{5}}{7}$

problem set 62

1. Fourteen big ones could be purchased for 9 crowns. Write the two rates (ratios) implied by this statement. How many big ones could be purchased for 360 crowns?

2. Forty-three bottles could be filled in 2 minutes. Write the two rates (ratios) implied. How long would it take to fill 860 bottles?

3. The first was fourteen thousand, fifty-eight. The second was twenty-one thousand, fifty-two. The third was forty-four thousand, forty-four. How much greater was the third than the sum of the first two?

4. The first measurement was one hundred forty-one ten-thousandths. The second measurement was one thousand, three hundred forty-two hundred-thousandths. What was the sum of the measurements?

5. Complete the table. Begin by inserting the reference numbers.

6. (a) Write 3.09×10^{-5} in standard form.
   (b) Write 19,000 in scientific notation.

Fraction	Decimal	Percent
(a)	0.12	(b)
$\frac{5}{6}$	(c)	(d)

7. 0.4 of what number is 316?

8. What fraction of $\frac{45}{6}$ is $7\frac{1}{4}$?

9. What decimal of 640 is 560?

10. $2\frac{1}{4}$ of what number is $6\frac{1}{3}$?

11. $3\frac{1}{2}$ of $1\frac{1}{10}$ is what number?

12. What is the volume of a right solid whose base is the figure shown on the left and whose height is 2 meters? Dimensions are in meters. Corners that look square are square.

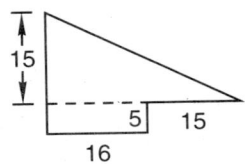 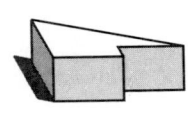

13. Find the least common multiple of 12, 16, and 30.

14. Use two unit multipliers to convert 16,000 square inches to square feet.

Solve:

15. $\dfrac{\frac{1}{3}}{\frac{2}{5}} = \dfrac{6}{x}$

16. $\dfrac{\frac{2}{3}}{\frac{3}{4}} = \dfrac{\frac{5}{12}}{p}$

17. $3\frac{5}{6}x = 7\frac{1}{2}$

18. $3\frac{2}{5}p = 1\frac{2}{5}$

Simplify:

19. $64 - 3[(6 - 2^2)(3^2 - 2^2) + 1]$

20. $\sqrt[3]{8} + \sqrt{16}[2(3 - 1) + 7]$

21. $2\frac{1}{4} \cdot 3\frac{2}{3} - \frac{5}{6}$

22. $17\frac{3}{4} - 4\frac{1}{5} + \frac{7}{20}$

23. 361.4×0.0012

24. $18,191.8 - 19.762$

25. $\dfrac{1762.3}{0.6}$

26. $\frac{1}{4}\left(\frac{1}{2} + 3\frac{1}{4}\right) - \frac{5}{8}$

27. $\frac{1}{3}\left(\frac{1}{2} + 3\frac{1}{3}\right) - \frac{5}{6}$

28. $\dfrac{4\frac{2}{3}}{2\frac{1}{6}}$

29. $1\frac{1}{2} \times 6\frac{2}{3} \div 3\frac{1}{6} \times 1\frac{2}{3}$

30. Evaluate: $x^2 + y^2 + 2xy + \sqrt[x]{z}$ if $x = 3$, $y = 5$, and $z = 27$

LESSON 63   *Circles · Circumference and pi*

63.A

circles   Every point on a circle is the same distance from the center of the circle. We call this distance the **radius** of the circle. We call a line that connects two points on a circle a **chord** of the circle. A **diameter** of a circle is a chord that passes through the center of the circle. From the circles shown here, we see that a diameter of a circle is twice the radius of the same circle.

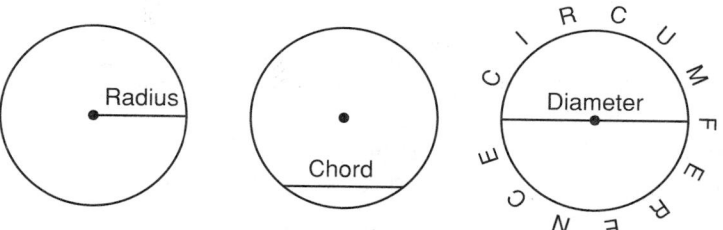

63.B
circumference and pi

The circumference of a circle is the total distance around the circle. Many ancients thought that the circle was the perfect geometric figure. They were especially interested in the relationship between the diameter of a circle and the circumference of the same circle. They found that three diameters would not go all the way around the circle. There is a little extra left over no matter how large or how small the circle is.

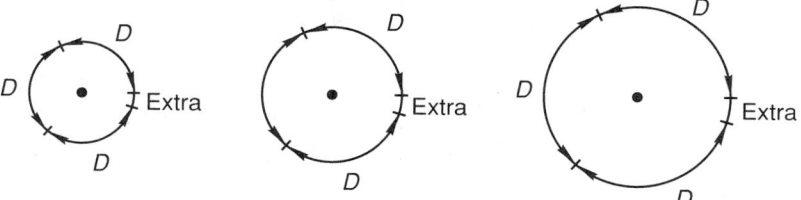

Now we know that the number of times the diameter will go around a circle is approximately

3.14 times

This is not exact because the exact numeral for this number has more digits than can ever be counted. We use the symbol π to represent this number and write it as *pi* (pronounced "pie"). We will use 3.14 as an approximation of π. If we multiply the diameter D by π, we can find the circumference of the circle. If we multiply the square of the radius R by π, we can find the area of the circle.

$$\text{Circumference} = \pi D \qquad \text{Area} = \pi R^2$$

example 63.1   The radius of a circle is 5 centimeters (cm).
(a) What is the circumference of the circle?
(b) What is the area of the circle?

solution   (a) If the radius is 5 cm, the diameter is 10 cm. So

$$\text{Circumference} = \pi D$$
$$= 3.14(10)$$
$$= \textbf{31.4 cm}$$

(b) The <u>area is π times the radius squared</u>. So

$$\text{Area} = \pi R^2$$
$$= 3.14(5)^2$$
$$= \textbf{78.5 cm}^2$$

practice The radius of a big circle is 100 meters.

    (a) What is the diameter of the circle?

    (b) What is the circumference of the circle?

    (c) What is the area of the circle?

problem set 63

1. The bus labored up the hill at 400 yards per minute. If it was 28,000 yards to the top, how long would it take the bus to get there?

2. The skiers could traverse 7000 feet of rough terrain in 20 minutes. How long would it take them to traverse 26,950 feet of rough terrain?

3. The new machine bottled 8000 bottles in 4 hours. How long would it take the machine to bottle 40,000 bottles?

4. The tram could cover 400 centimeters in 50 seconds. How far could the tram go in 20,000 seconds?

5. Complete the table. Begin by inserting the reference numbers.

6. (a) Write 3.09×10^{-4} in standard form.
   (b) Write 6,103,000 in scientific notation.

Fraction	Decimal	Percent
(a)	0.16	(b)
$\frac{2}{5}$	(c)	(d)

7. What decimal part of 720 is 80?

8. What fraction of 625 is 75?

9. $6\frac{2}{5}$ of what number is $6\frac{1}{4}$?

10. $3\frac{1}{5}$ of $1\frac{1}{2}$ is what number?

11. The radius of a circle is 10 centimeters.
    (a) What is the circumference of the circle?
    (b) What is the area of the circle?

12. The base of a right solid is shown on the left. If the sides of the solid are 5 inches high, what is the volume of the solid? Dimensions are in inches. Corners that look square are square.

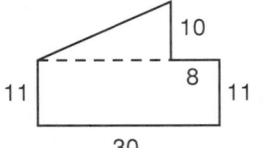 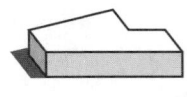

13. Use two unit multipliers to convert 28,000 square miles to square feet.

14. The radius of a big circle is 40 centimeters.
    (a) What is the diameter of the circle?
    (b) What is the circumference of the circle?
    (c) What is the area of the circle?

Solve:

15. $\dfrac{\frac{3}{4}}{\frac{2}{5}} = \dfrac{6}{x}$

16. $\dfrac{\frac{5}{12}}{\frac{5}{3}} = \dfrac{\frac{5}{2}}{k}$

17. $6\frac{2}{7}x = 2\frac{1}{3}$ **18.** $x - 3\frac{6}{7} = 7\frac{16}{21}$

Simplify:

19. $56 - 3[3(3 - 1)(2^2 - 2) - 5]$ **20.** $\sqrt{4} + 5[3(2 - 1) + \sqrt{9}]$

21. $\frac{4}{5} + 2\frac{1}{5} \cdot 1\frac{3}{4}$ **22.** $13\frac{7}{10} - 3\frac{2}{5} + \frac{4}{15}$

23. 621.3×0.0014 **24.** $21.62 - 18.9261$

25. $\dfrac{123.45}{0.006}$ **26.** $\frac{1}{2}\left(6\frac{2}{3} + \frac{1}{4}\right) - \frac{5}{24}$

27. $\dfrac{3\frac{1}{4}}{6\frac{5}{7}}$ **28.** $3\frac{6}{7} \times 2\frac{1}{3} \div 4\frac{1}{2} \div \frac{1}{3}$

29. Evaluate: $xyz + x^y + \sqrt[y]{x} - x$ if $x = 16$, $y = 2$, and $z = \frac{1}{4}$

30. Round 41,190,364 to three digits and write the number in scientific notation.

LESSON 64   *Using both rules to solve equations*

The answer to all four of these equations is 6.

(a) $\frac{x}{3} = 2$ (b) $3x = 18$ (c) $x + 2 = 8$ (d) $x - 3 = 3$

We use the multiplication-division rule to solve (a) and (b) and use the addition-subtraction rule to solve (c) and (d).

(a) $3 \cdot \frac{x}{3} = 2 \cdot 3$ (b) $\frac{3x}{3} = \frac{18}{3}$

$x = 6$ $x = 6$

(c) $x + 2 - 2 = 8 - 2$ (d) $x - 3 + 3 = 3 + 3$

$x = 6$ $x = 6$

Often it is necessary to use both rules to solve the same equation. When it is necessary to use both rules, we always use the addition-subtraction rule first. Then we use the multiplication-division rule.

example 64.1   Solve: $3x + 2 = 7$

solution   First we subtract 2 from both sides.

$$3x + 2 - 2 = 7 - 2$$
$$3x = 5$$

Now we divide both sides by 3.

$$\frac{3x}{3} = \frac{5}{3} \quad \longrightarrow \quad x = \frac{5}{3}$$

example 64.2   Solve: $\frac{2}{3}x - \frac{1}{2} = \frac{10}{3}$

solution   First we add $\frac{1}{2}$ to both sides.

$$\frac{2}{3}x - \frac{1}{2} + \frac{1}{2} = \frac{10}{3} + \frac{1}{2}$$

$$\frac{2}{3}x = \frac{23}{6} \qquad \text{simplified}$$

Now we multiply both sides by $\frac{3}{2}$.

$$\frac{3}{2} \cdot \frac{2}{3}x = \frac{23}{6} \cdot \frac{3}{2} \qquad \text{multiplied by } \frac{3}{2}$$

$$x = \frac{23}{4} \qquad \text{simplified}$$

example 64.3   Solve: $2\frac{1}{3}x - 1\frac{1}{5} = \frac{23}{10}$

solution   We begin by writing the mixed numbers as improper fractions.

$$\frac{7}{3}x - \frac{6}{5} = \frac{23}{10}$$

Next we eliminate the $-\frac{6}{5}$ by adding $\frac{6}{5}$ to both sides.

$$\frac{7}{3}x - \frac{6}{5} + \frac{6}{5} = \frac{23}{10} + \frac{6}{5}$$

$$\frac{7}{3}x = \frac{35}{10} \qquad \text{simplified}$$

$$\frac{7}{3}x = \frac{7}{2} \qquad \text{reduced}$$

Now we multiply both sides by $\frac{3}{7}$.

$$\frac{3}{7} \cdot \frac{7}{3}x = \frac{7}{2} \cdot \frac{3}{7} \qquad \text{multiplied by } \frac{3}{7}$$

$$x = \frac{3}{2} \qquad \text{simplified}$$

practice   Solve:

a. $\frac{2}{3}x - \frac{1}{6} = \frac{1}{2}$

b. $2\frac{1}{2}x + \frac{1}{8} = \frac{1}{4}$

c. $1\frac{1}{4}x - \frac{1}{3} = \frac{3}{5}$

d. $\frac{2}{7}x + \frac{1}{5} = 2\frac{1}{10}$

problem set 64

1. Roland heard the clarion from afar and increased the speed of the column to 4 miles per hour. How far could the column go in 16 hours?

2. Charlemagne's column covered the first 12 miles in 4 hours. He then increased the speed by 2 miles per hour. If the total distance of the trip was 52 miles, how long did it take to finish the trip?

3. Alison traveled at 5 miles per hour for 12 hours. Then she traveled at 8 miles per hour for 8 hours. How far did she go in all?

4. The roses had to be packed 48 to a shipping box. If 4800 boxes were available, how many roses could be shipped?

5. Complete the table. Begin by inserting the reference numbers.

6. (a) Write 10,300 in scientific notation.
   (b) Write 6.019×10^8 in standard form.

Fraction	Decimal	Percent
(a)	(b)	60

7. What decimal part of 240 is 160?   8. What fraction of 576 is 36?

9. $5\frac{2}{3}$ of what number is $3\frac{1}{4}$?

10. The radius of a circle is 7 feet.
    (a) What is the circumference of the circle?
    (b) What is the area of the circle?

11. The base of a right solid is shown on the left. If the sides of the solid are 1 foot high, what is the volume of the solid in cubic feet? Dimensions are in feet. Corners that look square are square.

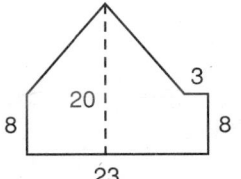 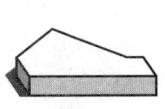

12. Find the least common multiple of 12, 20, and 30.

13. Find the surface area of a cube whose edges are 7 meters long.

Solve:

14. $\dfrac{\frac{1}{5}}{\frac{3}{10}} = \dfrac{5}{x}$

15. $\dfrac{\frac{7}{12}}{\frac{5}{24}} = \dfrac{5}{y}$

16. $4\frac{1}{2}x - 1\frac{1}{3} = \dfrac{17}{12}$

17. $2\frac{1}{2}x - 2\frac{1}{4} = 3\frac{2}{5}$

18. $3x + 7 = 25$

Simplify:

19. $22 + 2[(6 + 1)(3 - 2)3 + 1]$

20. $2^2 + 2[2^2(4^2 - 3^2) - 3^3]$

21. $\dfrac{3}{7} + 3\frac{1}{2} \cdot \dfrac{2}{3}$

22. $6\frac{7}{10} - 4\frac{4}{15} + \dfrac{7}{30}$

23. 17.82×0.0011

24. $178.22 - 19.621$

25. $\dfrac{192.61}{0.005}$

26. $\dfrac{1}{3}\left(2\frac{1}{3} + 6\frac{1}{2}\right) - \dfrac{7}{24}$

27. $\dfrac{2\frac{1}{3}}{3\frac{1}{5}}$

28. $2\frac{1}{4} \times 3\frac{1}{2} \div \dfrac{3}{4} \times 1\frac{1}{4}$

29. Evaluate: $xy + \sqrt[y]{z} + xyz + x^y$ if $x = 3$, $y = 3$, and $z = 1$

30. Round 109,376 to two digits and write the number in scientific notation.

LESSON 65 *Fractional-part word problems*

The equation

$$F \times of = is$$

is one of the most important equations in all mathematics. This equation is used to solve many problems that are encountered in everyday life. Also, this equation is almost exactly the same as one form of the percent equation, as we will see in a later lesson.

example 65.1 Five-eighths of the gnomes who lived in the magic forest had happy faces. If 840 gnomes had happy faces, how many gnomes lived in the magic forest?

solution We can change the wording of this problem to

$$\frac{5}{8} \text{ of what number is 840?}$$

which we can solve by using the fractional-part-of-a-number equation.

$$F \times of = is$$

We substitute and solve.

$$\frac{5}{8} \cdot WN = 840$$

$$\frac{8}{5}\left(\frac{5}{8} \cdot WN\right) = 840 \cdot \frac{8}{5} \qquad \text{multiplied by } \frac{8}{5}$$

$$WN = \mathbf{1344} \qquad \text{simplified}$$

So 1344 gnomes lived in the forest. If 840 had happy faces, then 504 did not have happy faces.

example 65.2 On Monday, $2\frac{4}{5}$ times the acceptable number took refuge in the mountain caves. If 640 was the acceptable number, how many took refuge in the mountain caves on Monday?

solution We can restate the problem as

$$2\frac{4}{5} \text{ of 640 is what number?}$$

We will use the fractional-part-of-a-number equation, substitute, and solve.

$$F \times of = is$$

$$\frac{14}{5} \cdot 640 = WN$$

$$\mathbf{1792 = WN}$$

Thus 640 was an acceptable number, but 1152 more took refuge for a total of 1792.

example 65.3 When the fog lifted, 400 ghosts were spied skulking near the outskirts. If 240 ghosts were not spied, what fraction of the ghosts was not spied?

solution Since there are 640 ghosts in all, we can restate the problem as

What fraction of 640 is 240?

We use the fractional-part-of-a-number equation, substitute, and solve.

$$F \times of = is$$

$$WF \cdot 640 = 240$$

$$\frac{WF \cdot 640}{640} = \frac{240}{640}$$

$$WF = \frac{3}{8}$$

Thus $\frac{3}{8}$ of the ghosts were not spied.

practice When the gun sounded, only two-fifths of the racers began to run. If 460 racers began to run, how many racers were there in all?

problem set 65

1. Four-fifths of the pixies in the kingdom had sad faces. If 840 pixies had sad faces, how many pixies lived in the kingdom?

2. On Monday, $2\frac{1}{5}$ times the acceptable number hid under the front porch. If 880 was the acceptable number, how many hid under the front porch?

3. When the siren sounded, 440 were seen walking in the forest. If 360 were not seen, what fraction of the total was seen?

4. The cutters cut out 400 in 8 hours. How many could they cut out in the next 160 hours?

5. Complete the table. Begin by inserting the reference numbers.

6. Use one unit multiplier to convert 12,000 feet to inches.

Fraction	Decimal	Percent
(a)	0.18	(b)

7. What decimal part of 360 is 300?

8. What fraction of 360 is 300?

9. $2\frac{1}{3}$ of $3\frac{1}{4}$ is what number?

10. The radius of a circle is 3 inches.
    (a) What is the circumference of the circle?
    (b) What is the area of the circle?

11. The base of a right solid is shown on the left. If the solid is 20 centimeters high, what is the volume of the solid in cubic centimeters? Dimensions are in centimeters. Corners that look square are square.

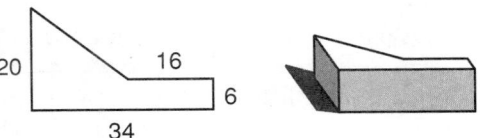

12. Find the surface area of a cube whose edges are 16 inches long.

13. Write $\dfrac{3211}{7}$ as a mixed number.

Solve:

14. $\dfrac{\frac{1}{3}}{\frac{7}{10}} = \dfrac{6}{x}$

15. $\dfrac{\frac{6}{7}}{\frac{12}{21}} = \dfrac{4}{x}$

16. $3\frac{1}{4}x - 2\frac{1}{3} = \dfrac{17}{12}$

17. $3\frac{1}{6}x - \dfrac{3}{8} = 1\frac{17}{24}$

18. $17x - 12 = 19$

Simplify:

19. $12 + 2[(6 - 2)(4 - 3)2 - 5]$

20. $3^2 + 2^3[2(3^2 - 4) - 3]$

21. $\dfrac{2}{7} + 2\frac{1}{3} \cdot \dfrac{1}{2}$

22. $13\frac{2}{3} - 4\frac{1}{2} + 2\frac{5}{6}$

23. 181.2×0.013

24. $1921.61 - 19.897$

25. $\dfrac{175.61}{0.7}$

26. $\dfrac{1}{4}\left(3\frac{1}{5} + 2\frac{1}{2}\right) - \dfrac{3}{10}$

27. $3\frac{1}{3} \times 2\frac{1}{2} \div \dfrac{2}{3} \div 2\frac{5}{6}$

28. Evaluate: $xyz + \sqrt{y} + x^z$ if $x = 1$, $y = 36$, and $z = 2$

29. Write 621723131.72 in words.

30. Round 61,237,899,721.2 to the nearest thousand.

LESSON 66 *Changing rates*

We know that a rate is a ratio. Either number can be on top. If 30 jars can be filled in 1 minute, we can write two rates.

(a) $\dfrac{30 \text{ jars}}{1 \text{ minute}}$ (b) $\dfrac{1 \text{ minute}}{30 \text{ jars}}$

In everyday usage the word *rate* means speed, which is the number per unit time. This

is the rate with the time on the bottom. If the original rate in this problem were doubled, the two new rates would be

$$\frac{60 \text{ jars}}{1 \text{ minute}} \quad \text{and} \quad \frac{1 \text{ minute}}{60 \text{ jars}}$$

If the original rate were increased by 7 jars per minute, the two new rates would be

$$\frac{37 \text{ jars}}{1 \text{ minute}} \quad \text{and} \quad \frac{1 \text{ minute}}{37 \text{ jars}}$$

example 66.1   Prince Charming traveled 60 leagues in 2 days. Then he doubled his rate. How long would it take him to go 300 leagues at this new speed?

solution   The first two rates are

$$\frac{60 \text{ leagues}}{2 \text{ days}} = \frac{30 \text{ leagues}}{1 \text{ day}} \quad \text{and} \quad \frac{2 \text{ days}}{60 \text{ leagues}} = \frac{1 \text{ day}}{30 \text{ leagues}}$$

If he doubled his rate, **he doubles the distance** he can travel in 1 day. Thus the new rates are

$$\frac{(2 \times 30) \text{ leagues}}{1 \text{ day}} = \frac{60 \text{ leagues}}{1 \text{ day}}$$

and

$$\frac{1 \text{ day}}{(2 \times 30) \text{ leagues}} = \frac{1 \text{ day}}{60 \text{ leagues}}$$

To find the time required to go 300 leagues at the new rate, we choose the rate with time on the top.

$$\frac{1 \text{ day}}{60 \text{ leagues}} \cdot 300 \text{ leagues} = \frac{300 \text{ days}}{60} = \textbf{5 days}$$

example 66.2   The machine could cap 500 bottles in 2 hours. If the rate of the machine was tripled, how many bottles could be capped in 10 hours at the new rate?

solution   The given rates are

$$\frac{500 \text{ bottles}}{2 \text{ hours}} = \frac{250 \text{ bottles}}{1 \text{ hour}}$$

and

$$\frac{2 \text{ hours}}{500 \text{ bottles}} = \frac{1 \text{ hour}}{250 \text{ bottles}}$$

If the rate is tripled, the machine could cap 3 times the number of bottles in the same time. The new rates would be

$$\frac{(3 \times 250) \text{ bottles}}{1 \text{ hour}} \quad \text{and} \quad \frac{1 \text{ hour}}{(3 \times 250) \text{ bottles}}$$

To find the number of bottles, we choose the rate with bottles on top.

$$\frac{750 \text{ bottles}}{1 \text{ hour}} \times 10 \text{ hours} = \textbf{7500 bottles}$$

example 66.3   Vacuous ran 6 miles in 2 hours. Then he ran 20 miles in 4 hours. By how much did his rate increase?

solution   Unless otherwise specified, rate means speed and time is on the bottom.

$$\text{Rate 1} = \frac{6 \text{ mi}}{2 \text{ hr}} = \frac{3 \text{ mi}}{1 \text{ hr}} \qquad \text{Rate 2} = \frac{20 \text{ mi}}{4 \text{ hr}} = \frac{5 \text{ mi}}{1 \text{ hr}}$$

His rate increased from 3 miles per hour to 5 miles per hour. His rate increased **2 miles per hour.**

practice   **a.** Jim drove the 120 miles to grandmother's house in 3 hours. He increased his speed by 10 miles per hour on the way home. How long did it take him to drive home?

b. Margot could pickle 500 cucumbers in 2 hours. She got tired and reduced her rate by one-half. How long would it take her to pickle 2000 cucumbers at the new rate?

problem set 66

1. Ephemera was running out of time. She traveled the first 100 leagues in 4 days, but she found that she would have to double her pace in order to avoid disaster. How long would it take her to complete the last 200 leagues at the new pace?

2. Tenacious ran 12 miles in 2 hours. Then he ran 36 miles in 3 hours. By how much did his rate increase?

3. When Aristotle proclaimed his new theory, 87 responded. If this was only $\frac{3}{5}$ the number of responses expected, how many responses were expected?

4. The apothecary could press 104 packages of medicine in 4 hours. How many packages could be pressed in 10 hours?

5. Complete the table. Begin by inserting the reference numbers.

6. Use two unit multipliers to convert 25,000 square centimeters to square meters.

Fraction	Decimal	Percent
(a)	(b)	73

7. What decimal part of 450 is 300?

8. What fractional part of 450 is 300?

9. $5\frac{1}{2}$ of $2\frac{1}{3}$ is what number?

10. The radius of a circle is 7 centimeters.
(a) What is the circumference of the circle?
(b) What is the area of the circle?

11. Find the least common multiple of 18, 30, and 54.

12. The base of a right solid is shown on the left. If the solid is 5 feet high, what is the volume of the solid in cubic feet? Dimensions are in feet. All angles are right angles.

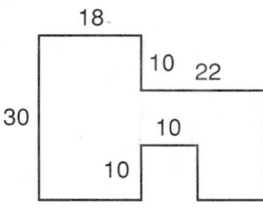
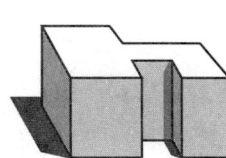

13. (a) Write 0.00392 in scientific notation.
  (b) Write 6.03×10^{-9} in standard form.

Solve:

14. $\dfrac{\frac{1}{4}}{\frac{4}{5}} = \dfrac{3}{x}$

15. $\dfrac{\frac{3}{4}}{\frac{8}{9}} = \dfrac{2}{x}$

16. $5\frac{1}{3}x - 3\frac{3}{4} = \dfrac{19}{32}$

17. $2\frac{1}{2}x - \frac{1}{4} = 2\frac{3}{16}$

18. $22x - 9 = 57$

Simplify:

19. $15 + 3[(5 - 1)(7 - 2)2 - 4]$

20. $4^2 + 3^3[2(4^2 - 15) - 2]$

21. $\frac{1}{5} + 3\frac{1}{2} \cdot \frac{2}{7}$

22. $11\frac{3}{4} - 5\frac{1}{5} + 3\frac{1}{3}$

23. 201.6×0.021

24. $2653.19 - 5.319$

25. $\dfrac{226.14}{0.003}$

26. $\frac{1}{2}\left(2\frac{1}{4} + 1\frac{3}{16}\right) - \dfrac{31}{32}$

27. $2\frac{1}{5} \times 3\frac{1}{4} \div 5\frac{1}{2} \div 6\frac{1}{2}$

28. Evaluate: $kmx + \sqrt[m]{k} + m^x$ if $k = 64$, $m = 3$, and $x = 4$

29. Use two unit multipliers to convert 1,000,000 square inches to square feet.

30. Round 1,789,322.137 to the nearest tenth.

LESSON 67 *Semicircles*

Half a circle is called a **semicircle.** The length of the arc of a semicircle is one-half the perimeter of a whole circle. The area of a closed semicircle is one-half the area of a whole circle.

$$\text{Arc length of a semicircle} = \frac{\pi D}{2}$$

$$\text{Area of a closed semicircle} = \frac{\pi R^2}{2}$$

Sometimes is it necessary to find the perimeter or the area of a figure that contains one or more semicircles.

example 67.1 Find (a) the perimeter and (b) the area of this figure. Dimensions are in meters.

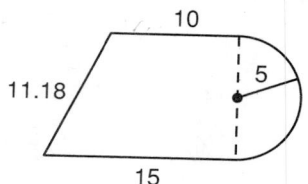

solution (a) The perimeter includes the length of the semicircle.

$$\text{Perimeter} = 11.18 + 15 + 10 + \frac{3.14(10)}{2}$$

$$= 11.18 + 15 + 10 + 15.7 = \textbf{51.88 m}$$

(b) The area is the sum of the areas of the triangle, the rectangle, and the semicircle.

$$\text{Area} = \frac{10 \cdot 5}{2} + 10 \cdot 10 + \frac{3.14(5)^2}{2}$$

$$= 25 + 100 + 39.25 = \textbf{164.25 m}^2$$

example 67.2 Find (a) the perimeter and (b) the area of this figure. Dimensions are in centimeters.

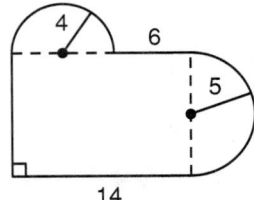

solution (a) The perimeter includes two semicircles.

$$\text{Perimeter} = 10 + 14 + 6 + \frac{3.14(8)}{2} + \frac{3.14(10)}{2}$$

$$= 10 + 14 + 6 + 12.56 + 15.7 = \textbf{58.26 cm}$$

(b) The area includes two semicircles.

$$\text{Area} = 10 \cdot 14 + \frac{3.14(4)^2}{2} + \frac{3.14(5)^2}{2}$$

$$= 140 + 25.12 + 39.25 = \textbf{204.37 cm}^2$$

practice **a.** Find the perimeter of this figure. Dimensions are in centimeters.

b. Find the area of this figure.

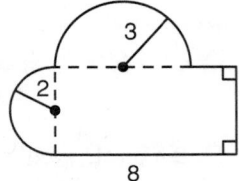

problem set 67

1. Three-fourths of the fans did not carry banners. If 1000 fans attended the game, how many fans carried banners?

2. Seven-sixteenths of the children smirked when they heard the joke. If 210 children smirked, how many children heard the joke?

3. When the score was announced over the PA system, 480 fans were happy. If 3360 fans attended the game, what fraction of the fans did the announcement make happy?

4. The hikers slogged 10 miles through the muck in 5 hours. Then they came to the pavement and were able to triple their rate. How long would it take them to cover 24 miles on the pavement?

5. Complete the table. Begin by inserting the reference numbers.

Fraction	Decimal	Percent
$\frac{2}{5}$	(a)	(b)

6. What number is $3\frac{3}{5}$ of $2\frac{1}{2}$?

7. 0.6 of what number is 1.44?

8. What fraction of 35 is 21?

9. What decimal part of 810 is 270?

10. Find (a) the perimeter and (b) the area of this figure. Dimensions are in meters.

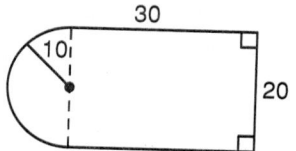

11. The figure shown on the left is the base of a right solid whose sides are 10 centimeters high. What is the volume of the solid in cubic centimeters? Dimensions are in centimeters.

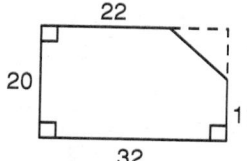 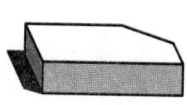

12. Find the least common multiple of 12, 16, and 30.

13. Use one unit multiplier to convert 10,000 feet to inches.

Solve:

14. $\dfrac{\frac{1}{4}}{\frac{7}{8}} = \dfrac{6}{x}$

15. $\dfrac{\frac{5}{7}}{\frac{10}{21}} = \dfrac{6}{p}$

16. $4\frac{1}{7}x - 2\frac{2}{3} = 3\frac{1}{4}$

17. $4\frac{1}{3}x - \frac{5}{8} = 1\frac{5}{24}$

18. $18x + 14 = 27$

Simplify:

19. $13 + 3[(3 - 1)(6 - 4)2 - 1]$

20. $2^3 + 3[3(3^2 - 5) + \sqrt{4}]$

21. $1\frac{2}{7} + 3\frac{3}{4} \cdot 2\frac{1}{3}$

22. $12\frac{4}{5} - 4\frac{2}{3} + 11\frac{7}{15}$

23. 191.4×0.0012

24. $9218.98 - 178.621$

25. $\dfrac{161.82}{0.06}$

26. $\frac{1}{5}\left(2\frac{1}{6} - 1\frac{1}{4}\right) - \frac{1}{30}$

27. $4\frac{3}{4} \times 3\frac{7}{12} \div 2\frac{1}{3} \times \frac{1}{6}$

28. Use two unit multipliers to convert 144 square feet to square miles.

29. (a) Write 1,390,000 in scientific notation.
    (b) Write 4.26×10^{17} in standard form.

30. Evaluate: $xyz + y^2 + x^y - xy$ if $x = 6$, $y = 3$, and $z = \frac{1}{9}$

LESSON 68   *Proportions with mixed numbers*

To solve proportions that contain mixed numbers, the first step is to write the mixed numbers as improper fractions.

example 68.1   Solve: $\dfrac{\frac{4}{9}}{x} = \dfrac{2\frac{1}{5}}{1\frac{3}{4}}$

solution   As the first step, we change the mixed numbers to improper fractions. Then we have

$$\dfrac{\frac{4}{9}}{x} = \dfrac{\frac{11}{5}}{\frac{7}{4}}$$

Now we cross multiply and get

$$\frac{4}{9} \cdot \frac{7}{4} = \frac{11}{5}x \qquad \text{cross multiplied}$$

$$\frac{7}{9} = \frac{11}{5}x \qquad \text{simplified}$$

Now we will solve by multiplying both sides by $\frac{5}{11}$.

$$\frac{7}{9} \cdot \frac{5}{11} = \frac{5}{11} \cdot \frac{11}{5}x$$

$$\frac{35}{99} = x$$

practice   Solve:

a. $\dfrac{1\frac{3}{4}}{\frac{2}{3}} = \dfrac{5\frac{1}{4}}{x}$

b. $\dfrac{3\frac{1}{4}}{p} = \dfrac{2\frac{1}{6}}{3\frac{2}{10}}$

problem set 68

1. The ancient Sumerians did not have paper. They wrote by pressing wedges into clay. Their writing is called *cuneiform*, which means wedge-shaped. Four-fifths of the Sumerians in the town could not read cuneiform. If 16,000 Sumerians lived in the town, how many could read cuneiform?

2. Monoliths were erected at a rate of 39 every 3 days. However, this was not enough to clear the back orders. If the original rate was quadrupled, how many days would it take to erect 208 monoliths?

3. Enrique drove the 240 miles to the park in 4 hours. If his return trip took only 3 hours, what was the difference in his rate?

4. Three-fourths of all the skate-boarders at the competition could execute a "boneless." If 14 could not execute a boneless, how many skate-boarders were at the competition?

5. Complete the table. Begin by inserting the reference numbers.

Fraction	Decimal	Percent
(a)	0.06	(b)

6. What decimal part of 930 is 558?

7. What fractional part of 930 is 558?

8. $4\frac{3}{5}$ of $9\frac{1}{2}$ is what number?

9. Find (a) the perimeter and (b) the area of this figure. Dimensions are in feet.

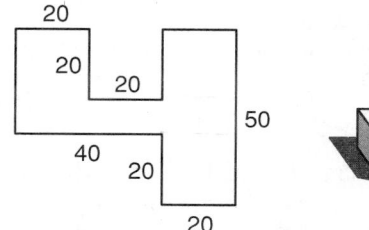

10. The figure shown on the left is the base of a right solid whose sides are 30 meters high. What is the volume of the solid in cubic meters? Dimensions are in meters. Remember that volume equals the area of the base times the height. All angles are right angles.

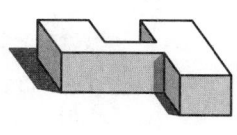

11. Find the least common multiple of 7, 9, and 11.

12. Round 0.000893 to the nearest hundred-thousandth and write the rounded number in scientific notation.

13. Use two unit multipliers to convert 100,000 square feet to square miles.

Solve:

14. $\dfrac{\frac{1}{6}}{\frac{4}{5}} = \dfrac{10}{x}$

15. $\dfrac{2\frac{2}{3}}{\frac{8}{14}} = \dfrac{\frac{1}{2}}{m}$

16. $\dfrac{2\frac{1}{5}}{m} = \dfrac{1\frac{1}{5}}{\frac{1}{3}}$

17. $6\frac{1}{2}x - \frac{1}{4} = 2\frac{1}{13}$

18. $3\frac{1}{5}x - 7\frac{2}{3} = 1\frac{3}{4}$

19. $35x - 19 = 51$

Simplify:

20. $11 + 3[(5 - 2)(7 - 3)4 - 36]$

21. $3^2 + 2[2(2^3 - 7) + \sqrt{16}]$

22. $1\frac{2}{3} + 2\frac{1}{2} \cdot 3\frac{1}{5}$

23. $11\frac{2}{3} - 6\frac{1}{2} + 9\frac{1}{12}$

24. 310.8×0.00005

25. $6842.96 - 41.096$

26. $\dfrac{230.95}{0.00025}$

27. $\frac{1}{4}\left(3\frac{1}{3} + 2\frac{1}{2}\right) - \frac{11}{12}$

28. $6\frac{1}{2} \times 1\frac{3}{5} \div 1\frac{1}{4} \times \frac{7}{10}$

29. Use two unit multipliers to convert 75,000 centimeters to kilometers.

30. Evaluate: $mp + \sqrt[p]{x} + p^m$ if $m = 2$, $p = 4$, and $x = 16$

LESSON 69 *Ratio word problems*

Often it is desirable to maintain a fixed ratio between two quantities. The conditions are stated in ratio word problems. Three of the four parts are usually given. We use the statement of the problem to help us write a conditional proportion. Then we solve the conditional proportion to find the missing part.

example 69.1 The ratio of the number of parrots to the number of macaws was 5 to 7. How many macaws were there when the parrots numbered 750?

solution The two ratios are between P for parrots and M for macaws. **If P is on top on one side, P must be on top on the other side.** We decide to put P on top. So we have

$$\frac{P}{M} = \frac{P}{M}$$

We were told that the ratio of the number of parrots to the number of macaws was 5 to 7. The first number given (5) goes with the first word (parrots), and the second number given (7) goes with the second word (macaws). Thus we write

$$\frac{5}{7} = \frac{P}{M}$$

These numbers are the ratio numbers. The other numbers are the problem numbers. Since parrots numbered 750, we replace P with 750.

$$\frac{5}{7} = \frac{750}{M}$$

Now we cross multiply and solve for M.

$$5M = 7 \cdot 750 \qquad \text{cross multiplied}$$

$$\frac{5M}{5} = \frac{7 \times 750}{5} \qquad \text{divided both sides by 5}$$

$$M = 1050$$

Thus, there were 1050 macaws when there were 750 parrots.

example 69.2 The ratio of the number of wrigglers to the number of squirmers was 13 to 2. When 26 students had the wriggles, how many were squirming?

solution We begin by writing

$$\frac{W}{S} = \frac{W}{S}$$

Now on the left we put in the ratio numbers 13 for W and 2 for S.

$$\frac{13}{2} = \frac{W}{S}$$

On the right we put in the given number 26 for W.

$$\frac{13}{2} = \frac{26}{S}$$

Now we cross multiply and solve for S.

$$13S = 2 \cdot 26 \qquad \text{cross multiplied}$$

$$\frac{13S}{13} = \frac{2 \cdot 26}{13} \qquad \text{divided by 13}$$

$$S = 4 \qquad \text{simplified}$$

Thus, when 26 students were wriggling, 4 were squirming.

practice   The ratio of winners to losers in the valley was 22 to 3. If there were 3000 losers gathered in the valley, how many winners were gathered in the valley?

problem set 69

1. The ratio of the number of birds to the number of beasts was 3 to 7. How many beasts were there when the birds numbered 750?

2. The ratio of the number of smilers to the number of frowners was 13 to 2. When 52 students were smiling, how many were frowning?

3. Two-thirds of the costumes were polychromatic. If 4800 costumes were polychromatic, how many were not polychromatic?

4. Matildabelle traveled the first 100 miles in 4 hours. Then she quadrupled her speed. How long would it take her to travel the next 1400 miles?

5. Complete the table. Begin by inserting the reference numbers.

Fraction	Decimal	Percent
(a)	(b)	25

6. (a) Write 0.000913 in scientific notation.
    (b) Write 6.14×10^{-5} in standard form.

7. 0.3 of what number is 123?

8. What fraction of 36 is 32?

9. What decimal of 360 is 300?

10. $3\frac{1}{2}$ of what number is $4\frac{2}{3}$?

11. Find (a) the perimeter and (b) the area of this figure. Dimensions are in feet.

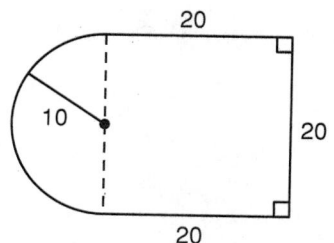

12. The figure shown on the left is the base of a solid whose sides are 3 meters high. What is the volume of the solid in cubic centimeters? Dimensions are in centimeters. Remember to convert the height to centimeters and multiply by the area of the base.

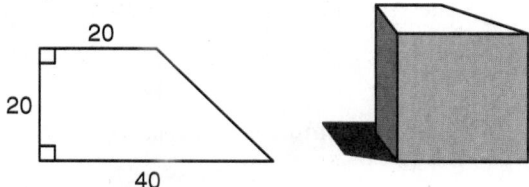

13. Find the least common multiple of 12, 20, and 36.

14. Write $\frac{6131}{5}$ as a mixed number.

Solve:

15. $\dfrac{2\frac{3}{4}}{3\frac{7}{8}} = \dfrac{8}{x}$

16. $\dfrac{\frac{6}{7}}{\frac{10}{21}} = \dfrac{9}{x}$

17. $5\frac{1}{2}x + 3\frac{1}{3} = 10\frac{1}{7}$

18. $4\frac{6}{7}x - \frac{4}{21} = 11\frac{5}{14}$

19. $18x + 14 = 72$

Simplify:

20. $14 + 2[2^2(2 - 1)(3^2 - 2^3) + 1]$

21. $3^2 + 3[5(3^2 - 2^3) + \sqrt{4}]$

22. $1\frac{3}{7} + 2\frac{1}{3} \cdot 6\frac{1}{2}$

23. $13\frac{4}{5} - 6\frac{1}{8} + 2\frac{3}{40}$

24. 291.8×0.0013

25. $10{,}818.17 - 689.891$

26. $\dfrac{1921.1}{0.004}$

27. $\frac{1}{10}\left(2\frac{1}{5} + 7\frac{1}{2}\right) - \frac{3}{10}$

28. $3\frac{1}{4} \times 11\frac{5}{6} \div 2\frac{1}{8} \div \frac{1}{4}$

29. Evaluate: $xyz + x^2 + y^2 + z^2$ if $x = 3$, $y = 3$, and $z = 2$

30. Round $182.\overline{157}$ to eight decimal places.

LESSON 70   *Using ratios to compare*

We can use ratios to determine unit prices. The unit price is the price for one item.

example 70.1   If 20 pounds of beans sold for $1.20, what was the price per pound of the beans?

solution   We get two ratios from this statement.

$$\frac{120 \text{ cents}}{20 \text{ pounds}} = \frac{\textbf{6 cents}}{\textbf{1 pound}} \qquad \frac{20 \text{ pounds}}{120 \text{ cents}} = \frac{1 \text{ pound}}{6 \text{ cents}}$$

We call the first ratio the **unit price** because it tells the cost for 1 pound. The second ratio tells us the number of pounds we can buy for 1 cent.

example 70.2 The big can held 16 ounces and cost 80 cents. The small can held 12 ounces and cost 72 cents. Which can was the better buy?

solution To tell which price is the best price, we usually compare unit prices. This is the cost for 1 item. We compute the cost per ounce for both cans.

$$\text{Big can} = \frac{80 \text{ cents}}{16 \text{ ounces}} \qquad \text{Small can} = \frac{72 \text{ cents}}{12 \text{ ounces}}$$

$$= 5 \frac{\text{cents}}{\text{ounce}} \qquad\qquad = 6 \frac{\text{cents}}{\text{ounce}}$$

Thus, the **big can** at 5 cents per ounce is a better buy than the small can at 6 cents per ounce.

practice One store sold packages of 1 dozen (12) for $1.44 a package. The other store sold packages of 3 dozen (36) for $4.68. Use unit prices to tell which price was the best price.

problem set 70

1. The teacher's first guess was twenty-two million, seven hundred forty-six thousand, nine hundred seventy-two. The teacher's second guess was twenty-one million, three hundred sixty-five thousand, two hundred forty. By how much was the first guess greater?

2. The big can held 18 ounces and cost $2.52. The small can held only 12 ounces and cost $1.08. Find both unit prices. Which can was the better buy?

3. The ratio of painted faces to unpainted faces was 3 to 5. If 1800 had painted their faces, how many had not painted their faces?

4. Seven-eighths of the worker's income was used to pay for necessities. If $5600 was spent on necessities, how much was the worker's income?

5. Complete the table. Begin by inserting the reference numbers.

6. Round 0.000009153 to the ten-millionths' place and write in scientific notation.

Fraction	Decimal	Percent
(a)	0.37	(b)

7. 0.4 of what number is 216?

8. What fraction of 39 is 24?

9. What decimal part of 480 is 450?

10. $4\frac{2}{3}$ of what number is $5\frac{4}{5}$?

11. Find (a) the perimeter and (b) the area of the figure on the left. Dimensions are in inches.

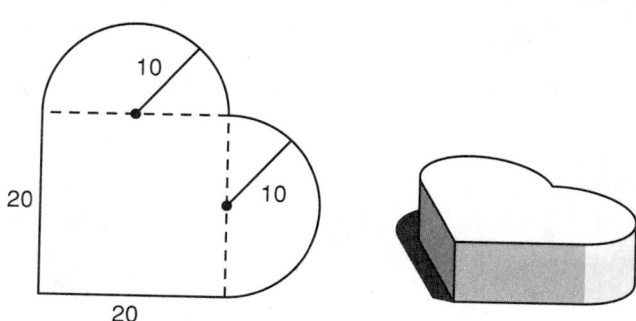

12. What is the volume of a right solid whose base is the figure shown in Problem 11 and whose sides are 7 inches tall?

13. Find the least common multiple of 24, 36, and 40.

14. Find the surface area of this triangular prism. Dimensions are in centimeters.

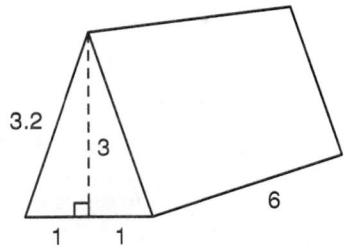

Solve:

15. $\dfrac{3\frac{1}{5}}{2\frac{1}{3}} = \dfrac{4}{x}$

16. $\dfrac{\frac{3}{4}}{\frac{10}{21}} = \dfrac{8}{x}$

17. $6\frac{1}{3}x - 4\frac{7}{8} = 3\frac{3}{4}$

18. $5\frac{6}{7}x - \frac{5}{21} = 11\frac{3}{14}$

19. $19x + 6 = 121$

Simplify:

20. $13 + 2[2(3 - 1)(2^2 - 1) + 1]$

21. $2^2 + 3[2^2(2^4 - 3^2) - \sqrt{4}]$

22. $2\frac{2}{3} + \frac{1}{6} \cdot 2\frac{1}{4}$

23. $16\frac{11}{12} - 3\frac{5}{6} + 2\frac{7}{18}$

24. 12.21×0.0017

25. $11{,}716.181 - 891.7891$

26. $\dfrac{136.18}{0.03}$

27. $\frac{1}{3}\left(2\frac{1}{3} + 3\frac{1}{4}\right) - \frac{13}{36}$

28. $2\frac{1}{3} \times 12\frac{1}{6} \div 1\frac{5}{12} \div \frac{1}{4}$

29. Evaluate: $xy + xyz - x + y^2$ if $x = 6$, $y = 8$, and $z = \frac{1}{24}$

30. Write 621321611.01 in words.

LESSON 71 *Forms of the percent equation · Percents less than 100*

71.A
forms of the percent equation

There are three forms of the percent equation that are commonly used. They are the fractional form (a), the ratio form (b), and the rate form (c).

(a) $\dfrac{P}{100} \times of = is$ (b) $\dfrac{P}{100} = \dfrac{is}{of}$ (c) Rate $\times of = is$

We will look closely at forms (a) and (b) now and save form (c) until later. In forms (a) and (b), the letter P stands for percent, and the words *of* and *is* are used the same way as in the fractional-part-of-a-number problem. We will use these forms in this lesson to investigate percents less than 100. In a later lesson, we will look at percents greater than 100.

71.B
percents less than 100

When the percent is less than 100, the problem always divides a number into two parts. A two-part diagram that shows this division is helpful. We will use one of these diagrams to discuss the statement

30 percent of 140 is 42

On the left we show the number 140. This is the "before" diagram and represents 100 percent.

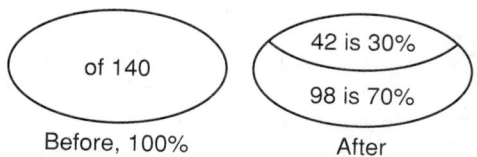

Before, 100%   After

On the right we show 140 divided into two parts. One of the parts is 42, which is 30 percent. The other part is 98, which is 70 percent. The sum of the numbers 42 and 98 is 140. The sum of 30 percent and 70 percent is 100 percent. **In the diagram, the sum of the two "after" numbers must always equal the "before" number. The sum of the two "after" percents must equal the "before" percent, which is always 100 percent.** We will practice by working percent problems and then drawing the two-part diagrams that give us a picture of the solution.

Learning to draw the percent diagrams is very important. The diagrams let you visualize the percent problem.

example 71.1   Complete the percent diagram by finding a and b.

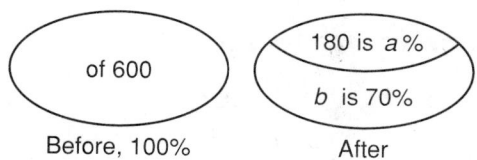

Before, 100%   After

solution   The sum of the two percents in the diagram on the right is always 100 percent. Thus, the sum of a% and 70 percent must be 100 percent. This means that a% is **30 percent.** The sum of the two numbers on the right must equal the number on the left. Thus 180 plus b must be 600. This means that b is **420.**

example 71.2   Twenty percent of what number is 240? Draw a diagram of the problem.

solution   We see that the percent is 20, that *of* goes with "what number," and that *is* goes with 240. We will work the problem twice. We will use the fractional form of the percent equation and then rework the problem using the ratio form of the percent equation to demonstrate that both forms will give the same answer.

(a)   FRACTIONAL FORM

$$\frac{P}{100} \times of = is$$

(b)   RATIO FORM

$$\frac{P}{100} = \frac{is}{of}$$

$$\frac{20}{100} \times WN = 240$$

$$\frac{100}{20} \times \frac{20}{100} \times WN = \frac{100}{20} \times 240$$

$$WN = 1200$$

$$\frac{20}{100} = \frac{240}{WN}$$

$$20WN = 240(100)$$

$$\frac{20WN}{20} = \frac{240(100)}{20}$$

$$WN = 1200$$

The diagram shows 1200 as the "before" number. One part is 240, which is 20 percent. Thus the other part must be 960, which is 80 percent.

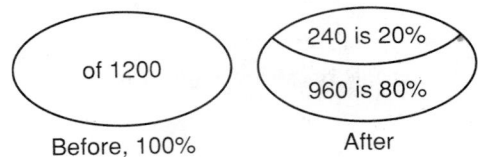

Before, 100% After

example 71.3 Twenty is what percent of 400? Draw a diagram of the problem.

solution This time we will use the fractional form of the percent equation.

$$\frac{P}{100} \times of = is$$

We replace *percent* with *WP*, replace *is* with 20, and replace *of* with 400.

$$\frac{WP}{100} \times 400 = 20$$

To solve we multiply both sides by $\frac{100}{400}$.

$$\frac{100}{400} \cdot \frac{WP}{100} \cdot 400 = 20 \cdot \frac{100}{400}$$

$$WP = 5$$

Now we draw the diagram (not exactly to scale).

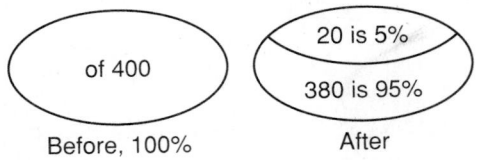

Before, 100% After

The "after" percents must add to 100 percent, so the other percent is 95 percent. **The sum of the "after" numbers must equal the "before" number.** Thus, the other "after" number must be 380, as shown.

practice **a.** Complete the percent diagram by finding *a* and *b*.

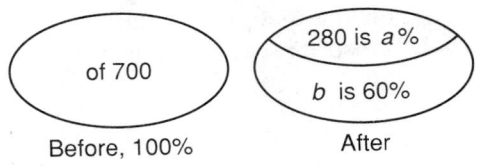

Before, 100% After

b. Complete the percent diagram by finding *a*, *b*, and *c*.

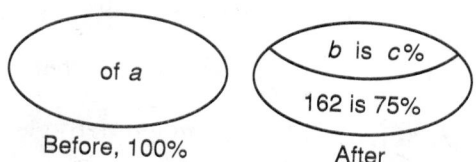

Before, 100%   After

c. Forty percent of what number is 160? Work the problem and then draw a diagram of the problem.

d. Seventy percent of 700 is what number? Work the problem and then draw a diagram of the problem.

problem set 71

1. The ratio of students who rode horses to school to students who walked to school was 2 to 9. If 18 students walked to school, how many rode horses to school?

2. Three-and-one-half times as many came as were invited. If 840 came, how many had been invited?

3. Computer disks could be purchased for $10.90 a dozen or could be purchased in cases of 60 for $80 a case. Use unit prices to tell which price was the best price.

4. Hyunah managed the first 40 miles in 8 hours. If she then doubled her speed, how long did it take her to manage the next 20 miles?

5. Complete the percent diagram by finding *a*, *b*, and *c*.

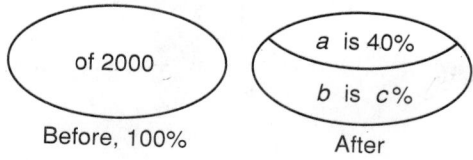

Before, 100%   After

6. Twenty percent of what number is 400? Draw a diagram of the problem.

7. Fifty-five is what percent of 275? Draw a diagram of the problem.

8. Round 0.000735492 to the hundred-thousandths' place and write in scientific notation.

9. Complete the table. Begin by inserting the reference numbers.

10. What decimal part of 300 is 120?

11. $2\frac{2}{3}$ of what number is $6\frac{1}{7}$?

Fraction	Decimal	Percent
(a)	0.30	(b)

12. Find (a) the perimeter and (b) the area of the figure on the left. Dimensions are in feet.

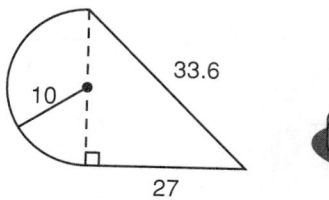
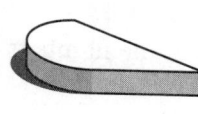

13. What is the volume of a solid whose base is the figure shown in Problem 12 and whose sides are 2 feet tall?

14. Write $\dfrac{6211}{5}$ as a mixed number.

15. Find the least common multiple of 20, 24, and 36.

Solve:

16. $\dfrac{4\frac{1}{3}}{2\frac{6}{7}} = \dfrac{7}{x}$

17. $\dfrac{\frac{4}{5}}{\frac{16}{15}} = \dfrac{x}{3}$

18. $7\frac{1}{8}x + 2\frac{3}{10} = 3\frac{2}{5}$

19. $8\frac{1}{4}x - \dfrac{5}{11} = 8\dfrac{13}{22}$

20. $18x + 7 = 122$

Simplify:

21. $14 + 2[2^2(3 - 1)(5 - 2) + 1]$

22. $2^3 + 2^2[3(2^2 - 1) + \sqrt{16}]$

23. $6\frac{1}{3} + \dfrac{2}{3} \cdot 3\frac{3}{4}$

24. $17\dfrac{5}{12} - 4\dfrac{5}{6} + 1\dfrac{3}{4}$

25. 18.21×0.0018

26. $18{,}217.81 - 876.897$

27. $\dfrac{161.17}{0.004}$

28. $\dfrac{1}{4}\left(3\frac{1}{3} + 2\frac{1}{4}\right) - \dfrac{11}{24}$

29. $3\frac{1}{4} \times 4\dfrac{5}{12} \div 3\dfrac{7}{8} \div \dfrac{1}{4}$

30. Evaluate: $x^y + y^x + \sqrt[y]{z}$ if $x = 2$, $y = 3$, and $z = 125$

LESSON 72   *Negative numbers · Absolute value · Adding signed numbers*

72.A

negative numbers   The number ray introduced in Lesson 8 is a positive number ray. It begins at the origin and points to the right.

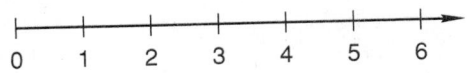

All the numbers of arithmetic can be graphed on a positive number ray.

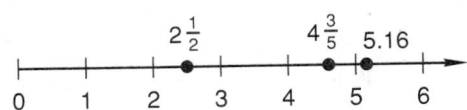

To make a **number line**, we draw the positive number ray and also draw a ray that points to the left. The two number rays form a number line.

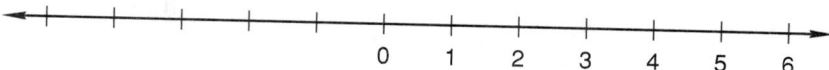

The graph of a positive number is to the right of the origin. Every positive number has an opposite whose graph is the same distance to the left of the origin. We call these numbers the **negative numbers.**

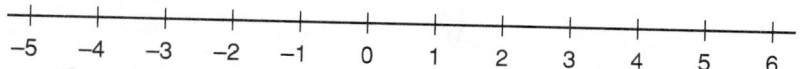

We can see that the graph of +4 is 4 units to the right of the origin.

We must use a minus sign every time we write a negative number. It is not necessary to use a plus sign to write a positive number.

$$-4 \qquad \text{means negative 4}$$
$$+4 \quad \text{and} \quad 4 \qquad \text{both mean positive 4}$$

72.B
absolute value

Numbers that have either a plus sign or a minus sign are called **signed numbers.** Signed numbers have two qualities. One of the qualities is designated by the + or − sign. This tells us if the graph of the number is to the left of the origin or to the right of the origin. The other quality is indicated by the numeral and tells us how far the graph of the number is from the origin. We call this quality the **absolute value** of the number. We use two vertical lines to designate the absolute value, as we show here.

$$|-4| = 4 \qquad |4| = 4$$

The absolute value of −4 is 4 because the graph of −4 is 4 units from the origin. The absolute value of 4 is also 4 because the graph of 4 is 4 units from the origin.

72.C
adding signed numbers

There are rules for adding signed numbers. Many authors use arrows to help explain the rules for adding signed numbers. Positive numbers are represented by arrows that point to the right. Negative numbers are indicated by arrows that point to the left.

To use these arrows to demonstrate the addition of +2 and −3, we begin at 0 on the number line and draw the +2 arrow. From the head of this arrow, we draw the −3 arrow. The head of the −3 arrow is over −1 on the number line.

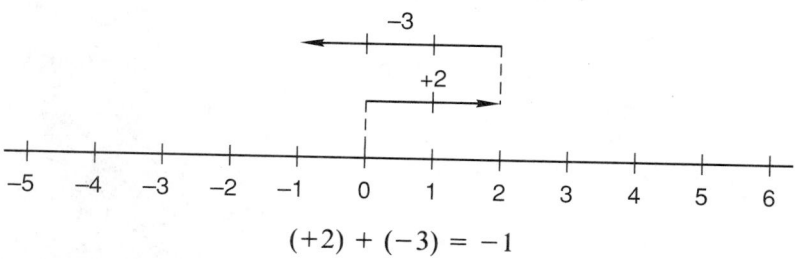

Thus,
$$(+2) + (-3) = -1$$

example 72.1   Use arrows and a number line to add -3 and $+1$.

solution   Either arrow can be drawn first. We will draw the -3 arrow first.

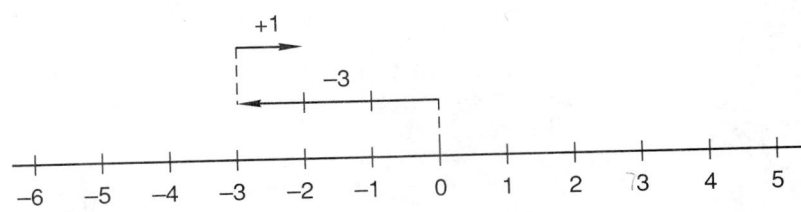

The head of the $+1$ arrow is above -2 on the line, so we say that the answer is -2.

$$(-3) + (+1) = -2$$

example 72.2   Use arrows and a number line to add $(+2) + (+1)$ and $(-2) + (-1)$.

solution   We will use a number line for each addition.

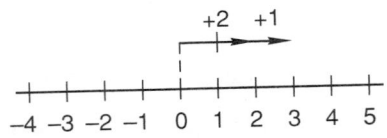   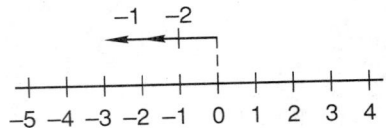

From these diagrams, we see that

$$(+2) + (+1) = +3 \quad \text{and} \quad (-2) + (-1) = -3$$

practice   Quickly sketch four number lines. Then draw arrows to add these numbers.

a. $(+2) + (-4)$ **b.** $(-2) + (+4)$

c. $(-2) + (-3)$ **d.** $(+2) + (+3)$

problem set 72

1. Twenty dozen starters cost a total of $5. What was the cost of 1 dozen starters? What would it cost to buy only 10 dozen starters?

2. Three-fifths of the clowns had solid red noses. If there were 150 clowns in all, how many had solid red noses?

3. The ratio of big spenders to the penurious ones was 2 to 5. If 1400 were big spenders, how many were penurious?

4. The entourage wended slowly through the forest. If it covered 400 yards in the first 20 minutes, how far would it go in the next hour?

5. Health insurance was offered in three plans. One year of coverage costs $762. Three years of coverage costs $1850, and five years of coverage costs $3000. Find the unit cost per year for each plan. Which insurance plan is least expensive?

6. Use arrows and number lines to add -6 and $+2$.

7. Complete the percent diagram by finding a, b, and c.

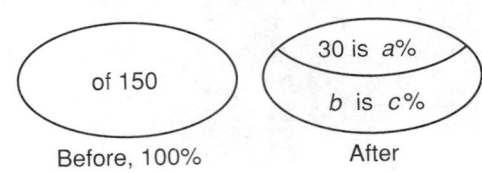

8. Thirty percent of what number is 360? Draw a diagram of the problem.

9. Seventy percent of 420 is what number? Draw a diagram of the problem.

10. Complete the table. Begin by inserting the reference numbers.

Fraction	Decimal	Percent
(a)	(b)	40

11. Round 216,348,219 to the millions' place and write in scientific notation.

12. What decimal part of 420 is 147?

13. Find (a) the perimeter and (b) the area of this figure. Dimensions are in inches.

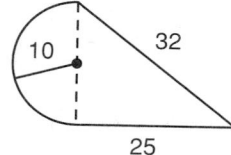 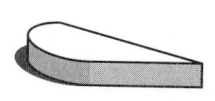

14. What is the volume of a solid whose base is the figure shown on the left in Problem 13 and whose sides are 5 inches high?

15. Write $36\frac{11}{12}$ as an improper fraction.

16. Find the least common multiple of 16, 24, and 36.

Solve:

17. $\dfrac{6\frac{1}{4}}{2\frac{1}{3}} = \dfrac{5}{x}$

18. $\dfrac{\frac{3}{5}}{\frac{9}{25}} = \dfrac{p}{3}$

19. $6\frac{2}{3}x - 4\frac{1}{4} = 3\frac{1}{12}$

20. $20x + 18 = 132$

Simplify:

21. $13 + 2[2^3(2^2 - 1)(3 - 1) + 1]$

22. $2^3 + 2^2[3(2^2 - 1) + \sqrt{16}]$

23. $2\frac{11}{48} - \frac{1}{6} \cdot 3\frac{3}{8}$

24. $18\frac{5}{12} - 12\frac{1}{6} + 1\frac{3}{4}$

25. 19.31×0.0021

26. $61,131.812 - 817.819$

27. $\dfrac{181.31}{0.005}$

28. $\frac{1}{5}\left(3\frac{1}{4} - 2\frac{1}{3}\right) + \frac{13}{60}$

29. $6\frac{1}{4} \times 2\frac{1}{3} \div 1\frac{5}{12} \div \frac{1}{6}$

30. Evaluate: $x^x + y^y + x^y + y^x$ if $x = 3$ and $y = 2$

LESSON 73 *Visualizing signed numbers*

We need to have a quick way to visualize the process of adding signed numbers. This will allow us to add signed numbers quickly and accurately. To do this we will think of two numbers the same distance from the origin as being a **pair of opposites.** Here we show that the graph of $+2$ and -2 are the same distance from the origin.

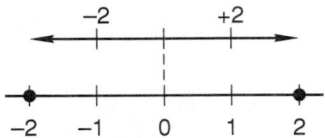

If we add a pair of opposites, the sum is zero.

$$(+2) + (-2) = 0$$

We can add two numbers whose signs are not alike by breaking one of the numbers into two parts so that we can get a pair of opposites. To add $(+4) + (-1)$, we can mentally break $+4$ into two parts and think of the problem this way.

$$\underbrace{(+3) + (+1)}_{(+4)} + (-1)$$

The $(+1)$ and (-1) are a pair of opposites. Their sum is zero. All that is left is $+3$.

$$+3 + 0 = \mathbf{3}$$

example 73.1 Visualize $(+4) + (-6)$ and add.

solution We can write -6 in two parts with one of the parts being -4 so we will have a pair of opposites.

$$(+4) + \underbrace{(-4) + (-2)}_{(-6)}$$

The sum of $+4$ and -4 is 0 and all that is left is -2.

$$(+4) + (-4) + (-2) = \mathbf{-2}$$

example 73.2 Visualize the addition of $(+2) + (4)$.

solution $(+2) + (+4) = ?$

There are no pairs of opposites because there are no negative numbers. Thus the sum of the numbers is $+6$.

$$(+2) + (+4) = +6$$

example 73.3 Visualize the addition of $(-2) + (-3)$.

solution $(-2) + (-3) = ?$

There are no pairs of opposites to cancel themselves, so the sum is -5.

$$(-2) + (-3) = \mathbf{-5}$$

example 73.4 Visualize the sum of $(+2) + (-6)$.

solution To get a pair of opposites, we think of -6 as the sum of (-2) and (-4).

$$(+2) + (-2) + (-4)$$

The first sum is zero because the $+2$ and -2 are a pair of opposites. This leaves (-4). Now we have

$$0 + (-4) = \mathbf{-4}$$

practice Find the following sums:

   a. $(-14) + (+2)$ **b.** $(+256) + (-3)$ **c.** $(41) + (-8)$

   d. $(-3) + (-2)$ **e.** $(-18) + (9)$ **f.** $(-4) + (-15)$

problem set 73

1. The result was 5400. This was $2\frac{1}{2}$ times the expected number. What was the expected number?

2. The ratio of unusual students to normal students was 15 to 2. If 3000 students were unusual, how many were normal?

3. Seven thousand, seven hundred made it on time. If the ratio of those who were on time to those who were late was 7 to 2, how many were late?

4. The rate of production of the machine was increased to 1400 bottles per hour. How many bottles could the machine produce if it operated for 8 hours?

5. Add: (a) $(-5) + (-3)$ (b) $(-8) + (4)$

6. Use two unit multipliers to convert 1,000,000 square meters to square centimeters.

7. Complete the percent diagram by finding a, b, and c.

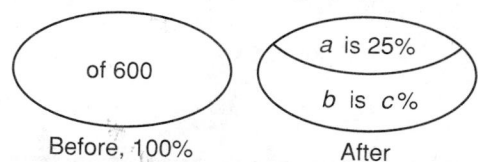

8. Eighty percent of 280 is what number? Draw a diagram of the problem.

9. Sixty is what percent of 150? Draw a diagram of the problem.

10. Complete the table. Begin by inserting the reference numbers.

11. Write 6.04×10^{-6} in standard form.

Fraction	Decimal	Percent
(a)	(b)	23

12. What decimal part of 600 is 360?

13. Find (a) the perimeter and (b) the area of the figure on the left. Dimensions are in meters. All angles are right angles.

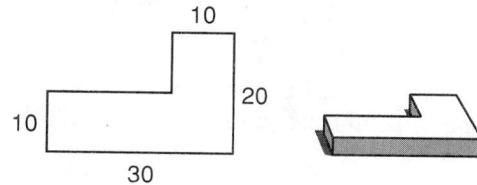

14. What is the volume of a right solid whose base is the figure shown on the left in Problem 13 and whose sides are 200 centimeters high?

15. Write $\dfrac{6213}{30}$ as a mixed number.

16. Find the least common multiple of 18, 24, and 30.

Solve:

17. $\dfrac{5\frac{1}{3}}{3\frac{1}{6}} = \dfrac{6}{x}$

18. $\dfrac{\frac{4}{5}}{\frac{8}{25}} = \dfrac{x}{5}$

19. $5\frac{1}{3}x - 2\frac{1}{4} = 4\frac{7}{12}$

20. $16x + 19 = 131$

Simplify:

21. $14 + 2[2^2(2^3 - 2^2)(3 - 1) - 1]$

22. $\sqrt[3]{8} + \sqrt{4}[(2^3 - 4)3 + 1]$

23. $3\frac{7}{24} - \frac{1}{8} \cdot 2\frac{5}{6}$

24. $19\frac{6}{7} - \frac{3}{14} + 1\frac{4}{21}$

25. 18.87×0.0032

26. $18{,}123.18 - 619.98$

27. $\dfrac{192.18}{0.006}$

28. $\frac{1}{4}\left(3\frac{1}{5} - 2\frac{1}{3}\right) + \frac{7}{60}$

29. $5\frac{1}{3} \times 2\frac{1}{2} \div 2\frac{1}{6} \div 1\frac{1}{12}$

30. Evaluate: $x^y + y^x + \sqrt[y]{x}$ if $x = 8$ and $y = 3$

LESSON 74 *Rules for addition*

When we add two numbers that have the same sign, we always get an answer that has the same sign as the sign of the numbers.

$$(2) + (3) = 5 \qquad (-2) + (-3) = (-5)$$

> The value of the sum of two numbers of the same sign is the sum of the absolute values of the numbers. The sign of the sum is the same as the sign of the numbers.

When we add two numbers whose signs are different, sometimes the answer is a positive number and sometimes the answer is a negative number.

$$(-3) + (5) = 2 \qquad (3) + (-5) = -2$$

> The value of the sum of two numbers of opposite signs
> is the difference of the absolute values of the numbers.
> The sign of the sum is the same as the sign of the
> number with the greatest absolute value.

When we must add more than two signed numbers, we can add from left to right.

example 74.1 Add these signed numbers mentally from left to right.

$$(-4) + (+3) + (-5) + (+7)$$

solution We add from left to right.

$(-4) + (+3) + (-5) + (+7)$	problem
$(-1) + (-5) + (+7)$	added (-4) and $(+3)$
$(-6) + (+7)$	added (-1) and (-5)
+1	added (-6) and $(+7)$

example 74.2 Add these signed numbers mentally from left to right.

$$(-4) + (+3) + (-2) + (-6) + (5)$$

solution We begin by adding (-4) and $(+3)$ to get (-1).

$(-1) + (-2) + (-6) + (5)$	added (-4) and $(+3)$
$(-3) + (-6) + (5)$	added (-1) and (-2)
$(-9) + (5)$	added (-3) and (-6)
−4	added (-9) and (5)

In a given problem, it is customary to write all positive numbers with a plus sign, as $(+3)$, or without the plus sign, as (3). In this example we used both notations to emphasize that both notations have the same meaning.

practice Find the following sums by adding mentally from left to right:

a. $(-4) + (-3) + (2) + (-5)$ b. $(-8) + (13) + (-5) + (-8)$

c. $(3) + (-8) + (-9) + (2) + (-4)$ d. $(-5) + (8) + (9) + (-5) + (-7)$

problem set 74

1. The white cell count was only 980. The patient was in trouble as the count should have been at least $5\frac{1}{2}$ times this number. What was the least the white cell count should have been?

2. Doctor Hal and his explorer scouts traveled 32 miles in 8 hours. The next day they had to travel 49 miles, so they increased their speed by 3 miles per hour. How long did it take them to travel the 49 miles?

3. The ratio of winners to losers in the throng was 7 to 2. If 4200 were losers, how many winners were in the throng?

4. Three-fourths of the bats were in the belfry. If 42 bats were in the belfry, how many bats were there in all?

5. Complete the percent diagrams by finding a, b, and c.

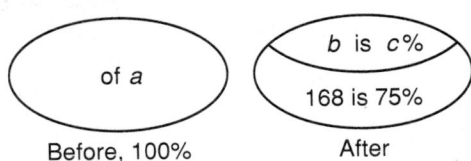

Before, 100%   After

6. Thirty-five is what percent of 175? Draw a diagram of the problem.

7. Thirty percent of what number is 123? Draw a diagram of the problem.

8. (a) What is 1 percent of 34?
   (b) What is 115 percent of 34?

9. Complete the table. Begin by inserting the reference numbers.

10. Round 0.0030796 to the ten-thousandths' place and write the resulting number in scientific notation.

Fraction	Decimal	Percent
$\frac{37}{100}$	(a)	(b)

11. What decimal part of 500 is 125?

12. Find (a) the perimeter and (b) the area of the figure on the left. Dimensions are in meters.

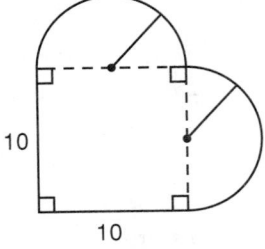
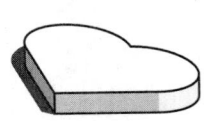

10

10

13. What is the volume of a right solid whose base is the figure shown on the left above and whose sides are 50 centimeters high? Remember to find the area of the base in square centimeters before multiplying by the height.

14. Write $16\frac{11}{12}$ as an improper fraction.

15. Find the least common multiple of 12, 15, and 25.

Solve:

16. $\dfrac{4\frac{2}{3}}{1\frac{5}{6}} = \dfrac{8}{x}$

17. $\dfrac{\frac{3}{5}}{\frac{9}{20}} = \dfrac{10}{x}$

18. $2\frac{1}{4}x + 4\frac{1}{3} = 5\frac{7}{12}$

Simplify:

19. $13 + 2[2^2(3^2 - 2^3)(2^2 + 1) - 15]$

20. $(-4) + (5) + (-10) + (+2)$

21. $(-11) + (-2) + (20) + (-7)$

22. $(12) + (-7) + (-10) + (1)$

23. $(-21) + (7) + (-2) + (6)$

24. $16\frac{7}{8} - 2\frac{1}{4} \cdot 1\frac{1}{2} + 1\frac{7}{16}$

25. 166.5×0.0024

26. $19,621.81 - 698.971$

27. $\dfrac{213.19}{0.005}$

28. $\frac{1}{5}\left(4\frac{2}{5} - 2\frac{1}{3}\right) + \frac{4}{15}$

29. Write 625361811.01 in words.

30. Evaluate: $x^y + y^x - \sqrt[x]{z}$ if $x = 3$, $y = 2$, and $z = 64$

LESSON 75   *Powers and roots of fractions*

75.A
powers of fractions

We remember that an expression that has an exponent is called an **exponential expression** or a **power**. If we write

$$2^3$$

the number 2 is the base. The number 3 is the exponent. Thus, 2^3 means

$$2 \cdot 2 \cdot 2 = 8$$

Sometimes we use the word *power* to describe the value of the exponential expression. We can say that the third power of 2 is 8. Sometimes the word power means the exponent. If we do this, we say that

$$2^3$$

is 2 raised to the third power. We can also use fractions as the base of a power.

$$\left(\frac{1}{2}\right)^3 \quad \text{means} \quad \frac{1}{2} \cdot \frac{1}{2} \cdot \frac{1}{2} = \frac{1}{8}$$

example 75.1   Simplify: (a) $\left(\frac{1}{3}\right)^2$ (b) $\left(\frac{1}{4}\right)^3$

solution   (a)   The exponent 2 tells us to use the base $\frac{1}{3}$ as a factor twice.

$$\left(\frac{1}{3}\right)^2 \quad \text{means} \quad \frac{1}{3} \cdot \frac{1}{3} = \frac{1}{9}$$

(b)   The exponent 3 tells us to use the base $\frac{1}{4}$ as a factor 3 times.

$$\left(\frac{1}{4}\right)^3 = \frac{1}{4} \cdot \frac{1}{4} \cdot \frac{1}{4} = \frac{1}{64}$$

75.B
roots of fractions

When we use the radical

$$\sqrt[3]{\text{number}}$$

we ask a question. The question is, "What number used as a factor 3 times equals the given number?" This expression

$$\sqrt[3]{\frac{1}{64}}$$

asks us to find the number that used as a factor 3 times equals $\frac{1}{64}$. This number is $\frac{1}{4}$

because

$$\frac{1}{4} \cdot \frac{1}{4} \cdot \frac{1}{4} = \frac{1}{64}$$

Thus, we can write

$$\sqrt[3]{\frac{1}{64}} = \frac{1}{4}$$

example 75.2   Simplify: $\sqrt{\frac{1}{16}}$

solution   This notation asks us to find the number which used as a factor twice gives an answer of $\frac{1}{16}$. The answer is $\frac{1}{4}$ because

$$\frac{1}{4} \cdot \frac{1}{4} = \frac{1}{16}$$

The answer is not $\frac{1}{4} \cdot \frac{1}{4}$. The answer is just $\frac{1}{4}$.

$$\sqrt{\frac{1}{16}} = \frac{1}{4}$$

example 75.3   Simplify: $\sqrt[3]{\frac{1}{27}}$

solution   If we use $\frac{1}{3}$ as a factor 3 times, the product is $\frac{1}{27}$.

$$\frac{1}{3} \cdot \frac{1}{3} \cdot \frac{1}{3} = \frac{1}{27}$$

The answer is not $\frac{1}{3} \cdot \frac{1}{3} \cdot \frac{1}{3}$. The answer is just $\frac{1}{3}$.

$$\sqrt[3]{\frac{1}{27}} = \frac{1}{3}$$

practice   Simplify:

a. $\left(\frac{1}{3}\right)^3$ b. $\sqrt[3]{\frac{1}{8}}$ c. $\left(\frac{1}{2}\right)^4$ d. $\sqrt{\frac{1}{64}}$

problem set 75

1. The receipts for the day totaled $5200. This was only three-fifths of the money needed to pay the bills. How much money was needed to pay the bills?

2. Edsel traveled the 350 miles from the factory to the River Rouge plant in 7 hours. The next day he was in a hurry to return. If he must complete the trip back to the factory in 5 hours, what should be his speed?

3. The ratio of believers to doubters at the meeting was 8 to 3. If 2400 of those in attendance were believers, how many doubters were present?

4. Four-fifths of the bees were not in the hive. If 1350 bees were in the hive, how many were not in the hive?

5. Sixty-five is what percent of 325? Draw a diagram of the problem.

6. Complete the percent diagrams by finding *a*, *b*, and *c*.

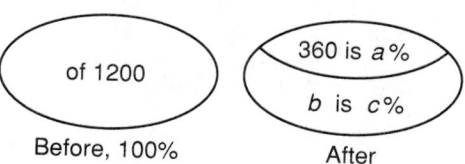

7. Complete the table. Begin by inserting the reference numbers.

8. Write 26,900,000,000 in scientific notation.

9. What decimal part of 790 is 474?

Fraction	Decimal	Percent
(a)	0.71	(b)

10. Find (a) the perimeter and (b) the area of the figure on the left. Dimensions are in yards.

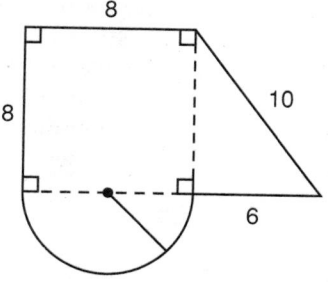

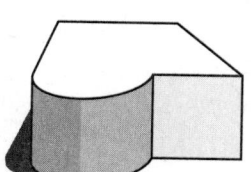

11. What is the volume of a right solid whose base is the figure shown on the left above and whose sides are 4 feet high?

12. Write 8.03×10^5 in standard form.

13. Use two unit multipliers to convert 1,000,000 inches to miles.

Simplify:

14. $\left(\dfrac{2}{3}\right)^3$

15. $\sqrt[3]{\dfrac{8}{27}}$

16. $\left(\dfrac{1}{5}\right)^3$

17. $\sqrt[4]{\dfrac{16}{256}}$

Solve:

18. $\dfrac{3\frac{1}{2}}{1\frac{2}{5}} = \dfrac{10}{m}$

19. $\dfrac{\frac{4}{7}}{\frac{11}{12}} = \dfrac{4}{p}$

20. $5\frac{1}{4}y - 2\frac{1}{2} = 4\frac{5}{12}$

Simplify:

21. $10 + 3[3^2(1^5 + 2^3)(3^2 + 1)]$

22. $(-3) + (7) + (-9) + (-1)$

23. $(-21) + (-4) + (30) + (-1)$

24. $(8) + (-9) + (6) + (-14)$

25. $(-30) + (14) + (-1) + (17)$

26. $11\frac{2}{8} - 5\frac{1}{2} \cdot 1\frac{2}{3} + 1\frac{1}{4}$

27. 312.063×0.013 **28.** $132{,}313.04 - 78.788$

29. $\dfrac{657.12}{0.0012}$ **30.** $\dfrac{1}{4}\left(3\dfrac{2}{3} - 1\dfrac{1}{4}\right) + \dfrac{6}{7}$

LESSON 76 *Graphing inequalities*

In Lesson 8 we used the positive number ray to help us understand how the positive numbers are arranged in order. Now we have added the negative number ray to form the number line. The number line can be used to help us understand how all positive and negative numbers can be arranged in order. Before algebra books began using the number line in the 1960s, many students had difficulty in understanding how -2 could be greater than -5. Now all books use the number line to help define "greater than," and we can see on the number line why we say that -2 is greater than -5.

One number is greater than another number if its graph on the number line is to the right of the graph of the other number.

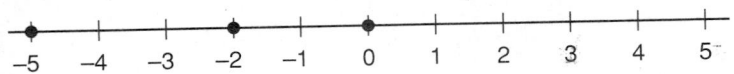

We see that -2 is greater than -5 because the graph of -2 is to the right of the graph of -5. For the same reason, we see that 0 is greater than either -5 or -2. The symbol

$$<$$

is a greater than/less than symbol. The open end is read as "greater than." The closed or pointed end is read as "less than." We read the end we come to first. We do not read the other end. The pointed end points to the smaller number. If we read this

$$2 < 4$$

from left to right, we say "2 is less than 4." If we read it from right to left, we say "4 is greater than 2." We can turn the symbol around if we turn the numbers around. This is read

$$4 > 2$$

from left to right as "4 is greater than 2" or from right to left as "2 is less than 4."

If we add half an equals sign, we indicate that **equals to** can also be part of the solution. We read

$$-4 \geq -12$$

from left to right as "-4 is greater than or equal to -12" or from right to left as "-12 is less than or equal to -4." The statement

$$4 \geq 2 + 2$$

is read from left to right as "4 is greater than or equal to $2 + 2$." The statement is read from right to left as "2 plus 2 is less than or equal to 4."

Some statements of inequality are false.

$$4 > 10 \qquad \text{false}$$

This is a false statement because 4 is not greater than 10. Some statements of inequality are conditional inequalities.

$$x > 2 \qquad \text{conditional}$$

The truth or falsity of this inequality depends on the number that is used to replace x. If the replacement number makes the inequality a true inequality, we say the number is a **solution of the inequality.** In the following examples, we show how graphs on the number line can be used to indicate the numbers that satisfy an inequality.

example 76.1 Graph: $x > 2$

solution We are asked to graph all numbers that are greater than 2.

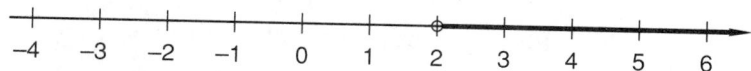

The open circle at 2 tells us that 2 is not a solution because 2 is not greater than 2. The arrow means that all numbers to the right of 2 are solutions.

example 76.2 Graph: $x \le 2$

solution This time we are asked to indicate all numbers that are less than or equal to 2.

This time the circle at 2 is a solid circle because 2 is a solution to this inequality, as well as all numbers to the left of 2.

practice Sketch four number lines quickly. Then graph each inequality.

   a. $x > -2$ **b.** $x \ge -2$ **c.** $x \le 4$ **d.** $x < -4$

problem set 76

1. Rockabilly was old hat, but some of the fans still enjoyed it. If there were 29,000 fans at the festival, what fraction enjoyed rockabilly if 3000 fans were rockabilly fans?

2. When the trumpet sounded the charge, seven-eights of the malingerers disappeared. If there were 16,000 malingerers in all, how many disappeared?

3. The stagecoach clattered and rattled but still made the 40-mile trip in 5 hours. If the speed for the next leg was increased by 4 miles per hour, how long would it take to travel the 36 miles that remained?

4. The ratio of players to onlookers was 7 to 5. If 112 were players, how many were onlookers?

Graph:

5. $x > -4$

6. $x \le 4$

7. Sixty percent of 480 is what number? Draw a diagram of the problem.

8. Forty percent of what number is 220? Draw a diagram of the problem.

9. Forty is what percent of 200? Draw a diagram of the problem.

10. Complete the table. Begin by inserting the reference numbers.

11. Simplify:

    (a) $\left(\dfrac{2}{3}\right)^4$ (b) $\sqrt{\dfrac{64}{81}}$

Fraction	Decimal	Percent
(a)	0.39	(b)

12. 0.4 of what number is 148?

13. Find (a) the perimeter and (b) the area of the figure on the left. Dimensions are in feet.

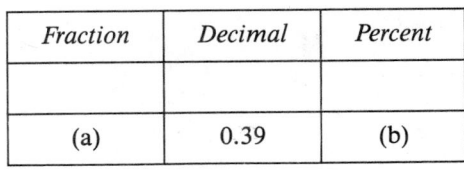

14. What is the volume of a solid whose base is the figure shown in Problem 13 and whose sides are 1 yard high?

15. Find the surface area of a cube whose edges are 8 meters long.

16. Find the least common multiple of 12, 18, and 30.

Solve:

17. $\dfrac{6\frac{1}{2}}{7\frac{1}{3}} = \dfrac{4}{x}$

18. $\dfrac{\frac{3}{7}}{\frac{9}{21}} = \dfrac{z}{3}$

19. $6\frac{1}{2}x + 3\frac{1}{4} = 5\frac{1}{8}$

Simplify:

20. $4^2 + \sqrt{16}[2^2(3^2 - 2^2)(4^2 - 3^2) - 10]$

21. $(-6) + (4) + (-10) + (+2)$

22. $(-11) + (-2) + (6) + (+3)$

23. $(-36) + (21) + (-6) + (-3)$

24. $13\frac{3}{4} - 1\frac{4}{5} \cdot \frac{3}{2} + \frac{7}{20}$

25. 124.3×0.0018

26. $19{,}611.62 - 687.91$

27. $\dfrac{218.31}{0.006}$

28. $\dfrac{1}{6}\left(2\frac{1}{3} \cdot \frac{1}{2} - \frac{3}{4}\right) + \frac{5}{24}$

29. Round 162,189,921.618 to the nearest hundred million.

30. Evaluate: $xy + x^2 + y^2 + x^y$ if $x = 4$ and $y = 2$

LESSON 77 *Right circular cylinders*

The right circular cylinder is important in science because of its structural strength. For example, the big steel oil storage tanks are all right circular cylinders. Have you ever noticed that there are no square oil storage tanks or square water towers?

Many people memorize the formulas for the volume and the surface area of a right circular cylinder. Then they forget the formulas. Problems involving right circular cylinders are easy to learn how to work without special formulas. Suppose the base of a right circular cylinder that is 20 meters tall has a radius of 2 meters. Then the area of the base is $\pi(2)^2$, which is approximately 12.56 m².

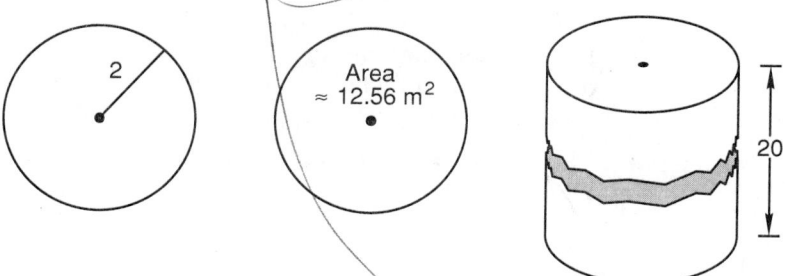

This means that if we crush 12.56 one-meter sugar cubes and pour them in the cylinder, the cylinder will be filled to a depth of 1 meter. Since the cylinder is 20 meters tall, it will take 20 × 12.56 crushed 1-meter sugar cubes to fill it completely. This leads to our definition of the volume of any right solid.

> **The volume of a right solid equals the area of the base times the height of the solid.**

To find the surface area of a right circular cylinder that is 20 meters tall and has a radius of 3 meters, we treat it like a tin can. First we cut off the top and the bottom. The area of the top equals the area of the bottom. Each area equals $\pi r^2 = \pi(3)^2 = 9\pi$.

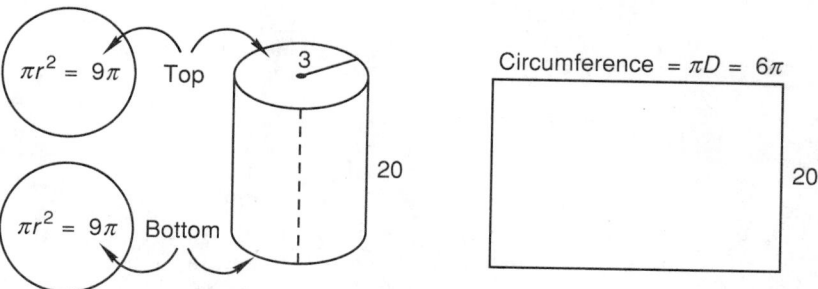

Then we cut the can down the dotted line and unroll it. We get a rectangle. The length of the rectangle is the circumference of the cylinder. The height of the rectangle is the height of the cylinder. The area of the rectangle is called the lateral surface area (the area of the sides). The total surface area is the area of the top and the bottom plus the area of the rectangle.

Surface area = area of top + area of bottom + area of rectangle

$$= \pi(3)^2 + \pi(3)^2 + (6\pi)(20)$$

$$= 9\pi + 9\pi + 120\pi = \mathbf{138\pi \ m^2}$$

practice **a.** Find the volume of a right circular cylinder that has a height of 30 feet and whose base has a radius of 10 feet.

b. Find the surface area of the same cylinder.

problem set 77 **1.** A harsh law is called a *draconian law* after the Greek law-giver Draco. Two-fifths of the laws were draconian. If 42 laws were draconian, how many laws were there in all?

2. A neophyte is a beginner. On the first day of classes 28,000 students arrived at the university campus. If three-sevenths were neophytes, how many were not neophytes?

3. Patton drove his column north and covered 60 miles in 3 hours. By how much did he need to increase the column's rate in order to complete the last 200 miles in the remaining 5 hours of daylight?

4. The ratio of the timorous to the undaunted was 5 to 17. If 200 were timorous, how many were undaunted?

Graph:

5. $x > -1$

6. $x \le 0$

7. Complete the percent diagram by finding a and b.

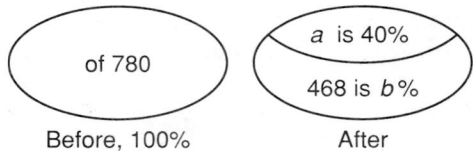

8. Forty percent of 310 is what number? Draw a diagram of the problem.

9. Seventy percent of what number is 350? Draw a diagram of the problem.

10. Complete the table. Begin by inserting the reference numbers.

11. Simplify:

    (a) $\left(\dfrac{3}{4}\right)^3$ (b) $\sqrt[4]{\dfrac{16}{81}}$

Fraction	Decimal	Percent
$\dfrac{26}{50}$	(a)	(b)

12. 0.65 of what number is 780?

13. Find the volume of this right circular cylinder in cubic meters. Dimensions are in meters. Remember that the volume is equal to the area of the base times the height.

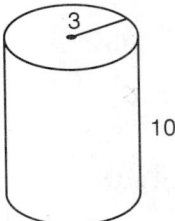

14. Find the surface area of the right circular cylinder in Problem 13. Remember that surface area is equal to the area of the top plus the area of the bottom plus the rectangular area of the unrolled side surface.

15. Round 1,039,296,119.4 to the nearest ten million and write the resulting number in scientific notation.

16. Write $6\dfrac{11}{13}$ as an improper fraction.

Solve:

17. $\dfrac{4\frac{1}{4}}{6\frac{4}{5}} = \dfrac{15}{x}$

18. $\dfrac{\frac{5}{8}}{\frac{15}{22}} = \dfrac{x}{2}$

19. $7\dfrac{1}{5}x + 2\dfrac{1}{9} = 4\dfrac{1}{3}$

Simplify:

20. $5^2 + \sqrt{64}[2^2(2 - \sqrt[9]{1})(5^2 - 4^2) + (+11)]$

21. $(-11) + (3) + (-8) + (+9)$ **22.** $(-15) + (-3) + (+14) + (5)$

23. $(-29) + (11) + (-5) + (-3)$ **24.** $10\frac{1}{6} - 2\frac{5}{6} \cdot \frac{5}{4} + \frac{11}{12}$

25. 416.09×0.0011 **26.** $15{,}342.16 - 91.091$

27. $\dfrac{419.42}{0.004}$ **28.** $\frac{1}{5}\left(3\frac{1}{2} \cdot \frac{4}{7} - \frac{11}{14}\right) + \frac{3}{28}$

29. Use one unit multiplier to convert 80,000 feet to inches.

30. Evaluate: $\sqrt[m]{p} + \dfrac{x}{\sqrt{p}}$ if $p = 16$, $m = 4$, and $x = 3$

LESSON 78 *Inserting parentheses · Order of addition*

78.A
inserting
parentheses

In arithmetic a minus sign always means to subtract.

$$5 - 3 = 2$$

This expression tells us to subtract 3 from 5. In algebra this expression means to add -3 to 5. It means this:

$$5 + (-3) = 2$$

The answer is the same. The thought process is different. It is important to learn the thought process of algebraic addition because it will help us to solve hard problems. If a problem does not have parentheses, the parentheses can be inserted as the first step.

If we have this problem

$$-4 + 3 - 2 - 6$$

we consider that the signs tell if the numbers are positive or negative. We enclose each number and its sign in parentheses and insert plus signs in between the parentheses.

$$(-4) + (+3) + (-2) + (-6)$$

Now we simplify by adding the signed numbers.

$$
\begin{array}{ll}
(-1) + (-2) + (-6) & \text{added } (-4) \text{ and } (+3) \\
(-3) + (-6) & \text{added } (-1) \text{ and } (-2) \\
\mathbf{-9} & \text{added } (-3) \text{ and } (-6)
\end{array}
$$

example 78.1 Add: $-2 + 3 - 5 + 7$

solution First we use parentheses and insert plus signs between the parentheses.

$$(-2) + (+3) + (-5) + (+7)$$

Now we add.

$$(+1) + (-5) + (+7) \qquad \text{added } (-2) \text{ and } (+3)$$
$$(-4) + (+7) \qquad \text{added } (+1) \text{ and } (-5)$$
$$+3 \qquad \text{added } (-4) \text{ and } (+7)$$

example 78.2 Add: $-3 - 2 - 6 + 2 - 7$

solution We insert parentheses and plus signs as necessary.

$$(-3) + (-2) + (-6) + (+2) + (-7)$$

Now we add.

$$(-5) + (-6) + (+2) + (-7) \qquad \text{added } (-3) \text{ and } (-2)$$
$$(-11) + (+2) + (-7) \qquad \text{added } (-5) \text{ and } (-6)$$
$$(-9) + (-7) \qquad \text{added } (-11) \text{ and } (+2)$$
$$-16 \qquad \text{added } (-9) \text{ and } (-7)$$

78.B
order of addition

Numbers that are enclosed in parentheses can be added in any order. Some people begin by adding the positive numbers together, and they then add the negative numbers together. Then they add the two sums.

example 78.3 Add: $-3 - 2 + 3 - 4 + 5$

solution We begin by inserting the necessary parentheses.

$$(-3) + (-2) + (+3) + (-4) + (+5)$$

Now we add numbers whose signs are the same and get

$$(-9) + (8) \qquad \text{combined positive numbers and negative numbers}$$
$$-1 \qquad \text{added } (-9) \text{ and } (8)$$

practice Use parentheses to enclose each number and its sign. Then insert plus signs between the parentheses. Then add.

   a. $-4 + 6 - 1 + 7 - 3 + 2$ **b.** $-2 - 4 - 3 + 2 + 3 + 2 + 4$

problem set 78

1. Three-sevenths of the cherubs had angelic faces. If there were 420,000 cherubs in the firmament, how many did not have angelic faces?

2. The ratio of chimeras to gargoyles was 2 to 11. If there was a total of 380 chimeras, what was the total number of gargoyles?

3. Jean traveled the first 100 miles in 5 hours. Then she doubled her speed for the last part of the trip. If the total trip was 300 miles, how many hours did it take Jean to travel the last part?

4. Four dozen roses cost $24. Homeratho could afford only 6 roses. How much did he pay for them?

Graph:

5. $x \le 3$

6. $x > -3$

7. Complete the percent diagram by finding a, b, and c.

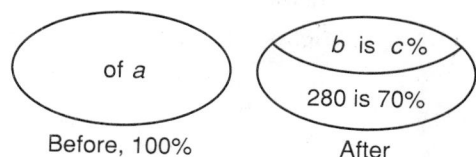

8. Eighty percent of 620 is what number? Draw a diagram of the problem.

9. Sixty is what percent of 300? Draw a diagram of the problem.

10. Complete the table. Begin by inserting the reference numbers.

11. Write 1.03×10^{10} in standard form.

Fraction	Decimal	Percent
(a)	(b)	47

12. $2\frac{1}{4}$ of what number is $5\frac{3}{8}$?

13. Find the volume in cubic inches of a right solid whose base is the figure shown on the left and whose sides are 1 foot high. Dimensions are in inches.

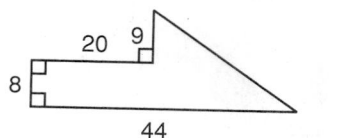 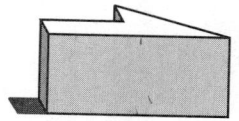

14. What is the surface area of a rectangular solid whose length is 10 feet, whose width is 5 feet, and whose height is 2 feet?

15. Find the volume of this right circular cylinder. Dimensions are in feet.

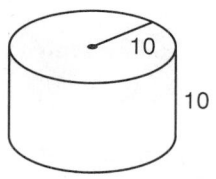

16. Find the least common multiple of 10, 12, and 18.

Solve:

17. $\dfrac{1\frac{2}{3}}{6\frac{3}{4}} = \dfrac{\frac{3}{4}}{x}$

18. $\dfrac{\frac{4}{7}}{\frac{8}{28}} = \dfrac{p}{6}$

19. $5\frac{1}{2}x + 2\frac{1}{3} = 4\frac{5}{6}$

Simplify:

20. $3^2 + \sqrt{25}[2^2(2^2 - 3)(5^2 - 4^2) - 20]$

21. $-8 + 13 - 6 + 9$

22. $-4 - 6 + (-3) + 1 - 6$

23. $-6 + (-3) + 4 - 5 + 10$

24. $-2 + (-4) + 5 - 6 + (-7)$

25. $16\frac{9}{20} - 2\frac{1}{2} \cdot 3\frac{2}{5} + \frac{9}{10}$

26. 19.81×0.0061

27. $3689.87 - 98.176$ 28. $\dfrac{198.71}{0.005}$

29. $\dfrac{1}{8}\left(3\dfrac{1}{2} \cdot \dfrac{1}{2} - \dfrac{4}{5}\right) + \dfrac{9}{40}$

30. Evaluate: $\sqrt{x} + x^y + xy$ if $x = 4$ and $y = 3$

LESSON 79 *Implied ratios*

We remember that a ratio is a comparison of two numbers. Ratios are often written in the form of fractions. We also remember that a proportion is a statement that two ratios are equal. These are equal ratios.

$$\frac{3}{4} = \frac{9}{12}$$

The equation is called a **proportion**. Most ratio word problems do not use the word ratio. When we read the problem, we must recognize that the problem is a ratio problem. We must also be able to pick out the **implied ratio.**

example 79.1 It took $2\frac{1}{2}$ eggs to make 140 cookies. Jenny wanted to make 1680 cookies. How many eggs did she need?

solution This problem is a ratio problem about eggs and cookies. We decide to put eggs on top.

$$\frac{E}{C} = \frac{E}{C}$$

The first sentence in the problem gives us the implied ratio. It tells us that the ratio of eggs to cookies is $2\frac{1}{2}$ to 140. We make the substitution.

$$\frac{2\frac{1}{2}}{140} = \frac{E}{C}$$

We want to make 1680 cookies, so we use 1680 for C and solve for E.

$$\frac{2\frac{1}{2}}{140} = \frac{E}{1680} \qquad \text{substituted}$$

$$\frac{5}{2} \cdot 1680 = 140E \qquad \text{cross multiplied}$$

$$4200 = 140E \qquad \text{simplified}$$

$$30 = E \qquad \text{divided by 140}$$

It will take **30 eggs** to make 1680 cookies.

example 79.2 It takes 3 tons of fertilizer to fertilize 170 acres. Farmer Brown wants to fertilize 1870 acres. How many tons of fertilizer does he need?

solution This problem concerns ratios of tons and acres. We decide to put tons on top.

$$\frac{T}{A} = \frac{T}{A}$$

The first sentence in the problem gives us the implied ratio. It says that the ratio of tons to acres is 3 to 170. We substitute these numbers on the left. On the right we substitute 1870 for A. Then we solve for T.

$$\frac{3}{170} = \frac{T}{1870} \qquad \text{substituted}$$

$$3 \cdot 1870 = 170T \qquad \text{cross multiplied}$$

$$\frac{3 \cdot 1870}{170} = T \qquad \text{divided by 170}$$

$$33 = T \qquad \text{simplified}$$

It will take **33 tons** of fertilizer to fertilize 1870 acres.

practice The baker found that it took 4 huge measures of sugar to make 13 confections. The baker needed to make 143 confections. How many huge measures of sugar were needed?

problem set 79

1. The shop uses 65 gallons of oil every 5 weeks. At this rate, how many gallons will the shop need for 1 year (52 weeks)?

2. The recipe for 4 cakes required 6 eggs. How many eggs will be used to make 86 cakes?

3. The ratio of the number that were mundane to the number that were exotic was 11 to 9. If there were 121,000 mundane ones, how many were exotic?

4. Five criterion team bikes cost $11,300. How much would the club have to pay for a dozen criterion team bikes?

Graph:

5. $x \leq -4$

6. $x > 0$

7. Complete the percent diagram by finding a, b, and c.

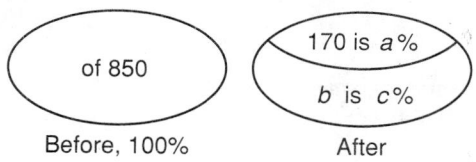

of 850		170 is a%
Before, 100%		b is c%
		After

8. Twenty-six percent of 900 is what number? Draw a diagram of the problem.

9. Eighty is what percent of 1600? Draw a diagram of the problem.

10. Complete the table. Begin by inserting the reference numbers.

11. Write 103,000,000,000 in scientific notation.

Fraction	Decimal	Percent
(a)	(b)	73

12. $3\frac{1}{2}$ of what number is 4200?

13. Find the volume in cubic meters of a right solid whose base is the figure shown on the left and whose sides are 5 meters high. Dimensions are in meters.

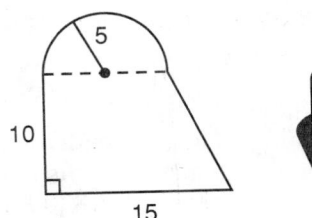 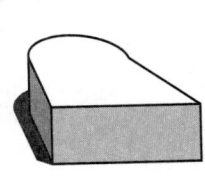

14. Find the surface area of a right circular cylinder whose diameter is 4 meters and whose height is 10 meters.

15. Write $\dfrac{1621}{7}$ as a mixed number.

16. Find the least common multiple of 5, 25, and 80.

Solve:

17. $\dfrac{2\frac{1}{2}}{5\frac{1}{3}} = \dfrac{8}{x}$

18. $\dfrac{\frac{3}{8}}{5\frac{2}{5}} = \dfrac{\frac{1}{9}}{p}$

19. $4\frac{1}{8}m + 1\frac{1}{5} = 5\frac{2}{3}$

Simplify:

20. $2^3 + \sqrt{25}\,[3^2(2 - 1^{11})(2^2 - \sqrt[5]{1}) + (-9)]$

21. $-4 + 9 - 11 + 3$

22. $-7 - 5 + (-3) + 1 - 2$

23. 203.641×0.00055

24. $\dfrac{78.045}{0.06}$

25. $83,041.76 - 76.0915$

26. $\dfrac{1}{4}\left(2\frac{1}{3} \cdot \frac{1}{4} + \frac{5}{6}\right) + \dfrac{7}{9}$

27. $\left(\dfrac{3}{4}\right)^3$

28. $\sqrt{\dfrac{169}{225}}$

29. Use one unit multiplier to convert 1,000,000 meters to centimeters.

30. Evaluate: $\sqrt[m]{p} + p^x - x^m$ if $m = 3$, $p = 27$, and $x = 1$

LESSON 80 *Multiplication with scientific notation*

To multiply

$$4. \times 1000 \times 100$$

we begin by multiplying 4 by 1000. When we do this, we move the decimal point 3 places to the right. Now we have

$$4000. \times 100$$

When we do the next multiplication, we move the decimal point 2 more places to the right and get

400,000

Multiplying by 1000 moved the decimal point 3 places. Multiplying by 100 moved it 2 more places. We moved the decimal point a total of 5 places to the right. If we use scientific notation to do the same problem, we get

$$4 \times 10^3 \times 10^2$$

We can think of 10^3 and 10^2 as **decimal point movers**. The 10^3 moves the decimal point 3 places to the right. The 10^2 moves the decimal point 2 places to the right. This is a total of 5 places to the right. So this means that

$$4 \times 10^3 \times 10^2 \qquad \text{equals} \qquad 4 \times 10^5$$

> To multiply numbers in scientific notation, we multiply the numbers and we add the exponents of the decimal point movers algebraically.

---

example 80.1   Multiply: $(4 \times 10^5) \times (2 \times 10^{-2})$

solution   As the first step we put the numbers together and put the decimal point movers together. We now have

$$(4 \times 2) \times (10^5 \times 10^{-2})$$

Now we multiply 4 by 2 and get 8.

$$8 \times (10^5 \times 10^{-2})$$

The first decimal point mover says move the decimal point to the right 5 places. The second decimal point mover says move the decimal point to the left 2 places. This is a net move of 3 places to the right. Our answer is

$$\mathbf{8 \times 10^3}$$

example 80.2   Multiply: $(3 \times 10^8) \times (3 \times 10^3)$

solution   First we rearrange the factors and get

$$3 \times 3 \times 10^8 \times 10^3$$

Now we multiply the numbers and add the exponents of the decimal point movers. We get

$$\mathbf{9 \times 10^{11}}$$

example 80.3   Multiply: $(4 \times 10^{-6}) \times (2 \times 10^4)$

solution   We multiply the two numbers. Then we add the exponents of the decimal point movers algebraically. We get

$$\mathbf{8 \times 10^{-2}}$$

practice   Multiply and simplify.

a. $(4 \times 10^{-4}) \times (2 \times 10^{14})$ b. $(3 \times 10^6) \times (3 \times 10^{-2})$

c. $(2 \times 10^{10}) \times (2 \times 10^3)$ d. $(2 \times 10^{-5}) \times (3 \times 10^{-8})$

**problem set
80**

1. Ingrid used 2 eggs to make 153 cookies. How many eggs would she need to make 1071 cookies?

2. The ratio of noisy ones to quiet ones was 13 to 5. If 3900 were noisy, how many were quiet?

3. Twain traveled the first 200 miles to Calaveras in 4 hours. Then, in the mountains, he slowed to half that speed. If the total trip was 300 miles, how many hours did it take Twain to complete the trip?

4. When the spelunkers entered the cave, four-fifths of them got excited. If there were 1650 spelunkers in all, how many did not get excited?

Graph:

5. $x \geq -2$ 6. $x < 3$

7. Forty-eight percent of what number is 110,592? Draw a diagram of the problem.

8. Sixty-eight is what percent of 340? Draw a diagram of the problem.

9. Complete the percent diagram by finding a, b, and c.

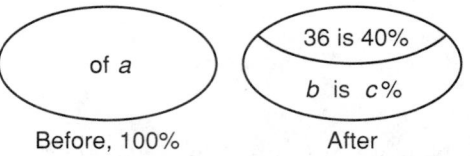

10. Complete the table. Begin by inserting the reference numbers.

Fraction	Decimal	Percent
(a)	0.28	(b)

Simplify:

11. $(1 \times 10^{-2}) \times (6 \times 10^{8})$

12. $(3 \times 10^{-15}) \times (3 \times 10^{3})$

13. $(3 \times 10^{7}) \times (2 \times 10^{2})$ 14. $(4 \times 10^{-9}) \times (2 \times 10^{5})$

15. $\left(\dfrac{3}{8}\right)^{2}$ 16. $\sqrt[5]{\dfrac{1}{32}}$

17. Use two unit multipliers to convert 90,000 square miles to square feet.

18. Write 108,000,000,000,000 in scientific notation.

19. Find the volume in cubic feet of a right solid whose base is the figure shown on the left and whose sides are 8 feet high. Dimensions are in feet.

20. Find the volume and surface area of the right circular cylinder shown. Dimensions are in inches.

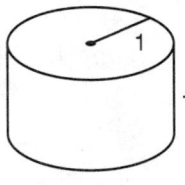

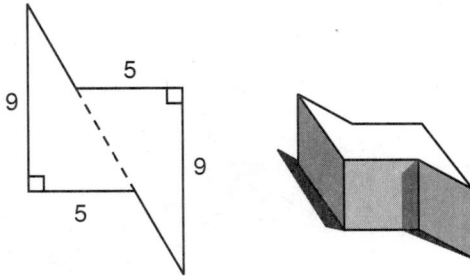

Solve:

21. $\dfrac{3\frac{1}{5}}{1\frac{1}{2}} = \dfrac{\frac{8}{5}}{x}$

22. $\dfrac{\frac{4}{7}}{1\frac{2}{14}} = \dfrac{m}{\frac{2}{9}}$

23. $11\frac{1}{4}x - 3\frac{1}{2} = \frac{1}{9}$

Simplify:

24. $3^2 + \sqrt{121}[4^3(5 - 3)(3^2 - 5) + (-400)]$

25. $-41 + (-9) + 6 + 14$

26. $-9 + (-4) + 3 + 14$

27. 114.09×0.03

28. $\dfrac{623.89}{0.02}$

29. $11{,}314.091 - 91.3621$

30. Evaluate: $x^m + xm + \sqrt{m}$ if $x = 2$ and $m = 4$

LESSON 81 *Percents greater than 100*

In percent problems the initial amount always equals 100 percent. If the amount increases, then the final percent will be greater than 100 percent. **In these problems, the "after" diagram will always be larger than the "before" diagram and will not be divided.** The diagram will look like this:

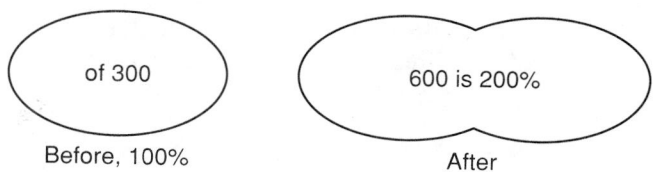

Before, 100%   After

We get this type of diagram whenever:

1. The percent is greater than 100%.
2. The *is* number, which is the "after" number, is greater than the *of* number.

example 81.1 For which of these problems will the "after" diagram be a two-part diagram? For which will the "after" diagram be a big one-part diagram?

(a) Forty percent of what number is 160?
(b) What percent of 140 is 60?
(c) What percent of 60 is 140?
(d) One hundred forty percent of 20 is what number?
(e) What percent of 160 is 280?

solution (a) The after diagram is a **two-part diagram** because the percent is less than 100 percent.

(d) The after diagram is a **one-part diagram** because the percent is greater than 100 percent.

(b, c, e)   These are the troublesome statements because the percent is not stated.

(b)   The *is* number, which is the after number, is less than the *of* number. The after diagram is a **two-part diagram.**

(c)   The *is* number, which is the after number, is greater than the *of* number. The after diagram is a **one-part diagram.**

(e)   The *is* number, which is the after number, is greater than the *of* number. The after diagram is a **one-part diagram.**

example 81.2   What number is 140 percent of 80? Find the number and then draw a diagram that depicts the problem.

solution   This time we will work the problem twice. On the left we use the fractional format and on the right we use the ratio format.

FRACTIONAL FORMAT   RATIO FORMAT

$$\frac{P}{100} \times of = is \qquad\qquad \frac{P}{100} = \frac{is}{of}$$

Now we make the necessary substitutions and solve.

$$\frac{140}{100} \cdot 80 = WN \qquad\qquad \frac{140}{100} = \frac{WN}{80}$$

$$\frac{11,200}{100} = WN \qquad\qquad 140 \cdot 80 = 100\,WN$$

$$112 = WN \qquad\qquad \frac{11,200}{100} = \frac{100\,WN}{100}$$

$$\mathbf{112 = WN}$$

The after diagram must be larger than the before diagram because 112 is greater than 80. Thus, the after diagram has only one part.

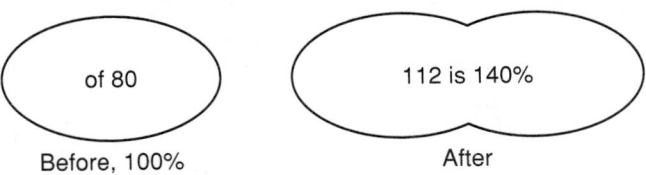

Before, 100%   After

example 81.3   What percent of 40 is 160?

solution   **It is very important to remember that the *of* number is the before number and that the *is* number is the after number. Whenever the *is* number is greater, we will get a percent greater than 100% on the right and a one-part diagram on the right.** We will use the fractional format this time.

$$\frac{P}{100} \times of = is$$

We substitute *WP* for *P*, 40 for *of*, and 160 for *is*.

$$\frac{WP}{100} \times 40 = 160$$

To solve, we multiply both sides by 100 over 40 and cancel on the left.

$$\frac{\cancel{100}}{\cancel{40}} \cdot \frac{WP}{\cancel{100}} \times \cancel{40} = 160 \times \frac{100}{40}$$

$$\mathbf{WP = 400\%}$$

Our diagram is

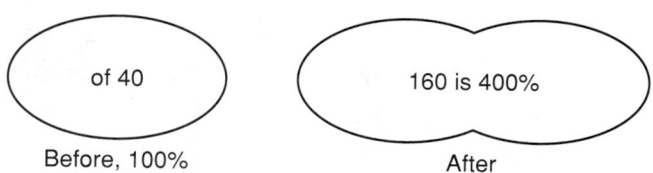

We note that when the second number is greater, the percent is greater than 100 percent.

example 81.4   One hundred fifty percent of what number is 300? Work the problem and draw the diagram.

solution   The percent is greater than 100 percent, so the diagram on the right will be larger than the one on the left. We will use the ratio format.

$$\frac{P}{100} = \frac{is}{of}$$

We replace *P* with 150, replace *of* with *WN*, and replace *is* with 300.

$$\frac{150}{100} = \frac{300}{WN} \qquad \text{substitution}$$

Now we cross multiply and solve.

$$150\,WN = (100)(300) \qquad \text{cross multiplied}$$

$$WN = \frac{100 \times 300}{150} \qquad \text{divided both sides by 150}$$

$$\boldsymbol{WN = 200} \qquad \text{simplified}$$

```
   of 200          300 is 150%
 Before, 100%          After
```

practice   **a.**   Which of these problems will have a one-part diagram on the right?

1.   Forty percent of what number is 240?
2.   One hundred fifty percent of what number is 240?
3.   What percent of 40 is 240?
4.   What percent of 240 is 40?
5.   What percent of 20 is 80?

b.   What percent of 20 is 80? Work the problem and draw the diagram.

c.   One hundred forty percent of what number is 364? Work the problem and draw the diagram.

problem set 81   **1.**   From her turret the chatelaine could see 440 sheep in the meadow. If she could only see four-ninths of the sheep, how many sheep were there in all?

2.   The ratio of brigands to highway robbers skulking in the shadows was 5 to 7. If 450 brigands were skulking in the shadows, how many highway robbers were there?

3. Falstaff ate 160 ounces of comestibles and bragged that he could have eaten $3\frac{1}{2}$ times that much. How many ounces did Falstaff think he could have eaten?

4. Jana had to travel 480 kilometers before nightfall. Her speed would be 20 kilometers per hour. How long would the trip take?

Graph:

5. $x \geq -2$ 6. $x < 5$

7. Complete the percent diagram by finding a.

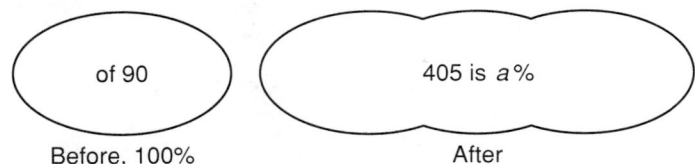

8. What number is 160 percent of 90? Find the number and then draw a diagram of the problem.

9. Two hundred fifty percent of what number is 80?

10. Complete the table. Begin by inserting the reference numbers.

11. Use two unit multipliers to convert 12 square miles to square feet.

Fraction	Decimal	Percent
(a)	0.12	(b)

12. Find the volume of a right solid whose base is the figure shown on the left and whose sides are 2 yards high. Dimensions are in feet.

13. What is the surface area of this right prism? Dimensions are in feet.

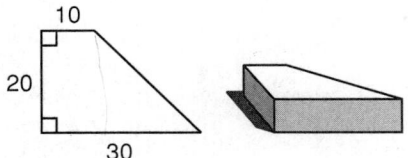

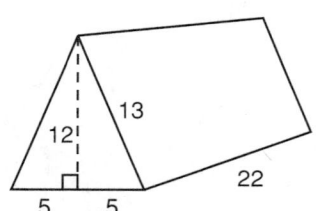

14. Write $6\frac{12}{29}$ as an improper fraction.

15. Find the least common multiple of 12, 18, and 27.

Solve:

16. $\dfrac{1\frac{3}{4}}{3\frac{3}{8}} = \dfrac{\frac{1}{3}}{x}$ 17. $\dfrac{\frac{5}{6}}{\frac{25}{36}} = \dfrac{p}{12}$ 18. $6\frac{1}{3}x + 1\frac{5}{6} = 7\frac{2}{3}$

Simplify:

19. $2^3 + \sqrt{36}[3^2(3^2 - 2^3)(4^2 - 2^4) + 3]$

20. $-5 + 6 + (-5) + 7$ 21. $-4 - 6 + 10 + (+6)$

22. $-2 - 2 + (-3) - 6 + (+3)$ 23. $-6 + (-3) - 6 - 10 + 1$

24. $17\frac{3}{5} - 1\frac{3}{5} \cdot \frac{1}{2} + 3\frac{1}{5}$ 25. 0.1315×0.0012

26. $(5 \times 10^{-15}) \times (1 \times 10^{8})$

27. $(1 \times 10^{-27}) \times (4 \times 10^{-20})$

28. $(2 \times 10^{-8}) \times (3 \times 10^{-8})$

29. Evaluate: $xyz + yz + y^z$ if $x = \frac{1}{3}$, $y = 6$, and $z = 2$

LESSON 82 *Multiplication and division of signed numbers*

It is easy to memorize the rules for multiplying and dividing signed numbers. These rules are difficult to explain but are not difficult to use. In this book, we will concentrate on learning how to use the rules.

The rules for the signs of the answers are the same for both multiplication and division.

RULES FOR BOTH MULTIPLICATION AND DIVISION

1. **The answer is a positive number if the signs of the numbers are alike.**
2. **The answer is a negative number if the signs of the numbers are different.**

example 82.1 Simplify: (a) $(4)(2)$ (b) $(-4)(-2)$ (c) $(-4)(2)$ (d) $(4)(-2)$

solution The answer to (a) and (b) is $+8$ because both numbers have the same sign. **Note that (-4) times (-2) equals $+8$.**

$$\text{(a)} \quad (4)(2) = 8 \qquad \text{(b)} \quad (-4)(-2) = 8$$

The answers to (c) and (d) are -8 because the numbers have different signs.

$$\text{(c)} \quad (-4)(2) = -8 \qquad \text{(d)} \quad (4)(-2) = -8$$

example 82.2 Simplify: (a) $\frac{8}{2}$ (b) $\frac{-8}{-2}$ (c) $\frac{8}{-2}$ (d) $\frac{-8}{2}$

solution The answers to (a) and (b) are both positive because the numbers have the same sign. **Note that a negative number divided by a negative number equals a positive number.**

$$\text{(a)} \quad \frac{8}{2} = 4 \qquad \text{(b)} \quad \frac{-8}{-2} = 4$$

The answers to (c) and (d) are both negative because the signs of the numbers are different.

$$\text{(c)} \quad \frac{8}{-2} = -4 \qquad \text{(d)} \quad \frac{-8}{2} = -4$$

practice Simplify:

a. $(4)(-3)$ b. $(-3)(4)$ c. $(-4)(-3)$ d. $(4)(3)$

e. $\dfrac{4}{-8}$ f. $\dfrac{-14}{-7}$ g. $\dfrac{-14}{7}$ h. $\dfrac{14}{-7}$

problem set 82

1. Four-fifths of the artichokes had disfiguring growths. If 500 did not have these growths, how many artichokes were there in all?

2. Chicanery was rampant at the medicine show. If seven-eighths of the customers were chicaned and there were 3200 customers, how many were not chicaned?

3. The demagogue pandered to five-sixths of those who were prejudiced. If he pandered to 1200 who were prejudiced, how many were prejudiced in all?

4. After traveling 120 miles in 3 hours, Rebecca reduced the speed of the sled by 10 miles per hour. How long did it take her to travel the next 450 miles?

Graph:

5. $x \geq -3$

6. $x < 4$

7. Complete the percent diagram by finding a.

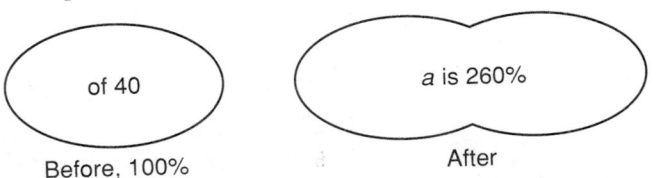

of 40

Before, 100%

a is 260%

After

8. What number is 180 percent of 60? Draw a diagram of the problem.

9. What number is 65 percent of 300? Draw a diagram of the problem.

10. What percent of 80 is 200? Draw a diagram of the problem.

11. Complete the table. Begin by inserting the reference numbers.

12. Write 108,000 in scientific notation.

13. What decimal part of 480 is 360?

Fraction	Decimal	Percent
$\dfrac{29}{100}$	(a)	(b)

14. Express in cubic centimeters the volume of a right solid whose base is the figure shown on the left and whose sides are 1 meter high. Dimensions are in centimeters.

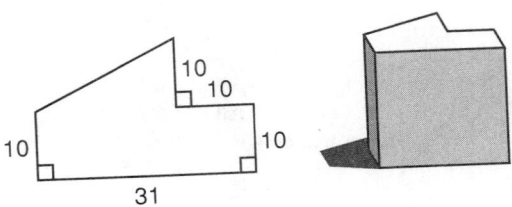

15. Find the perimeter of the figure shown. Dimensions are in feet.

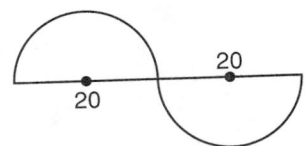

16. Write $\dfrac{12,311}{5}$ as a mixed number.

17. Find the least common multiple of 12, 20, and 30.

Solve:

18. $\dfrac{2\frac{6}{7}}{1\frac{1}{14}} = \dfrac{\frac{1}{2}}{y}$

19. $\dfrac{\frac{6}{5}}{\frac{12}{25}} = \dfrac{5}{p}$

20. $2\frac{1}{5}x + 2\frac{1}{10} = 6\frac{3}{20}$

Simplify:

21. $2^4 + \sqrt{25}\,[2^2(3^2 - 2^3)(2^2 + 1) - 1]$

22. (a) $(6)(3)$ (b) $(-6)(-3)$ (c) $(-6)(3)$ (d) $(6)(-3)$

23. (a) $(6)(2)$ (b) $(-6)(-2)$ (c) $(-6)(2)$ (d) $(6)(-2)$

24. (a) $\dfrac{6}{2}$ (b) $\dfrac{-6}{-2}$ (c) $\dfrac{6}{-2}$ (d) $\dfrac{-6}{2}$

25. $-6 + (-2) - 5 - 4 + 3$

26. $-7 + 3 - 31 + (+3)$

27. $(7 \times 10^{14}) \times (1 \times 10^{-12})$

28. 0.1319×0.0014

29. $\dfrac{1}{6}\left(2\frac{1}{3} \cdot \frac{1}{2} - \frac{1}{6}\right) + \dfrac{13}{36}$

30. Express 2540 as the product of primes.

31. Evaluate: $\sqrt[3]{x} + \sqrt{y} + xy$ if $x = 8$ and $y = 16$

LESSON 83 *Algebraic addition · Using mental parentheses*

83.A
algebriac addition

If we have to simplify the expression

$$8 - 5$$

there are two ways to do it. On the left below we let the minus sign indicate subtraction. We simplify by subtracting 5 from 8.

$$8 - 5 = 3 \quad \text{subtracted} \qquad (8) + (-5) = 3 \quad \text{added}$$

On the right we let the minus sign indicate that -5 is a negative number. We inserted parentheses and then we added. We call this procedure **algebraic addition.** It might seem that we are making a hard problem out of an easy problem. This is not so because in a later lesson we will discuss the simplification of complicated expressions such as

$$-(-4) + \{-[-(-2)]\}$$

These problems are easy to simplify if we use algebraic addition but are much more difficult if we use subtraction. In problems that involve signed numbers, we should try to avoid the use of the word subtraction because using the word subtraction often leads to trouble. Algebraic subtraction is defined, however, and some people use algebraic subtraction. The definition of algebraic subtraction is given here.

> ALGEBRAIC SUBTRACTION
>
> To subtract algebraically, we change the sign of the subtrahend and add.

To use this rule to simplify

$$(8) - (+5)$$

we change the sign of the 5 from + to − and add instead of subtracting.

$$(8) + (-5) = 3$$

This is exactly the same process we use for algebraic addition except that this time we used the definition of subtraction to explain what we did.

83.B
using mental parentheses

Writing the parentheses is not necessary. When we see an expression such as

$$-4 + 2 - 5 - 7$$

we consider that the signs go with the numbers. We can look at this problem and think

$$(-4) + (+2) + (-5) + (-7)$$

but not write down the parentheses.

example 83.1   Simplify: $-3 - 2 + 6 - 4$

solution   We insert the parentheses mentally and do not write them down. Then we add.

$$-5 + 6 - 4 \qquad \text{added } -3 \text{ and } -2$$
$$1 - 4 \qquad \text{added } -5 \text{ and } +6$$
$$-3 \qquad \text{added } 1 \text{ and } -4$$

practice   Simplify by using algebraic addition. Do not write the parentheses.

a. $-4 + 2 - 3 - 7 - 8 + 1$ **b.** $-7 - 3 - 2 - 4 + 8$

problem set 83

1. The ratio of the bumptious to the gracious was 7 to 3. If 1400 were bumptious, how many were gracious?

2. The Visigoths executed only $\frac{3}{11}$ of the number the Vandals executed. If the Visigoths executed 2202, how many did the Vandals execute?

3. Five hours was allotted for the trip. If Apocrypha's top speed was 77 miles per hour, how far could she go in the time allotted?

4. There was not enough food to go around because gate crashers had swelled the total to $2\frac{1}{2}$ times the number expected. If 1200 came, how many had been expected?

Graph:

5. $x > -2$

6. $x \leq 4$

7. What number is 175 percent of 220? Draw a diagram of the problem.

8. Complete the percent diagram by finding a.

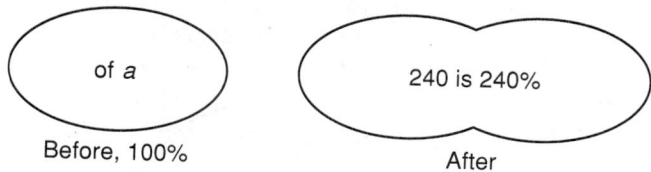

of a

240 is 240%

Before, 100%   After

9. What percent of 60 is 105? Draw a diagram of the problem.

10. Complete the table. Begin by inserting the reference numbers.

11. Write 9.1×10^{-11} in standard form.

Fraction	Decimal	Percent
(a)	0.37	(b)

12. $3\frac{1}{2}$ of what number is $4\frac{1}{4}$?

13. Find the volume in cubic centimeters of a right solid whose base is shown on the left and whose sides are 2 meters high. Dimensions are in centimeters.

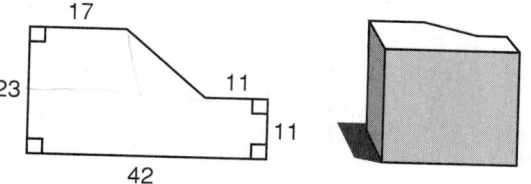

17

23   11

11

42

14. What is the surface area of a rectangular solid whose length, width, and height are 5 meters, 10 meters, and 3 meters, respectively?

15. Simplify: $(2 \times 10^{-14}) \times (3 \times 10^{-13})$

16. Find the least common multiple of 16, 24, and 32.

Solve:

17. $\dfrac{6\frac{1}{4}}{2\frac{1}{3}} = \dfrac{\frac{5}{12}}{x}$

18. $\dfrac{\frac{7}{12}}{\frac{21}{24}} = \dfrac{p}{12}$

19. $2\frac{1}{2}x + 2\frac{1}{2} = 6\frac{3}{8}$

Simplify:

20. $2^3 + \sqrt{36}[2^3(2^2 - 1)(5^2 - 22) - 1]$

21. (a) $(3)(-2)$ (b) $(-3)(2)$ (c) $(-3)(-2)$

22. (a) $\dfrac{12}{-4}$ (b) $\dfrac{-12}{4}$ (c) $\dfrac{-12}{-4}$

23. $-8 - 12 + 14 - 4$

24. $-5 - 6 + 3 - 2$

25. $-2 + 3 - 6 + (-2) - 5$

26. 1.218×0.0016

27. $16\frac{2}{7} - 2\frac{1}{2} \cdot 1\frac{1}{7} + \dfrac{9}{14}$

28. $\dfrac{1}{3}\left(1\frac{1}{4} \cdot \dfrac{7}{8} - \dfrac{1}{4}\right) + \dfrac{17}{32}$

29. Round $0.00\overline{516}$ to the nearest ten-thousandth.

30. Evaluate: $xy + x^y + y^x + \sqrt{x}$ if $x = 4$ and $y = 2$

LESSON 84 *Increases in percent*

The left-hand diagram in a percent problem holds the *of* number. We treat this number as the original number. Then the number either divides into two smaller numbers or the number gets larger. If the number gets larger, the percent is greater than 100 percent and the right-hand diagram is larger.

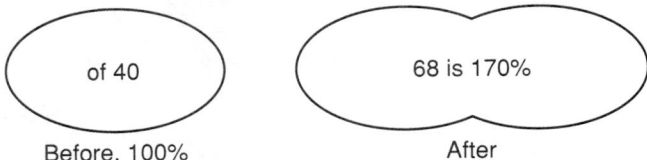

of 40   68 is 170%

Before, 100%   After

We note that we began with 40. Forty is 100 percent of 40. We ended up with 68. Sixty-eight is 100 percent of 40 plus 28, and 28 is 70 percent of 40. **This is an increase of 70 percent.**

If a problem says that the final number was 70 percent greater, the percent in the problem is 170 percent. If the problem talks about a 170 percent increase, the percent in the problem is 270 percent. An increase of 500 percent means a total percent of 600.

example 84.1 To their dismay the townspeople found that the rodent population had increased 140 percent. If the rodent population had been 400 before, what was the rodent population now?

solution An increase of 140 percent means the final percent is 240 percent. What number is 240 percent of 400?

$$\frac{P}{100} \times of = is$$

We replace percent P with 240, replace *is* with WN, and replace *of* with 400.

$$\frac{240}{100} \times 400 = WN \qquad \text{substituted}$$

$$960 = WN \qquad \text{simplified}$$

of 400   960 is 240%

Before, 100%   After

The 140 percent increase means that the rodent population increased from 400 rodents to **960 rodents.**

example 84.2 The number of cheering fans increased from 60 to 240. What percent increase was this?

solution We replace P with WP, *of* with 60, and *is* with 240.

$$\frac{WP}{100} \times 60 = 240 \qquad \text{equation}$$

To solve, we multiply both sides by 100 over 60 and cancel.

$$\frac{\cancel{100}}{\cancel{60}} \cdot \frac{WP}{\cancel{100}} \times \cancel{60} = 240 \cdot \frac{100}{60}$$

$$WP = 400\%$$

<table>
<tr><td>of 60</td><td>240 is 400%</td></tr>
<tr><td>Before, 100%</td><td>After</td></tr>
</table>

A final percent of 400 means that there was an increase of 300 percent.

practice **a.** The number of acorns on the ground increased 240 percent during the storm. If 5400 acorns were on the ground before the storm, how many were on the ground after the storm?

b. The number increased from 40 to 280. What percent increase was this?

problem set **1.** Between 1960 and 1997 the number of school-age children will increase 225
84 percent. If there were 10,000,000 school-age children in 1960, how many will there be in 1997?

2. The ratio of the loquacious to the taciturn was 9 to 5. If 25,000 were taciturn, how many were loquacious?

3. Only two-thirds of the philosophers were peripatetic. If 49 philosophers were not peripatetic, how many philosophers were there in all?

4. Thunderheads rolled across the plains at 20 miles per hour. At the beginning of the third hour the storm's speed quadrupled. If it covered 600 miles in all, how many hours did it take for the storm to cover that distance?

Graph:

5. $x \geq -7$ **6.** $x \leq 0$

7. What number is 230 percent of 80? Draw a diagram of the problem.

8. Ninety is 225 percent of what number? Draw a diagram that depicts the number.

9. $6\frac{1}{4}$ of what number is $8\frac{4}{7}$?

10. Write 3.13×10^4 in standard form.

11. Find the least common multiple of 7, 28, and 32.

12. Simplify: $(2 \times 10^{-14}) \times (2 \times 10^{-6})$

13. Find the volume in cubic feet of a right solid whose base is shown on the left and whose sides are 4 feet high. Dimensions are in feet.

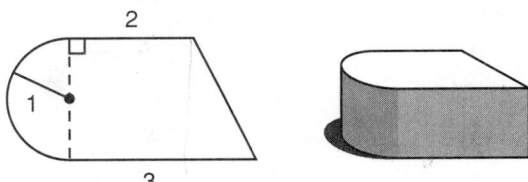

14. Find the volume and surface area of the right circular cylinder shown. Dimensions are in centimeters.

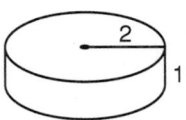

Solve:

15. $6a + 15 = 33$

16. $7\frac{1}{2}m + 2\frac{1}{3} = 9\frac{5}{6}$

17. $\dfrac{3\frac{1}{2}}{2\frac{1}{5}} = \dfrac{\frac{1}{4}}{p}$

18. $\dfrac{\frac{2}{3}}{\frac{1}{7}} = \dfrac{s}{21}$

Simplify:

19. $3^3 + \sqrt{144}[4^2(3^2 - 8)(5^2 - \sqrt{225}) + (-110)]$

20. (a) $(4)(-5)$ (b) $(-4)(5)$ (c) $(-4)(-5)$

21. (a) $\dfrac{16}{-2}$ (b) $\dfrac{-16}{2}$ (c) $\dfrac{-16}{-2}$

22. $-9 - 12 + 16 - 1$

23. $-4 - 3 + 3 - 4$

24. 3.049×0.0021

25. $\dfrac{17.658}{0.009}$

26. $11\frac{1}{4} - 3\frac{2}{3} \cdot 1\frac{1}{5} + \sqrt{\dfrac{4}{121}}$

27. $\dfrac{1}{4}\left(\dfrac{1}{2} \cdot \dfrac{1}{8} - \dfrac{1}{3}\right) + \dfrac{11}{24}$

28. Write 1.63×10^9 in standard form.

29. Round $0.00\overline{673}$ to the nearest ten-millionth.

30. Evaluate: $p^x + \sqrt[m]{x}$ if $m = 3$, $p = 2$, and $x = 8$

LESSON 85 *Opposites*

The graph of the number 2 is 2 units to the right of the origin. The graph of the number -2 is 2 units to the left of the origin. Both graphs are the same distance from the origin, but the graphs are on opposite sides of the origin.

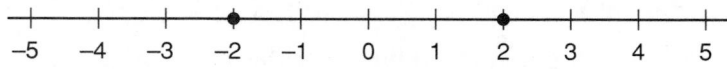

We say that these numbers are a pair of **opposites.** The number -2 is the opposite of 2 and the number 2 is the opposite of -2.

Sometimes it is helpful if we say the words **the opposite of** instead of saying **negative.** If we do this, we read

$$-2$$

as the opposite of 2. And we would read

$$-(-2)$$

as the opposite of the opposite of 2.

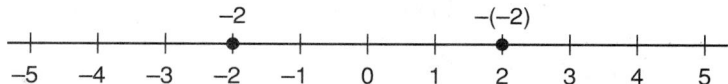

From this diagram, we see that -2 is to the left of the origin, and thus the opposite of this is on the right of the origin. Thus,

$$-(-2) = \text{the opposite of}$$
$$\qquad\quad \text{the opposite of 2} \quad \text{which is 2}$$

If we record another minus sign and write $-[-(-2)]$, we move back to the left side.

$$-[-(-2)] = \text{the opposite of}$$
$$\qquad\qquad \text{the opposite of}$$
$$\qquad\qquad \text{the opposite of 2} \quad \text{which is } -2$$

Every time we record another minus sign, we switch sides again. We can see this in the pattern developed here.

	READ AS	WHICH IS
2	2	2
-2	the opposite of 2	-2
$-(-2)$	the opposite of the opposite of 2	2
$-[-(-2)]$	the opposite of the opposite of the opposite of 2	-2
$-\{-[-(-2)]\}$	the opposite of the opposite of the opposite of the opposite of 2	2

The expressions in the left-hand column are all equivalent expressions for 2 or for -2. If we look at the right-hand column, we see that every time an additional $(-)$ is included in the left-hand expression, the right-hand expression changes sign.

A similar alternation in sign occurs whenever a particular number is multiplied by a negative number. For instance,

$$-2 = -2$$
$$(-2)(-2) = +4$$
$$(-2)(-2)(-2) = -8$$
$$(-2)(-2)(-2)(-2) = +16$$
$$(-2)(-2)(-2)(-2)(-2) = -32$$

From the pattern above, we can induce that when there is an even number of negative factors, the answer is positive and when there is an odd number of negative factors, the answer is negative.

In the following examples we use braces { }, which are another symbol of inclusion.

example 85.1 Simplify: $-(-\{-[-(-7)]\})$

solution There are five $-$ signs. Five is an odd number, so

$$-(-\{-[-(-7)]\}) = \mathbf{-7}$$

example 85.2 Simplify: $-\{-[-(-6)]\}$

solution There are four − signs. Four is an even number, so
$$-\{-[-(-6)]\} = \mathbf{6}$$

example 85.3 Simplify: $(-4)(-2)(-5)$

solution The product of three negative numbers is a negative number.
$$(-4)(-2)(-5) = \mathbf{-40}$$

example 85.4 Simplify: $(-2)(-2)(-3)(-3)(-1)(-1)$

solution There are six factors that are negative numbers. Six is an even number, and the product of an even number of negative factors is a positive number. So
$$(-2)(-2)(-3)(-3)(-1)(-1) = \mathbf{+36}$$

practice Simplify:

a. $-[-(-3)]$ b. $-\{-[-(-2)]\}$

c. $(-2)(-2)(-2)(-2)$ d. $-(-\{-[-(-5)]\})$

e. Is the sign of this product positive or negative?

$(-242)(41)(-17)(-22)(-3)(4)(-7)(5)(-3)(-2)$

problem set 85

1. The Braves started the season by hitting 1.875 times their last season's average. If their last season's average was .368, what was their beginning average this season?

2. Fortunately, the ratio of lovers to misanthropes was 13 to 2. If 3900 were lovers, how many were misanthropes?

3. For 4 hours Sam traveled at 40 miles per hour. Then for the next 3 hours, he increased his speed to 60 miles per hour. How far did he go in the 7 hours he traveled?

4. Only a vestige of the original remained. The vestige weighed 14 ounces and represented one thirty-fifth ($\frac{1}{35}$) of the original. What did the original weigh?

Graph:

5. $x > -3$ 6. $x \le 2$

7. What number is 160 percent of 120? Draw a diagram of the problem.

8. If 220 is increased by 120 percent, what is the resulting number? Draw a diagram of the problem.

9. What percent of 50 is 125? Draw a diagram that depicts the problem.

10. Complete the table. Begin by inserting the reference numbers.

11. Write 0.000000913 in scientific notation.

Fraction	Decimal	Percent
(a)	(b)	43

12. What decimal part of 800 is 280?

13. Find the volume in cubic centimeters of a right solid whose base is shown on the left and whose sides are 150 centimeters high. Dimensions are in centimeters.

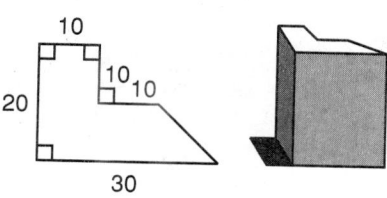

14. What is the perimeter of the figure shown? Dimensions are in yards.

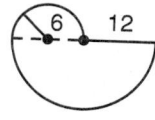

15. Write $\dfrac{1811}{15}$ as a mixed number.

16. Find the least common multiple of 15, 18, and 27.

Solve:

17. $\dfrac{7\frac{1}{4}}{2\frac{1}{2}} = \dfrac{\frac{1}{6}}{x}$

18. $\dfrac{\frac{3}{4}}{\frac{9}{16}} = \dfrac{z}{6}$

19. $3\frac{1}{5}x + 2\frac{1}{3} = 5\frac{2}{5}$

Simplify:

20. $3^2 + 2^3[\sqrt{36}(2^2 - 3)(2^2 + 1) - 20]$

21. (a) $(6)(-3)$ (b) $(-6)(-3)$ (c) $(-6)(3)$

22. (a) $\dfrac{24}{-6}$ (b) $\dfrac{-24}{6}$ (c) $\dfrac{-24}{-6}$

23. $-\{-[-(-3)]\}$

24. $-(-\{-[-(-9)]\})$

25. $-5 - 4 - (-2) + (6) + (-3)$

26. $-2 - (2) + (-4) - (-2)$

27. $3\frac{3}{7} - 1\frac{1}{7} \cdot 3\frac{1}{3} + \dfrac{8}{21}$

28. $(3 \times 10^{26}) \times (2 \times 10^{-9})$

29. Write 136121134.5 in words.

30. Evaluate: $tv + t^v + v^t + v^2$ if $t = 3$ and $v = 3$

LESSON 86 *Evaluation with signed numbers*

The order of operations for simplifying expressions that contain signed numbers is the same as the order of operations for simplifying expressions that contain the numbers of arithmetic. We always do the multiplications before we do the algebraic additions.

example 86.1 Simplify: $2(-4) - 3 - (-6)(-3)$

solution First we will do the multiplications and get

$$-8 - 3 - 18$$

We finish by adding algebraically and get -29 for our answer.

$$-8 - 3 - 18 = \mathbf{-29}$$

example 86.2 Evaluate: $-x - xy$ if $x = -2$ and $y = -4$

solution Some people find that using parentheses helps prevent making mistakes in signs. We will write parentheses for each variable.

$$-(\ \ ) - (\ \ )(\ \ )$$

Now we insert the proper numbers in the parentheses.

$$-(-2) - (-2)(-4)$$

We remember to multiply before we add, so the final result is -6,

$$2 - 8 = \mathbf{-6}$$

example 86.3 Evaluate: $-a - xa$ if $x = -3$ and $a = 2$

solution We will use parentheses and write

$$-(2) - (-3)(2)$$

Now we simplify, remembering that we always multiply first.

$$-2 - (-6) \qquad \text{multiplied } (-3) \text{ and } (2)$$
$$-2 + 6 \qquad \text{simplified } -(-6)$$
$$\mathbf{4} \qquad \text{added}$$

practice Evaluate:

a. $ab - a$ if $a = -2$ and $b = 3$

b. $x - xy$ if $x = -3$ and $y = -2$

c. $-xy - x$ if $x = -3$ and $y = -2$

problem set 86

1. Harry traveled the last 240 miles of the trip at 60 miles per hour. The entire trip covered 480 miles and took 10 hours. How fast did Harry travel for the first 240 miles?

2. Forty tons cost $20,000. What would be the cost for only 12 tons?

3. The ratio of screaming fans in box seats to fans in bleacher seats was 2 to 17. If 68,000 screamed in the bleachers, how many screaming fans were in box seats?

4. One thousand, four hundred remained standing after the crash. If this was one-seventh of the total, how many were no longer standing?

Graph:

5. $x > -1$

6. $x \le -2$

7. What number is 140 percent of 140? Draw a diagram of the problem.

8. If 230 is increased by 160 percent, what is the resulting number? Draw a diagram of the problem.

9. What percent of 70 is 98? Draw a diagram of the problem.

10. Complete the table. Begin by inserting the reference numbers.

Fraction	Decimal	Percent
$\frac{4}{5}$	(a)	(b)

11. Simplify: $(8 \times 10^{10}) \times (1 \times 10^{-10})$

12. Six-tenths of what number is 750?

13. Find the volume of a right solid whose base is shown on the left and whose sides are 3 inches high. Dimensions are in inches.

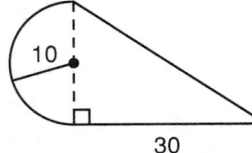

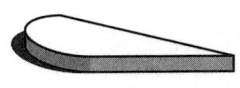

14. Use two unit multipliers to convert 4 miles to inches.

15. Write $\frac{1357}{12}$ as a mixed number.

16. Find the least common multiple of 10, 15, and 25.

Solve:

17. $\dfrac{3\frac{3}{4}}{6\frac{3}{8}} = \dfrac{\frac{1}{4}}{x}$

18. $\dfrac{\frac{3}{5}}{\frac{27}{25}} = \dfrac{z}{5}$

19. $2\frac{1}{3}x + 3\frac{1}{2} = 6\frac{1}{6}$

Simplify:

20. $3^2 + 2^2\left[2^3(\sqrt{36} - 2^2)(1 + \sqrt{16}) - \sqrt[3]{\dfrac{27}{125}}\right]$

21. (a) $3(-6)$ (b) $(-2)(-5)$ (c) $(-1)(+7)$

22. (a) $\dfrac{15}{-3}$ (b) $\dfrac{-20}{4}$ (c) $\dfrac{-36}{-12}$

23. $-(-\{-[-(-1)]\})$

24. $-(-\{-[-(+2)]\})$

25. $-6 - 3 + (-2) - (-3)$

26. $2(-4) - (-6)(-3)$

Evaluate:

27. $-x - xy$ if $x = -2$ and $y = 4$

28. $-a + ya$ if $a = -1$ and $y = -2$

29. $ay - a - y$ if $a = -7$ and $y = -3$

30. $x^y + y^x + 2xy$ if $x = 4$ and $y = 3$

LESSON 87 *Rate problems as ratio problems*

We remember that a rate is a ratio of two quantities. If 7 apples can be purchased for 2 dollars, we can write two rates or ratios.

$$\text{(a)}\quad \frac{7 \text{ apples}}{2 \text{ dollars}} \qquad \text{(b)}\quad \frac{2 \text{ dollars}}{7 \text{ apples}}$$

If we want to know how many apples could be purchased for 10 dollars, we would use rate (a) and multiply by 10 dollars.

$$\frac{7 \text{ apples}}{2 \text{ \sout{dollars}}} \times 10 \text{ \sout{dollars}} = 35 \text{ apples}$$

If we want to know how much 10 apples would cost, we would use rate (b) and multiply by 10 apples.

$$\frac{2 \text{ dollars}}{7 \text{ \sout{apples}}} \times 10 \text{ \sout{apples}} = \frac{20}{7} \text{ dollars}$$

The advantage of using this method is that we are able to keep track of the units. When two units cancel and the unit we need is still there, we know we have worked the problem correctly.

We can also work this problem as a proportion problem. A proportion is a statement that two ratios are equal. The ratios are between A for apples and D for dollars.

$$\frac{A}{D} = \frac{A}{D}$$

On the left we replace A with 7 and D with 2.

$$\frac{7}{2} = \frac{A}{D}$$

To find how many apples we could buy with 10 dollars, we replace D with 10, cross multiply, and solve.

$$\frac{7}{2} = \frac{A}{10} \qquad \text{substituted}$$

$$70 = 2A \qquad \text{cross multiplied}$$

$$35 = A \qquad \text{divided by 2}$$

To find how much 10 apples could cost, we replace A with 10 and solve for D.

$$\frac{7}{2} = \frac{10}{D} \qquad \text{substituted}$$

$$7D = 20 \qquad \text{cross multiplied}$$

$$D = \frac{20}{7} \qquad \text{divided by 7}$$

This method has the disadvantage that the units are not used, and we must be careful. The advantage of this method is that it emphasizes that the problem deals with equal ratios. Both methods should be practiced because some science books use one method and some science books use the other method.

example 87.1 Fourteen bags could be filled in 3 hours. How many hours would it take to fill 84 bags? Work this problem two ways: once as a rate problem and once as a ratio problem.

solution We get two rates.

$$(a) \quad \frac{14 \text{ bags}}{3 \text{ hours}} \qquad (b) \quad \frac{3 \text{ hours}}{14 \text{ bags}}$$

We want to know how many hours, so we use rate (b) because it has hours on top.

$$\frac{3 \text{ hours}}{14 \text{ bags}} \times 84 \text{ bags} = \textbf{18 hours}$$

To use the ratio method, we can put either bags or hours on top. We decide to put bags on top.

$$\frac{B}{H} = \frac{B}{H}$$

On the left we use 14 for bags and 3 for hours. On the right we use 84 for bags. Then we solve.

$$\frac{14}{3} = \frac{84}{H} \qquad \text{substituted}$$

$$14H = 252 \qquad \text{cross multiplied}$$

$$H = \textbf{18 hours}$$

practice Jimmy could cap 60 bottles in 7 hours. How long would it take to cap 420 bottles? Work once using rates. Then work the problem again using a proportion.

problem set 87

1. Beatrice traveled the last 220 miles of the trip at 55 miles per hour. The entire trip covered 430 miles and took 7 hours. What was her rate for the first 210 miles?

2. Fifty designer dresses cost $15,000. What would be the cost of only 19 designer dresses?

3. The ratio of masterpieces in the museum to reproductions in the shop was 2 to 120. If 228,000 reproductions were offered in the shop, how many masterpieces could be viewed in the museum?

4. One thousand, one hundred drenched fans awaited the outcome of the final quarter. If this was one-twentieth of the total number of fans who had come to the game, how many fans had come to the game?

5. Jack could check the chemicals in 25 pools in 7 hours. How long would he take to check the chemicals in 275 pools? Use both the rate and ratio methods to solve the problem.

Graph:

6. $x > 1$

7. $x \leq 1$

8. What number is 260 percent of 90? Draw a diagram of the problem.

9. If 650 is increased 120 percent, what is the resulting number? Draw a diagram of the problem.

10. What percent of 90 is 144? Draw a diagram of the problem.

11. Simplify: $(2 \times 10^{-27})(3 \times 10^{-10})$

12. Write this product in scientific notation: $20,000 \times 0.000004$

13. Three one-hundredths of what number is 9.3?

14. Use two unit multipliers to convert 10,000,000 square inches to square feet.

15. Use two unit multipliers to convert 10,000,000 square centimeters to square meters.

16. Find the volume of a right solid whose base is the figure shown on the left and whose sides are 2 meters high. The upper and lower sides of the base are parallel. Dimensions are in meters.

17. Find the volume in cubic yards of a right circular cylinder whose diameter measures 3 yards and whose height is 3 yards.

Solve:

18. $9x + 11 = 92$

19. $\dfrac{4\frac{1}{5}}{3\frac{1}{2}} = \dfrac{1\frac{1}{8}}{x}$

20. $\dfrac{\frac{6}{7}}{\frac{54}{42}} = \dfrac{\frac{2}{3}}{x}$

21. $8\frac{3}{14}x + 1\frac{1}{5} = 1\frac{1}{30}$

Simplify:

22. $2^3 + 2^2[2^1(\sqrt{64} - 7)(1 + 3^2) + (-3)]$

23. (a) $4(-7)$ (b) $(-4)(7)$ (c) $(-4)(-7)$

24. (a) $\dfrac{27}{-9}$ (b) $\dfrac{-27}{9}$ (c) $\dfrac{-27}{-9}$

25. $-(-\{-[-(-4)]\})$ 26. $-(-\{-[-(+5)]\})$

Evaluate:

27. $x + xy$ if $x = -1$ and $y = -9$

28. $-m - mx$ if $m = -2$ and $x = -3$

29. $ap - p + a$ if $a = -4$ and $p = -3$

30. Use one unit multiplier to convert 0.6 centimeter to meters.

LESSON 88   *Formats for equations · Negative coefficients*

88.A
formats for the addition rule

The addition rule for equations tells us we can add the same number to both sides of an equation. The number can be a positive number or a negative number.

ADDITION RULE FOR EQUATIONS

The same quantity can be added to both sides of an equation without changing the solution of the equation.

To solve this equation,

$$x + 2 = 6$$

we can add -2 to both sides of the equation:

$$x + 2 - 2 = 6 - 2$$

We have been working problems like this by subtracting 2 from both sides. Now we say that we are adding -2 instead of subtracting $+2$. We simplify and get

x = 4

The rule does not say that the -2s have to be on the same line. If we wish, we can use another format and write

$$
\begin{array}{lll}
x + 2 = & 6 & \text{equation} \\
\underline{\quad -2 \quad -2} & & \text{add } -2 \text{ to both sides} \\
x \quad = \quad 4 & & \text{solution}
\end{array}
$$

example 88.1   Use the new format and algebraic addition to solve

$$x + 5 = 7$$

solution   We will solve by adding -5 to both sides.

$$
\begin{array}{lll}
x + 5 = & 7 & \text{equation} \\
\underline{\quad -5 = -5} & & \text{add } -5 \text{ to both sides} \\
x \quad = \quad 2 & & \text{solution}
\end{array}
$$

example 88.2   Solve: $2x + \dfrac{1}{2} = \dfrac{3}{4}$

solution   First we add $-\frac{1}{2}$ to both sides.

$$
\begin{array}{lll}
2x + \dfrac{1}{2} = & \dfrac{3}{4} & \text{equation} \\[2mm]
\underline{\quad -\dfrac{1}{2} \quad -\dfrac{1}{2}} & & \text{add } -\dfrac{1}{2} \text{ to both sides} \\[2mm]
2x = & \dfrac{1}{4} & \text{added}
\end{array}
$$

$$\frac{1}{2} \cdot 2x = \frac{1}{4} \cdot \frac{1}{2} \qquad \text{multiplied both sides by } \frac{1}{2}$$

$$x = \frac{1}{8} \qquad \text{simplified}$$

88.B

negative coefficients Often we encounter variables with negative coefficients. In the following equation, the coefficient of x is -2:

$$-2x = 4$$

This equation can be solved by dividing both sides by -2.

$$\frac{-2x}{-2} = \frac{4}{-2}$$

$$\mathbf{x = -2}$$

Beginners often make mistakes when they divide by negative numbers. This kind of mistake can be avoided by using an extra step. Before we divide, we will multiply both sides of the equation by -1. This changes the signs on both sides. Then we can solve by dividing both sides by a positive number.

example 88.3 Solve: $-3x = 12$

solution **As the first step, we mentally multiply both sides by -1. This changes both signs, and** we get

$$3x = -12$$

Now we solve by dividing by $+3$.

$$\frac{3x}{3} = \frac{-12}{3} \qquad \text{divided by 3}$$

$$\mathbf{x = -4} \qquad \text{simplified}$$

example 88.4 Solve: $-3x + 4 = 10$

solution We always use the addition rule first. Thus, we begin by adding -4 to both sides.

$$\begin{array}{rl} -3x + 4 = & 10 \qquad \text{equation} \\ \underline{-4 \quad -4} & \qquad \text{add } -4 \text{ to both sides} \\ -3x \quad = & 6 \end{array}$$

Next we multiply mentally by -1 and get

$$3x = -6$$

Now we divide by 3 to solve.

$$\frac{3x}{3} = \frac{-6}{3} \qquad \text{divided by 3}$$

$$\mathbf{x = -2} \qquad \text{simplified}$$

practice Solve:

a. $-2x + 5 = 9$

b. $-3x + \dfrac{1}{2} = \dfrac{3}{8}$

problem set 88

1. Wanataxa peered into the gloom and could positively identify only 2000. If there was a total of 16,000 in the gloom, what fraction of the total did she identify?

2. The ratio of defective components to nondefective components was 2 to 19. If 380 parts were nondefective, how many were defective?

3. The delegates crowded into the hall until they numbered $1\frac{1}{4}$ times the limit set by the fire marshall. If the fire marshall's limit was 800, how many delegates were in the hall?

4. For the first 50 miles, Hannah traveled at 25 miles per hour. Then she doubled her speed. If the entire trip was 450 miles, how long did the trip take?

Graph:

5. $x \geq -2$

6. $x < 2$

7. What number is 240 percent of 120? Draw a diagram of the problem.

8. If 160 is increased by 185 percent, what is the resulting number? Draw a diagram of the problem.

9. What percent of 60 is 96? Draw a diagram of the problem.

10. Complete the table. Begin by inserting the reference numbers.

Fraction	Decimal	Percent
(a)	0.41	(b)

11. Find the surface area of a cube whose edges are 6 meters long.

12. What decimal part of 620 is 217?

13. Express in cubic meters the volume of a solid whose base is shown on the left and whose sides are 10 centimeters high. Dimensions are in meters.

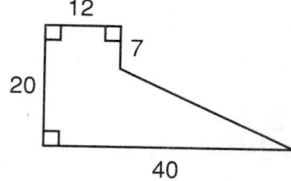 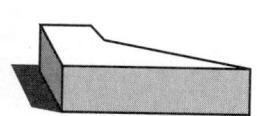

14. Simplify: $(3 \times 10^{-11}) \times (2 \times 10^{15})$

15. Convert 1385.31 kilometers to centimeters.

16. Find the least common multiple of 60, 120, and 180.

Solve:

17. $-8x + 4 = 28$

18. $-5x + 7 = 12$

19. $\dfrac{2\frac{2}{3}}{\frac{4}{5}} = \dfrac{\frac{3}{4}}{x}$

20. $-2\frac{1}{4}x + \frac{1}{2} = 3\frac{3}{4}$

Simplify:

21. $2^3 + 2^2\left[2(\sqrt{49} - \sqrt{36})(1 + \sqrt{9}) - \sqrt[3]{\frac{8}{64}}\right]$

22. $1\frac{3}{4} \cdot 2\frac{1}{3} - \frac{5}{12}$

23. (a) $(4)(-2)$ (b) $\dfrac{16}{-2}$ (c) $(-6)(-2)$

24. $-(-\{-[-(5)]\})$

25. $4 - 3(5) - 7(-6) - 4(-5)$

26. $\frac{1}{2}\left(1\frac{2}{3} \cdot 1\frac{4}{5} - \frac{1}{2}\right) + \frac{5}{6}$

Evaluate:

27. $-x + xy$ if $x = -1$ and $y = -3$

28. $xyz + xy$ if $x = -1$, $y = -2$, and $z = -3$

29. $xy + y^x + x^x$ if $x = 2$ and $y = 3$

30. Write 6256183122 in words.

LESSON 89 *Algebraic phrases*

In algebra we learn to write conditional algebraic equations that have the same meanings as statements made by using words. Then we find the solutions to the equations. These solutions give us the answers to the questions asked. When we write the conditional equations, we use algebraic phrases. These algebraic phrases have the same meanings as the word phrases. There are several keys to writing algebraic phrases. The word **sum** means things are added, as do the words **greater than** or **increased by**.

WORD PHRASE	ALGEBRAIC PHRASE
The sum of a number and 7	$N + 7$
The sum of a number and -7	$N - 7$
7 greater than a number	$N + 7$
A number increased by 7	$N + 7$

The words **less than** or **decreased by** mean to subtract (add the opposite of).

7 less than a number	$N - 7$
A number decreased by 7	$N - 7$

The word **product** means things are multiplied.

The product of a number and 7	$7N$

If we use N to represent an unknown number, then we will use $-N$ to represent the opposite of the unknown number. Thus twice a number and 9 times the opposite of a number could be written as follows:

Twice a number	$2N$
9 times the opposite of a number	$9(-N)$

Sometimes two or more operations are designated by a single phrase.

The sum of twice a number and -8	$2N - 8$
5 times the sum of twice a number and -8	$5(2N - 8)$

Cover the answers in the right-hand column on the next page and see if you can write the algebraic phrase that is indicated.

The sum of a number and 9	$N + 9$
A number decreased by 9	$N - 9$
The opposite of a number, decreased by 6	$-N - 6$
The sum of the opposite of a number and -3	$-N - 3$
The product of 5 times a number and 6	$6(5N)$
The sum of twice a number and -3	$2N - 3$
4 times the sum of twice a number and -3	$4(2N - 3)$
The product of -2 and a number increased by 5	$-2(N + 5)$

practice   Write the algebraic phrases that correspond to each statement.

a.   The sum of a number and -4

b.   A number reduced by 4

c.   The sum of the opposite of a number and -3

d.   The product of the opposite of a number and 6

e.   5 times the sum of a number and -6

f.   The product of -4 and a number increased by 6

problem set 89

1.   Frances got the first 40 pictures framed for a total of $120. How much would it cost her to get 800 pictures framed?

2.   The ratio of roses to snapdragons was 4 to 5. If there were 26,000 roses on the float, how many snapdragons were there?

3.   If $7\frac{4}{5}$ of the total came to 109,200, what was the total?

4.   The last leg of the trip was traveled at 680 miles per hour. If the last leg took 10 hours to travel, what was the length of the last leg?

5.   Write the algebraic phrase that is described:
   (a)   The sum of twice a number and -16
   (b)   The product of -3 and a number increased by 5
   (c)   The opposite of a number, decreased by 4
   (d)   5 times the sum of twice a number and 6

6.   Graph: $x \geq -1$

7.   What number is 190 percent of 340? Draw a diagram of the problem.

8.   If 200 is increased by 160 percent, what is the resulting number? Draw a diagram of the problem.

9.   What percent of 80 is 128? Draw a diagram of the problem.

10.   Complete the table. Begin by inserting the reference numbers.

11.   Find the surface area of a cube whose edges measure 0.25 centimeter.

Fraction	Decimal	Percent
$\frac{3}{4}$	(a)	(b)

12. Express in cubic yards the volume of a right solid whose base is shown on the left and whose sides are 2 yards high. Dimensions are in yards.

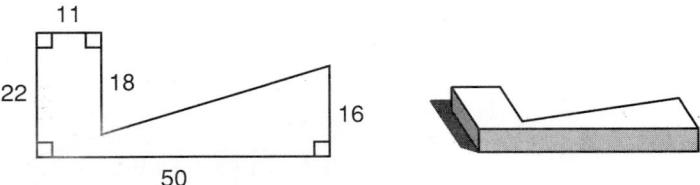

13. Use four unit multipliers to convert 612 square inches to square yards.

14. Find the least common multiple of 25, 36, and 45.

15. Express 720 as a product of primes.

Solve:

16. $-6x - 6 = -3$

17. $-7x - 3 = 12$

18. $\dfrac{3\frac{1}{3}}{6\frac{1}{2}} = \dfrac{\frac{4}{13}}{x}$

19. $-3\frac{1}{3}x + \frac{3}{4} = 2\frac{5}{12}$

Simplify:

20. $2^2 + 2^3[3(\sqrt{49} - \sqrt{36})(\sqrt{16} + \sqrt{4}) - \sqrt{9}] + \sqrt{81}$

21. $2\frac{2}{3} \cdot 3\frac{1}{4} - \frac{5}{12}$

22. (a) $(3)(-2)$ (b) $\dfrac{-6}{-3}$ (c) $(-1)(-1)$

23. $-(-\{-[-(+6)]\})$

24. $-4(-3) + (-2)(-5)$

25. $\dfrac{621.21}{0.004}$

26. $\dfrac{1}{3}\left(2\frac{1}{4} \cdot \frac{3}{4} - \frac{15}{16}\right) + \frac{11}{48}$

27. 118.21×0.0014

Evaluate:

28. $yz - xz$ if $x = -1$, $y = 1$, and $z = 2$

29. $z + y + yz$ if $y = -1$ and $z = -2$

30. $y^y + x^x + y^x + x^y$ if $x = 2$ and $y = 3$

LESSON 90 *Surface area of a right solid*

We remember that a right solid is a solid whose sides are perpendicular to the top and bottom. Until now we have concentrated on finding the surface area of right circular cylinders and right triangular prisms. We have also noted that the **lateral** surface area of a right circular cylinder can be computed by cutting the cylinder lengthwise and unrolling it flat to make a rectangle.

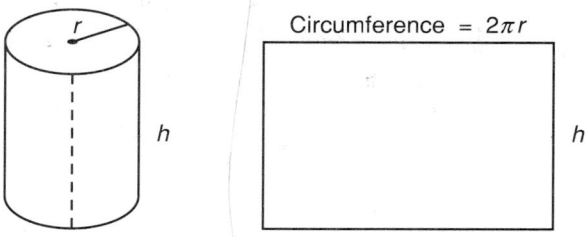

The lateral surface area equals the height times the circumference. We can use this procedure to find the lateral surface area of any right solid.

example 90.1   Find the surface area of the right triangular solid shown. Dimensions are in meters.

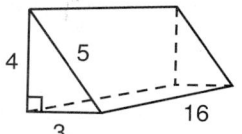

solution   The total surface area equals the area of the two triangular ends plus the lateral surface area. **The lateral surface area equals the perimeter of the base times the height of the solid.**

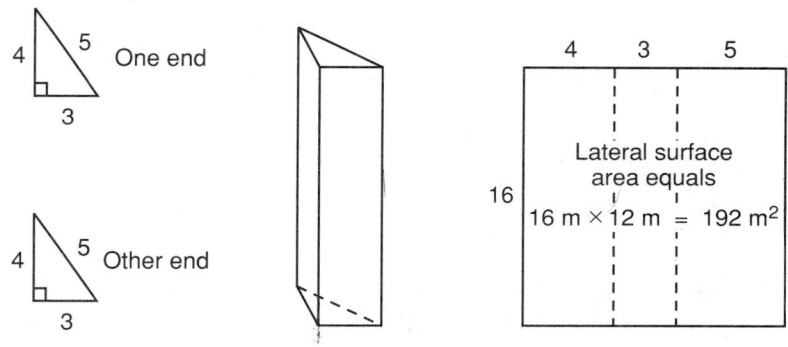

On the right above we cut the prism down one edge and find it makes a rectangle. One side is 16, which is the height, and the other side is the perimeter of the base, which is 4 + 3 + 5, or 12.

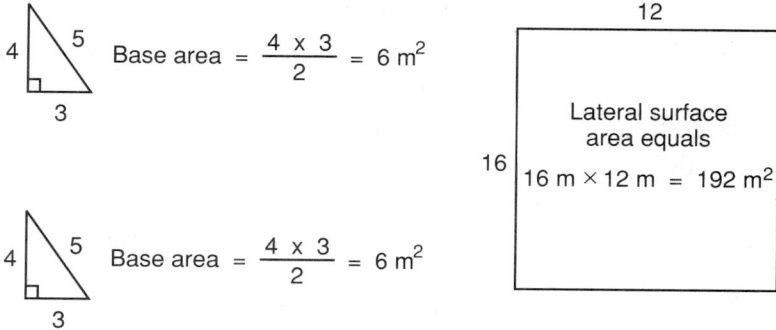

Thus

$$\text{Total surface area} = 6m^2 + 6m^2 + 192 \text{ m}^2 = \mathbf{204 \text{ m}^2}$$

example 90.2   Find the lateral surface area of a right solid 30 feet high whose base is figure shown. Dimensions are in feet.

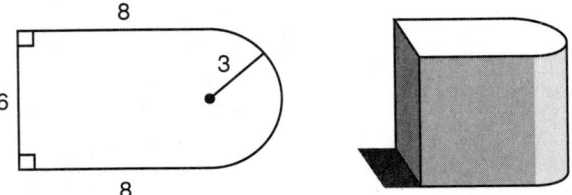

solution   The thought process of the preceding example can be extended to help us find the lateral surface of any right solid. **The lateral surface area of a right solid equals the perimeter of the base times the height of the solid.**
The perimeter is

$$6 + 8 + 8 + \frac{2\pi(3)}{2} = 22 + 3\pi$$

If we use 3.14 for π, the perimeter is

$$22 + 3(3.14) = 31.42 \text{ feet}$$

The lateral surface area is the height times the perimeter.

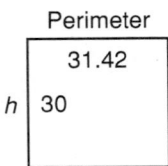

Lateral surface area = 30(31.42) = **942.6 ft²**

practice   The figure shown is the base of a right solid 40 meters high. What is the lateral surface area of the right solid? Dimensions are in meters.

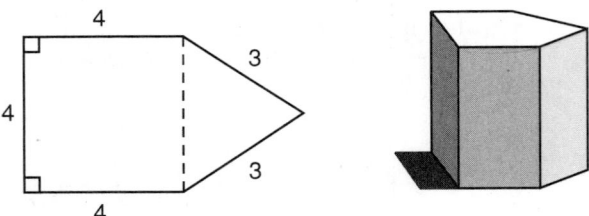

**problem set
90**

1. Adrian purchased 20 ballet tickets for a total of $1110. How much would it have cost him for 150 ballet tickets?

2. The ratio of *Sequoia gigantias* to *Sequoia semivirons* was 6 to 1. If there were 24,600 *Sequoia gigantias* in the forest, how many *Sequoia semivirons* were there?

3. Business improved by 260 percent this year. If the previous year's receipts totaled $240,000, what was the amount of this year's receipts?

4. The large aircraft traveled at 1140 miles per hour for the first 2 hours of the flight. For the next 6 hours it traveled at $1\frac{1}{2}$ times the original speed. What was the total distance covered?

5. Write the algebraic phrase that is described:
   (a) The sum of 3 times a number and -11
   (b) The product -5 and a number decreased by 2
   (c) The opposite of 3 times a number decreased by 4.

6. Graph: $x \geq 3$

7. What number is 270 percent of 250? Draw a diagram of the problem.

8. If 78 is increased by 150 percent, what is the resulting number? Draw a diagram of the problem.

9. What percent of 90 is 360? Draw a diagram of the problem.

10. Find the volume in cubic feet of a right solid whose base is shown and whose sides are 4 feet high.

11. Find the lateral surface area of the right solid in Problem 10.

12. Find the total surface area of the right solid in Problem 10.

13. Simplify: $(1 \times 10^{20})(5 \times 10^{24})$

14. Use two unit multipliers to convert 0.75 square meter to square centimeters.

15. Seven-tenths of what number is 9.1?

16. Write 8.51×10^{-11} in standard form.

Solve:

17. $-5x - 4 = 16$

18. $-11x + 9 = 42$

19. $\dfrac{5\frac{1}{2}}{\frac{2}{3}} = \dfrac{3}{x}$

20. $-1\frac{1}{2}x + 2\frac{3}{5} = 4\frac{3}{10}$

Simplify:

21. $5^2 + 4^2[2(\sqrt{121} - \sqrt{81})]$

22. $2\frac{1}{5} \cdot 1\frac{1}{2} - \frac{7}{10}$

23. (a) $(-5)(-3)$ (b) $\dfrac{14}{-2}$ (c) $\dfrac{-8}{-4}$

24. $-(-\{-[-(-6)]\})$

25. $2 - 5(2) - 4(-2) - 5(-3)$

26. 314.14×0.00002

27. $\dfrac{78.99}{0.00003}$

Evaluate:

28. $km + m(-k)$ if $k = 3$ and $m = 9$

29. $-(km) - k(m)$ if $k = 5$ and $m = 3$

30. $a^b + bax + \sqrt[x]{a}$ if $a = 8$, $b = 2$, and $x = 3$

LESSON 91   *Trichotomy · Symbols of negation*

91.A
trichotomy

We have discussed how we compare numbers by using the words **less than, equal to,** or **greater than.** We recall that we use the number line to define what we mean by these words. On this number line, we have graphed the numbers -4, -2, 0, and 2.

We remember that a number is greater than another number if its graph is to the right of the graph of the other number. By looking at the graphs above, we can verify that all of the following inequalities are true inequalities.

$$0 > -4 \qquad -2 < 0 \qquad 2 > -2 \qquad -4 < 2$$

It is interesting to note that if we have two numbers, then one and only one of the following statements can be true.

(a)   The first number is greater than the second number.
(b)   The first number is equal to the second number.
(c)   The first number is less than the second number.

Because there are only three possibilities, this is often called the **trichotomy axiom** from the Greek word *tricha*, which means "in three parts."

91.B
symbols of negation

The symbols shown here

$$< \qquad > \qquad = \qquad \geq \qquad \leq$$

are used to denote that things are equal or are not equal. We can negate each of these symbols by drawing a slash through the symbol. Thus, if we are asked to read the following inequalities:

(a) $4 \nless 1$ (b) $4 \neq 2 + 6$ (c) $7 \ngeq 10$

we can read them from left to right as

(a) $4 \nless 1$ 4 is not less than 1   true

(b) $4 \neq 2 + 6$ 4 is not equal to 2 + 6   true

(c) $7 \ngeq 10$ 7 is not greater than or equal to 10   true

Negated inequalities can also be false inequalities.

$-4 \nless 10$ -4 is not less than 10   false

$7 \ngeq 3$ 7 is not greater than or equal to 3   false

When we graph conditional inequalities, we indicate the numbers that will make them true inequalities.

example 91.1   Graph: $x \nless 2$

solution   **As the first step, we will rewrite the negated inequality.** If x is not less than 2, then x must be greater than or equal to 2.

$$x \geq 2$$

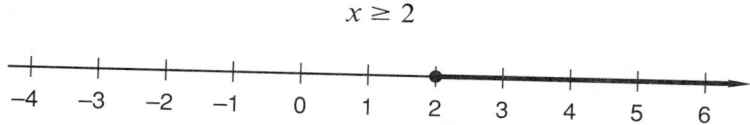

example 91.2   Graph: $x \ngeq -1$

solution   **As the first step, we will rewrite the negated inequality.** If x is not greater than or equal to -1, then x must be less than -1.

$$x < -1$$

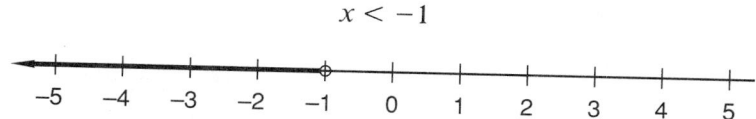

example 91.3   Graph: $x \nless -2$

solution   If x is not less than -2, then x must be greater than or equal to -2.

$$x \geq -2$$

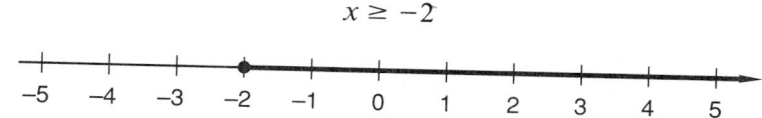

Note that in this example, we used a solid circle at -2 because of the $\geq$ sign. In the preceding example, the circle at -1 was not solid because the inequality $<$ did not contain part of an equals sign.

practice   Graph:

a. $x \nleq -1$ **b.** $x \ngtr 2$ **c.** $x \ngeq 2$

problem set 91

1. There was no water in the lake. Then the rain came down in torrents. When three-eighths of the rain had fallen, the water in the lake was 21 inches deep. How deep was the water in the lake when all the rain had fallen?

2. The ratio of fishermen to fish caught was 2 to 23. How many fish were caught by 460 fishermen?

3. Forty of the expensive ones could be purchased for $28,000. What would 72 of the expensive ones cost?

4. After traveling the first 480 miles in 12 hours, the traveler increased her speed by 20 miles per hour. How long did it take her to travel the last 300 miles?

5. Write the algebraic phrase that is described.
    (a) Four times the sum of 3 times a number and -6
    (b) The product of -2 and a number increased by 5
    (c) The opposite of a number, decreased by 25

Graph:

6. $x \nless -9$ 7. $x \ngeq -4$

8. What number is 210 percent of 250? Draw a diagram of the problem.

9. If 250 is increased by 170 percent, what is the resulting number? Draw a diagram of the problem.

10. What percent of 90 is 126? Draw a diagram of the problem.

11. Convert to scientific notation and simplify:

$$(100,000,000)(0.00000004)$$

12. Express in cubic centimeters the volume of a right solid whose base is shown on the left and whose sides are 1 meter high. Dimensions are in centimeters.

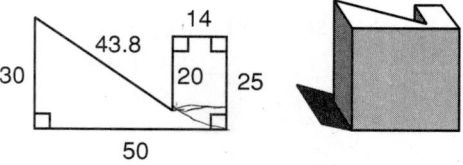

13. Find the lateral surface area for the solid in Problem 12.

14. Use two unit multipliers to convert 2 miles to inches.

Solve:

15. $-7x - 7 = -2$

16. $-2x + 3 = 7$

17. $\dfrac{2\frac{1}{4}}{1\frac{1}{2}} = \dfrac{1\frac{5}{6}}{x}$

18. $-2\frac{1}{3}x - \frac{3}{4} = 3\frac{7}{12}$

Simplify:

19. $2^3 + 2^2[3(\sqrt{64} - \sqrt{49})(\sqrt[3]{8} + 1) - 1]$

20. $1\frac{2}{3} \cdot 3\frac{1}{4} - \frac{7}{12}$

21. (a) $2(-3)$ (b) $\dfrac{-12}{-3}$ (c) $(-1)(-2)(3)$

22. $-2(-3) + (-2)(-4) - (-2)$

23. $(-2)(3) - (-2)(-4) - 3(-2)$

24. $\frac{1}{2}\left(3\frac{2}{5} \cdot \frac{3}{4} - 1\frac{1}{4}\right)$

25. $3\frac{1}{2} \times 2\frac{1}{3} \div \frac{5}{6} \times \frac{1}{3}$

26. $\dfrac{118.02}{0.004}$

27. $672.8913 - 19.8761$

Evaluate:

28. $-xy + y$ if $x = -2$ and $y = -3$

29. $xy + yz + xz$ if $x = -1$, $y = -2$, and $z = -3$

30. $x^3 + y^3 + 3x^2y + 3xy^2$ if $x = 2$ and $y = 3$

LESSON 92 *Number word problems*

To solve word problems, we look for statements that describe equal quantities. Then we use an equals sign and algebraic phrases to write an equation that makes the same statement of equality.

We will avoid the use of the meaningless variables x and y when we write these equations. We will try to use variables whose meanings are easy to remember. The problems in this lesson discuss some unknown number. We will use the letter N to represent the unknown number.

example 92.1 The sum of twice a number and 42 is 128. Find the number.

solution The word **is** means **equals.** Thus the sum of twice a number and 42 equals 128.

$$
\begin{array}{rll}
2N + 42 &= 128 & \text{equation} \\
-42 & -42 & \text{add } -42 \text{ to both sides} \\
\hline
2N & = 86 & \\
N &= 43 & \text{divided by 2}
\end{array}
$$

We will use 43 for N in the original equation to check.

$$2(43) + 42 = 128 \longrightarrow 86 + 42 = 128 \longrightarrow 128 = 128 \qquad \text{check}$$

example 92.2 Twenty-seven less than 3 times a number is 144. What is the number?

solution If we use N for the number, then 3 times the number is $3N$.

$$
\begin{array}{rll}
3N - 27 &= 144 & \text{equation} \\
+27 & +27 & \text{add 27 to both sides} \\
\hline
3N & = 171 & \\
\end{array}
$$

$$
\frac{3N}{3} = \frac{171}{3} \qquad \text{divided by 3}
$$

$$N = 57 \qquad \text{simplified}$$

We will use 57 for N in the original equation to check.

$$3(57) - 27 = 144 \longrightarrow 171 - 27 = 144 \longrightarrow 144 = 144 \qquad \text{check}$$

example 92.3 Four more than 33 times a number equals -95. What is the number?

solution First we write the equation. Then we solve.

$$
\begin{array}{rll}
33N + 4 &= -95 & \text{equation} \\
-4 & -4 & \text{add } -4 \text{ to both sides} \\
\hline
33N & = -99 & \\
N &= -3 & \text{divided by 33}
\end{array}
$$

Now we use -3 for N in the original equation to check.

$$33(-3) + 4 = -95 \longrightarrow -99 + 4 = -95 \longrightarrow -95 = -95 \qquad \text{check}$$

practice **a.** The sum of 3 times a number and -4 is 32. What is the number?

b. Four more than 9 times a number is -77. What is the number?

c.   Five less than 5 times a number is -55. What is the number?

problem set   **1.**   The sum of twice a number and 42 is 126. Find the number.

92   **2.**   Twenty less than 3 times a number is 223. What is the number?

   3.   Four more than 33 times a number equals -260. What is the number?

   4.   For the first 20 miles, the stagecoach crawled along at 4 miles per hour. When the robbers were sighted, the speed was quadrupled. How long did it take to travel the last 80 miles at the new speed?

Graph:

   5. $x \nleq 3$ **6.** $x \ngtr -2$ **7.** $x \geq -2$

   8.   What number is 225 percent of 140? Draw a diagram of the problem.

   9.   If 300 is increased by 125 percent, what is the resulting number? Draw a diagram of the problem.

   10.   What percent of 640 is 896? Draw a diagram of the problem.

   11.   Complete the table. Begin by inserting the reference numbers.

Fraction	Decimal	Percent
(a)	(b)	40

   12.   Find the volume in cubic meters of a solid whose base is shown on the left and whose sides are 50 centimeters high. Dimensions are in meters.

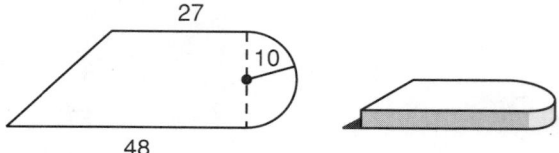

   13.   Find the volume of a rectangular solid whose length, width, and height measure 12 inches, 10 inches, and 3 inches, respectively.

   14.   Find the lateral surface area of the rectangular solid in Problem 13.

Solve:

   15. $-6x - 5 = -1$ **16.** $-2x + 6 = 1$

   17. $\dfrac{3\frac{1}{6}}{2\frac{1}{3}} = \dfrac{\frac{1}{2}}{x}$ **18.** $-3\frac{1}{3}x - \frac{1}{4} = 6\frac{1}{2}$

Simplify:

   19. $2^2 + 2[\sqrt{9}(2^3 - 2 \cdot 3)(3^2 - 4 \cdot 2)]$ **20.** $2\frac{1}{3} \cdot 1\frac{2}{3} - \frac{7}{9}$

   21.   (a) $3(-2)$ (b) $\dfrac{-18}{3}$ (c) $(-2)(-2)(5)$

22. $-(-3) - (-4) + 6 - (-5)$

23. $-[-(-2)] - (-1) - 5 + \left(\sqrt[3]{\dfrac{64}{27}} \right)$

24. $\dfrac{1}{3} \left(2\dfrac{1}{4} \cdot 1\dfrac{1}{5} - \dfrac{17}{20} \right)$

25. $4\dfrac{1}{3} \div 1\dfrac{2}{3} \times 2\dfrac{3}{4} \div \dfrac{1}{4}$

26. $\dfrac{117.03}{0.005}$

27. $1182.6218 - 13.6211$

Evaluate:

28. $xy + yz$ if $x = -1$, $y = -2$, and $z = -3$

29. $zy - zx$ if $x = -2$, $y = -3$, and $z = -1$

30. $x^2 + y^2 + 2xy$ if $x = 2$ and $y = 3$

LESSON 93 *Operations with signed numbers*

We always multiply before we add algebraically. If an expression contains symbols of inclusion, we simplify within these symbols first, always beginning with the innermost symbols of inclusion.

example 93.1 Simplify: $-2(-2 - 3 + 1) - [(-3 - 2) \cdot 4]$

solution We first simplify within the parentheses.

$$-2(-4) - [(-5) \cdot 4] \qquad \text{simplified within parentheses}$$

Then we simplify within the brackets.

$$-2(-4) - [-20] \qquad \text{simplified within brackets}$$

Now we multiply and get

$$8 - [-20] \qquad \text{multiplied } (-2) \text{ and } (-4)$$

Then we add to get

$$28 \qquad \text{added}$$

example 93.2 Simplify: $-2(-2 - 3 \cdot 5) - 2[(3 - 5)2 + 2]$

solution First we simplify within the parentheses.

$$-2(-17) - 2[(-2)2 + 2] \qquad \text{simplified within parentheses}$$

Now we simplify within the brackets.

$$-2(-17) - 2[-2] \qquad \text{simplified within brackets}$$

Now we multiply.

$$34 - [-4] \qquad \text{multiplied twice}$$

Then we add.

$$38 \qquad \text{added}$$

practice   Simplify:

   a. $-2(-3 - 5) + [6(-2 + 8) + 2]$

   b. $3(-8 - 2) + 2 - [4(3 - 2)]$

problem set
93

1. The product of a number and 15 is increased by 4. The result is 49. What is the number?

2. Fifteen less than 10 times a number is -105. What is the number?

3. Hartzler multiplied his magic number by 14. Then he added 8. The result was 50. What was Hartzler's magic number?

4. The ratio of straights to bents was $2\frac{1}{2}$ to 3. If 300 were bent, how many were straight?

Graph:

5. $x \nleq 1$ 6. $x \ngtr -1$

7. What number is 230 percent of 350? Draw a diagram of the problem.

8. If 200 is increased by 130 percent, what is the resulting number? Draw a diagram of the problem.

9. What percent of 80 is 108? Draw a diagram of the problem.

10. Thirty percent of what number is 240? Draw a diagram of the problem.

11. Complete the table. Begin by inserting the reference numbers.

12. Find the surface area in square inches of a right circular cylinder with a diameter of 1 foot and height of 1 foot.

Fraction	Decimal	Percent
(a)	0.36	(b)

13. Express in cubic yards the volume of a solid whose base is shown on the left and whose sides are 2 yards high. Dimensions are in yards.

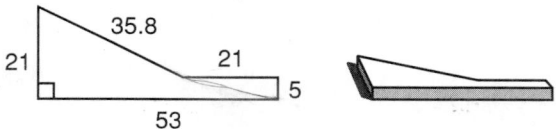

14. Find the lateral surface area of the solid in Problem 13.

15. Use one unit multiplier to convert 1234.8792 meters to kilometers.

Solve:

16. $-3x - 10 = -1$ 17. $-6x + 4 = 6$

18. $\dfrac{3\frac{1}{3}}{2\frac{1}{4}} = \dfrac{1\frac{1}{2}}{x}$ 19. $-3\frac{1}{4}x - \frac{3}{8} = 4\frac{1}{2}$

Simplify:

20. $2^3 + 2^3[2^2(\sqrt{16} - \sqrt{9})(\sqrt{9} + 2^2)]$ 21. $-2(-2 - 3 \cdot 5) - [2(3 - 5) + 2]$

22. $-3(-2 - 6 \cdot 2) - [4(2 - 4) - 2]$

23. $-(-2)(-3) + (-6)(-1) - 2(-1 - 3 \cdot 5)$

24. $\frac{1}{3}\left(2\frac{1}{4} \cdot \frac{2}{3} - \frac{1}{6} \cdot \frac{1}{2}\right)$ **25.** $2\frac{1}{3} \times 1\frac{1}{4} \div \frac{3}{8} \div \frac{1}{2}$

26. 21.62×0.0012 **27.** $\dfrac{615.01}{0.003}$

Evaluate:

28. $yy - xy$ if $x = -1$ and $y = -2$

29. $abc + ab$ if $a = -2$, $b = -1$, and $c = 3$

30. $a^2 + b^2 + c^2 + 2ab$ if $a = 1$, $b = 2$, and $c = 3$

LESSON 94 *Roots by cut and try*

Thus far we have restricted our investigation of roots of numbers to roots that are whole numbers. The cube root of 27 is a whole number. The fourth root of 16 is also a whole number.

$$\sqrt[3]{27} = 3 \qquad \sqrt[4]{16} = 2$$

The cube root of 5 is not a whole number. Neither is the fourth root of 5.

$$\sqrt[3]{5} = ? \qquad \sqrt[4]{5} = ?$$

The easiest way to simplify these expressions is to use the inv y^x key on a pocket calculator. This key is often used incorrectly, and many beginners lack the understanding necessary to recognize that the answer is incorrect. In this lesson we will use a method that is less likely to produce errors. The method is called "cut and try." We make a guess. Then we try our guess. Then we change our guess and try again and again. . . .

example 94.1   Use the method of cut and try to estimate $\sqrt[3]{5}$ to one decimal place.

solution   We want to find the number which, when used as a factor 3 times, equals 5.

$$(?)(?)(?) = 5$$

We know that $(1)(1)(1) = 1$ and $(2)(2)(2) = 8$. This tells us the number we are searching for is between 1 and 2 and is probably closer to 2. Let's try 1.5 and use a calculator.

$$(1.5)(1.5)(1.5) = 3.375$$

This number is not great enough. Let's try 1.8.

$$(1.8)(1.8)(1.8) = 5.832$$

Too great. Let's try 1.7.

$$(1.7)(1.7)(1.7) = 4.913$$

Closer, but too low. Let's try 1.73.

$$(1.73)(1.73)(1.73) = 5.18$$

Now we have

$$(1.70)^3 = 4.913 \qquad (1.73)^3 = 5.18$$

The cube root of 5 is a number between 1.7 and 1.73. Thus, to one decimal place,

$$\sqrt[3]{5} \approx 1.7$$

The symbol $\approx$ means approximately equal to. We could have continued to refine our guesses to find the cube root of 5 correct to any number of decimal places that we wish.

example 94.2   Use the method of cut and try to estimate $\sqrt{7}$ to one decimal place.

solution   We are searching for the number that multiplied by itself equals 7.

$$(?)(?) = 7$$

Well, $(2)(2) = 4$ and $(3)(3) = 9$, so 2.8 seems like a reasonable guess.

$$(2.8)(2.8) = 7.84$$

Too great. Let's try 2.7.

$$(2.7)(2.7) = 7.29$$

Too great. Let's try 2.6.

$$(2.6)(2.6) = 6.76$$

Too low. Let's try 2.64.

$$(2.64)(2.64) = 6.9696$$

Still too low. Let's try 2.65.

$$(2.65)(2.65) = 7.0225$$

The square root of 7 is between 2.64 and 2.65, so to one decimal place our estimate is 2.6.

$$\sqrt{7} \approx 2.6$$

practice   Use a calculator and the cut-and-try method to estimate to one decimal place.

a. $\sqrt[3]{15}$ \qquad\qquad\qquad\qquad\qquad b. $\sqrt{23}$

problem set 94

1.   The product of a number and -4 is increased by 12. The result is 36. What is the number?

2.   Brodsky multiplied his number by 8. Then he added 12. The result was 84. What was Brodsky's original number?

3.   The ratio of the number that were translucent to the number that were opaque was $4\frac{1}{2}$ to 6. If 675 were translucent, how many were opaque?

4.   For the first 4 hours of the tour, the cyclists idled along at 18 miles per hour. If they increased their speed $1\frac{1}{2}$ times and traveled at the new speed for the next 4 hours, what was the total distance they traveled?

5.   Karen purchased 40 collectables at the auction for $5400. If they were of equal value, how many could she have purchased for $13,500?

6. One prognosticator predicted that from the late 1980s to the late 1990s the number of automobiles on the roads will have risen by 215 percent. According to this prediction, how many automobiles will there be in the late 1990s if, in the late 1980s, there were 400,000,000 automobiles?

Graph:

7. $x \not< 0$ **8.** $x \ne 2$ **9.** $x \not> 0$

10. What number is 315 percent of 160? Draw a diagram of the problem.

11. What percent of 600 is 2280? Draw a diagram of the problem.

12. Express in cubic yards the volume of a right solid whose base is shown on the left and whose sides are 3 feet high. All angles in the base are right angles. Dimensions are in yards.

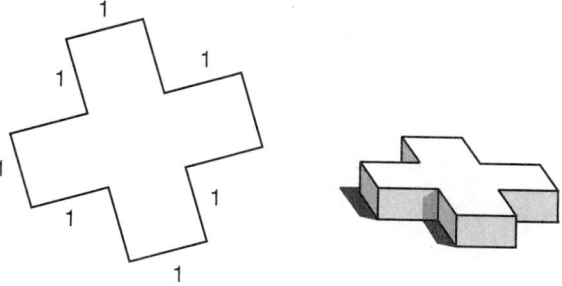

13. Find the lateral surface area of the solid in Problem 12. Remember that lateral surface area equals perimeter times height.

14. Find the volume in cubic meters of a right circular cylinder with a diameter of 20 meters and a height of 6 meters.

Solve:

15. $-9x - 4 = 5$

16. $-3x + 7 = 2$

17. $\dfrac{2\frac{1}{11}}{5\frac{6}{8}} = \dfrac{\frac{1}{11}}{p}$

18. $-5\frac{2}{3}x - \frac{1}{5} = 4\frac{7}{15}$

Simplify:

19. $(3 \times 10^{-17}) \times (3 \times 10^{-19})$

20. $3^2\left[\sqrt{25}(4^2 - 5 \cdot 2)\sqrt{\dfrac{1}{169}}\right]$

21. $5\frac{3}{4} \cdot 1\frac{7}{9} - \dfrac{35}{36}$

22. (a) $-5(-5)$ (b) $\dfrac{21}{-7}$ (c) $(-3)(5)(-2)$

23. $-(-4) - (-9) + 7 + (-9)$

24. $-[-(+3)] - (-2) - 4 + (-6)$

25. $\frac{1}{4}\left(3\frac{1}{2} \cdot 5\frac{1}{5} - \dfrac{11}{20}\right)$

26. $\dfrac{78.016}{0.0008}$

27. Write 8.17×10^{-4} in standard form.

28. Use two unit multipliers to convert 0.85 mile to inches.

Evaluate:

29. $-zm - (-m)(-z)$ if $z = 3$ and $m = 5$

30. $zp - mp + z^p$ if $z = -3$, $p = 4$, and $m = 5$

LESSON 95 *Order of division*

The procedure for simplifying expressions that require multiplication, division, and algebraic addition is as follows:

1. Do the multiplications and divisions from left to right in the order that they are encountered.
2. Do the algebraic additions.

To use these steps to simplify

$$32 \div 2 \cdot 4 - 3 \cdot 5 + 2$$

we move from left to right, doing only multiplication and division.

$$16 \cdot 4 - 3 \cdot 5 + 2 \qquad \text{divided 32 by 2}$$

$$64 - 15 + 2 \qquad \text{multiplied 16 by 4 and 3 by 5}$$

Now we do the algebraic additions.

$$49 + 2 \qquad \text{added 64 and } -15$$

$$\mathbf{51} \qquad \text{added 49 and 2}$$

The notation used in this problem is unduly complicated. If symbols of inclusion are used and if a fraction line is used to designate division, the same expression can be written in any way that clearly indicates what we must do.

$$\left(\frac{32}{2} \cdot 4 \right) - (3 \cdot 5) + 2 \qquad \text{expression}$$

$$64 - 15 + 2 \qquad \text{simplified within parentheses}$$

$$\mathbf{51} \qquad \text{added algebraically}$$

For this reason we will avoid the use of the symbol $\div$, and instead we will use a fraction bar to indicate division. We will also use parentheses, brackets, and braces to clarify our expressions. We will begin by simplifying within these symbols of inclusion and will remember to multiply before performing algebraic addition.

example 95.1   Simplify: $\dfrac{4 - (3 - 7) + 3 \cdot 2 + 6}{2(-4 - 1)}$

solution   We will simplify above and below and will divide as the last step.

$$\frac{4 - (-4) + 3 \cdot 2 + 6}{2(-5)} \qquad \text{simplified within parentheses}$$

$$\frac{4 - (-4) + 6 + 6}{-10} \qquad \text{multiplied}$$

$$\frac{20}{-10} \qquad \text{added}$$

$$-2 \qquad \text{divided}$$

example 95.2 Simplify: $\dfrac{6(-4 + 3) - (-2 - 6)(2)}{3(2 - 4)}$

solution Again we simplify above and below.

$$\frac{6(-1) - (-8)(2)}{3(-2)} \qquad \text{simplified within parentheses}$$

$$\frac{-6 + 16}{-6} \qquad \text{multiplied}$$

$$\frac{10}{-6} \qquad \text{added}$$

$$-\frac{5}{3} \qquad \text{simplified}$$

We leave the answer as an improper fraction.

practice Simplify: $\dfrac{6 - (2 - 5) + 4 \cdot 2 + 3}{2(-1 - 1)}$

problem set 95

1. A number was multiplied by 4. Then this product was increased by 4. If the result was 24, what was the number?

2. A number was multiplied by -4. Then this product was increased by -4. If the result was -24, what was the number?

3. Seventy-five less than 30 times a number is -225. What is the number?

4. The container overflowed because the kids tried to pour in $3\frac{3}{5}$ times the amount it would hold. If they tried to pour in 288 gallons, how much would the container hold?

Graph:

5. $x \not< 2$

6. $x \not\geq -2$

7. What number is 130 percent of 220? Draw a diagram of the problem.

8. If 230 is increased by 60 percent, what is the resulting number? Draw a diagram of the problem.

9. What percent of 70 is 112? Draw a diagram of the problem.

10. Seventy-six is what percent of 95? Draw a diagram of the problem.

11. Complete the table. Begin by inserting the reference numbers.

12. Simplify: $(3 \times 10^{-17})(2 \times 10^{-15})$

Fraction	Decimal	Percent
(a)	0.53	(b)

13. Express in cubic inches the volume of a right solid whose base is shown on the left and whose sides are 2 feet high. Dimensions are in inches.

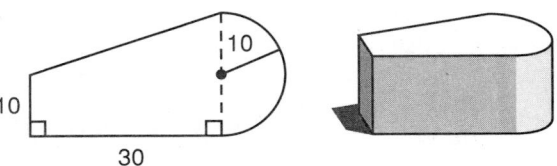

14. Find the perimeter of the figure shown. Dimensions are in yards. All angles are right angles.

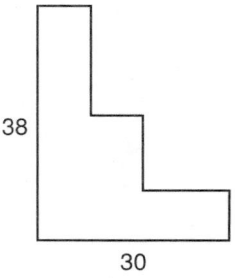

15. Use two unit multipliers to convert 12.687 kilometers to centimeters.

Solve:

16. $-2x - 12 = -2$

17. $-3x + 6 = 4$

18. $\dfrac{4\frac{1}{5}}{2\frac{1}{3}} = \dfrac{1\frac{1}{2}}{x}$

19. $-2\frac{1}{3}x - \frac{3}{4} = 1\frac{3}{8}$

Simplify:

20. $3^2 + 2^3[2^2(\sqrt{16} - \sqrt{9}) - (\sqrt{25} - \sqrt{4})]$

21. $-2(-6 - 2 \cdot 5) + 2(-2 - 1)$

22. $-1(6 - 2 \cdot 2) - 3(2 \cdot 1 - 3)$

23. $\dfrac{5 - (8 - 6) + 3 \cdot 4 + 6}{5(4 - 5)}$

24. $\dfrac{5 - (4 - 6)2 + 4 \cdot 2}{3(-2 - 1)}$

25. $\dfrac{3 - (6 - 3)2 + 3 \cdot 2}{4(-1 + 3)}$

26. $\dfrac{1}{4}\left(2\frac{1}{3} \cdot \frac{1}{4} - \frac{5}{6} \cdot \frac{1}{2}\right)$

27. $3\frac{1}{3} \times 2\frac{1}{4} \div \frac{1}{8} \times \frac{1}{2}$

Evaluate:

28. $-bc - ac$ if $a = -1$, $b = -3$, and $c = -5$

29. $a^2 + b^2 + c^2 + 2abc + ab$ if $a = 3$, $b = 2$, and $c = 4$

30. Use the method of cut and try to find $\sqrt[4]{20}$ to one decimal place. Show your work.

LESSON 96   *Adding like terms, Part 1*

We know that multiplication is shorthand for repeated addition. If we write

$$(5 + 5 + 5 + 5) + (5 + 5 + 5) + 4 + 2$$

We could combine the 5s and write

$$4(\text{fives}) + 3(\text{fives}) + 4(\text{ones}) + 2(\text{ones})$$

Now 4 fives + 3 fives is 7 fives. Four ones plus 2 ones is 6 ones.

$$7(\text{fives}) + 6(\text{ones})$$

We can make no further simplifications by combining like numbers because 5 and 1 are different numbers. Now suppose we write x in place of each 5. We get

$$(x + x + x + x) + (x + x + x) + 4(\text{ones}) + 2(\text{ones})$$

We can combine the x's and get

$$4x + 3x + 4(1) + 2(1)$$

Now we add $4x$ and $3x$ and $4(1) + 2(1)$.

$$7x + 6$$

We cannot add $7x$ and 6 because this is a general expression and x could represent any number. We could add the x's because in any problem every x must represent the same number everywhere in the problem. We say that

$$4x \qquad \text{and} \qquad 3x$$

are **like terms** because the variable x represents the same number in each expression.

example 96.1   Simplify by adding like terms: $-3x + 2y + 2x + 3 + 7$

solution   We can combine $-3x$ and $2x$ to get $-x$. These terms are like terms because x must represent the same number everywhere it appears in a problem. We can add 3 and 7 to get 10. The $2y$ term cannot be added to the x's because y often represents some number other than x. The answer is

$$-x + 2y + 10$$

example 96.2   Simplify: $3x + 2m - 14x + 2x - 4m + 4 - 2$

solution   By adding like terms, we can combine 4 and -2. We can combine the m terms and the x terms. We get

$$-9x - 2m + 2$$

practice   Simplify by combining like terms:

a. $-4x + 7x + 4 - y - x - 8 + 4y$

b. $y + 3y - 4 + 10 - 6y + 2m - 6m$

c. $a + 3b - 6a + 4 - 10 + 7a - 2 + 8b$

problem set 96

1. A number was multiplied by 6. Then this product was increased by 3. If the result was 33, what was the number?

2. The product of a number and −3 is increased by −6. If the result is −51, what is the number?

3. Sixty-three less than 15 times a number is 72. What is the number?

4. Five-eighths of the ranch animals were equine. If 330 ranch animals were not equine, what was the total number of ranch animals?

5. The ratio of idle onlookers to serious spectators in the crowd was 2 to 29. If there were 8410 spectators, how many idle onlookers were in the crowd?

6. The number who attended this year's conference was $2\frac{3}{8}$ times the number who attended the conference last year. If 4180 people attended this year, how many attended last year?

Graph:

7. $x \not> 3$ 8. $x \not\leq 3$

9. What number is 160 percent of 350? Draw a diagram of the problem.

10. If 95 is increased by 40 percent, what is the resulting number? Draw a diagram of the problem.

11. What percent of 80 is 360? Draw a diagram of the problem.

12. Seventy-five is what percent of 125? Draw a diagram of the problem.

13. Write in scientific notation and simplify: $600,000 \times 0.0000000001$

14. Find the volume and surface area of a right circular cylinder with a diameter of 200 centimeters and a height of 100 centimeters.

15. Express in cubic feet the volume of a right solid whose base is shown on the left and whose sides are 5 feet high. Dimensions are in feet.

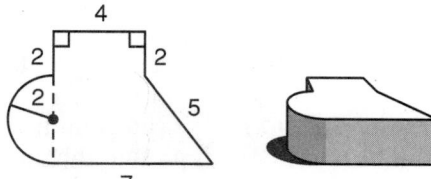

16. Find the lateral surface area of the solid in Problem 15.

17. Use two unit multipliers to convert 1,000,000 square meters to square kilometers.

Simplify by combining like terms:

18. $-6m + 2m + 8 - p - m$ 19. $5x + 5 - 11x - 9 + 2x$

Solve:

20. $-5x + 9 = 6$ 21. $\dfrac{2\frac{1}{4}}{5\frac{2}{5}} = \dfrac{x}{4}$ 22. $-4\frac{2}{3}x - 1\frac{1}{2} = 1\frac{1}{4}$

Simplify:

23. $\sqrt{81} + 4^3[7^0(2^3 - 3^2) - (\sqrt{4} - 1^{10})]$

24. $-3(-5 - 2 \cdot 3) + 3[5 + 2(-3)]$ 25. $\dfrac{-6 - (3 - 4)\,3 + 2\,(-2)}{3(-2 - 1)}$

26. $(7.81 \times 10^{-3}) \times (1 \times 10^{15})$

Evaluate:

27. $-[bp - (-b)]$ if $b = -3$ and $p = 2$

28. $a^2b + \sqrt{ab}$ if $a = 2$ and $b = 8$

29. Write in scientific notation and simplify: $4,800,000 \times 10,000$

30. Use the method of cut and try to find $\sqrt[5]{18}$ to one decimal place. Show your work.

LESSON 97 *Variables on both sides*

Some equations have a variable on both sides of the equation. To solve these equations, we first add as necessary to eliminate the variable on one side or on the other side.

example 97.1   Solve: $3x + 3 = x - 5$

solution   We can eliminate either the $3x$ or the x. We decide to eliminate the x, so we add $-x$ to both sides.

$$
\begin{array}{ll}
3x + 3 = x - 5 & \text{equation} \\
\underline{-x -x} & \text{add } -x \text{ to both sides} \\
2x + 3 = -5 &
\end{array}
$$

Now we eliminate the 3 by adding -3 to both sides. Then we divide both sides by 2.

$$
\begin{array}{ll}
2x + 3 = -5 & \\
\underline{ -3 -3} & \text{add } -3 \text{ to both sides} \\
2x = -8 & \\
\\
x = -4 & \text{divided by 2}
\end{array}
$$

Check:

$$3(-4) + 3 = (-4) - 5 \longrightarrow -12 + 3 = -9 \longrightarrow -9 = -9 \quad \text{check}$$

example 97.2   Solve: $3x + 3 = x - 5$

solution   This is the same problem as the preceding problem. This time we will eliminate the $3x$ by adding $-3x$ to both sides.

$$
\begin{array}{ll}
3x + 3 = x - 5 & \\
\underline{-3x -3x} & \text{add } -3x \text{ to both sides} \\
3 = -2x - 5 &
\end{array}
$$

Now we will eliminate the -5 by adding $+5$ to both sides.

$$
\begin{array}{ll}
3 = -2x - 5 & \\
\underline{5 +5} & \text{add } + 5 \text{ to both sides} \\
8 = -2x &
\end{array}
$$

Next we change the sign of $-2x$ by multiplying both sides by -1. Then we divide both sides by $+2$.

$$
\begin{array}{ll}
-8 = 2x & \text{multiplied by } -1 \\
\\
-4 = x & \text{divided by 2}
\end{array}
$$

example 97.3   Solve: $4x + 2 = 2x + 5$

solution   We can either eliminate the $4x$ or the $2x$. We decide to eliminate the $4x$, so we add $-4x$ to both sides.

$$
\begin{array}{rl}
4x + 2 = 2x + 5 & \text{equation} \\
\underline{-4x -4x } & \text{add } -4x \text{ to both sides} \\
2 = -2x + 5 &
\end{array}
$$

Now we add -5 to both sides. Then we change signs and divide by 2.

$$
\begin{array}{rl}
2 = -2x + 5 & \\
\underline{-5 -5} & \text{add } -5 \text{ to both sides} \\
-3 = -2x & \\
3 = 2x & \text{multiplied both sides by } -1 \\
\dfrac{3}{2} = x & \text{divided by 2}
\end{array}
$$

Check:

$$4\left(\frac{3}{2}\right) + 2 = 2\left(\frac{3}{2}\right) + 5 \quad \longrightarrow \quad 6 + 2 = 3 + 5 \quad \longrightarrow \quad 8 = 8 \quad \text{check}$$

example 97.4   Solve: $4x - 2 = -x$

solution   This time we will eliminate the $-x$ on the right by adding $+x$ to both sides.

$$
\begin{array}{rl}
4x - 2 = -x & \text{equation} \\
\underline{+x +x} & \text{add } +x \text{ to both sides} \\
5x - 2 = 0 &
\end{array}
$$

Now we add $+2$ to both sides and then divide by 5.

$$
\begin{array}{rl}
5x - 2 = 0 \\
\underline{+2 +2} \\
5x = 2 \\
x = \dfrac{2}{5}
\end{array}
$$

Check:

$$4\left(\frac{2}{5}\right) - 2 = -\left(\frac{2}{5}\right) \quad \longrightarrow \quad \frac{8}{5} - \frac{10}{5} = -\frac{2}{5} \quad \longrightarrow \quad -\frac{2}{5} = -\frac{2}{5} \quad \text{check}$$

practice   Solve:

a. $-4x + 3 = 7x - 8$ **b.** $2x - 5 = 6x + 4$

problem set 97

1. The product of a number and 5 was decreased by 4 and the result was 96. What was the number?

2. The sum of twice a number and 7 equals 27. What is the number?

3. In 4 hours the big horses pulled the wagon 16 miles. Then they tired and slowed down to half speed. How long did it take them to pull the wagon the last 10 miles?

4. The ratio of red hats to polka-dotted hats was 5 to 13. If 260 hats were polka-dotted, how many hats were red?

Graph:

5. $x \not> -1$

6. $x \not\leq 3$

7. What number is 160 percent of 240? Draw a diagram of the problem.

8. If 270 is increased by 80 percent, what is the resulting number? Draw a diagram of the problem.

9. What percent of 60 is 96? Draw a diagram of the problem.

10. Twenty percent of what number is 60? Draw a diagram of the problem.

11. Write in scientific notation and simplify: $640{,}000{,}000{,}000 \times 100{,}000$

12. Express in cubic meters the volume of a right solid whose base is shown on the left and whose sides are 2 meters high. Dimensions are in meters.

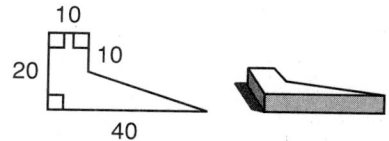

13. Use the method of cut and try to find $\sqrt[3]{11}$ to one decimal place.

Solve:

14. $-3x - 11 = -3$

15. $-2x + 3 = 2$

16. $4x + 6 = x - 5$

17. $2x - 5 = -x$

18. $6x + 6 = 2x - 4$

19. $-3\frac{1}{4}x + \frac{2}{3} = 1\frac{5}{6}$

Simplify:

20. $-2(6 - 3 \cdot 2) - 2(3 \cdot 1 - 2)$

21. $-3(6 - 2 \cdot 3) - 3(2 \cdot 7 - 6)$

22. $-4(-2 - 3 \cdot 4) - 2(3 \cdot 4 - 7)$

23. $\dfrac{6 - (2 - 3)3 + 4 \cdot 1}{3(-1 + 2)}$

24. $\dfrac{6 - (3 - 5)2 + 1 \cdot 2}{6(-2 + 3)}$

25. $\frac{1}{3}\left(2\frac{1}{2} \cdot \frac{1}{3} - \frac{3}{4} \cdot \frac{1}{2}\right)$

26. $2^2 + 2^3[3^2(\sqrt{36} - 5)(2^4 - 3 \cdot 5)]$

27. $3\frac{1}{2} \times 2\frac{1}{3} \div 6\frac{1}{3} \div \frac{1}{2}$

Evaluate:

28. $bc - ab$ if $a = -3$, $b = 2$, and $c = -1$

29. $xy^2 + yx^2 + x^3y - xy$ if $x = 2$ and $y = 3$

30. Write 67821132.3 in words.

LESSON 98 *Two-step problems*

Some of the problems that have appeared in the problem sets thus far have required two steps for their solutions. In this lesson we will look at two-step problems that require the solution of an equation as the first step. These problems are easy to recognize because they almost always contain the word **if**.

example 98.1   If $2x + 4 = 6$, what is the value of $3x - 7$?

solution   The word **if** tells us that the problem is a two-step problem. We begin by solving the equation to find the value of x.

$$
\begin{array}{rll}
2x + 4 = & 6 & \text{equation} \\
\underline{-4 \quad -4} & & \text{add } -4 \text{ to both sides} \\
2x \quad = & 2 & \\
x = & 1 & \text{divided}
\end{array}
$$

Now we use 1 for x to find the value of $3x - 7$.

$$
\begin{array}{rl}
3(1) - 7 & \text{substituted} \\
3 - 7 & \text{multiplied} \\
\mathbf{-4} & \text{added}
\end{array}
$$

example 98.2   If $4x - 2 = 3$, what is the value of $\frac{2}{5}x - \frac{1}{4}$?

solution   As the first step, we solve the equation to find the value of x.

$$
\begin{array}{rll}
4x - 2 = & 3 & \text{equation} \\
\underline{+2 \quad +2} & & \text{add } +2 \text{ to both sides} \\
4x \quad = & 5 & \\
x = & \dfrac{5}{4} &
\end{array}
$$

Now we use $\frac{5}{4}$ for x to find the value of $\frac{2}{5}x - \frac{1}{4}$.

$$
\begin{array}{rl}
\dfrac{2}{5}\left(\dfrac{5}{4}\right) - \dfrac{1}{4} & \text{substituted} \\[2mm]
\dfrac{1}{2} - \dfrac{1}{4} & \text{multiplied} \\[2mm]
\mathbf{\dfrac{1}{4}} & \text{added}
\end{array}
$$

example 98.3   If $\frac{4}{3}x - 2 = 4$, what is the value of $6x - \frac{2}{5}$?

solution   We will solve the equation for x.

$$
\begin{array}{rll}
\dfrac{4}{3}x - 2 = & 4 & \text{equation} \\
\underline{+2 \quad +2} & & \text{add } +2 \text{ to both sides} \\
\dfrac{4}{3}x \quad = & 6 &
\end{array}
$$

$$\frac{3}{4} \cdot \frac{4}{3}x = 6 \cdot \frac{3}{4}$$

$$x = \frac{9}{2} \qquad \text{solved}$$

Now we use $\frac{9}{2}$ for x to find the value of $6x - \frac{2}{5}$.

$$6\left(\frac{9}{2}\right) - \frac{2}{5} \qquad \text{substituted}$$

$$27 - \frac{2}{5} \qquad \text{multiplied}$$

$$26\frac{3}{5} \qquad \text{added}$$

practice **a.** If $2x - 4 = 6$, what is the value of $3x + 2$?

b. If $5x - 2 = 5$, what is the value of $10x - 3$?

c. If $\frac{1}{5}x = \frac{3}{10}$, what is the value of $15x + 2$?

problem set 98

1. The sum of 7 times a number and 42 was -98. What was the number?

2. A number was multiplied by -4. Then -6 was added. If the final result was -34, what was the number?

3. If 140 of the new ones cost \$980, what would 200 of the new ones cost?

4. In an attempt to cut them off at the pass, the posse rode 14 miles in only 2 hours. The horses were tired so they reduced their speed by 2 miles per hour. If it was still 15 miles to the pass, how much longer did it take to get there?

Graph:

5. $x \not\geq 3$

6. $x \not\leq 2$

7. What number is 230 percent of 270? Draw a diagram of the problem.

8. If 260 is increased by 170 percent, what is the resulting number? Draw a diagram of the problem.

9. What percent of 80 is 116? Draw a diagram of the problem.

10. Thirty percent of what number is 180? Draw a diagram of the problem.

11. Complete the table. Begin by inserting the reference numbers.

Fraction	Decimal	Percent
$\frac{9}{10}$	(a)	(b)

12. Express in cubic centimeters the volume of a right solid whose base is shown on the left and whose sides are 12 centimeters high. Dimensions are in centimeters.

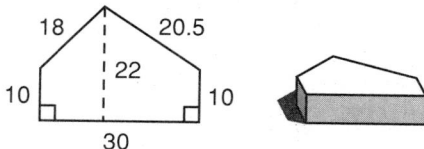

13. Find the lateral surface area of the solid in Problem 12.

14. Find the perimeter of the figure shown. Dimensions are in feet. All angles are right angles.

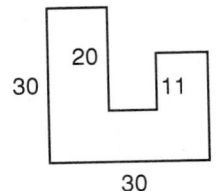

15. Use two unit multipliers to convert 1287.321 centimeters to kilometers.

16. If $3x - 9 = 6$, what is the value of $2x - 6$?

17. If $6x - 2 = 10$, what is the value of $\frac{1}{2}x + 3$?

18. If $\frac{3}{4}x - 2 = 1$, what is the value of $3x - 4$?

Solve:

19. $-4x - 12 = -1$

20. $4x + 2 = 2x - 4$

21. $-2\frac{1}{3}x + \frac{2}{3} = 1\frac{4}{6}$

22. $\dfrac{3\frac{1}{2}}{2\frac{1}{3}} = \dfrac{\frac{1}{2}}{x}$

Simplify:

23. $-3(6 - 1 \cdot 4) - 3(2 \cdot 6 - 4)$

24. $-2(8 - 2 \cdot 3) - 4(3 \cdot 1 - 2)$

25. $\dfrac{7 - (2 - 4)2 + 5 \cdot 1}{4(-2 + 3)}$

26. $\dfrac{8 - (4 - 7)3 + 4 \cdot 2}{5(-3 + 5)}$

27. $3^2 + 3[2^3(\sqrt{49} - 2^2)(3^2 - 2^3) - 2^2]$

28. $3\frac{1}{3} \times 2\frac{1}{5} \div 1\frac{2}{3} \div \frac{1}{4}$

29. Evaluate: $bcb - ac$ if $a = -1$, $b = -2$, and $c = -5$

30. Use the method of cut and try to find $\sqrt[4]{31}$ to one decimal place.

LESSON 99 *Unequal quantities*

Some word problems tell us that one quantity is a certain amount greater than another quantity. Other word problems tell us that a quantity is a certain amount less than another quantity. To solve these problems, we must use equations. We add to both sides of the equations as necessary to make the quantities equal.

example 99.1   Twice a number is 56 less than -72. What is the number?

solution   **We must be careful because the problem tells us about things that are not equal. We begin by writing an equation that we know is incorrect.**

$$2N = -72 \qquad \text{incorrect}$$

The problem said that $2N$ is 56 less than -72. To make the equation correct, we must add 56 to $2N$ or add -56 to -72.

ADDING 56 TO $2N$

$$2N + 56 = -72$$
$$\underline{-56 \quad\quad -56}$$
$$2N \quad\quad = -128$$
$$N = \;\; -64$$

correct

ADDING -56 TO -72

$$2N = -72 - 56$$
$$2N = -128$$
$$N = -64$$

correct

Check:

$$2(-64) + 56 = -72 \;\longrightarrow\; -128 + 56 = -72 \;\longrightarrow\; -72 = -72 \quad \text{check}$$

example 99.2 Six times a number is 14 less than the opposite of the number. What is the number?

solution **Again we begin by writing an equation that we know is incorrect.**

$$6N = -N \quad \text{incorrect}$$

We know that $6N$ is 14 less than $-N$. We can make the equation a correct equation by adding 14 to $6N$ or by adding -14 to $-N$.

ADDING 14 TO $6N$

$$6N + 14 = -N$$
$$\underline{+N \quad\quad\quad +N}$$
$$7N + 14 = 0$$

ADDING -14 TO $-N$

$$6N = -N - 14$$
$$\underline{+N \quad +N}$$
$$7N = -14$$

On the left we add -14 to both sides before dividing. On the right we divide.

$$7N + 14 = \quad\; 0$$
$$\underline{\quad\quad -14 \quad -14}$$
$$\dfrac{7N}{} \quad\quad = -14$$
$$N = \;\; -2$$

$$7N = -14$$
$$N = \;\; -2$$

practice **a.** Seven times a number is 8 greater than the sum of the number and 10. What is the number?

b. Three times a number is 15 less than twice the opposite of the number. What is the number?

problem set 99

1. Twice a number is 34 less than -14. What is the number?

2. Six times a number is 14 less than the opposite of the number. What is the number?

3. Five times a number is 49 greater than the product of 2 and the opposite of the number. What is the number?

4. Seven times a number is 30 greater than the sum of twice the number and 5. What is the number?

Graph:

5. $x \not\leq -1$

6. $x \not\geq -3$

7. What number is 320 percent of 250? Draw a diagram of the problem.

8. If 250 is increased by 190 percent, what is the resulting number? Draw a diagram of the problem.

9. What percent of 90 is 162? Draw a diagram of the problem.

10. Forty percent of what number is 62? Draw a diagram of the problem.

11. Use the method of cut and try to find $\sqrt[5]{43}$ to one decimal place.

12. Find the least common multiple of 20, 24, and 30.

13. Find the volume of a solid whose base is shown on the left and whose sides are 10 centimeters high. Dimensions are in centimeters.

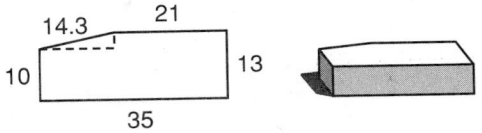

14. Find the lateral surface area of the solid in Problem 13.

15. Use two unit multipliers to convert 3 miles to inches.

16. If $6x - 2 = 16$, what is the value of $x - 3$?

17. If $3x + 2 = 8$, what is the value of $\frac{1}{3}x + \frac{1}{2}$?

18. If $\frac{3}{2}x - 3 = 1$, what is the value of $2x + 3$?

Solve:

19. $-3x + 12 = 1$

20. $6x + 2 = 3x - 5$

21. $-6x + 2 = 3x - 6$

22. $-3\frac{1}{4}x + \frac{3}{4} = 1\frac{3}{8}$

Simplify:

23. $-6(2 \cdot 3 - 2 \cdot 4) - 2(6 \cdot 1 - 2)$

24. $-2(3 \cdot 2 - 7) - 3(2 \cdot 1 - 5) + \sqrt[4]{\frac{16}{81}}$

25. $\dfrac{6 - 3(2 - 1) + 4 \cdot 3}{5(-3 + 6)}$

26. $\dfrac{8 - 2(3 - 2) + 4 \cdot 2}{4(2 - 3)}$

27. $2^3 + 3[2^2(\sqrt{36} + \sqrt{4})(\sqrt[3]{8} - 1) + 1]$

28. $2\frac{1}{4} \times 3\frac{1}{3} \div 2\frac{1}{3} \times 1\frac{1}{4}$

Evaluate:

29. $d + c - ac$ if $a = -1$, $c = -1$, and $d = 1$

30. $abc - a$ if $a = -2$, $b = -1$, and $c = -3$

LESSON 100 Exponents and signed numbers

We must be careful when we simplify exponential expressions (powers) whose bases are signed numbers. The exponential expression

$$4^2$$

indicates that 4 is to be used as a factor twice.

$$(4)(4) = 16$$

The notation

$$-4^2$$

indicates the opposite of "4 squared," which is

$$-(4)(4) = -16$$

If we wish to indicate that -4 is to be used as a factor twice, we must enclose the -4 in parentheses and write

$$(-4)^2$$

This means -4 times -4 so

$$(-4)(-4) = 16$$

example 100.1 Simplify: $-4^2 - (-3)^2 - 3^3$

solution First we simplify the exponential expressions,

$$-16 - 9 - 27$$

and then we add and get

$$-52$$

example 100.2 Simplify: $(-3)^3 - (-2)^3 - 2^2$

solution First we simplify the exponential expressions.

$$-27 - (-8) - 4$$

Then we add.

$$-27 + 8 - 4 \qquad \text{simplified within parentheses}$$
$$-23 \qquad \text{added}$$

practice Simplify:

a. $-(-2)^3 - (-2)^2$

b. $-2^2 - (-2)^2 - (-2)^3$

problem set 100

1. Five times a number is 35 greater than 2 times the opposite of the number. What is the number?

2. Seven times the opposite of a number is 50 less than 3 times the number. What is the number?

3. Only $\frac{3}{5}$ of the students in the school came to hear the distinguished scholar. If there were 1200 students in the school, how many came to hear the distinguished scholar?

4. The ratio of those who stayed awake to those who dozed was $2\frac{1}{2}$ to 7. If 1400 stayed awake, how many dozed?

Graph:

5. $x \nless -2$

6. $x \ngtr -1$

7. What number is 160 percent of 230? Draw a diagram of the problem.

8. If 180 is increased by 180 percent, what is the resulting number? Draw a diagram of the problem.

9.   What percent of 60 is 111? Draw a diagram of the problem.

10. Twenty-five percent of what number is 61? Draw a diagram of the problem.

11. Write 7.34×10^{-5} in standard form.

12. Write $\dfrac{3121}{7}$ as a mixed number.

13. Use the method of cut and try to find $\sqrt[3]{215}$ to one decimal place.

14. Find the volume of a solid whose base is shown on the left and whose sides are 6 feet high. Dimensions are in feet.

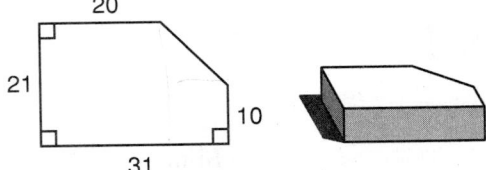

15. Use two unit multipliers to convert 7892.321 centimeters to kilometers.

16. If $3x - 5 = 7$, what is the value of $3x - 2$?

17. If $4x + 2 = 11$, what is the value of $2x - 1$?

18. If $\frac{1}{2}x - 2 = 3$, what is the value of $\frac{1}{2}x + 3$?

Solve:

19. $-4x + 3 = x + 4$

20. $3x - 4 = -x + 6$

21. $-3\frac{1}{3}x - \frac{1}{6} = 1\frac{1}{3}$

22. $\dfrac{6\frac{1}{2}}{2\frac{1}{3}} = \dfrac{\frac{1}{2}}{x}$

Simplify:

23. $-2^2 - (-3)^2 - 3^4$

24. $-3^2 + (-2)^2$

25. $-2^2(2 \cdot 2 - 3 \cdot 2) - 3(2 \cdot 3 - 2^2)$

26. $\dfrac{3 - 2(3 - 2) + 3 \cdot 2}{4(-2 + 5)}$

27. $\dfrac{-3 - 2(3 \cdot 2 - 1) + 2^2}{2^2(-3 - 2^2)}$

28. $3\frac{1}{2} \times 2\frac{1}{3} \div \frac{1}{4} \div \frac{1}{6}$

Evaluate:

29. $abc + ab$ if $a = -2$, $b = -3$, and $c = -4$

30. $-ab + bc$ if $a = 2$, $b = -3$, and $c = -4$

LESSON 101   *Advanced ratio problems*

In this lesson we will consider advanced ratio problems. We begin with an example.

example 101.1   The ratio of nuts to bolts was 11 to 2. If there were 260 nuts and bolts in the pile, how many were bolts?

solution We begin confidently by writing

$$\frac{N}{B} = \frac{11}{2}$$

Now where does the 260 go? What do we do now? The answer is that this problem is more difficult than the ratio problems we have been working and we have begun the wrong way. This problem has nuts and bolts but also has the **total**. There is a trick to working these problems. **The trick is to begin by listing the given information to include the total.**

$$\text{Nuts} = 11$$
$$\text{Bolts} = 2$$
$$\text{Total} = 13$$

There are three proportions hidden here. We can see the proportions if we cover up the lines one at a time with a finger. First we cover the top line. We see that what remains concerns bolts and total.

$$\text{Bolts} = 2 \quad \longrightarrow \quad \frac{B}{T} = \frac{2}{13} \qquad \text{(a)}$$
$$\text{Total} = 13$$

Now we cover the middle line and see that what remains concerns nuts and total.

$$\text{Nuts} = 11$$
$$\longrightarrow \quad \frac{N}{T} = \frac{11}{13} \qquad \text{(b)}$$
$$\text{Total} = 13$$

Now we cover the bottom line and see that what remains concerns nuts and bolts.

$$\text{Nuts} = 11$$
$$\text{Bolts} = 2 \quad \longrightarrow \quad \frac{N}{B} = \frac{11}{2} \qquad \text{(c)}$$

We see that the given information was enough to write three proportions. Ratio (a) is between bolts and total. Ratio (b) is between nuts and total. Ratio (c) is between nuts and bolts. We were told that we had a total of 260 and were asked for the number of bolts. So we want to use ratio (a).

$$\frac{B}{T} = \frac{2}{13} \qquad \text{(a)}$$

$$\frac{B}{260} = \frac{2}{13} \qquad \text{substituted}$$

$$B = \frac{2 \cdot 260}{13} \qquad \text{multiplied both sides by 260}$$

$$B = 40 \qquad \text{simplified}$$

In ratio problems that discuss the total, it is helpful to begin by writing all three equations. Then reread the problem to see which equation should be used.

example 101.2 Farmer Dunncowski wanted to plant his farm in wheat and corn in the ratio of 7 to 9. If he had 640 acres, how many should be planted in wheat?

solution First we write down the figures given and also write the total.

$$W = 7$$
$$C = 9$$
$$T = 16$$

With this information, we can write 3 proportions.

(a) $\dfrac{W}{C} = \dfrac{7}{9}$ (b) $\dfrac{W}{T} = \dfrac{7}{16}$ (c) $\dfrac{C}{T} = \dfrac{9}{16}$

We were given the total and asked for wheat, so we will use proportion (b). In (b) we replace total with 640.

$$\dfrac{W}{T} = \dfrac{7}{16} \qquad \text{proportion (b)}$$

$$\dfrac{W}{640} = \dfrac{7}{16} \qquad \text{replaced } T \text{ with } 640$$

$$16W = 7 \cdot 640 \qquad \text{cross multiplied}$$

$$\dfrac{16W}{16} = \dfrac{7 \cdot 640}{16} \qquad \text{divided both sides by 16}$$

$$\boldsymbol{W = 280} \qquad \text{simplified}$$

If he plants 280 acres in wheat, then $640 - 280 = 360$ acres will be planted in corn.

example 101.3 When the race began, the ratio of fats to leans was 2 to 17. If 3800 racers were in the race, how many were lean?

solution First we write down the numbers for fats and leans. We add these two numbers to get the total.

$$F = 2$$
$$L = 17$$
$$T = 19$$

Now we can write the three proportions.

(a) $\dfrac{F}{L} = \dfrac{2}{17}$ (b) $\dfrac{F}{T} = \dfrac{2}{19}$ (c) $\dfrac{L}{T} = \dfrac{17}{19}$

We were given a total of 3800 and asked for the number of leans. Thus, we will use proportion (c) and replace T with 3800.

$$\dfrac{L}{T} = \dfrac{17}{19} \qquad \text{proportion (c)}$$

$$\dfrac{L}{3800} = \dfrac{17}{19} \qquad \text{replaced } T \text{ with } 3800$$

$$19L = 17 \cdot 3800 \qquad \text{cross multiplied}$$

$$\dfrac{19L}{19} = \dfrac{17 \cdot 3800}{19} \qquad \text{divided both sides by 19}$$

$$\boldsymbol{L = 3400} \qquad \text{simplified}$$

If 3400 out of 3800 were lean, then 400 must have been fat.

practice   The ratio of big ships to small ships in the harbor was 2 to 9. If there were 242 ships in the harbor, how many were big ships? How many were small ships?

problem set 101

1. Farmer Brown wanted to plant her farm in rye and barley in the ratio of 7 to 5. If she had 240 acres, how many acres should she plant in rye? How many acres should she plant in barley?

2. The ratio of green ones to purple ones balanced on the ledge was 3 to 17. If there were 400 balanced on the ledge, how many were green?

3. Four times a number was 22 greater than twice the same number. What was the number?

4. Five-sixteenths of the teenies swooned as the star approached the microphone. If 2200 did not swoon, how many teenies attended the concert?

5. Graph: $x \not\le -1$

6. What number is 140 percent of 75? Draw a diagram of the problem.

7. If 150 is increased by 30 percent, what is the resulting number? Draw a diagram of the problem.

8. What percent of 70 is 91? Draw a diagram of the problem.

9. Thirty-five is what percent of 140? Draw a diagram of the problem.

10. Use the method of cut and try to find $\sqrt[4]{180}$ to one decimal place.

11. Find the least common multiple of 16, 18, and 20.

12. Complete the table. Begin by inserting the reference numbers.

Fraction	Decimal	Percent
(a)	(b)	85

13. Find the volume in cubic feet of a solid whose base is shown on the left and whose sides are 2 yards $_{6 \, ft}$ high. Dimensions are in feet.

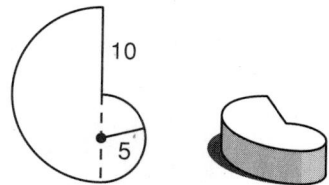

14. Find the total surface area of the solid in Problem 13.

15. Use two unit multipliers to convert 61.131121 kilometers to centimeters.

16. If $2x - 5 = 7$, what is the value of $6x - 1$?

17. If $5x - 3 = 22$, what is the value of $\frac{1}{5}x + 2$?

18. If $\frac{3}{4}x + 2 = 23$, what is the value of $\frac{1}{2}x + 2$?

Solve:

19. $-3x + 6 = 2x + 1$

20. $2x - 4 = -x + 5$

21. $-2\frac{1}{2}x - \frac{1}{3} = 2\frac{2}{3}$

22. $\dfrac{\frac{3}{5}}{\frac{6}{25}} = \dfrac{\frac{1}{3}}{x}$

Simplify:

23. $-2^2 - (-2)^3$

24. $-3^2 - (-3)^2$

25. $-8(2 \cdot 3 - 4) - 3(8 - 9)$

26. $\dfrac{4 - 2(6 - 4) + 2^2 \cdot 3}{3(-3 + 5)}$

27. $\dfrac{-2 + 3(2 \cdot 4 - 3) + 2^2}{3(2^2 - 1)}$

28. $2\frac{1}{2} \div 3\frac{1}{3} \times \sqrt{\frac{9}{16}} \div \frac{1}{3}$

Evaluate:

29. $-bc - ab$ if $a = -2$, $b = -2$, and $c = -5$

30. $ab - ac$ if $a = -1$, $b = -2$, and $c = -1$

LESSON 102 *Multiplication of exponential expressions*

102.A
multiplication of exponential expressions

We remember that we use exponential notation to indicate how many times a number is to be used as a factor. To review, we consider the notation

$$3^5$$

The base of this exponential expression is 3 and the exponent is 5. This means that 3 is to be used as a factor 5 times.

$$3^5 = 3 \cdot 3 \cdot 3 \cdot 3 \cdot 3$$

Now we will consider the notation

$$3^5 \cdot 3^2$$

The first part tells us that 3 is a factor 5 times, and the second part tells us that 3 is a factor twice.

$$(3 \cdot 3 \cdot 3 \cdot 3 \cdot 3)(3 \cdot 3)$$

This means that 3 is a factor 7 times, which is written

$$3^7$$

We see from this that the number of times 3 is to be used as a factor can be found by adding the exponents.

$$3^5 \cdot 3^2 = 3^{5+2} = 3^7$$

example 102.1   Simplify: $5^2 \cdot 5^3 \cdot 5^4$

solution   This means

$$(5 \cdot 5)(5 \cdot 5 \cdot 5)(5 \cdot 5 \cdot 5 \cdot 5)$$

and 5 is used as a factor 9 times. Thus we can write

$$5^2 \cdot 5^3 \cdot 5^4 = \mathbf{5^9}$$

example 102.2   Simplify: $10^{15} \cdot 10^{17}$

solution   The first part has 10 as a factor 15 times, and the second part has 10 as a factor 17 times. Thus,

$$10^{15} \cdot 10^{17} = 10^{15+17} = 10^{32}$$

If we use 10 as a factor 32 times, we get

$$100,000,000,000,000,000,000,000,000,000,000$$

This numeral is cumbersome and difficult to work with as it is easy to miscount the number of zeros. Thus the exponential expression

$$\mathbf{10^{32}}$$

is a better way to write this number and is preferred for most applications.

example 102.3   Simplify: $2^3 \cdot 5^2$

solution   If we write the exponential expressions in expanded form, we get

$$(2 \cdot 2 \cdot 2)(5 \cdot 5)$$

We have 2 as a factor 3 times and 5 as a factor twice. This shows us why the expression

$$2^3 \cdot 5^2$$

cannot be simplified by adding the exponents. The bases are not the same.

102.B
variable bases

Since letters stand for numbers, all the rules for numbers also apply to letters. Thus if we write

$$x^2 \cdot x^3$$

we mean

$$(x \cdot x)(x \cdot x \cdot x)$$

Here we have x used as a factor 5 times, and we can write this as

$$x^5$$

We call the rule demonstrated by this example the **product rule for exponents.**

PRODUCT RULE FOR EXPONENTS

If m and n are real numbers and $x \neq 0$,

$$x^m \cdot x^n = x^{m+n}$$

This is a very formal way of saying that

> We multiply exponential expressions that have the same base by adding the exponents.

example 102.4   Simplify: $x \cdot x^2 \cdot y^{17} \cdot x^4 \cdot y^6$

solution   **When we write x, we mean x^1.** We rearrange the expression and get

$$xx^2x^4y^{17}y^6$$

Now we add exponents of like bases and get

$$x^7 y^{23}$$

example 102.5   Simplify: $aa^2 mm^5 a^4 m^6$

solution   We will rearrange mentally. Then we add the exponents of like bases and get

$$a^7 m^{12}$$

practice   Simplify:

a. $3 \cdot 3^2 \cdot 3^5$

b. $xxx^5 x^2 yy^3$

c. $ab^2 ba^3 ba^4$

d. $yx^2 y^3 yy^4$

problem set 102

1.   In the very long hallway the ratio of red hats to green hats was 2 to 5. If 4900 hats were stored in the hallway, how many were red?

2.   Five times the opposite of a number was 56 less than twice the number. What was the number?

3.   Four hundred umbrellas cost $2000. What would 140 umbrellas cost?

4.   Wilson bought 60 items for $40 each and 40 items for $150 each. What was the average cost of the items?

5.   Graph: $x \not\le -2$

6.   What number is 130 percent of 80? Draw a diagram of the problem.

7.   If 160 is increased by 40 percent, what is the resulting number? Draw a diagram of the problem.

8.   What percent of 60 is 108? Draw a diagram of the problem.

9.   Forty percent of what number is 240? Draw a diagram of the problem.

10.   Find the volume in cubic centimeters of a right circular cylinder whose radius is 2 meters and whose height is 1 meter.

11.   Find the perimeter of the figure shown. Dimensions are in inches. All angles are right angles.

12.   Use the method of cut and try to find $\sqrt[4]{550}$ to one decimal place.

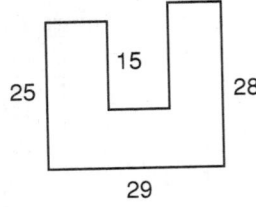

13.   If $3x - 6 = 9$, what is the value of $3x - 4$?

14.   If $4x - 4 = 28$, what is the value of $2x + 2$?

15.   If $3x + 4 = 31$, what is the value of $\frac{1}{4}x + 1$?

Solve:

16. $-2x + 4 = 3x + 2$

17. $3x - 2 = -2x + 5$

18. $-3\frac{1}{2}x - \frac{1}{2} = 2\frac{1}{4}$

19. $\dfrac{-\frac{2}{5}}{\frac{4}{25}} = \dfrac{-\frac{1}{2}}{x}$

Simplify:

20. $-2^3 - (-2)^3$

21. $-(-3)^2$

22. $(-3)^2 + (-2)^3$

23. $x^3 \cdot x^2 \cdot y^{15} \cdot x^6 \cdot y^6$

24. $5^{15} \cdot 5^{17}$

25. $aa^3mm^3am^4a^2$

26. $xx^3x^2yy^5y^6xy^2$

27. $\dfrac{2^3 - 2(3 - 1)(2^3 - 3)}{2^2(2^2 - 1)}$

28. $\dfrac{3 - 2^2(2^2 - 1)(2^3 - 1)}{3^2(3^2 - 2^3)}$

29. $3\frac{1}{2} \div \frac{1}{4} \times 2\frac{1}{3} \div \frac{1}{6}$

30.   Evaluate: $-ab + bc$ if $a = -2$, $b = -1$, and $c = 1$

LESSON *103* *Terms · Adding like terms, part 2*

103.A
terms

We remember that the terms of an algebraic expression are separated by plus and minus signs. This expression has 6 terms.

$$x + 2y + 3x + 5 - 7y - 2$$

We can simplify this expression by combining the x terms, the y terms, and the numbers. If we do this, we get

$$4x - 5y + 3$$

Terms can have more than one letter and the letters can be written as indicated multiplications or divisions. The expression

$$4xy + 6yx$$

has two terms. The coefficients of the terms are the numbers 4 and 6, respectively.

103.B
like terms

To see if

$$4xy \qquad \text{and} \qquad 6yx$$

are **like terms,** we consider only the letters, so we discard the coefficients and get

$$xy \qquad \text{and} \qquad yx$$

If these both represent the same number regardless of the numbers that are used for x and y, we say that the expressions are **like terms.** To investigate, we let $x = -2$ and $y = 3$. Then

$$xy = (-2)(3) = -6 \qquad \text{and} \qquad yx = (3)(-2) = -6$$

Since the order of factors does not affect a product, we see that this will hold true for any numbers that we use for x and y. **So we say that 4xy and 6yx are like terms because xy and yx will have the same values no matter what numbers are used as replacements for the variables.** If we have four (-6)s plus six (-6)s, we would have ten (-6)s.

$$4(-6) + 6(-6) = 10(-6)$$

This shows why we can add $4xy$ and $6yx$ and get $10xy$.

$$4xy + 6yx = 10xy$$

To add like terms, we add the coefficients of the terms.

example 103.1   Are $2xy^2$ and $3y^2x$ and $7yxy$ like terms?

solution   We first discard the coefficients and get

$$xy^2 \quad \text{and} \quad y^2x \quad \text{and} \quad yxy$$

If we expand these expressions, we get

$$xyy \quad \text{and} \quad yyx \quad \text{and} \quad yxy$$

Each of these has as factors two y's and one x. Since the order of multiplication does not affect a product, these expressions will always represent the same number. To demonstrate we will use -2 for x and 3 for y.

$$
\begin{array}{ccc}
xyy & yyx & yxy \\
(-2)(3)(3) & (3)(3)(-2) & (3)(-2)(3) \\
-18 & -18 & -18
\end{array}
$$

Thus the terms 2xy^2 and 3y^2x and 7yxy are like terms. Since xy^2, y^2x, and yxy are like terms we can add them by adding the coefficients.

$$2xy^2 + 3y^2x + 7yxy = 12y^2x \quad \text{or} \quad 12xy^2$$

example 103.2   Simplify by adding like terms: $4x + 3 + 6x + 4 + 2xy + 3yx + 2$

solution   We add the numbers for a sum of 9.

$$3 + 4 + 2 = 9$$

We add the x terms and get $10x$.

$$4x + 6x = 10x$$

And we add the xy terms and get $5xy$.

$$2xy + 3yx = 5xy \quad \text{(or 5yx if you prefer)}$$

So the sum is

$$\textbf{9 + 10}\boldsymbol{x}\textbf{ + 5}\boldsymbol{xy}$$

example 103.3   Add like terms: $4 + 3x^2 + 2xx + 4x + 7 + 3x$

solution   We add like terms and get

$$\textbf{11 + 5}\boldsymbol{x}^2\textbf{ + 7}\boldsymbol{x}$$

practice   Simplify by adding like terms:

a. $3yxy + 2xyx - 6xy^2 + 3y^2x$

b. $xxx + 3x^2x + 4x^3 + 2xy - 3yx$

c. $y^2xy + x^2yy - 5x^3y$

**problem set
103**

1.   The ratio of red marbles to blue marbles was 2 to 5. If the bowl contained 350 marbles, how many were blue?

2.   The ratio of flooded land to dry land was 2 to 9. If the plantation consisted of 1210 acres, how many acres were flooded?

3.   The product of a number and 2 is 20 less than the product of the number and -3. What is the number?

4.   The tree harvest was followed by seed planting. The forester was amazed because the number of seeds that sprouted was $2\frac{3}{4}$ times the number that had been planned on. If 140,000 sprouts had been planned on, how many sprouts did the forester get?

5.   Graph: $x \geq 1$

6.   What number is 75 percent of 220? Draw a diagram of the problem.

7.   If 150 is increased by 60 percent, what is the resulting number? Draw a diagram of the problem.

8.   What percent of 50 is 80? Draw a diagram of the problem.

9.   Use the method of cut and try to find $\sqrt[3]{31}$ to one decimal place.

10.   Find the area of the figure shown. Dimensions are in feet.

11.   Find the least common multiple of 12, 30, and 36.

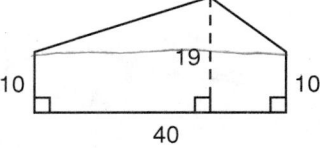

12.   If $6x - 5 = 31$, what is the value of $2x - 3$?

13.   If $5x - 3 = 22$, what is the value of $6x - 2$?

Solve:

14. $2x - 5 = 5x - 6$

15. $6x - 2 = 2x + 4$

16. $-4\frac{1}{3}x - \frac{1}{4} = 2\frac{1}{3}$

17. $\dfrac{-\frac{1}{4}}{\frac{1}{2}} = \dfrac{\frac{1}{3}}{x}$

Simplify by adding like terms:

18. $6x + 8 + 6x + 5 + 2xy + 9yx + 2$

19. $3 + 2ab^2 + 3a + 2abb - 4a$

20. $6x + 2y + 3xy + 2 + 3x + 4y - xy$

Simplify:

21. $-2^3 - (-3)^2$

22. $-(-3)^2 - [-(-1)]$

23. $2^3 - 2[3(2 - 3) + 2(2 - 4)]$

24. $aa^3ba^2a^4b^3b^2$

25. $xxyyx^2y^3x^3y^4$

26. $ababa^2b^2a^2b^2a^3b^3$

27. $\dfrac{2^4 - 3(2^2 - 1)(3^2 - 2^3)}{2^3(2^2 - 1)}$

28. $\dfrac{2^3 - 3^2(2^2 - 1)(3 - 2^2)}{2(3^2 - 2^3)}$

29. $3\dfrac{1}{3} \div \dfrac{1}{4} \times 2\dfrac{1}{3} \div \dfrac{1}{3}$

30. Evaluate: $a + bc + ca$ if $a = -1$, $b = -2$, and $c = 3$

LESSON 104 *Distributive property · Estimating higher-order roots*

104.A
distributive property

The notation

$$4(2 + 3)$$

tells us to multiply 4 by what is inside the parentheses. There are two ways we can do this.

$4(2 + 3)$	$4(2 + 3)$
$4(5)$	$4 \cdot 2 + 4 \cdot 3$
20	$8 + 12$
	20

On the left we added 2 and 3 to get 5. Then we multiplied 4 by 5 to get 20. On the right we multiplied 4 by 2 and multiplied 4 by 3 and then added 8 and 12. We get an answer of 20 both times. We call this peculiarity or property of numbers the **distributive property**. This property is especially helpful when we simplify expressions that have variables. If we look at

$$4(x + y)$$

we cannot add x and y and then multiply. But we can multiply x by 4 and then multiply y by 4 and then add the two products.

$$4(x + y) = 4x + 4y$$

example 104.1   Use the distributive property to multiply: $4(x + 2y - 8)$

solution   The distributive property also applies to the algebraic sum of three or more terms. We multiply each of the terms in the parentheses by 4.

$$4(x + 2y - 8) = \mathbf{4x + 8y - 32}$$

example 104.2   Use the distributive property to multiply: $3x(x^2 - 2 + 2x)$

solution   We multiply each of the terms in the parentheses by $3x$.

$$3x(x^2 - 2 + 2x) = 3x^3 - 6x + 6x^2$$

104.B
estimating higher-order roots

We can use the cut-and-try method to find roots to any accuracy we wish. However, the multiplication required is often time-consuming and untidy. Therefore, we will just practice finding whole number approximations of higher-order roots.

example 104.3 Estimate $\sqrt[3]{250}$ to the nearest whole number.

solution We want the cube root, so there are three factors.

$$(\ )(\ )(\ ) = 250$$

Let's try 3 as our first guess.

$$(3)(3)(3) = 27$$

That was a poor guess. Let's try 6.

$$(6)(6)(6) = 216$$

This is a little small, so let's try 7.

$$(7)(7)(7) = 343$$

Now we have a guess that is too large and one that is too small.

$$(6)(6)(6) = 216$$
$$(\ )(\ )(\ ) = 250$$
$$(7)(7)(7) = 343$$

Since 216 is closer to 250 than 343, we say that the whole number approximation of $\sqrt[3]{250}$ is 6.

$$\sqrt[3]{250} \approx 6$$

example 104.4 Estimate $\sqrt[4]{390}$ to the nearest whole number.

solution The fourth root is used as a factor four times so the problem can be stated as

$$(\ )(\ )(\ )(\ ) = 390$$

Let's try 5.

$$(5)(5)(5)(5) = 625$$

That was too large, so we will try 4.

$$(4)(4)(4)(4) = 256$$

We now have the answer bracketed.

$$(4)(4)(4)(4) = 256$$
$$(\ )(\ )(\ )(\ ) = 390$$
$$(5)(5)(5)(5) = 625$$

The number we are looking for is between 4 and 5 and is closer to 4. So to the nearest whole number the 4th root of 390 is 4.

$$\sqrt[4]{390} \approx 4$$

practice   Multiply:

a. $4(3x^2 + 2x)$ b. $3x(2 + x^2)$ c. $x(-x^2 - x - 4)$

Estimate to the nearest whole number:

d. $\sqrt[4]{750}$ e. $\sqrt[5]{200}$

problem set 104

1. The football coach shuddered when he realized that the ratio of the dextrous to the inept was 3 to 7. If 140 players were trying to make the team, how many were dextrous?

2. When the score was announced, 420 students were elated and the rest were dejected. If the ratio of elated to dejected was 2 to 9, how many students were there in all?

3. When Lafayette led the column, the troops marched 24 miles in 6 hours. Then he left and the troops reduced their speed by 1 mile per hour. How long did it take them to march the last 18 miles?

4. Four times a number is 36 greater than the product of the number and -2. What is the number?

5. What number is 45 percent of 200? Draw a diagram of the problem.

6. If 80 is increased by 80 percent, what is the resulting number? Draw a diagram of the problem.

7. What percent of 60 is 93? Draw a diagram of the problem.

8. Find the least common multiple of 20, 25, and 30.

9. Find the surface area of this right prism.   Dimensions   are   in centimeters.

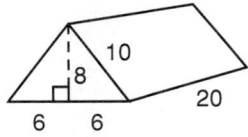

10. If $6x - 3 = 12$, what is the value of $\frac{2}{3}x + 5$?

11. If $3x + 5 = 22$, what is the value of $2x + 1$?

Use the distributive property to multiply:

12. $5x(x^2 - 3 + 3x)$ 13. $2ab(a + b)$ 14. $5a(6a + 2b + ab)$

Simplify by adding like terms:

15. $3x + 5 + 3y + 2xy + 6x + 5xy$ 16. $a^2 + 5a^2 + 3b^2 + 2a^2 - 5b^2$

Solve:

17. $2x - 6 = 3x - 4$ 18. $6x - 10 = -3 + 2x$

19. $-2x + 6 = -4x + 6$ 20. $\dfrac{-\dfrac{2}{3}}{\dfrac{1}{4}} = \dfrac{\dfrac{1}{6}}{x}$

Simplify:

21. $-3^2 - (-2)^3$ 22. $-[-(-3)] + (-2)^3$

23. $2^3 - 2^4[2(3 - 5) + 2^2(3^2 - 2^2)]$ 24. $aa^2ba^3b^2a^2b^3$

25. $xy^2x^2x^3y^3y^2$

26. $m^2nn^3m^3n^2n^3$

27. $\dfrac{2^3 - 2(1 - 2^2)(3 - 2^2)}{2(2^3 - 3^2)}$

28. $\dfrac{1 - 3(2^2 - 3) + (3^3 - 2 \cdot 3)}{6(2 \cdot 3 - 2)}$

29. Use the method of cut and try to estimate $\sqrt[3]{350}$ to the nearest whole number.

30. Use the method of cut and try to estimate $\sqrt[4]{480}$ to the nearest whole number.

LESSON 105 *Powers with negative bases*

Let's take another look at powers with negative bases. Beginners often confuse the notations

$$\text{(a)} \quad (-2)^2 \quad \text{and} \quad \text{(b)} \quad -2^2$$

The expression on the left means to multiply -2 by -2

$$(-2)(-2) = +4$$

The expression on the right tells us to find the opposite of the value of 2^2.

$$-2^2 = -(2)(2) = -4$$

There is an easy way to remember this. If the $-$ sign is not protected by a parentheses, cover it with a finger. To simplify

$$-2^4$$

we cover the minus sign with a finger.

 2^4

The value of 2^4 is 16 so now we have

 16

Now we remove our finger and uncover the minus sign and get the final result.

$$\mathbf{-16}$$

example 105.1   Evaluate: (a) $(-3)^2$ (b) -3^2

solution   If we try to cover the minus sign with a finger, we find that the minus sign is protected by the parentheses in (a) but not in (b).

(a) 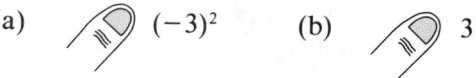 $(-3)^2$ (b) 3^2

So $(-3)^2$ means $(-3)(-3)$, and -3^2 means $-(3)(3)$.

$$\text{(a)} \quad (-3)^2 = \mathbf{9} \qquad \text{(b)} \quad -3^2 = \mathbf{-9}$$

example 105.2   Evaluate: (a) a^2 (b) $-a^2$ if $a = -2$

solution   (a)   We first write

$$(\ \ )^2$$

Then we write -2 inside the parentheses and simplify:

$$(-2)^2 = \mathbf{4}$$

(b)   This time we begin by writing

$$-(\ \ )^2$$

Then we write -2 inside the parentheses and simplify.

$$-(-2)^2 = \mathbf{-4}$$

example 105.3   Simplify: $a^2 - ab^2 - b^2$ if $a = -2$ and $b = -3$

solution   Parentheses are not absolutely necessary, but we will use them to help us with the second term. We substitute and get

$$4 - (-2)(-3)^2 - 9$$

We simplify this and get

$$
\begin{array}{ll}
4 - (-18) - 9 & \text{multiplied} \\
4 + 18 - 9 & \text{simplified} \\
\mathbf{13} & \text{added}
\end{array}
$$

practice   Evaluate:

a. $a^2b - b$ if $a = -2$ and $b = -3$

b. $b^2 - a^2b$ if $a = -2$ and $b = -3$

problem set 105

1. Seven-twelfths of the students mispronounced the word *chic*. If 1440 knew how to pronounce this word, how many mispronounced it?

2. One-fifth of the students were pertinacious in their pursuit of knowledge. If 840 students were not in this category, how many students were there in all?

3. The ratio of the number that paid the piper this month to the number that paid the piper last month was 12 to 5. If there were 170 in all, how many paid the piper this month?

4. The ratio of sycophants to bullies was 14 to 1. If there were 750 in all, how many were sycophants?

5. The Lilliputians marched 500 meters in 4 hours. For the next 4 hours they tripled their pace. What was the total distance traveled by the Lilliputians in 8 hours?

6. Three times a number is 12 less than the product of the number and 5. What is the number?

7. What number is 170 percent of 120? Draw a diagram of the problem.

8. If 80 is increased by 60 percent, what is the resulting number? Draw a diagram of the problem.

9. What percent of 70 is 77? Draw a diagram of the problem.

10. Find the volume in cubic centimeters of a right circular cylinder whose diameter is 4 centimeters and whose height is 1 meter.

11. The figure shown on the left is the base of a right solid 100 centimeters tall. Find the total surface area of the solid. Dimensions are in centimeters.

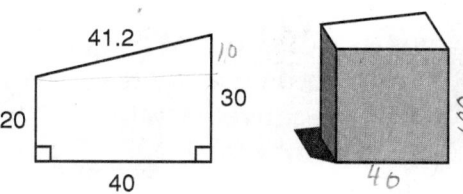

12. If $3x - 6 = 12$, what is the value of $\frac{1}{6}x + 3$?

13. If $4x - 3 = 21$, what is the value of $\frac{1}{2}x + 2$?

Use the distributive property to multiply.

14. $2bc(a + b + c)$

15. $ab(a + b + ab)$

Simplify by adding like terms:

16. $a^3 + 2abb + 3aa^2 + 2ab^2$

17. $yx^2 + xyx + 3y^2x + 2yxy$

Solve:

18. $4x - 3 = -x + 6$

19. $-5x - 2 = -x + 7$

20. $-3x + 5 = -2x - 3$

21. $\dfrac{-\frac{1}{3}}{2\frac{1}{3}} = \dfrac{\frac{1}{2}}{x}$

Simplify:

22. $-2^3 - [-(-3)]$

23. $-[-(-6)] + (-2)(-3)$

24. $2^3 - 2^2[2(3 - 2^2) + 2^2(2 \cdot 3 - 2^2)]$

25. $aba^2b^2a^3$

26. $xy^3x^2yy^2x$

27. $\dfrac{2^2 - 2^3(3^2 - 2^2) + 2^3}{6(5 - 2^2)}$

28. Use the method of cut and try to estimate $\sqrt[5]{250}$ to the nearest whole number.

Evaluate:

29. $ab^2 + \dfrac{b}{a^2}$ if $b = -8$ and $a = -2$

30. $a^3 + b^3 - ab$ if $a = -1$ and $b = -2$

LESSON 106 *Roots of negative numbers ·*
Negative exponents

106.A
roots of
negative
numbers

We remember that every positive number has a positive square root. Also, every positive number has a negative square root because

$$(2)(2) = 4 \qquad (-2)(-2) = 4$$

When we write

$$\sqrt{4}$$

we indicate the positive square root of 4, which is 2. When we write

$$-\sqrt{4}$$

we indicate the negative square root of 4, which is -2.

Negative numbers **do not have** positive square roots or fourth roots or sixth roots because the product of an even number of positive numbers is always a positive number.

$$(2)(2) = +4 \qquad (2)(2)(2) = 8 \qquad (2)(2)(2)(2) = +16$$

But negative numbers **do have** odd negative roots.

$$(-2)(-2)(-2) = -8 \qquad \text{so} \qquad \sqrt[3]{-8} = -2$$
$$(-3)(-3)(-3) = -27 \qquad \text{so} \qquad \sqrt[3]{-27} = -3$$

example 106.1 Find $\sqrt[3]{-8}$.

solution What number used as a factor 3 times equals -8? The answer is **-2** because

$$(-2)(-2)(-2) = \mathbf{-8}$$

example 106.2 Find $\sqrt[5]{-32}$.

solution What number used as a factor 5 times equals -32? The answer is **-2** because

$$(-2)(-2)(-2)(-2)(-2) = \mathbf{-32}$$

106.B
negative
exponents

It is convenient to have an alternative notation for the reciprocal form of a power, and so mathematicians invented one. They decided to let

$$\frac{1}{4^2} = 4^{-2} \qquad \text{and to let} \qquad \frac{1}{4^{-2}} = 4^2$$

It is impossible to evaluate

$$\frac{4^{-2}}{1}$$

as it is, because a negative exponent does not indicate an operation. To evaluate this expression, we turn it upside down and write its reciprocal, which requires that we change the exponent from $-$ to $+$.

$$4^{-2} = \frac{1}{4^2} = \mathbf{\frac{1}{16}}$$

example 106.3   Evaluate: $\dfrac{1}{4^{-2}}$

solution   To evaluate a power with a negative exponent, we turn the expression upside down (write the reciprocal) and change the $-$ sign in the exponent to a $+$ sign.

$$\frac{1}{4^{-2}} = 4^2 = \mathbf{16}$$

example 106.4   Evaluate: $\dfrac{1}{(-2)^{-4}}$

solution   We flip the expression (write the reciprocal) and change the sign of the exponent from $-$ to $+$. **We do not change the sign inside the parentheses.**

$$\frac{1}{(-2)^{-4}} = (-2)^4 = \mathbf{16}$$

example 106.5   Evaluate: $-\dfrac{1}{(-3)^{-2}}$

solution   The minus sign is not protected by a parentheses so we cover it with a finger.

$$\text{☞} \quad \frac{1}{(-3)^{-2}}$$

Now we simplify.

$$\text{☞} \quad (-3)^2 = \quad \text{☞} \quad 9$$

As the last step, we remove our finger and get

$$\mathbf{-9}$$

practice   Find:

a. $\sqrt[3]{-27}$

b. $\sqrt[3]{-343}$

Simplify:

c. 5^{-2}

d. $\dfrac{1}{5^{-2}}$

problem set 106

1. Six-thirteenths of the audiophiles could discriminate between the two products. If 28 of the audiophiles could not discriminate, how many could discriminate?

2. Six-sevenths of the party delegates were proclaiming their positions. If 1100 were not involved in this behavior, how many delegates were there in all?

3. The ratio of the number who were contumacious to the number who were affable was 2 to 17. If there were 7600 in all, how many were contumacious and how many were affable?

4. Seven times a number is 9 less than the product of the number and -20. What is the number?

5. Gawain and Lancelot drove their horses through the pounding rain and covered 16 miles in 2 hours. When the rain stopped, they slowed to half that pace. How long did it take them to travel the remaining 20 miles?

6. What number is 180 percent of 160? Draw a diagram of the problem.

7. If 90 is increased by 80 percent, what is the resulting number? Draw a diagram of the problem.

8. What percent of 60 is 78? Draw a diagram of the problem.

9. Find the volume in cubic meters of the right solid whose base is shown on the left and whose height is 2 meters. Dimensions are in meters.

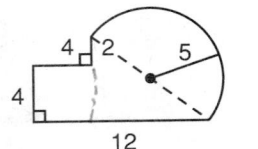 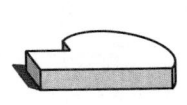

10. Find the lateral surface area of the solid in Problem 9.

Find:

11. $\sqrt[7]{-128}$

12. $\sqrt[3]{-512}$

Simplify:

13. (a) 4^{-3} (b) 3^{-2}

14. (a) $\dfrac{1}{2^{-4}}$ (b) $\dfrac{1}{5^{-3}}$

15. If $5x - 9 = 26$, what is the value of $\dfrac{1}{7}x + 49$?

16. If $9x - 9 = 153$, what is the value of $\dfrac{5}{18}x - 13$?

Use the distributive property to multiply:

17. $3px(p + x + 2px)$

18. $mn^2(m + mny + 3m^2)$

Simplify by adding like terms:

19. $m^2n^3 + 3nm^2nn - mnmn^2 + 2n^3mm$

20. $ap^3 + pap^2 - 5p^2ap$

Solve:

21. $5x - 4 = -3x + 20$

22. $-4x - 3 = -10x + 7$

23. $\dfrac{-\dfrac{2}{5}}{2\dfrac{1}{2}} = \dfrac{-\dfrac{11}{13}}{x}$

Simplify:

24. $-\dfrac{1}{3^{-3}} - [-(-2^2)] + \sqrt[3]{-8}$

25. $-[-(-6)^2] + (-3)(-4) + \sqrt[3]{-27}$

26. $a^3p^2ap^2a^4aa$

27. Use the cut-and-try method to estimate $\sqrt[3]{650}$ to the nearest whole number.

Evaluate:

28. $m^2 + \dfrac{n^2}{m}$ if $m = -8$ and $n = 4$

29. $p^3 + c^3$ if $p = -3$ and $c = -1$

LESSON 107 *Percent as a rate*

We have learned that a fraction, a decimal number, and a percent can be used to describe a part of a whole. The fraction $\frac{1}{5}$, the decimal number 0.2, and 20 percent are equivalent forms.

Fraction	Decimal	Percent
$\frac{1}{5}$	0.2	20

One-fifth of a number equals 0.2 of a number, which also equals 20 percent of the number.

$$\frac{1}{5} \text{ of } 40 = 8$$

$$0.2 \text{ of } 40 = 8$$

$$20\% \text{ of } 40 = 8$$

The two percent equations that we have been using are

$$\frac{P}{100} \times of = is \qquad \frac{P}{100} = \frac{is}{of}$$

When percent problems are worked in engineering or in business, the decimal equivalent is used instead of percent. The decimal equivalent is called **rate.** This rate is the percent divided by 100. The equation is

$$\text{Rate} \times of = is$$

example 107.1 Twenty percent of what number equals 160?

solution We use the equation

$$\text{Rate} \times of = is$$

We replace *rate* with 0.2, replace *of* with *WN*, and replace *is* with 160.

$$0.2\,WN = 160 \qquad \text{substituted}$$

$$\frac{0.2\,WN}{0.2} = \frac{160}{0.2} \qquad \text{divided both sides by 0.2}$$

$$WN = \textbf{800} \qquad \text{simplified}$$

example 107.2 What percent of 40 is 120?

solution We use the equation

$$\text{Rate} \times of = is$$

Now we substitute.

$$\text{Rate} \times 40 = 120 \qquad \text{substituted}$$

$$\frac{\text{Rate} \times 40}{40} = \frac{120}{40} \qquad \text{divided}$$

$$\text{Rate} = 3 \qquad \text{simplified}$$

Using rate is tricky. A rate of 3 means **300%**.

Decimal	Percent
3	300

example 107.3 Twenty percent of 400 is what number?

solution We use the equation

$$\text{Rate} \times of = is$$

Now we substitute.

$$0.2 \times 400 = WN \qquad \text{substituted}$$

$$\mathbf{80 = WN} \qquad \text{multiplied}$$

practice Use the rate equation to solve.

a. Thirty percent of what number is 63?

b. What percent of 4 is 80?

c. What percent of 400 is 0.8?

problem set 107

1. The number of errors in the manuscript was appalling. There were $3\frac{1}{2}$ times as many errors as were expected. If there were 84 errors, how many were expected?

2. The ratio of the number who were facile to the number who were inept was 11 to 3. If 28,000 were evaluated, how many were facile and how many were inept?

3. The product of a number and -5 is 14 less than the product of the number and -6. What is the number?

4. Three hundred twenty fedoras cost $7040. What would 215 fedoras cost?

5. The hikers packed their gear into the mountains. On their first day they hiked 21 miles in 7 hours. On the second day they increased their speed by 1 mile per hour. How long did it take them to hike the next 20 miles?

6. Graph: $x \le 3$

7. What number is 85 percent of 340? Draw a diagram of the problem.

8. If 260 is increased by 70 percent, what is the resulting number? Draw a diagram of the problem.

9. What percent of 80 is 132? Draw a diagram of the problem.

10. What is the volume in cubic centimeters of a right solid whose base is shown on the left and whose height is 3 meters? Dimensions are in centimeters.

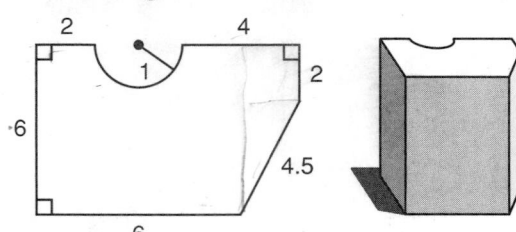

11. Find the lateral surface area in square centimeters of the solid in Problem 10.

12. Forty percent of what number is 92?

13. What percent of 200 is 2?

14. If $7x - 9 = 26$, what is the value of $\frac{3}{5}x + 19$?

15. If $5x + 6 = 46$, what is the value of $4x - 19$?

16. Use the distributive property to multiply: $3x(x^2 - 4 + 3x)$

Evaluate:

17. $x^3 + m^5 + \dfrac{m^2}{x}$ if $x = -4$ and $m = -2$

18. $\sqrt[x]{y}$ if $x = 5$ and $y = -32$

Simplify by adding like terms:

19. $5m + 4 + 2x + 4mx + 7mx$

20. $2p^2 + 2b^2 + 4m^2 + 7p^2 + 5b^2$

Solve:

21. $4x - 5 = 3x - 6$

22. $7x - 9 = -4 + 2x$

23. $-5x + 9 = -11x - 3$

24. $\dfrac{-\dfrac{1}{4}}{\dfrac{3}{5}} = \dfrac{\dfrac{1}{2}}{x}$

Simplify:

25. $-(4^2) - (3)^2 - \sqrt[3]{-64}$

26. $-[-(-4)] + [-(3^3)]$

27. $m^2 p^2 p m^3 p^3 p$

28. $\dfrac{3^3 - 1(2 - 3^2)(2^2 - 3)}{3(3^2 - 4^2)}$

29. Use the method of cut and try to estimate $\sqrt[3]{137}$ to the nearest whole number.

30. Use the method of cut and try to estimate $\sqrt[3]{740}$ to the nearest whole number.

LESSON 108 *Roman numerals*

The Romans had a hard time with numbers. They had only seven symbols they could use to write numbers. These symbols were

Symbol	I	V	X	L	C	D	M
Value	1	5	10	50	100	500	1000

The Romans did not use place value. They did not have a symbol for zero. The value of a Roman numeral equals the sum of the value of the symbols in the numeral. To write 3, 30, 300, or 3000, they wrote the same symbol 3 times.

III means 3   XXX means 30   CCC means 300   MMM means 3000

In this lesson we will use Roman numerals to write numbers between 1 and 50.

1	I	11	XI	21	XXI	31	XXXI	41	XLI
2	II	12	XII	22	XXII	32	XXXII	42	XLII
3	III	13	XIII	23	XXIII	33	XXXIII	43	XLIII
4	IV	14	XIV	24	XXIV	34	XXXIV	44	XLIV
5	V	15	XV	25	XXV	35	XXXV	45	XLV
6	VI	16	XVI	26	XXVI	36	XXXVI	46	XLVI
7	VII	17	XVII	27	XXVII	37	XXXVII	47	XLVII
8	VIII	18	XVIII	28	XXVIII	38	XXXVIII	48	XLVIII
9	IX	19	XIX	29	XXIX	39	XXXIX	49	XLIX
10	X	20	XX	30	XXX	40	XL	50	L

When we look at these numbers, we note that

1. No symbol appears more than 3 times in a row because the Romans used a trick to write 4, 9, 40, and 90.
2. To write 4, instead of writing IIII they wrote IV, which means 5 − 1. To write 9, instead of writing VIIII they wrote IX, which means 10 − 1. To write 40, instead of writing XXXX, they wrote XL, which means 50 − 10. To write 90, instead of writing LXXXX they wrote XC, which means 100 − 10.

We use these tricks whenever 4, 9, 40, and 90 appear in a number.

$$
\begin{array}{lll}
4 = & \text{IV} & 9 = & \text{IX} & 40 = & \text{XL} \\
14 = & \text{XIV} & 19 = & \text{XIX} & 44 = \text{XLIV} \\
24 = & \text{XXIV} & 29 = & \text{XXIX} & 49 = \text{XLIX} \\
34 = \text{XXXIV} & 39 = \text{XXXIX} &
\end{array}
$$

practice Write the Roman numeral for:

a. 4   **b.** 9   **c.** 34

d. 39   **e.** 40   **f.** 49

g. Use Roman numerals to write the numbers from 15 to 30.

problem set 108

1. Five times a number is 28 greater than the product of the number and −2. Find the number.

2. The ratio of ascetics to hedonists visiting the shrine was 4 to 7. If 4400 visited the shrine, how many were ascetics?

3. Enthusiasm for the new product ran high. There were $2\frac{3}{4}$ as many customers as there were on the previous day. If 2200 customers arrived to inspect the new product, how many were there on the previous day?

4. Fifty consultations cost $3150. What would 78 consultations cost?

5. It took the DC-9 four hours to go 1600 miles before it encountered a storm. The storm decreased the plane's speed by 45 miles per hour. How long would it take the plane to go the remaining 2485 miles at the new rate?

6. Graph: $x \geq -3$

7. What number is 78 percent of 900? Draw a diagram of the problem.

8. If 350 is increased by 90 percent, what is the resulting number? Draw a diagram of the problem.

9. What percent of 75 is 135? Draw a diagram of the problem.

10. Find the volume in cubic feet of a right solid whose base is the figure shown on the left and whose height is 1 yard. Dimensions are in feet.

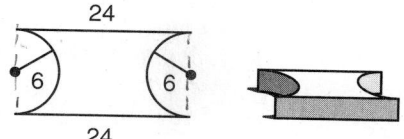

11. Find the total surface area of the solid in Problem 10.

Write the Roman numeral for each number:

12. 6   13. 29   14. 46

15. Use the rate equation to solve: What percent of 160 is 48?

16. If $11x - 4 = 29$, what is the value of $\frac{2}{3}x + 7$?

17. If $\frac{4}{5}x - 19 = 13$, what is the value of $\frac{3}{8}x + 4$?

18. Use the distributive property to multiply: $4ax(3a + 3z + 9ax)$

Evaluate:

19. $m^3 + \frac{x}{y} - x^2$ if $m = -3$, $x = -4$, and $y = -2$

20. $a^3 + b^3 + c^3$ if $a = -1$, $b = -2$, and $c = -3$

21. Simplify by adding like terms: $3m^2 + 2c^2 - 4cm + 6c^2 - 9m^2$

Solve:

22. $-7x + 4 = -x - 8$

23. $\dfrac{\frac{2}{3}}{-\frac{1}{4}} = \dfrac{x}{\frac{2}{5}}$

Simplify:

24. $-5^2 - (-3)^3 + \sqrt[3]{-64}$

25. $-[-(-3)^2] - (4)^2 - \sqrt[3]{-216}$

26. $a^3 m^3 a a^2 m^2 a$

27. $\dfrac{4^3 - 2^2(1^5 - 3^2)(3 - 2^2)}{2^2(2^3 - 3^2)}$

28. Use the method of cut and try to estimate $\sqrt[4]{216}$ to the nearest whole number.

29. Use the method of cut and try to estimate $\sqrt[3]{618}$ to the nearest whole number.

30. Use two unit multipliers to convert 1,000,000 square meters to square centimeters.

LESSON 109 *Percent word problems*

Many people have difficulty with percent word problems because they don't want to draw diagrams. They feel that somehow, somewhere, there exists a shortcut so that percent problems can be worked by rote and without understanding. No such shortcut exists, and that is the reason for drawing the diagrams. To work a percent word problem, we will first draw the diagram. Then we will put the given numbers into the diagram. Then we will use the percent equation to find the missing numbers.

example 109.1   Twenty percent of the clowns had red hair. If 40 clowns had red hair, how many did not have red hair?

solution   We draw diagrams and insert the numbers that we have.

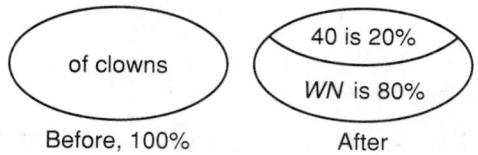

Now we can state the problem as

Forty is 20 percent of the clowns.

Now we use the percent equation to solve.

$$\frac{P}{100} \cdot of = is \qquad \text{equation}$$

$$\frac{20}{100} \cdot \text{clowns} = 40 \qquad \text{substituted}$$

$$\frac{100}{20} \cdot \frac{20}{100} \cdot \text{clowns} = 40 \cdot \frac{100}{20} \qquad \text{multiplied by } \frac{100}{20}$$

$$\text{Clowns} = 200$$

If there were 200 clowns in all and 40 had red hair, then **160 did not have red hair.**

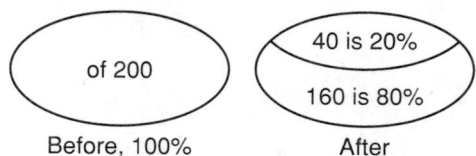

example 109.2   When the fog lifted, the number of frogs that could be seen was 140 percent greater than before the fog lifted. If 40 frogs could be seen before, how many could be seen after the fog lifted?

solution   We must be careful. A 140 percent increase means 240 percent total.

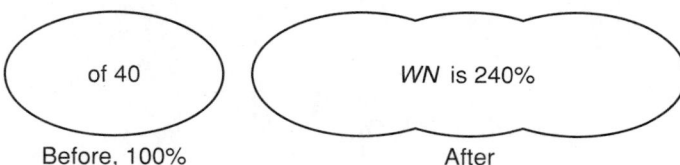

Now we can state the question as:

What number is 240 percent of 40?

Now we use the percent equation to solve.

$$\frac{P}{100} \cdot of = is \qquad \text{equation}$$

$$\frac{240}{100} \cdot 40 = WN \qquad \text{substituted}$$

$$\frac{9600}{100} = WN \qquad \text{multiplied}$$

$$\mathbf{96 = WN} \qquad \text{divided}$$

Thus, we find that 96 is 240 percent of 40. Now we can complete the diagram.

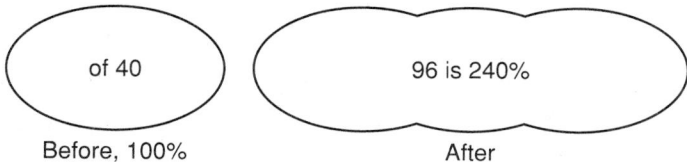

example 109.3   Forty percent of the trees were in bloom. If 1200 trees were not in bloom, how many trees were in bloom?

solution   First we draw the diagrams and use the numbers that were given.

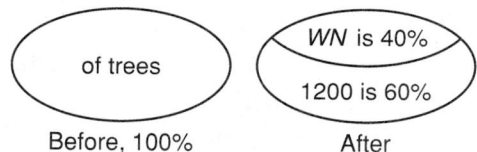

If 40 percent were in bloom, then 60 percent were not in bloom. We can state the problem as

1200 is 60 percent of the trees

Now we use the percent equation to solve.

$$\frac{P}{100} \cdot of = is \qquad \text{equation}$$

$$\frac{60}{100} \cdot trees = 1200 \qquad \text{substituted}$$

$$\frac{100}{60} \cdot \frac{60}{100} \cdot trees = 1200 \cdot \frac{100}{60} \qquad \text{multiplied by } \frac{100}{60}$$

$$Trees = 2000 \qquad \text{simplified}$$

Now we draw the final diagram.

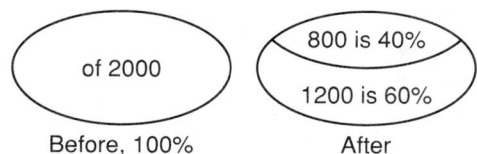

If there were 2000 trees in all and 1200 were not in bloom, then **800 were in bloom.**

practice

a. Thirty percent of the trees had red blossoms. If 600 trees had red blossoms, how many trees did not have red blossoms?

b. James found that the glob increased from 400 pounds to 640 pounds in just 3 hours. What was the percent increase in the weight of the glob?

problem set 109

1. Thirty percent of the clowns had blue hair. If 90 clowns had blue hair, how many clowns did not have blue hair?

2. When the fog lifted, the number of tadpoles that could be seen was 220 percent greater than before the fog lifted. If 80 tadpoles could be seen before, how many could be seen after the fog lifted?

3. Eighty percent of the trees were in bloom. If 2600 trees were not in bloom, how many trees were in bloom?

4. When all the extras had been counted, the number rose from 40 to 130. What percent increase was this?

5. Six times a number is 45 greater than the product of the number and -3. What is the number?

6. The ratio of sophists to pragmatists was 2 to 7. If there were 1800 sophists, how many pragmatists were there?

7. Ten credenzas cost $1240. What would be the cost of 7 credenzas?

8. What number is 130 percent of 230? Draw a diagram of the problem.

9. If 70 is increased by 60 percent, what is the resulting number? Draw a diagram of the problem.

10. What percent of 60 is 72? Draw a diagram of the problem.

11. Find the perimeter of the figure shown. Dimensions are in inches.

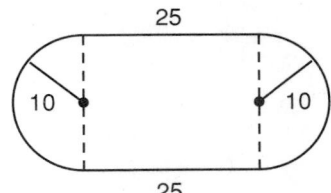

12. If $4x - 3 = 9$, what is the value of $\frac{1}{3}x + 2$?

13. If $3x + 5 = 17$, what is the value of $\frac{1}{2}x + 4$?

Use the distributive property to multiply:

14. $2a(3a + b + c)$

15. $3ab(a + b + c + d)$

Simplify by adding like terms:

16. $a^2 + b^2 + 3aa + 2bb - 3ab + 2ab$

17. $a^3 + 2aab + 2a^2b + 3b^2a + 3bba$

Solve:

18. $3x - 4 = 2x + 12$

19. $-4x - 5 = 3x + 4$

20. $-6x - 1 = -2x + 4$

21. $\dfrac{-\dfrac{2}{3}}{3\dfrac{1}{2}} = \dfrac{\dfrac{1}{4}}{x}$

Simplify:

22. $-3^2 - [-(2)] - \sqrt[5]{-1024}$

23. $-[-(-3)] + (-2)^3 + \sqrt[3]{-125}$

24. $ab^2ba^2b^3$

25. $2^2 - 2^3[2^2(3 - 2^2) + 2^2(3 - 1)]$

26. $xy^2x^3y^2x$

27. $m^3n^2m^2mn^3$

28. $\dfrac{2^3 - 2^2(3^2 - 2^3) + 2^2}{5(2^3 - 3^2)}$

29. Use the method of cut and try to estimate $\sqrt[3]{400}$ to the nearest whole number.

30. Evaluate: $ab + b^2$ if $a = -1$ and $b = -3$

LESSON 110 *Complex decimal numbers · Fractional percents*

110.A

complex decimal numbers

We can put a fraction in a decimal number. The value of the fraction depends on its place value. If we put the fraction $\frac{1}{2}$ in the ones' place, it has a value of one-half of 1, or 0.5.

$$\frac{1}{2} = 0.5$$

If we put the fraction $\frac{1}{2}$ in the tenths' place, it has a value of one-half of 0.1, or five-hundredths.

$$0.\frac{1}{2} = 0.05$$

If we put the fraction $\frac{1}{2}$ in the thousandths' place, it has a value of one-half of 0.001, or five ten-thousandths.

$$0.00\frac{1}{2} = 0.0005$$

If we write a fraction just after the hundredths' digit, as

$$0.03\frac{1}{2}$$

the fraction is not in the thousandths' place. **Both the 3 and the $\frac{1}{2}$ are in the hundredths' place.** This number has a value of 3 and $\frac{1}{2}$ hundredths, which equals thirty-five thousandths.

$$0.03\frac{1}{2} = 0.035$$

110.B

fractional percents

To write a fractional percent as a decimal number, we must remember the place value of the fraction.

example 110.1   Write each percent as a decimal number: (a) $4\frac{2}{5}\%$ (b) $0.7\frac{1}{8}\%$

solution   (a)   To change a percent to a decimal number, we move the decimal point two places to the left.

$$4\frac{2}{5}\% \quad \text{equals} \quad 0.04\frac{2}{5} \quad \text{equals} \quad \textbf{0.044}$$

(b)   To change $0.7\frac{1}{8}\%$ to a decimal number, we move the decimal point two places to the left.

$$0.7\frac{1}{8}\% \quad \text{equals} \quad 0.007\frac{1}{8} \quad \text{equals} \quad \textbf{0.007125}$$

practice   **a.**   Write $0.00046\frac{1}{5}$ as a decimal number without a fraction.

b.   Complete the boxes. Begin by inserting the reference numbers.

Fraction	Decimal	Percent
		$0.06\frac{2}{5}$
		$7\frac{1}{5}$

problem set 110

1.   Sixty percent of the populace were iconoclasts. If 16,000 were traditional, how many were iconoclasts?

2.   In the spring the number of flowers that could be seen was 360 percent greater than the year before. If 22,000 could be seen the year before, how many flowers could be seen this year?

3.   Originally there were 39,000 employees in the industry. Since the business began, the number of employees has grown to $2\frac{2}{13}$ of the original number. How many people are now employed in the industry?

4.   Three hundred new ones cost $2100. What would be the cost of 120 new ones?

5.   The ratio of goods to bads at the conference was 5 to 7. If 4800 had convened to weigh the issues, how many were good and how many were bad?

6.   Six times a number is 45 greater than the product of the number and -3. Find the number.

7.   Graph: $x > 0$

8.   What number is 84 percent of 1600? Draw a diagram of the problem.

9.   If 900 is increased by 35 percent, what is the resulting number? Draw a diagram of the problem.

10.   What percent of 89 is 445? Draw a diagram of the problem.

11. Find the volume in cubic centimeters of a right solid whose base is shown and whose height is 2 meters. Dimensions are in meters.

12. Find the lateral surface area of the solid described in Problem 11.

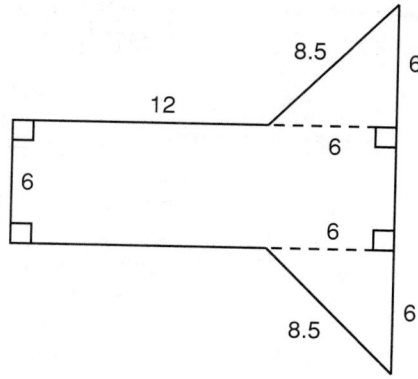

Write each Arabic number as a Roman numeral:

13. 17

14. 39

15. Use the rate equation to solve: What number is 230 percent of 70?

16. If $14x - 9 = 47$, what is the value of $\frac{11}{4}x - 12$?

17. If $\frac{3}{8}x - 36 = -12$, what is the value of $\frac{3}{4}x - 5$?

Use the distributive property to multiply:

18. $5c(3m + 2p + c)$

19. $7mp(m + p + x + 2mp)$

Simplify by adding like terms:

20. $m^2 + c^3 + 3mm + c^2c$

21. $p^2 + 3aam - 2p^2p + a^2m$

22. Write $0.0006\frac{1}{4}$ as a decimal number without a fraction.

23. Write each percent as a decimal number: (a) $4\frac{1}{2}\%$ (b) $6\frac{3}{8}\%$

Solve:

24. $-4x - 3 = -11x + 16$

25. $-7x - 4 = 3x + 3$

26. $\dfrac{-\dfrac{3}{8}}{2\dfrac{1}{4}} = \dfrac{\dfrac{1}{2}}{x}$

27. Simplify: $-[-(-4)] + (2)^2 - 3^2 + \sqrt[3]{-64} - \sqrt[3]{\dfrac{125}{1000}}$

28. Use the cut and try method to estimate $\sqrt[3]{800}$ to the nearest whole number.

29. Evaluate: $m^2p + p^2$ if $m = -4$ and $p = 5$

LESSON 111 *Simple interest · Compound interest*

111.A
simple interest

When you put money in a bank, the bank uses your money to make more money. They will pay you for the use of your money. The money they pay is called **interest.** The interest is a fixed percentage of the money you invest.

example 111.1 Jim put $5000 in the bank for 2 years at 8 percent simple interest. How much money did he get when he withdrew his money after 2 years?

solution The words *simple interest* tell us that Jim will get 8 percent of $5000 every year. We use the equation

$$\text{Rate} \times \text{amount} = \text{interest}$$

The interest is 8 percent, so the rate is 0.08.

$$0.08 \times \$5000 = \$400$$

The bank will pay him $400 for the use of his money each year. At the end of 2 years he will get back

$$\$5000 + \$400 + \$400 = \textbf{\$5800}$$

example 111.2 Maria deposited $800 at 8 percent simple interst. At the end of 3 years she will have how much money?

solution She will get 8 percent of $800 each year.

$$0.08 \times \$800 = \$64$$

At the end of 3 years she will have earned

$$3 \times \$64 = \$192 \text{ interest}$$

At the end of 3 years she will have

$$\$800 + \$192 = \textbf{\$992}$$

111.B
compound interest

At the end of the first year in the preceding example, Maria had earned $64 simple interest, so she had a total of $864. If the bank paid her **compound interest** the second year, they would pay her 8 percent interest on the total amount of $864. Thus in the second year, her interest would be

$$0.08 \times \$864 = \$69.12$$

Now she would have

$$
\begin{array}{r}
\$864.00 \\
+ \quad 69.12 \\
\hline
\$933.12
\end{array}
$$

As you can see, a calculator will help with one of these problems. Her interest the third year at 8 percent compound interest would be

$$0.08 \times \$933.12 = \$74.65$$

Now she would have

$$\begin{array}{r} \$933.12 \\ + \quad 74.65 \\ \hline \$1007.77 \end{array}$$

Thus, if the bank had paid compound interest, she would have had more money in her account than if the bank had paid simple interest.

3 years' compound interest	$1007.77
3 years' simple interest	$ 992.00
Difference	$ 15.77

example 111.3 How much would $700 be worth in 2 years if the money is deposited at 9 percent interest compounded annually?

solution We will use a calculator to do the arithmetic. The first year's interest would be

$$0.09 \times \$700 = \$63$$

The amount of money at the end of the first year would be

$$\begin{array}{r} \$700 \\ + \quad 63 \\ \hline \$763 \end{array}$$

The second year's interest would be

$$0.09 \times 763 = \$68.67$$

The amount of money at the end of the second year would be

End of first year	$763.00
Second year's interest	+ 68.67
Total	**$831.67**

practice **a.** Elvira put $7000 in the bank at 9 percent simple interest. How much money did she get when she withdrew her money after 2 years?

b. How much would Elvira's $7000 be worth in 2 years if it had been deposited at 9 percent compounded annually?

problem set 111

1. Consumption of health foods rose 110 percent in the year after the flood. If 500,000 units of health food were consumed the year before the flood, how many units were consumed the year after the flood?

2. The number of items recalled at the plant increased $2\frac{1}{8}$ times this year. If 8500 items were recalled this year, how many were recalled last year?

3. Eleven pedestals cost $74,800. What would be the cost of six pedestals?

4. The ratio of the number of beauties to the number of pretties was 7 to 13. If 2600 were present, how many were beauties and how many were pretties?

5. Akeem put $15,000 in the bank at 10 percent simple interest. How much money did he get when he withdrew his money after 3 years?

6. How much would Akeem's $15,000 be worth in 3 years if it had been deposited at 10 percent interest compounded annually?

7. Graph: $x \geq 1$

8. If 750 is increased by 80 percent, what is the resulting number? Draw a diagram of the problem.

9. What percent of 95 is 228? Draw a diagram of the problem.

10. What number is 16 percent of 9000? Draw a diagram of the problem.

11. Find the volume in cubic feet of a right solid whose base is shown and whose height is 2 feet. Dimensions are in feet.

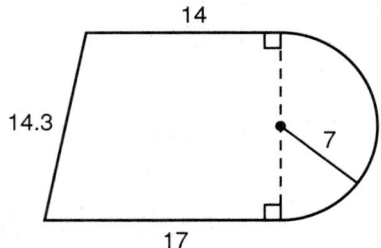

12. Find the total surface area of the figure in Problem 11.

13. Use Roman numerals to write each number: (a) 19 (b) 44

14. Use the rate equation to solve: What number is 80 percent of 90?

15. If $7x - 23 = 5$, what is the value of $\frac{6}{5}x + 100$?

16. If $\frac{7}{11}x - 40 = -19$, what is the value of $\frac{4}{13}x - 4$?

Use the distributive property to multiply:

17. $4x(3a + b + ab)$

18. $11p(11 + p + 2ab)$

Simplify by adding like terms:

19. $mp^2 + mpm + 2a^2b^2 + mp + abab$

20. $p^2ap + m^2nx + ap^3 + nmxm$

Write as a decimal number without a fraction:

21. $0.007\frac{1}{4}$

22. $0.00073\frac{3}{5}$

Write each percent as a decimal number:

23. $9\frac{1}{5}\%$

24. $86\frac{1}{4}\%$

Solve:

25. $6x + 3 = x - 22$

26. $-x - 9 = -11x + 91$

27. $\dfrac{\frac{4}{5}}{3\frac{1}{2}} = \dfrac{\frac{6}{7}}{x}$

Simplify:

28. $-[-(-4)(-4)] - [-(3^2)]$

29. $-[-(6)(3)] + (-9) - [-(-4)] - \sqrt[3]{-27}$

30. Evaluate: $a^2b^2 + b^2 + \dfrac{b}{a}$ if $a = -3$ and $b = -9$

LESSON *112 Markup · Markdown*

If a merchant is to make a living, goods must be sold for more than the merchant pays for them. It is customary to sell goods for the cost plus a percentage of the cost. This added price is called **markup**. If the price of an item is reduced for a sale, the amount it is reduced is called **markdown**.

example 112.1 Mr. Franklin bought new dresses for $40 each. He marked each dress up 20 percent. What was the price he charged his customers?

solution We can draw a picture of the problem.

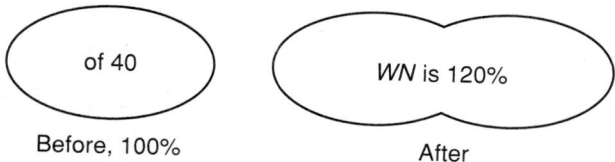

We want to answer the question

WN is 120 percent of 40?

We will use 1.2 for the rate instead of using 120 percent.

$$1.2 \times \$40 = WN \qquad \text{equation}$$
$$\$48 = \textbf{WN} \qquad \text{multiplied}$$

We could have found 20 percent of $40 and added this to $40.

$$(0.2)(\$40) = \$8 \qquad 20\% \text{ of } 40$$
$$\$8 + \$40 = \textbf{\$48} \qquad \text{added}$$

example 112.2 Sybil saw that the coat was marked down 30 percent. If the original price was $80, what was the price of the coat now?

solution The final price was less than the original price, so our diagram looks like this.

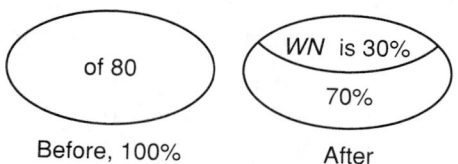

First we find out how much the coat was marked down.

30 percent of $80 is WN?

We will use 0.3 as the rate.

$$0.3 \times \$80 = \$24 \qquad \text{equation}$$

The coat was marked down $24. To find the sale price, we subtract.

$$\$80 - \$24 = \$56 \qquad \text{subtracted}$$

Thus, the sale price of the coat was **$56**.

example 112.3　The sale price of the coat was $120. It had been marked down 40 percent for the sale. What was the original price of the coat?

solution　The diagram of the problem looks like this.

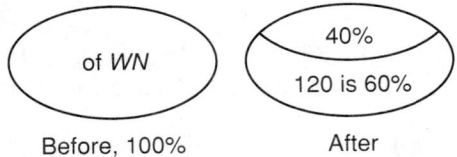

Before, 100%　　　　After

The original price was 100 percent. If the coat was marked down 40 percent, it was selling for 60 percent of the original price.

Sixty percent of what number is 120?

$$0.6WN = 120$$

$$\frac{0.6WN}{0.6} = \frac{120}{0.6}$$

$$WN = \$200$$

The original price of the coat was **$200.**

practice　**a.**　The original price of the tractor was $8000. If the store owner marked up the price 20 percent, at what price did the store owner sell the tractor?

b.　The suit was marked down 20 percent for the sale, and its sale price was $120. What was the original price of the suit?

problem set 112

1.　The ratio of the number of loquacious to the number of nonloquacious at the conference was 2 to 5. If 4900 people attended the conference, how many were loquacious?

2.　Fertilizer was purchased in large quantities. If 500 tons could be purchased for $3300, what would be the cost of 350 tons?

3.　This year the fertile valley produced a harvest 230 percent greater than average. If the average harvest produced 11,000,000 bushels, how many bushels were produced this year?

4.　Because the economy was good, $3\frac{3}{4}$ times as many people lived on the coast than before. If 12,000 people live on the coast now, how many people lived there before?

5.　If Jennifer put $5000 in the bank at 7 percent interest compounded annually, how much money would she have at the end of 2 years?

6.　The merchant has a standard 60 percent markup on her inventory. She pays $20 a pair for dress shoes. For how much does she sell the dress shoes?

7.　February is sale month at the clothing store. All items are marked down 20 percent. If the sale price for a purse is $42, what was the price before the markdown?

8.　Graph: $x \le -2$

9. If 90 is increased by 90 percent, what is the resulting number? Draw a diagram of the problem.

10. What percent of 68 is 170? Draw a diagram of the problem.

11. What number is 19 percent of 12,000? Draw a diagram of the problem.

12. Find the volume in cubic miles of a right solid whose base is shown and whose height is 0.5 mile. Dimensions are in miles. Corners that look square are square.

13. Find the volume in cubic feet and the lateral surface area in square feet of the right circular cylinder shown. Dimensions are in feet.

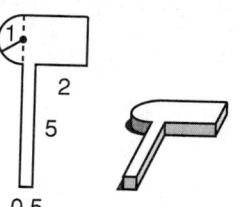

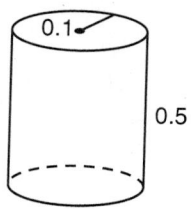

14. Use Roman numerals to write each number: (a) 26 (b) 39

15. If $3x - 9 = 18$, what is the value of $\frac{7}{3}x + 12$?

16. If $\frac{5}{4}x + 12 = -13$, what is the value of $\frac{8}{5}x - 3$?

Use the distributive property to multiply:

17. $3c(m + mn + n)$

18. $a^2(a + ab + b^2)$

Simplify by adding like terms:

19. $a^3p + mb^2m + apa^2 + m^2bb$

20. $p^2bc^2 + pbcpc + mx^3 + mp$

Write as a decimal number without a fraction:

21. $0.0091\frac{3}{4}$

22. $0.00007\frac{5}{8}$

Write each percent as a decimal number:

23. $8\frac{1}{5}\%$

24. $9\frac{1}{8}\%$

Solve:

25. $2x - 11 = -8x + 109$

26. $-x - 4 = -13x - 100$

27. $\dfrac{\frac{1}{4}}{-3\frac{1}{6}} = \dfrac{-1\frac{1}{2}}{x}$

Simplify:

28. $-[-(-3)(2)] - [-(-4)] - (2^2) + \sqrt[3]{-64}$

Evaluate:

29. $mx^3 + \dfrac{x}{m} + \dfrac{m}{x}$ if $m = -2$ and $x = -4$

30. $ab - a^2b$ if $a = -2$ and $b = -3$

LESSON 113   *Commission · Profit*

113.A
commission

Many companies pay their sales personnel a basic salary that is low. They also pay them a **commission,** which is a percentage of their sales. Many salespersons work hard and make big commissions.

example 113.1   A sales representative was paid a base salary of $300 a month. He was also paid a commission of 10 percent on everything he sold. If he sold $26,000 worth of merchandise in 1 month, what was his paycheck?

solution   His commission was 10 percent of $26,000.

$$(0.1)(\$26,000) = (\$2600)$$

To his commission we add his base salary.

Commission	$2600
Base salary	+ 300
	$2900

His paycheck for one month was **$2900.**

113.B
profit

Markup is not all profit. The store owner must pay overhead: rent, the water bill, the employees, insurance, and other costs. What the owner has left after all expenses are paid is **profit.**

example 113.2   A store owner marks up her merchandise 40 percent. Her expenses are $5200 a month. If she paid $16,000 for the goods sold in 1 month, what was her profit?

solution   She marked up the merchandise 40 percent, so her markup was

$$(0.4)(\$16,000) = \$6400$$

Her expenses were $5200, so we subtract.

$$
\begin{array}{r}
\$6400 \\
-\$5200 \\
\hline
\$1200
\end{array}
$$

Her profit for the month was **$1200.**

practice   **a.** A sales representative got a base salary of $274 a month. She was paid a 7 percent commission. If she sold $20,000 worth of merchandise, how much money did she make for the month?

b. A store owner purchased $200,000 of merchandise and sold it at a markup of 4 percent. If his expenses totaled $5500, what was his profit?

problem set 113

1. Twenty percent of the little people living in the forest suffered from agoraphobia, and they would not go into the clearing. If 1600 did not suffer from agoraphobia, how many little people lived in the forest?

2. The ratio of the number of loners living in the valley to the number who were gregarious was 3 to 14. If 15,300 lived in the valley, how many were gregarious?

3. The product of a number and -2 was increased by 7. This result was 27 greater than the product of the number and 3. What was the number?

4. Alfonso traveled 100 miles in 4 hours. Then he increased his speed by 5 miles per hour. How long did it take him to travel the last 120 miles?

5. A sales representative received a base salary of $350 a month. She was paid 10 percent commission on all sales. If she sold $45,000 worth of merchandise, how much money was she paid for the month?

6. A store owner purchased $14,800 of merchandise which he sold at a markup of 90 percent. If his expenses totaled $10,000, what was his profit?

7. Find the volume in cubic feet of a right solid whose base is shown on the left and whose height is 4 feet. Dimensions are in feet.

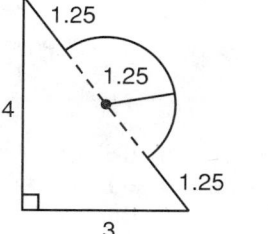 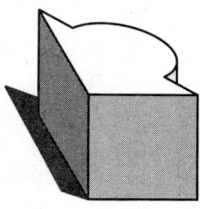

8. Find the lateral surface area in square feet of the solid in Problem 7.

Use Roman numerals to write each number:

9. 24

10. 19

11. If 120 is increased by 80 percent, what is the resulting number? Draw a diagram of the problem.

12. What percent of 50 is 60? Draw a diagram of the problem.

13. Write each percent as a decimal number: (a) $61\frac{3}{5}\%$ (b) $7\frac{4}{5}\%$

14. Use two unit multipliers to convert 12 miles to inches.

15. If $16x - 12 = 116$, what is $4x - 3$?

Use the distributive property to multiply:

16. $2ac(ab + a - b)$

17. $-am(a^2 + m - am)$

18. Simplify by adding like terms: $3xy^2 - 2xy + 3xyy + 3xy - y^2x$

Solve:

19. $3x + 6 = x + 7$

20. $-2\frac{1}{3}x - \frac{3}{4} = \frac{1}{2}$

21. $\dfrac{-\frac{1}{3}}{\frac{4}{9}} = \dfrac{\frac{1}{4}}{x}$

Simplify:

22. $-[-(-4)^2] + (-3)(4) + (-2)^3 + \sqrt[7]{-128}$

23. $2^2 + 2^3[-2(-3 + 2^2)(2^2 - 1) + 2]$ **24.** $6a^2bab^2b^3$

25. $\dfrac{-(-2)^2 + 3^2(2^2 - 5) + 3}{2(2^3 - 4)}$ **26.** $\dfrac{1}{4}\left(2\dfrac{1}{3} \cdot \dfrac{1}{4} - \dfrac{7}{12}\right)$

Evaluate:

27. $a^2 - 2a$ if $a = -2$

28. $a^2 + 3ab^2$ if $a = -1$ and $b = -3$

29. Graph: $x \not\geq 2$

30. Use the cut-and-try method to estimate $\sqrt{55}$ to one decimal place.

LESSON 114 *Complementary and supplementary angles · Measuring angles*

114.A
complementary and supplementary angles

We can name some angles by using the letter at the vertex. We can name the angles in the triangle on the left as angle A, angle B, and angle C. We also can use an angle symbol $\angle$ instead of writing the word *angle*.

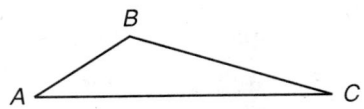

$\angle A$ is read "angle A"

$\angle B$ is read "angle B"

$\angle C$ is read "angle C"

When two or more angles have the same vertex, it is necessary to use three letters to name an angle. The center letter is the vertex and the other two letters name a point on each ray.

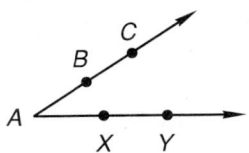

This angle has 9 names:

$\angle A, \angle CAX, \angle CAY,$

$\angle BAX, \angle BAY, \angle XAC,$

$\angle XAB, \angle YAB, \angle YAC$

All nine of the notations shown above name the same angle. If two angles have the same vertex and share the side between, then the angles are called **adjacent angles.**

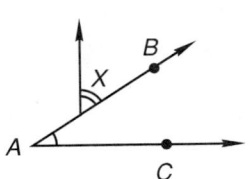
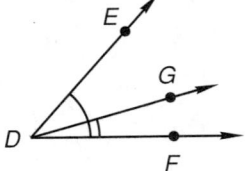
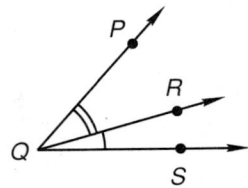

On the left, angles A and X are not adjacent because they do not have a common endpoint. In the center, angle GDE and angle FDE both have ray DE as a side, but it is not the side between the angles. The two angles $\angle PQR$ and $\angle RQS$ are adjacent angles. They both have ray QR as a side.

If the sum of the measures of two angles is 90°, the angles are called **complementary angles.** If the sum of the measure of two angles is 180°, the angles are called **supplementary angles.** Complementary angles and supplementary angles do not have to be next to each other.

Together angles *A* and *B* form a right angle, so these angles are complementary angles. Together angles *C* and *D* form a straight angle, so these angles are supplementary angles. Its easy to get the words complementary and supplementary confused. We can remember which is which if we associate the *C* in complementary with a picture of a right angle and the *S* in supplementary with two right angles.

example 114.1   What are the measures of (a) $\angle ABC$ and (b) $\angle DEF$?

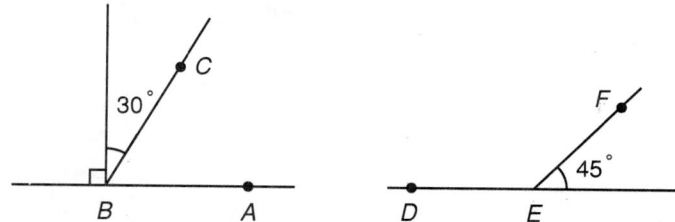

solution   (a)   Angle *ABC* and the 30° angle make a 90° angle. Thus angle *ABC* is a **60° angle.**

(b) $\angle DEF$ and the 45° angle make a 180° angle. Thus angle *DEF* is a **135° angle.**

114.B

measuring angles   To measure an angle by using a **protractor,** we align the baseline of the protractor with one side of the angle and move the protractor left or right as necessary to place the angle vertex under the origin at the center of the baseline. We read the measure of the angle where the other ray passes through the scale.

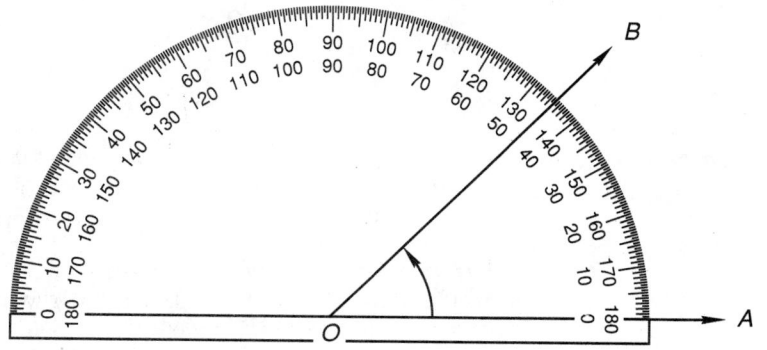

The scale on a protractor has two sets of numbers. One set is for measuring angles starting from the right-hand side, and the other is for measuring angles starting from the left-hand side. The easiest way to be sure we are reading from the correct scale is to decide if the angle we are measuring is acute or obtuse. Looking at ∠AOB we read the numbers 45° and 135°. Since the angle is less than 90° (acute), it must be **45° and not 135°.**

practice   **a.**   The supplement of an angle is 40°. What is the angle?
b.   The complement of an angle is 40°. What is the angle?

Given the angles shown:

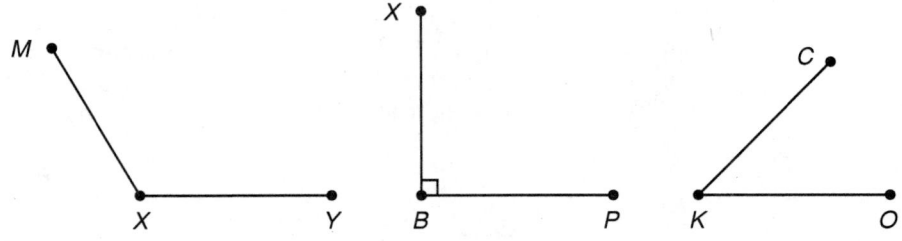

c.   Is angle *YXM* an acute, right, or obtuse angle?

d.   Is angle *PBX* an acute, right, or obtuse angle?

e.   Is angle *OKC* an acute, right, or obtuse angle?

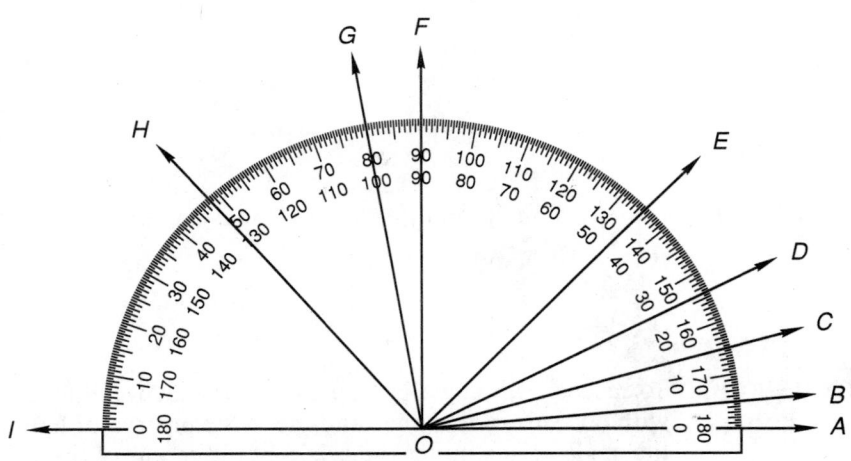

What is the measure of each angle?

f.   ∠AOC   **g.**   ∠AOD   **h.**   ∠AOG   **i.**   ∠AOH

j.   ∠IOG   **k.**   ∠IOF   **l.**   ∠IOB   **m.**   ∠IOA

problem set   **1.**   The wail of anguish turned into a paean of thanksgiving when the teacher
114   curved the grades. Forty percent of the students were saved. If 1440 students
were not saved, how many students were there in all?

2.   The number that eschewed unauthorized assistance rose 260 percent in 1 month. If the number was 400 last month, what was the number this month?

3. The ratio of froward students at the conference to those who were conciliatory was 2 to 11. If 390 students attended the conference, how many were froward?

4. The sum of a number and 40 was 13 greater than 10 times the number. What was the number?

5. Name:
   (a) The supplement of 65°
   (b) The complement of 65°

6. Label each angle as acute, right, or obtuse:

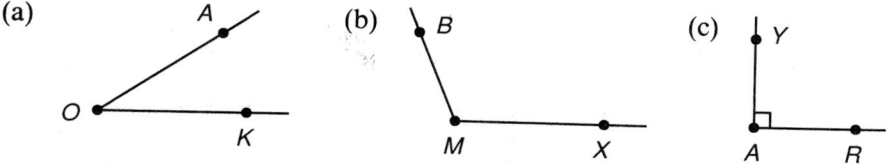

7. Find the measure of each angle:
   (a) ∠AOB   (b) ∠AOC   (c) ∠AOD

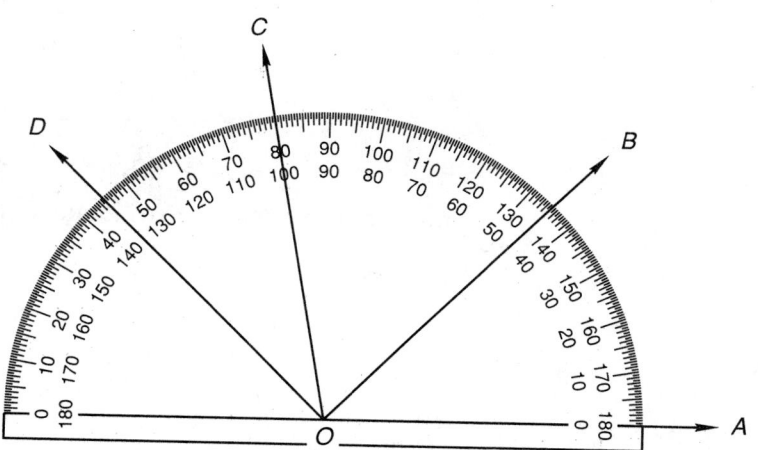

8. The sum of $8000 was deposited at 8 percent interest compounded yearly. How much money was in the bank at the end of 2 years?

9. What percent of 620 is 837? Draw a diagram of the problem.

10. Find the volume in cubic feet of a solid whose base is shown on the left and whose height is 2 yards. Dimensions are in feet.

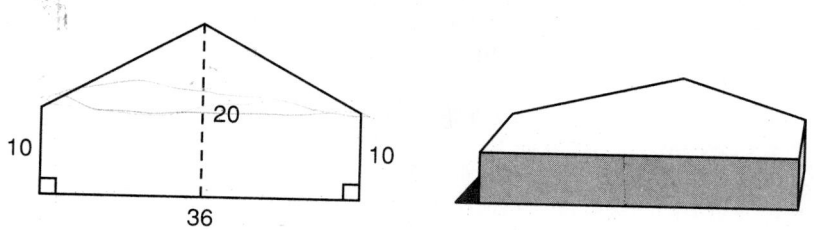

11. Use two unit multipliers to convert 2 square miles to square feet.

12. If $3x + 6 = 24$, what is the value of $\frac{1}{3}x - 2$?

13. If $7x + 6 = 27$, what is the value of $2x - 3$?

Use the distributive property to multiply:

14. $2ab(b + a + c)$ **15.** $mn(mn + n + m)$

Simplify by adding like terms:

16. $ab^2 + a^2b + 3abb + 2aab - a^2$ **17.** $xy^3 + x^2y + xyy^2 + xxy$

Solve:

18. $2x + 3 = 6x + 12$ **19.** $2\frac{1}{4}x - 2\frac{1}{2} = 3\frac{1}{4}$

20. $3x - 5 = 6x + 6$ **21.** $\dfrac{\frac{1}{3}}{\frac{9}{2}} = \dfrac{\frac{1}{4}}{x}$

Simplify:

22. $-[-(-3)] + (-2)(3)$ **23.** $-[-(-2)] + (-2) + (\sqrt[3]{-8})^2$

24. $ab^2b^3a^2b$ **25.** $xy^3y^2x^2x^3$

26. $\dfrac{2^4 - 2^3(2^3 - 4) + 2}{2^3(3^2 - 2^3)}$ **27.** $\frac{1}{3}\left(3\frac{1}{3} \cdot \frac{1}{3} - \frac{7}{9}\right)$

Evaluate:

28. $c^b - 2ab$ if $a = -2$, $b = +1$, and $c = -3$

29. $a + b^c + ac$ if $a = -1$, $b = -2$, and $c = 4$

30. Use the method of cut and try to estimate $\sqrt[3]{500}$ to the nearest whole number.

LESSON 115   *Copying angles · Construction*

115.A
copying angles

Sometimes we are asked to measure an angle whose sides are too short to permit the angle to be read on a protractor. We can use a piece of tracing paper to trace the given angle. Then we can use a straightedge to extend the sides of the angles so that a protractor can be used.

example 115.1   Measure this angle.

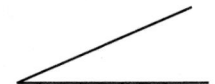

solution   First we trace the angle. Then we extend the sides of the angle and use a protractor to find the measure of the angle.

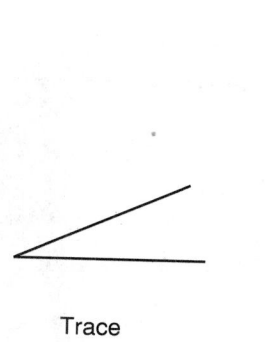

Trace

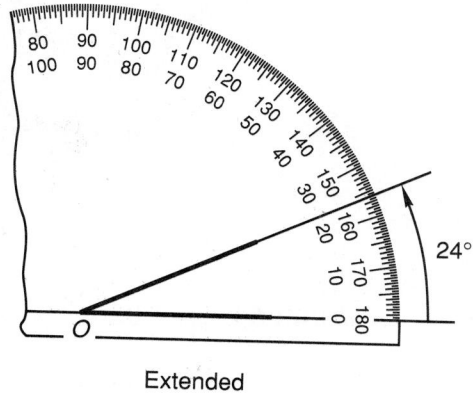

Extended

The measure of this angle is about **24°**.

The ancient Greeks did not have protractors to measure angles. But they could copy a given angle by using a straightedge and a compass.

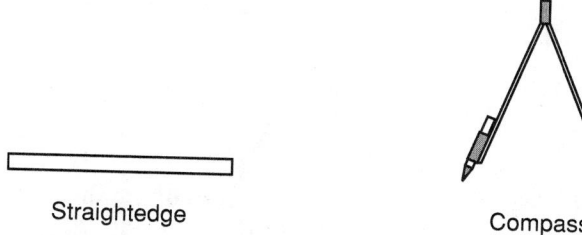

Straightedge

Compass

Of course, the Greeks did not have a modern metal compass. We imagine they used a forked stick or a piece of string instead. We can copy an angle by using a ruler, a compass, and five steps. Suppose we are given angle *A* shown here and asked to copy it. We use five steps as shown here.

Given angle

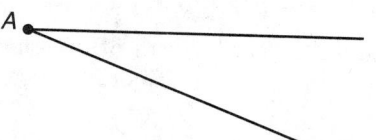

1. Draw one side.

2. Use a compass to draw an arc on the given angle.

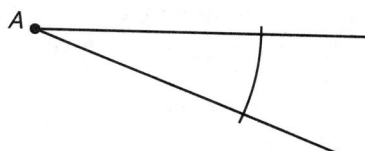

3. Draw the same arc on the side.

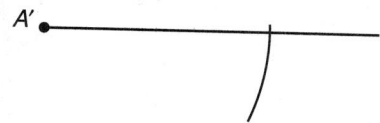

4. Use the compass to draw an arc on the arc that equals the opening.

5. Draw the same arc again and draw the ray from A' to B.

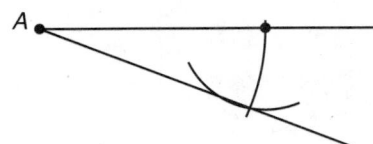

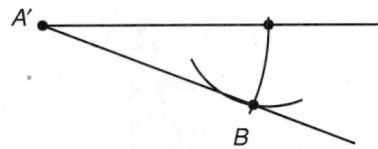

practice   Trace these angles. Extend the sides. Measure angles with a protractor.

a.

b.

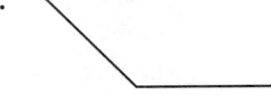

Copy these angles by using a compass and a straightedge.

c.

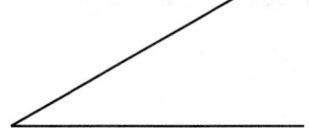

d.

problem set 115

1. Chad peeked around the corner and spied 440 of them lounging in the shade. If this was 120 percent more than David spied, how many did David spy?

2. Thirty percent of the natives were easily intimidated. If 450 natives were easily intimidated, how many natives were not easily intimidated?

3. The ratio of the squishy to the rock-hard in the room was 7 to 13. If 1400 total were in the room, how many were squishy?

4. When the $3400 was paid to the vendor, the students received 1700 items. If the vendor then reduced the price per item by 50 percent, how much did the students have to pay for the next 2000 items?

5. A store owner purchased a year's worth of merchandise for $200,000, which she sold at a markup of 80 percent. If her expenses were $110,000, what was her profit?

6. Use a protractor to draw a 66° angle. Then use a straightedge and a compass to copy the angle.

7. Use a protractor to draw a 50° angle. Then use a straightedge and a compass to copy the angle.

8. Describe each angle as acute, obtuse, or right:

(a) $\angle AOB < 90°$ (b) $\angle DOB > 90°$ (c) $\angle MOB = 90°$

9. What is the measure of:

(a) The complement of 34°
(b) The supplement of 35°

10. Use Roman numerals to write each number:
    (a) 16   (b) 22   (c) 29

11. Write each percent as a decimal number:

    (a) $5\frac{1}{5}\%$ (b) $17\frac{3}{4}\%$ (c) $11\frac{3}{8}\%$

12. Find the total surface area in square centimeters of a right circular cylinder whose radius is 10 centimeters and whose height is 7 centimeters.

13. There were about 156 million passenger cars in the United States in 1980. How many commercial vehicles were there?

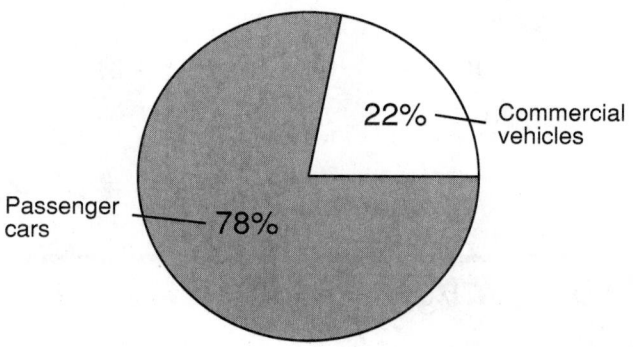

14. If 320 is increased by 15 percent, what is the resulting number? Draw a diagram of the problem.

15. Use six unit multipliers to convert 15 cubic yards to cubic inches.

16. Use two unit multipliers to convert 3 miles to inches.

17. Use the distributive property to multiply: $3ac(a + b - c + acb)$

18. Simplify by adding like terms: $2xxyyy - 6xy^2 + 3y^2x^2y + 3yxy$

19. Express in cubic centimeters the volume of the right solid whose base is shown on the left and whose sides are 2 meters high. Dimensions are in meters.

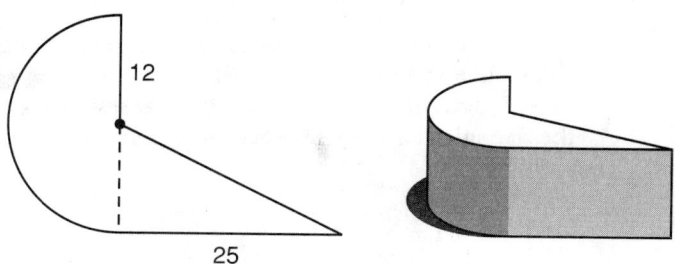

Solve:

20. $x + 5 = -3x - 6$

21. $-\frac{1}{6}x + 1\frac{1}{3} = -2\frac{5}{6}$

22. $\dfrac{-\dfrac{2}{3}}{\dfrac{4}{9}} = \dfrac{\dfrac{1}{2}}{-x}$

Simplify:

23. $-(-2)^3 + 2^3[-2(-3+5)(3-2^2)-7] + \sqrt[5]{-32}$

24. $2aba^2b^3a^2$

25. $\dfrac{-2^2 + 2^3(2^3 - 3^2) + 3}{3(2 \cdot 3 + 1)}$

26. $\dfrac{1}{2}\left(2\dfrac{1}{3} \cdot \dfrac{1}{4} - \dfrac{5}{6} \cdot \dfrac{1}{2}\right)$

Evaluate:

27. $xy^2 - x^2y$ 　if $x = -2$ and $y = -1$

28. $a - b^3a$ 　if $a = 1$ and $b = -5$

29. Graph: $x \not\geq 1$

30. If $6x + 7 = 28$, what is $6x - 4$?

LESSON 116 *Triangles · Measures*

116.A
classifying triangles

We can classify triangles by describing the angles. If one angle is a right angle, the triangle is a **right triangle**. If one angle is an obtuse angle, the triangle is an **obtuse triangle**. If all three angles are acute angles, the triangle is an **acute triangle**.

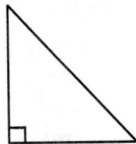
Right triangle

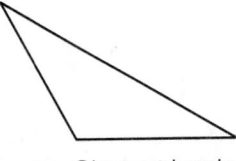
Obtuse triangle

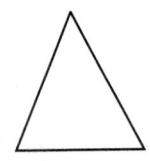
Acute triangle

　　　We can also classify triangles by describing the lengths of the sides. If no two sides have the same length, the triangle is a **scalene triangle**. If two sides have the same length, the triangle is an **isosceles triangle**. If all three sides have equal lengths, the triangle is an **equilateral triangle**. In the drawings, equal tick marks denote equal lengths.

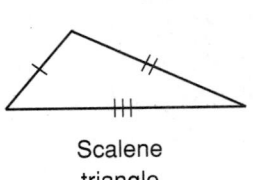
Scalene
triangle

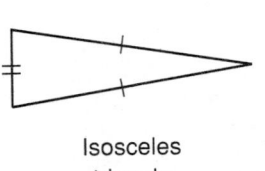
Isosceles
triangle

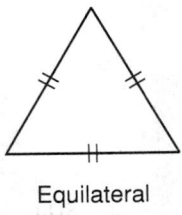
Equilateral
triangle

　　　If two sides of a triangle have equal lengths, the angles opposite these sides have equal measures. Conversely, if two angles have equal measures, the sides opposite

these angles have equal lengths.

 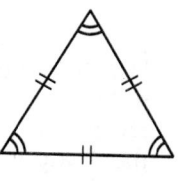

Isosceles
triangle

Equilateral
triangle

116.B
angles in triangles

The sum of the measures of the angles in any triangle is 180°.

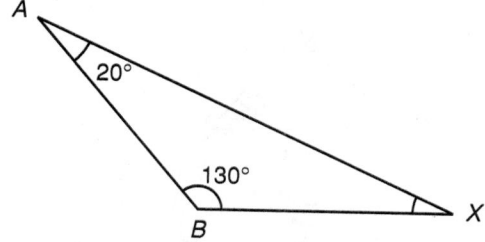

In this triangle the sum of the measures of angles *A* and *B* is 150°. Therefore, the measure of angle *X* must be 30° because

$$130° + 20° + 30° = 180°$$

example 116.1   Find the measure of angle *BCA*.

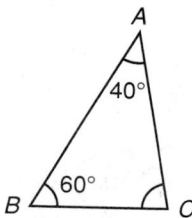

solution   The sum of the measures of the angles must be 180°. The measures of angles *A* and *B* add to 100°. Therefore angle *C*, or angle *BCA*, must have a measure of **80°**.

$$40° + 60° + 80° = 180°$$

We note that angle *BCA* can also be called angle *C* because only one angle has point *C* as its vertex.

example 116.2   What is the measure of angle *B*?

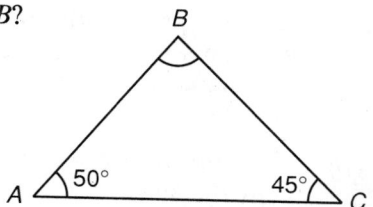

solution   The sum of the measures of the angles of any triangle is 180°. The two given angles are 50° and 45°. Their sum is 95°. Therefore, angle *B* must be an 85° angle because

$$95° + 85° = 180°$$

The measure of angle *B* is **85°**.

example 116.3   Angle *B* is a 110° angle. What are the measures of angles *B* and *C*?

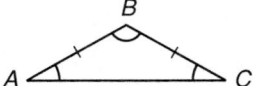

solution   The three angles must sum to 180°. Angle *B* is a 110° angle, so the sum of angles *A* and *C* must be 70° because 180° − 110° = 70°. Angles *A* and *C* must be equal angles because the sides opposite these angles have equal lengths. Since $\frac{70}{2} = 35$, angles *A* and *C* are **35° angles.**

practice   Find the measure of angle *A* and the measure of angle *B*:

a. *A*   **b.** *B*

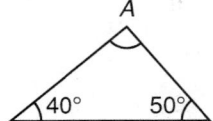   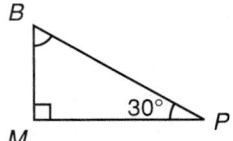

c. Angle *M* is a 50° angle. Find the measures of angles *D* and *E*.

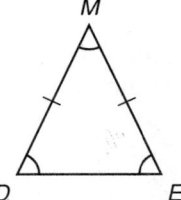

problem set 116

1. Sixty percent of the students were garrulous. If 2000 students were not garrulous, how many were garrulous?

2. The ratio of the number of loquacious students to the number of taciturn students was 7 to 3. If 750 students were in the school, how many were taciturn?

3. Five was added to the product of a number and 7. The result was 3 less than the product of the same number and 5. What was the number?

4. The caravan traveled 40 miles in 8 hours. Then the caravan doubled its speed. How long did it take the caravan to travel the last 60 miles?

5. A sales representative was paid a base salary of $400 a month. He was also paid a commission of 12 percent on everything he sold. If he sold $550,000 worth of merchandise in 1 year, what was his income? (Remember that he is paid a monthly salary.)

6. Use a protractor to draw a 40° angle. Then use a straightedge and a compass to copy the angle.

7. Use a protractor to draw a 70° angle. Then use a straightedge and a compass to copy the angle.

8. Name:
   (a) The supplement of 73°
   (b) The complement of 73°

9. Describe each angle as acute, obtuse, or right:

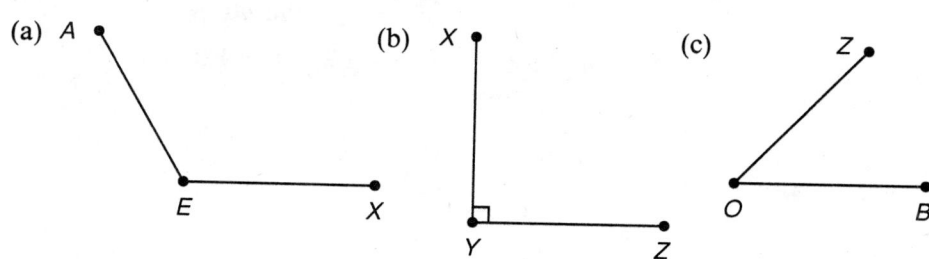

(a)   (b)   (c)

10. Use Roman numerals to write each number:
    (a) 49   (b) 45   (c) 43

11. Write each percent as a decimal number:

    (a) $16\frac{7}{8}\%$ (b) $4\frac{1}{4}\%$ (c) $29\frac{3}{5}\%$

12. Use a protractor to measure this angle.

13. Use three unit multipliers to convert 44 cubic meters to cubic centimeters.

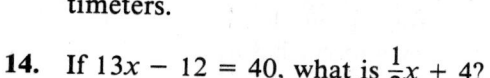

14. If $13x - 12 = 40$, what is $\frac{1}{2}x + 4$?

15. Use the distributive property to multiply: $2ab(ab + a^2 - b^2 + a)$

16. Simplify by adding like terms: $2xyy + 3xyx - 2x^2y + 5xy^2 - 3yxy$

17. Find the volume in cubic feet of a right solid whose base is shown on the left and whose height is 2 yards. Dimensions are in feet.

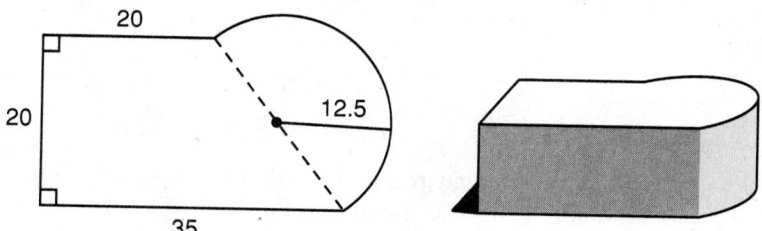

Solve:

18. $3x - 6 = 2x - 22$

19. $-1\frac{1}{3}x - \frac{5}{6} = \frac{2}{3}$

20. $\dfrac{-\frac{2}{5}}{\frac{4}{25}} = \dfrac{-\frac{1}{2}}{x}$

Simplify:

21. $-[-(-2)^2] + (-3)(-2) - \sqrt[5]{-243}$

22. $2^2 + 2^3[-3(-2 + 2^2)(2^3 - 2^2) + 3^2]$

23. $5a^3bab^2a^2b^3$

24. $\dfrac{-(-2)^2 + 2^3(3^2 - 2^3) + 2^2}{2^2(3^2 + 1)}$

25. $\frac{1}{5}\left(2\frac{1}{2} \cdot \frac{1}{6} - \frac{1}{3} \cdot \frac{5}{4}\right)$

Evaluate:

26. $-ab + 2a^3$ if $a = -1$ and $b = -2$

27. $-a^3b + b^c$ if $a = -2$, $b = -1$, and $c = 5$

28. Graph: $x \not\leq -1$

LESSON *117* *Roman numerals greater than 50*

We remember that the Romans used only seven symbols to write their numbers.

Symbol	I	V	X	L	C	D	M
Value	1	5	10	50	100	500	1000

We also remember that Romans avoided writing four symbols in a row by using a trick to write 4 and 9. They wrote 1 before 5 for 4 and wrote 1 before 10 for 9.

IV = 1 less than 5 = 4

IX = 1 less than 10 = 9

They also used a trick to write 40, 90, 400, and 900. They wrote 10 before 50 for 40. They wrote 100 before 500 for 400 and wrote 100 before 1000 for 900.

XL = 10 less than 50, which is 40

XC = 10 less than 100, which is 90

CD = 100 less than 500, which is 400

CM = 100 less than 1000, which is 900

Most people can remember what the symbols I, V, and X mean but have trouble remembering the meanings for L, C, D, and M. This is because we use Roman numerals so infrequently. So we suggest the following mnemonic to remember the order of L, C, D, and M.

Let Caesar Destroy Mountains

50   100   500   1000

L   C   D   M

This nonsense phrase can help us remember the order of the values of L, C, D, and M.

example 117.1 What number is represented by:
(a) DCCLXV   (b) MCMXLVII

solution First we write our key:

I   V   X   Let Caesar Destroy Mountains

1   5   10   50   100   500   1000

(a) With the key we can write that DCCLXV means

$$500 \cdots 100 \cdots 100 \cdots 50 \cdots 10 \cdots 5$$

Every number is equal to or less than the number to its left, so we add the numbers.

$$500 + 100 + 100 + 50 + 10 + 5 = \mathbf{765}$$

(b) With the key we write the numbers that correspond to MCMXLVII as

$$1000 \cdots 100 \cdots 1000 \cdots 10 \cdots 50 \cdots 5 \cdots 2$$

We circle pairs of numbers in which the first number is less than the second number. We also insert the necessary plus signs.

$$1000 + \boxed{100 \cdots 1000} + \boxed{10 \cdots 50} + 5 + 2$$

Now we take the difference of the numbers in each circle and write

$$1000 + 900 + 40 + 5 + 2 = \mathbf{1947}$$

example 117.2 Write the number 3493 in Roman numerals.

solution First we write our key.

I	V	X	L	C	D	M
1	5	10	50	100	500	1000

For 3000 we use three M's: MMM

For 400 we use 500 − 100: MMMCD

For 90 we use 100 − 10: MMMCDXC

And we have three 1s on the end.

MMMCDXCIII

example 117.3 Write the number 552,000 in Roman numerals.

solution To write numbers greater than MMM, or 3000, the Romans would put a bar over all or part of a number. The bar is called a **vinculum** and means to multiply by 1000.

DLII   means   552

$\overline{\text{DLII}}$ means   552,000

In the problem sets we will not consider Roman numerals for numbers greater than 4000.

practice Complete the chart. Begin by writing the key.

Roman Numeral	Arabic Numeral
DCCLXV	a.
MCMXLVII	b.
c.	888
d.	1990

problem set 117

1. Seventy percent of the population was gregarious. If 21,000 were not gregarious, how many were gregarious?

2. The number of fecund and the number of effete were in the ratio of 9 to 5. If 14,400 were fecund, how many were effete?

3. Ten was added to the product of a number and 4. The result was 4 less than the product of the same number and -10. What was the number?

4. The entourage flew 600 miles in 3 hours. Then they halved their speed. How long did it take to go the next 1400 miles?

5. There were $5\frac{3}{4}$ times as many participants as there were spectators. If there were 1200 spectators, how many participants were there?

6. A sales representative was paid a base salary of \$350 a month. She was also paid a commission of 20 percent on all sales. If she sold \$14,000 worth of merchandise in 1 month, what was her total paycheck?

7. Use Roman numerals to write 1435.

8. Write as an Arabic number: MMCMLVIII

9. Name:
   (a) The supplement of $83°$
   (b) The complement of $83°$

10. Use a protractor to draw a $65°$ angle. Then use a straightedge and a compass to copy the angle.

11. What number is 310 percent of 400? Draw a diagram of the problem.

12. What percent of 88 is 154? Draw a diagram of the problem.

13. Write as a decimal number: $6\frac{9}{10}\%$

14. Find the measures of angles A, B, and C:

    (a) 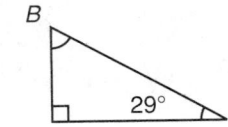   (b) 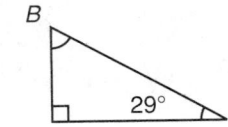   (c)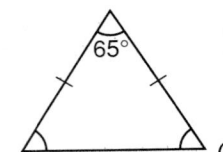

15. Use three unit multipliers to convert 160 cubic miles to cubic feet.

16. If $7x - 12 = 16$, what is $\frac{1}{4}x + 61$?

17. Use the distributive property to multiply: $4mn(mn + x^2 - z^2 + c)$

18. Simplify by adding like terms: $4mcm + 2cmc - 3m^2c + 5c^2m - bcmm$

19. Find the volume in cubic feet of a right solid whose base is shown on the left and whose height is 3 yards. Dimensions are in feet.

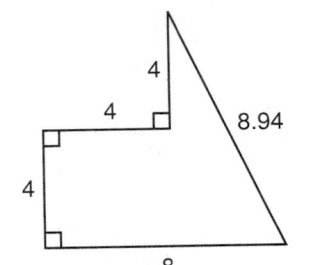

20. Find the lateral surface area of the solid in Problem 19.

Solve:

21. $3x - 9 = 7x + 12$

22. $\dfrac{-\dfrac{3}{4}}{\dfrac{9}{16}} = \dfrac{-\dfrac{8}{21}}{x}$

23. $-4\dfrac{1}{2}x - \dfrac{3}{7} = \dfrac{1}{4}$

Simplify:

24. $-[-(-9)^2] - \sqrt[3]{-125}$

25. $-4^2 + 3^3[-2(3 - 2^2)]$

26. $7c^5a^2cabc^2b^5$

27. $\dfrac{-(-3)^2 + 3^2}{-3^2}$

Evaluate:

28. $-m(-n)$ if $m = -3$ and $n = -4$

29. $(-b)^2 + a^5$ if $a = -2$ and $b = 3$

30. Graph: $x \not> 4$

LESSON *118 Probability*

The study of probability is based on the study of outcomes that have an equal chance of occurring. If we toss a fair coin many times, it should come up heads as many times as it comes up tails. We say the **probability** of one toss resulting in a head is the number of outcomes that are heads divided by the total number of outcomes.

$$P(H) = \frac{\text{number of outcomes that are heads}}{\text{total number of outcomes}} = \frac{1}{2}$$

The face of this spinner is divided into equal parts. If the spinner is spun many times, it should stop on each of the spaces an equal number of times. The probability of getting a 4 on any one spin is the number of outcomes that are 4 divided by the total number of outcomes.

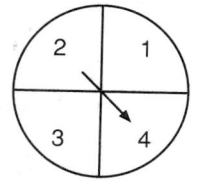

$$P(4) = \frac{\text{number of outcomes that are 4}}{\text{total number of outcomes}} = \frac{1}{4}$$

A single die has six faces.

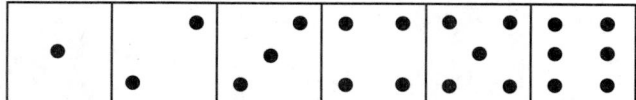

The probability of getting a number greater than 4 on one roll of the die is the number of outcomes that are greater than 4 divided by the total number of outcomes. There are two possible outcomes that are greater than 4.

$$P(>4) = \frac{\text{number of outcomes greater than 4}}{\text{total number of outcomes}} = \frac{2}{6} = \frac{1}{3}$$

example 118.1   A deck of cards contains 52 cards: 13 are spades, 13 are hearts, 13 are clubs, and 13 are diamonds. Jimmy tears up the ace of spades. Then he shuffles the deck and draws 1 card. What is the probability that the card is a spade?

solution   There are 51 cards and 12 are spades. So

$$P(S) = \frac{12}{51} = \frac{4}{17}$$

example 118.2   One die is red and the other die is green. Both are rolled. What is the probability of getting (a) a sum of 7? (b) a sum of 4?

solution   We make a table that shows every possible outcome.

Green die

	1	2	3	4	5	6
1	2	3	4	5	6	7
2	3	4	5	6	7	8
3	4	5	6	7	8	9
4	5	6	7	8	9	10
5	6	7	8	9	10	11
6	7	8	9	10	11	12

Red die

(a)   The total number of dots is indicated by the numbers in the center of each box in the table. Six of them are 7 and there are 36 squares in all. So

$$P(7) = \frac{6}{36} = \frac{1}{6}$$

(b)   There are only 3 ways to get a total of 4. Thus, the probability of getting a total of 4 is

$$P(4) = \frac{3}{36} = \frac{1}{12}$$

practice   **a.**   There are 16 marbles in a bowl: 3 are red, 6 are black, and 7 are green. Jimmy draws 1 marble from the bowl. What is the probability that the marble is either red or black?

b.   A red die and a green die are rolled at the same time. What is the probability of getting an 11?

problem set   **1.**   The ratio of red marbles to green marbles was 2 to 19. If there were 84,000
118   marbles in all, how many were red?

2.   A deck of cards contains 52 cards: 13 are spades, 13 are hearts, 13 are clubs, and 13 are diamonds. Joan tears up the ace of spades. Then she shuffles the deck and draws 1 card. What is the probability that the card is a heart?

3.   One die is rolled. What is the probability of getting a (a) 3? (b) 4?

4.   Thirty percent of the airplanes in the show were biplanes. If 120 were biplanes, how many were not biplanes?

5.   There were $3\frac{2}{5}$ times as many wet ones as there were dry ones. If there were 8500 dry ones, how many wet ones were there?

6.   A merchant purchased $16,000 worth of merchandise and then sold it at a 60 percent markup. Her expenses totaled $7000. What was her profit?

7.   Use a protractor to draw a 50° angle. Then use a straightedge and a compass to copy the angle.

8.   Write as a decimal number: $6\frac{1}{4}\%$

9.   Write MMMCDXXXIV using Arabic numerals.

10.   Use Roman numerals to write 3465.

11.   Forty-five is what percent of 225? Draw a diagram of the problem.

12.   Forty-two percent of what number is 126? Draw a diagram of the problem.

13.   Use three unit multipliers to convert 52 cubic meters to cubic centimeters.

14.   Use six unit multipliers to convert 4100 cubic miles to cubic inches.

15.   If $12x - 61 = 83$, what is $\frac{1}{6}x - 7$?

16.   Use the distributive property to multiply: $2ac(a + c - ac + a^2)$

17.   Simplify by adding like terms: $2xyy + 3xyx - 6xy^2 + 4x^2y$

18.   Find the volume in cubic meters of a solid whose base is shown and whose height is 3 centimeters. Dimensions are in centimeters.

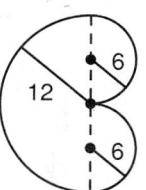

Solve:

19. $3x + 6 = 2x - 21$

20. $-2\frac{1}{3}x - \frac{5}{6} = \frac{1}{3}$

21. $\dfrac{-\dfrac{3}{4}}{\dfrac{9}{16}} = \dfrac{\dfrac{1}{3}}{x}$

Simplify:

22. $-[-(-1)^3] + (-4) + \sqrt[3]{-125}$

23. $2^2 a a^3 b a^2 b^3$

24. $\dfrac{2^2(2^2 - 2 \cdot 4)}{2^3(3^2 - 2^2)}$

25. $2\frac{1}{3}\left(2\frac{1}{2} \cdot \frac{1}{2} - \frac{1}{3} \cdot \frac{4}{5}\right)$

Evaluate:

26. $a^2 b + ab^2$ if $a = -1$ and $b = -2$

27. $a^2 - b^2$ if $a = -2$ and $b = -3$

28. Graph: $x \not\le 3$

29. Express 10,080 as a product of primes.

30. Round 62,987,134.32 to the nearest ten thousand.

LESSON *119* *Measuring lengths*

119.A

inch scale
We use a scale to measure lengths. The numbers on an inch scale tell the number of inches the long marks are from the left end of the scale. There are 16 marks between the inch marks. These marks are $\frac{1}{16}$ inch apart. Two of the spaces between these smaller marks equal $\frac{1}{8}$ inch. Four of the spaces between these shorter marks equal $\frac{1}{4}$ inch.

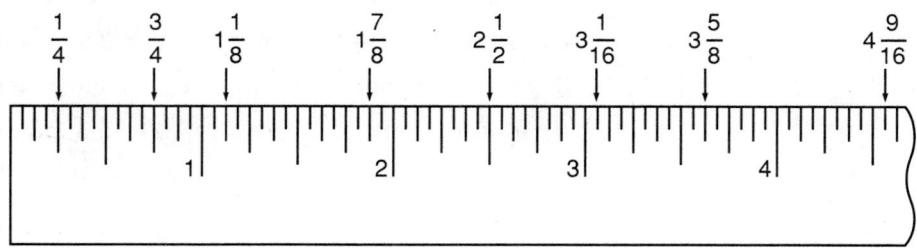

example 119.1 How far is arrow g from the left end of the scale?

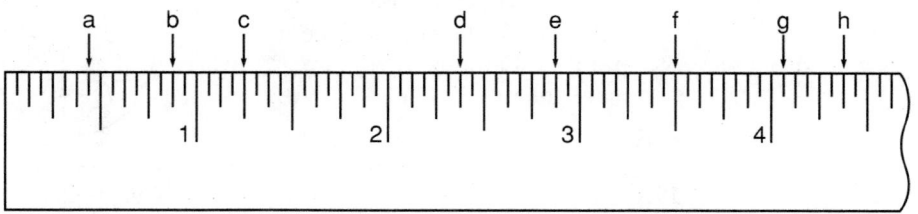

solution The small spaces are $\frac{1}{16}$ inch from the spaces on either side. Arrow g is **$4\frac{1}{16}$ inches** from the left end of the scale.

example 119.2   Use an inch scale to find the length of this line segment.

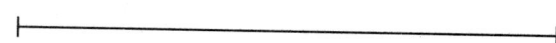

solution   We place the left end of the segment at the left end of an inch scale. Then we read the length on the right end.

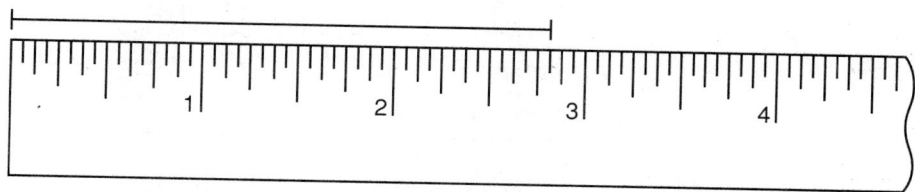

The line segment is about $2\frac{13}{16}$ **inches** long. No measurement is exact. This reading is to the nearest sixteenth of an inch.

119.B
metric scale   The numbers on a metric scale tell the number of centimeters the long lines are from the left end of the scale. There are 10 short lines between each pair of long lines. These short lines are 1 millimeter apart. A millimeter is a tenth of a centimeter. A metric scale is much easier to read than an inch scale.

example 119.3   How far is arrow d from the left end of the scale?

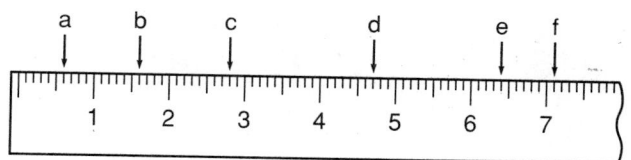

solution   The small spaces are 1 millimeter wide. A millimeter is a tenth of a centimeter. Arrow d is about **4.7 centimeters** from the left end of the scale.

example 119.4   Use a centimeter scale to measure the length of this line segment.

solution   We place the left end of the segment at the left end of the scale. Then we read the right-hand end.

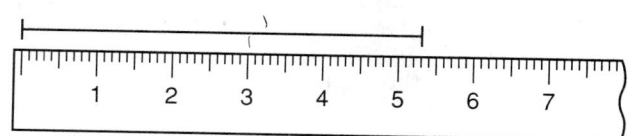

The segment is about **5.3 centimeters** long. This is the same as **53 millimeters**.

practice   **a.**   Use an inch scale and a centimeter scale to measure the length of this segment.

b.   Tell how far these arrows are from the left end of this inch scale.

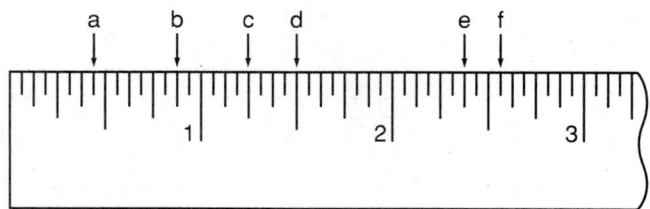

c.   Tell how far these arrows are from the left end of the centimeter scale.

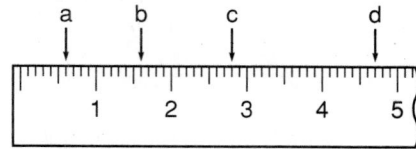

d.   Use an inch scale and a centimeter scale to find the approximate length of this line segment.

problem set   **1.**   There are 22 marbles in a bowl: 7 are black, 5 are red, and 10 are blue. Arcelia
119   draws a marble from the bowl. What is the probability that the marble is either
                      black or red?

2.   In a standard deck of 52 cards there are 13 of each suit: spades, hearts, clubs,
and diamonds. After the deck is fairly shuffled, what is the probability of
drawing a diamond?

3.   Forty percent of the airplanes were hydroplanes. If 660 were hydroplanes, how
many were not hydroplanes?

4.   There were $2\frac{2}{3}$ times as many that were cordial as those that were aloof. If 3000
were cordial, how many were aloof?

5.   Sixteen was added to the product of a number and 3. The result was 5 less than
the product of the same number and -4. What was the number?

6.   Find the ending balance in a bank account that began with $3000 and received
9 percent interest compounded annually for 4 years.

7.   Use a protractor to draw a 43° angle. Then use a straightedge and a compass to
copy the angle.

8.   Write as a decimal number: $19\frac{1}{8}\%$

9.   Write MMMXV as an Arabic numeral.

10. (a) Use an inch scale to measure this line segment to the nearest sixteenth of an inch.

(b) Use a metric scale to measure this line segment to the nearest millimeter.

11. Sixteen is what percent of 320? Draw a diagram of the problem.

12. Seventy-three percent of what number is 438? Draw a diagram of the problem.

13. What percent of 105 is 273? Draw a diagram of the problem.

14. Use three unit multipliers to convert 75 cubic feet to cubic inches.

15. If $5x - 203 = 22$, what is the value of $\frac{11}{30}x + 29$?

16. Use the distributive property to multiply: $3mp\left(2m + p + \dfrac{m^2}{p} + \dfrac{p^4}{m}\right)$

17. Simplify by adding like terms: $7d^2z + 7zdz + 7ddz + 7dz^2$

18. Find the volume in cubic miles of a right solid whose base is the figure shown and whose height is 0.5 mile. Dimensions are in miles. Angles that look square are square.

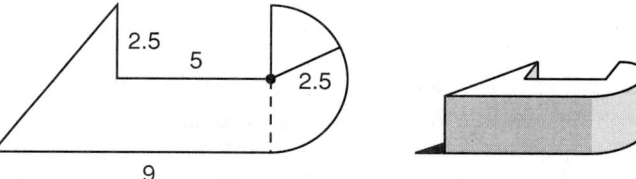

19. Find the surface area in square inches of a cube whose edges measure 3 inches.

Solve:

20. $5x + 6 = 14x - 32$

21. $-5\frac{5}{6}x - \frac{1}{2} = 1\frac{1}{4}$

22. $\dfrac{-\dfrac{5}{6}}{1\dfrac{2}{3}} = \dfrac{\dfrac{3}{4}}{x}$

23. Find the measures of angles A, B, and C:

(a)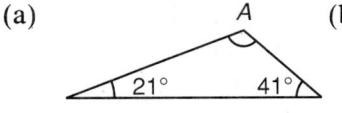

(b)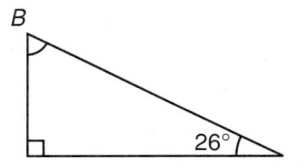

(c)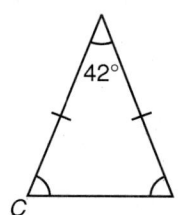

24. Graph: $x \not\geq -1$

Simplify:

25. $-[-(-2)^5] + (-9) - \sqrt[5]{-243}$

26. $3^2mn^5mn^2m^6(-1)^5$

27. $\dfrac{-(-3)^2 - 1^9(3^3 - 3 \cdot 2)}{3^3(3^2 - 2^3)}$

28. $3\frac{1}{2}\left(2\frac{1}{3} \cdot \frac{1}{3} - \frac{1}{2} \cdot \frac{3}{7}\right)$

29. (a) 2^{-5} (b) $\dfrac{1}{3^{-3}}$

Evaluate:

30. $m^3 n^{-2}$ if $m = -2$ and $n = -3$

31. $a^2 p^3 + \dfrac{1}{p^{-2}}$ if $a = -5$ and $p = -2$

LESSON 120 *Similar triangles*

If two triangles have the same angles, the triangles are similar triangles. We say they are similar because they have the same shapes and look alike.

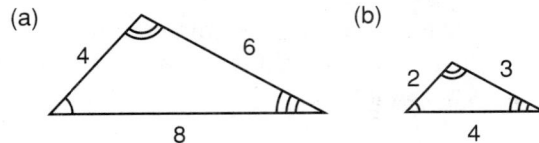

We use tick marks to denote equal angles. Angles with the same tick marks are equal angles. Sides opposite equal angles are called **corresponding sides.** Similar triangles have the same shape, but one of the triangles might be larger than the other triangle. Similar triangles are related by a **scale factor.** We can multiply one side of a triangle by the scale factor and find the length of the corresponding side in the other triangle. In the triangles above, the scale factor from (a) to (b) is $\frac{1}{2}$. The scale factor from (b) to (a) is 2.

example 120.1   Find x and y.

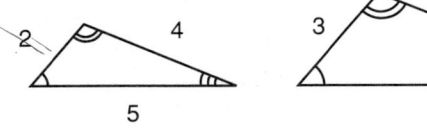 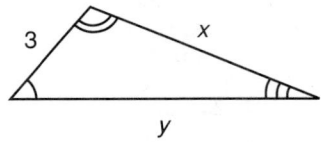

solution   The triangles are similar because the angles in one triangle are the same angles as the angles in the other triangle. The sides labeled 2 and 3 are corresponding sides because they are opposite equal angles. Let's find the scale factor from the triangle on the left to the triangle on the right.

$$2 \cdot \overrightarrow{SF} = 3 \qquad \text{2 times scale factor equals 3}$$

$$\overrightarrow{SF} = \frac{3}{2} \qquad \text{divided both sides by 2}$$

The arrow above SF shows us that this is the left-to-right scale factor. This means that if we multiply the lengths of each side in the triangle on the left by $\frac{3}{2}$, we will find the lengths of the corresponding sides in the triangle on the right.

$$3 \text{ equals } 2 \cdot \overset{\longrightarrow}{\frac{3}{2}} = 3 \qquad \text{check}$$

$$x \text{ equals } 4 \cdot \frac{\overrightarrow{3}}{2} = 6$$

$$y \text{ equals } 5 \cdot \frac{\overrightarrow{3}}{2} = \frac{15}{2} \text{ or } 7\frac{1}{2}$$

practice   What are the lengths of sides m and p? Begin by using the sides 5 and 4 to find $\overrightarrow{SF}$.

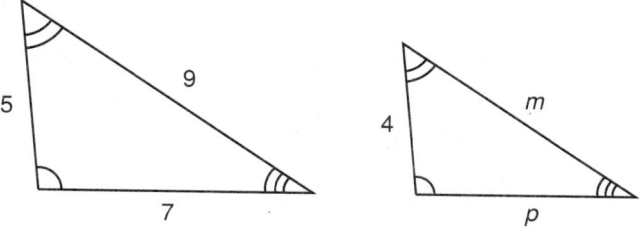

problem set 120

1. What is the probability of getting a number greater than 2 on one roll of a die?

2. Find the difference between simple interest and interest compounded yearly on $11,000 deposited in an account for 4 years at a yearly rate of 6 percent.

3. If the product of a number and −6 is increased by 12, the result is 4 less than the product of the same number and −2. What is the number?

4. The production line was 60 percent more efficient during the last month than the month before that. If they produced 1,600,000 units last month, how many units were produced in the month before?

5. The ratio of the number of melodic pieces to the number of atonal pieces was 7 to 2. If 3600 pieces were submitted, how many were atonal and how many were melodic?

6. Find the lengths of sides x and y:

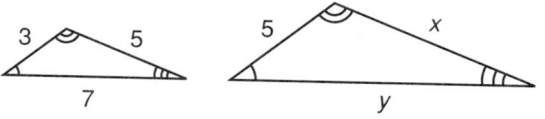

7. Use a protractor to draw a 59° angle. Then use a straightedge and a compass to copy the angle.

8. Write $9\frac{3}{8}\%$ as a decimal number.

9. Write 3964 as a Roman numeral.

10. (a) Use an inch scale to measure the distance between the marks to the nearest sixteenth of an inch.

   (b) Use a metric scale to measure the same length.

11. Fifty-three is what percent of 265? Draw a diagram of the problem.

12. Eighty-nine percent of what number is 178? Draw a diagram of the problem.

13. What percent of 94 is 235? Draw a diagram of the problem.

14. Use three unit multipliers to convert 1000 cubic kilometers to cubic meters.

15. If $13x - 130 = 39$, find the value of $\frac{7}{26}x + \frac{3}{13}$.

16. Use the distributive property to multiply: $6a^2z^3(a + z^3 + a^5)$

17. Simplify by adding like terms: $a^3m^{-2} + 5aa^2mm^{-3} + 2$

18. Find the volume in cubic meters of a right solid whose base is the figure shown on the left and whose height is 4 meters. Dimensions are in meters.

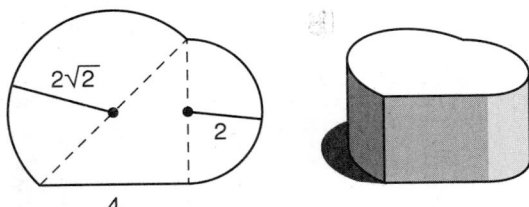

19. Find the perimeter of the base of the figure in Problem 18.

Solve:

20. $11x - 9 = -4x + 23$

21. $-4\frac{3}{11}x + \frac{1}{3} = \frac{16}{33}$

22. $\dfrac{-\frac{2}{3}}{2\frac{1}{2}} = \dfrac{x}{\frac{1}{3}}$

23. Find the measures of angles A, B, and C:

(a)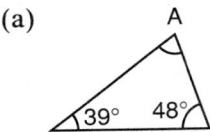
(b)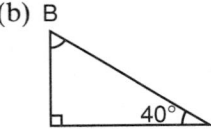
(c)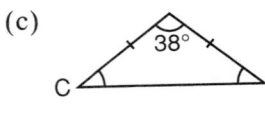

Simplify:

24. $-[-(-1)^{13}] + (6)(-2) - \sqrt[3]{-8}$

25. $4^{-2}mp^5m^2p^7$

26. $\dfrac{-(-3^2)(2) - (3^2 - 3)}{-2^3(2^3 - 3^2)}$

27. $5\frac{1}{3}\left(\frac{1}{2} \cdot \frac{1}{5} - \frac{1}{10} \cdot 2\right)$

28. (a) 4^{-3} (b) $\dfrac{1}{2^{-6}}$

Evaluate:

29. $a^3m^2 - am$ if $a = -3$ and $m = -2$

30. $a^4 - ab^3$ if $a = -2$ and $b = -1$

LESSON 121 *Probability, Part 2: Independent events*

121.A
independent events

We say that events that do not affect one another are **independent events.** If Danny flips a dime and Paul flips a penny, the outcome of Danny's flip does not affect the outcome of Paul's flip. Thus, we say that these events are independent events. **The probability of independent events occurring in a designated order is the product of the probabilities of the individual events.**

A tree diagram can always be used to demonstrate the probability of independent events occurring in a designated order. This diagram shows the possible outcomes if a coin is tossed twice.

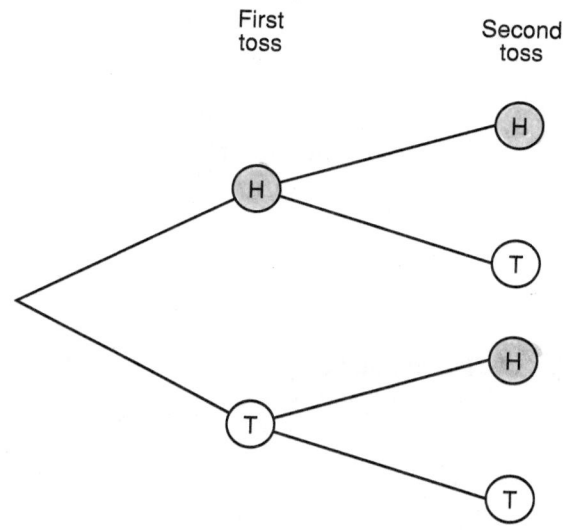

First toss

Second toss

On a single toss the probability of getting heads is $\frac{1}{2}$. The probability of getting tails is also $\frac{1}{2}$. There are only four possibilities on two tosses. If the first toss is heads, the second toss can be heads or tails.

$$(HT) \quad \text{or} \quad (HH)$$

Thus,

$$P(H, T) = \frac{1}{2} \cdot \frac{1}{2} = \frac{1}{4} \qquad P(H, H) = \frac{1}{2} \cdot \frac{1}{2} = \frac{1}{4}$$

If the first toss is tails, the second toss can be heads or tails.

$$(TH) \quad \text{or} \quad (TT)$$

Thus,

$$P(T, H) = \frac{1}{2} \cdot \frac{1}{2} = \frac{1}{4} \qquad P(T, T) = \frac{1}{2} \cdot \frac{1}{2} = \frac{1}{4}$$

There are no other possible outcomes for two tosses. The probability of getting any one of these outcomes is one-fourth.

example 121.1 A fair coin is tossed 3 times. What is the probability that it will come up heads every time?

solution Coin tosses are independent events because the result of one toss has no effect on the result of the next toss. Since the probability of independent events occurring in a designated order is the product of the individual probabilities, we have

$$P(H, H, H) = \frac{1}{2} \cdot \frac{1}{2} \cdot \frac{1}{2} = \frac{1}{8}$$

example 121.2   A fair coin is tossed 4 times and it comes up heads each time. What is the probability that it will come up heads on the next toss?

solution   The results of past coin tosses do not affect the outcome of future coin tosses. Thus, the probability of getting a head on the next toss is $\frac{1}{2}$.

$$P(H) = \frac{1}{2}$$

practice   **a.**   The spinner is spun 3 times. What is the probability it will stop on 3, 5, and 1 in that order?

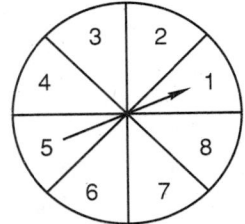

b.   The spinner is spun twice. What is the probability it will stop on either 3 or 5 the first time and will also stop on 2 the second time?

problem set 121

1.   The first urn contains 4 blue marbles and 2 white marbles. The second urn contains 4 blue marbles and 11 white marbles. A marble is picked from each of the urns. What is the probability that both marbles are white?

2.   A red die and a blue die are rolled. What is the probability of getting both a 6 on the red die and a 4 on the blue die?

3.   A fair coin is tossed five times. What is the probability of getting *HHTHH* in that order?

4.   The increase in the turnip crop was 260 percent. If last year's harvest was 230,000 tons, what was the harvest this year?

5.   The ratio of the number who were puissant to the number who were diffident was 2 to 5. If 2450 were present, how many were puissant?

6.   Use a protractor to draw a 50° angle. Then use a straightedge and a compass to copy the angle.

7.   Find sides *x* and *y*:

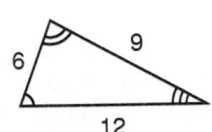
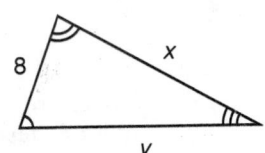

8.   (a)   Use an inch scale to measure the segment to the nearest sixteenth of an inch.

(b) Use a metric scale to measure the segment to the nearest millimeter.

9. Find the measures of angles *A*, *B*, and *C*:

(a)

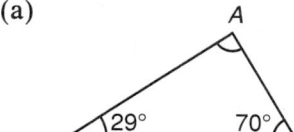

(b)

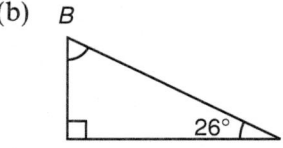

(c)

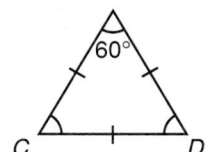

10. Simplify: (a) $2(4^{-2})$ (b) $2\left(\dfrac{1}{4^{-2}}\right)$

11. Write each percent as a decimal number: (a) $6\frac{7}{8}\%$ (b) $132\frac{3}{4}\%$

12. What percent of 90 is 162? Draw a diagram of the problem.

13. Sixty percent.of what number is 72? Draw a diagram of the problem.

14. Write MDXCIV in Arabic numerals.

15. Use three unit multipliers to convert 181 cubic miles to cubic feet.

16. Use the distributive property to multiply: $2ac(ab + bc - c)$

17. Simplify by adding like terms: $2xyx + 3yxy^2 - 3x^2y + 5y^3x - 6yx^2$

18. Express in cubic feet the volume of a right solid whose base is shown on the left and whose sides are 2 yards high. Dimensions are in feet.

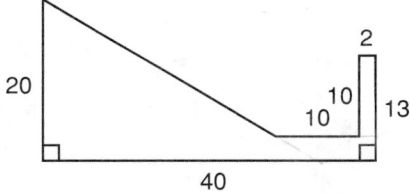

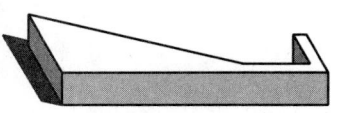

19. Find the surface area of a rectangular solid whose length, width, and height are 4 feet, 2 feet, and 10 feet, respectively.

Solve:

20. $3x - 3 = -8x + 5$

21. $-\dfrac{1}{4}x + \dfrac{2}{3} = 3\dfrac{5}{12}$

22. $\dfrac{-\dfrac{2}{3}}{\dfrac{4}{9}} = \dfrac{\dfrac{1}{3}}{x}$

Simplify:

23. $-[-(-1)^4] + (-3)(-2) - \sqrt[5]{-1}$

24. $5aba^2b^3a^2$

25. $3\dfrac{1}{2} \times 2\dfrac{1}{3} \div \dfrac{1}{3} \div \dfrac{1}{2}$

26. $139.287 - 19.876$

Evaluate:

27. $-xy + y$ if $x = -2$ and $y = -3$

28. $ab^2 - a^2b$ if $a = 2$ and $b = -1$

29.   If $3x - 12 = 12$, what is the value of $\frac{1}{4}x + 10$?

30.   Use the cut-and-try method to estimate $\sqrt[3]{330}$ to the nearest whole number.

LESSON 122   *Multiple-term equations*

When equations have like terms on either side of the equals sign, the first step in the solution is to combine the like terms.

example 122.1   Solve: $4x + 2 + 3x - 2x = 12$

solution   The first step is to combine the like terms.

$$5x + 2 = 12 \qquad \text{combined like terms}$$

Now we finish by adding -2 to both sides and then dividing by 5.

$$
\begin{array}{rl}
5x + 2 =& 12 \\
-2 & -2 \qquad \text{add} - 2 \text{ to both sides} \\
\hline
5x =& 10 \\
\end{array}
$$

$$x = \quad 2 \qquad \text{divided by 5}$$

Check:

$$4(2) + 2 + 3(2) - 2(2) = 12 \quad \longrightarrow \quad 8 + 2 + 6 - 4 = 12$$

$$\longrightarrow \quad 12 = 12 \quad \text{check}$$

example 122.2   Solve: $3x - 2 - x - 4x = 5x + 12$

solution   First we simplify the left side of the equation by adding like terms.

$$-2x - 2 = 5x + 12 \qquad \text{added like terms}$$

Next we decide to eliminate the variable on the left side, so we add $+2x$ to both sides.

$$
\begin{array}{rl}
-2x - 2 =& 5x + 12 \\
+2x & + 2x \qquad \text{add} +2x \text{ to both sides} \\
\hline
-2 =& 7x + 12 \\
\end{array}
$$

Now we finish by adding -12 to both sides and then dividing by 7.

$$
\begin{array}{rl}
-2 =& 7x + 12 \\
-12 & -12 \qquad \text{add} - 12 \text{ to both sides} \\
\hline
-14 =& 7x \\
\end{array}
$$

$$-2 = x \qquad \text{divided}$$

Check:

$$3(-2) - 2 - (-2) - 4(-2) = 5(-2) + 12$$
$$-6 - 2 + 2 + 8 = -10 + 12 \longrightarrow 2 = 2 \quad \text{check}$$

practice Solve:

a. $4x + 5 + 3x - x = -7x - 8$

b. $12x - 7 + 8x - 3x = -5x + 15$

problem set 122

1. Sixty percent of the boys did not like to euphemize. If 56 boys were not in this category, how many boys were there in all?

2. A 180 percent increase in 1 month could not be prevented. If the new total was 1120, what was the total before the increase?

3. When the rats invaded the town, some people ran and some stayed to fight the rats. The ratio of the runners to the fighters was 2 to 3. If 400 ran, how many people lived in the town before the rats came?

4. Four times a number was 24 greater than the product of the number and -2. What was the number?

5. A fair coin is tossed 4 times. What is the probability of getting *HHHH?*

6. A blue die and a green die are rolled. What is the probability of getting both a 5 on the blue die and a 3 on the green die?

7. If 65 is increased by 60 percent, what is the resulting number? Draw a diagram of the problem.

8. What percent of 960 is 1392? Draw a diagram of the problem.

9. Write $0.0012\frac{3}{4}$ as a decimal number without a fraction.

10. Complete the table. Begin by inserting the reference numbers.

Fraction	Decimal	Percent
$\frac{7}{20}$	(a)	(b)

11. Find the least common multiple of 24, 30, and 36.

12. If $6x + 4 = 28$, what is the value of $\frac{1}{4}x + 3$?

13. If $7x - 6 = 22$, what is the value of $\frac{1}{2}x - 3$?

Use the distributive property to multiply:

14. $2abc(a + b + c)$

15. $ab(abc + 2ab - 2bc)$

Simplify by adding like terms:

16. $ab^2 + a^3b + a^2ab - 3abb$

17. $m^2n^3 - mn^2 + mmnn^2 + 6mnn$

Solve:

18. $8x - 2 - x - 4x = -3x + 14$

19. $6\frac{1}{3}x - 2\frac{1}{3} = 3\frac{1}{6}$

20. $6x + 2 - 3x + 2x = 9x - 12$

21. Use the method of cut and try to estimate $\sqrt{33}$ to one decimal place.

Simplify:

22. $-2^3 - [-(-2)]$

23. $abb^2a^2bb^3$

24. $-[-(-2)] + (-3)(2) + (-1)^3 - \sqrt[3]{-1}$

25. $-2^3 - 2[2^3(2^2 - 3)] + \sqrt[5]{-243}$

26. $xy^2y^3x^2y$

27. $\dfrac{2^3 - 2^2(2^3 - 3^2) + 3}{3(2^3 - 1)}$

28. $\dfrac{1}{2}\left(2\dfrac{1}{3} \cdot \dfrac{1}{4} - \dfrac{1}{2}\right)$

Evaluate:

29. $a^2 + b^2 + 2ab$ if $a = -1$ and $b = 2$

30. $a^3 + b^3$ if $a = -2$ and $b = -2$

LESSON 123 *Polygons · Congruence*

123.A
polygons

In mathematics we call a flat surface a plane. A geometric figure drawn on a flat surface is called a planar figure. We remember that all mathematical lines are straight lines that have no ends. A part of a mathematical line is called a line segment. **A polygon is a closed planar geometric figure whose sides are line segments.** If a polygon has an indentation (a "cave"), it is called a **concave polygon.** If it does not have an indentation, it is called a **convex polygon.** Any two points in a convex polygon can be connected by a line that does not cross the boundary of the polygon.

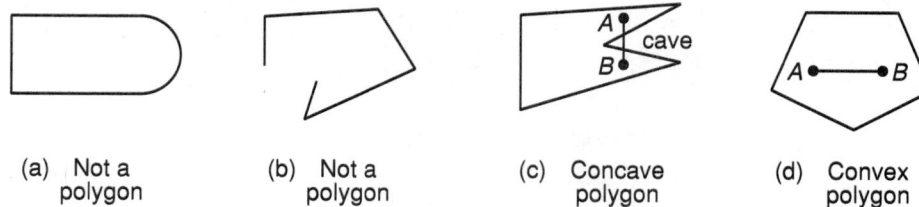

(a) Not a polygon (b) Not a polygon (c) Concave polygon (d) Convex polygon

Figure (a) is not a polygon because it has one side that is not straight. Figure (b) is not a polygon because it is not closed. Figure (c) is a concave polygon, and Figure (d) is a convex polygon. If all angles in a convex polygon have equal measures and all sides have equal lengths, the polygon is called a **regular polygon.** Polygons are named according to the number of sides they have. The polygon with the fewest number of sides is the triangle. A regular triangle is called an equilateral triangle. All the angles in an equilateral triangle are 60° angles. A quadrilateral has four sides. A regular quadrilateral is a square.

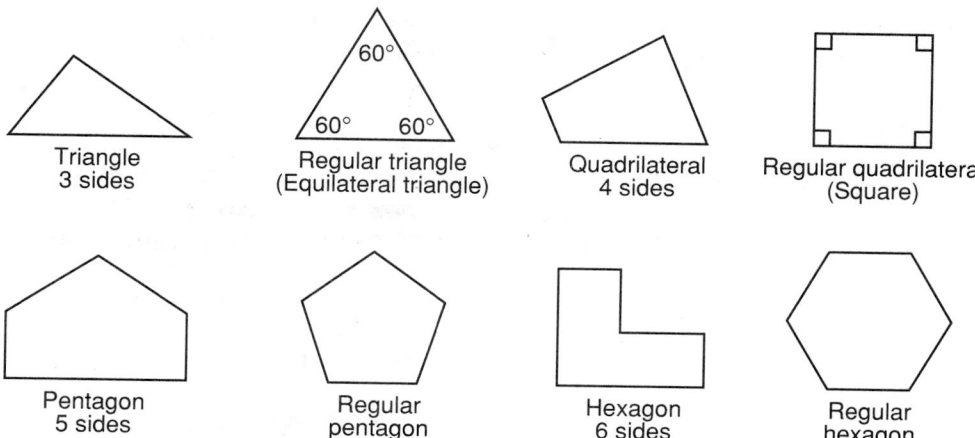

| Triangle
3 sides | Regular triangle
(Equilateral triangle) | Quadrilateral
4 sides | Regular quadrilateral
(Square) |

| Pentagon
5 sides | Regular
pentagon | Hexagon
6 sides | Regular
hexagon |

A polygon of 7 sides is a heptagon. A polygon of 8 sides is an octagon. A polygon of 9 sides is a nonogon. A polygon of 10 sides is a decagon. A polygon of 11 sides is an undecagon. A polygon of 12 sides is a dodecagon. We note that there is no special name for a polygon of more than 12 sides. When we speak of polygons of more than 12 sides, we use the word *polygon* and tell the number of sides or use the number of sides with the suffix *-gon*. Thus, if a polygon has 143 sides, we would call it a polygon with 143 sides, or a 143-gon.

A **diagonal** of a polygon is a line segment that connects two nonconsecutive vertices. We can name a polygon by listing the vertices in order. These polygons are quadrilaterals *MPRQ* and *AXCY*.

The diagonals of the polygon on the left are $\overline{QP}$ and $\overline{RM}$. Or we could call them $\overline{PQ}$ and $\overline{MR}$. If the diagonals of the polygon on the right were drawn in, they could be named $\overline{AC}$ and $\overline{YX}$.

A **trapezoid** is a quadrilateral that has exactly two sides parallel. A **parallelogram** is a quadrilateral that has two pairs of parallel sides. The sides that are parallel have the same lengths. In these figures equal tick marks denote equal lengths.

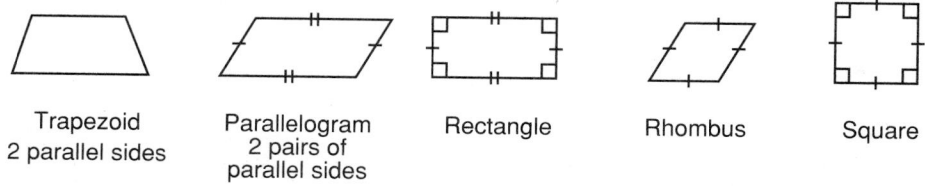

| Trapezoid
2 parallel sides | Parallelogram
2 pairs of
parallel sides | Rectangle | Rhombus | Square |

A **rectangle** is a parallelogram in which all angles have a measure of 90 degrees. A **rhombus** is a parallelogram that has four sides of equal lengths. A **square** is a rhombus in which all angles have a measure of 90°.

123.B
congruence

Consider this dialogue. Tom said, "Those two line segments are equal." Janice replied, "No, they are not. Only numbers can be equal. What you mean to say is that the number that describes the length of one segment equals the number that describes the length of the other segment." Tom then said, "But you will agree that those two triangles are equal." Janice replied, "No, they are not. The numbers that describe the lengths of the sides in one triangle equal the numbers that describe the lengths of the sides in the other triangle. Also, the numbers that describe the measures of the angles opposite the pairs of equal sides are also equal. We use the word **congruent** to mean geometrically equal." Janice is correct because **congruent means geometrically equal except for position and orientation.** If two geometric figures are congruent, then all corresponding measures of the figures are equal. Congruent is a term that is used often in geometry. In trigonometry and in other branches of mathematics we find it convenient to speak of equal angles and equal segments. When we do this, it is understood that we mean that the measures of the angles are equal and that the lengths of the segments are equal.

If two figures are congruent, we can place one figure over the other by sliding it over or rotating it or flipping it or some combination of these actions. If we slide the figure but do not rotate it, we call the action **translation.** A flip is called a **reflection** because a flipped figure looks like a mirror image.

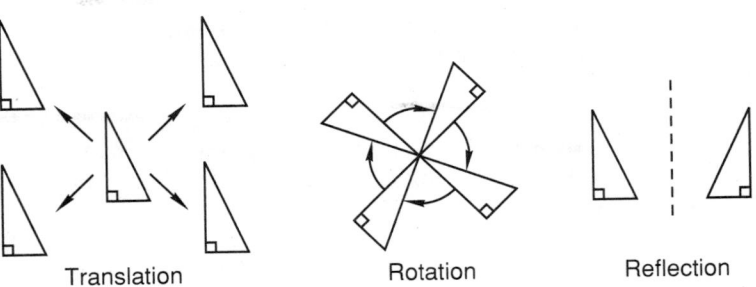

Translation   Rotation   Reflection

practice

All the figures shown below are polygons. Figure (a) is also a quadrilateral, a rectangle, a rhombus, and a square. Give as many names as you can for each of these figures.

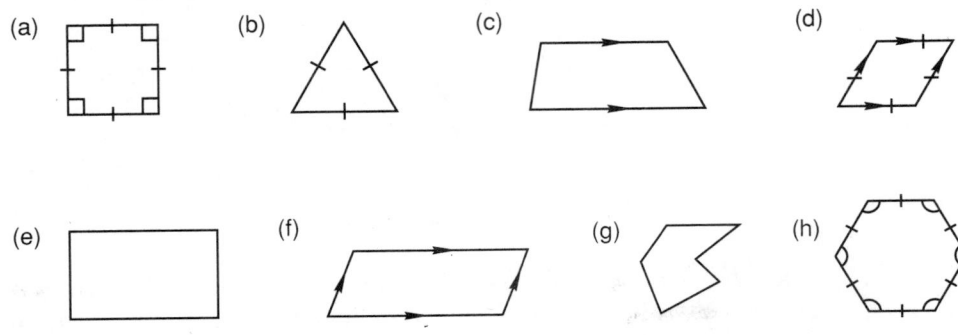

The pairs of figures shown below are congruent. Tell which movements are necessary to place figure *A* exactly on top of figure *B*.

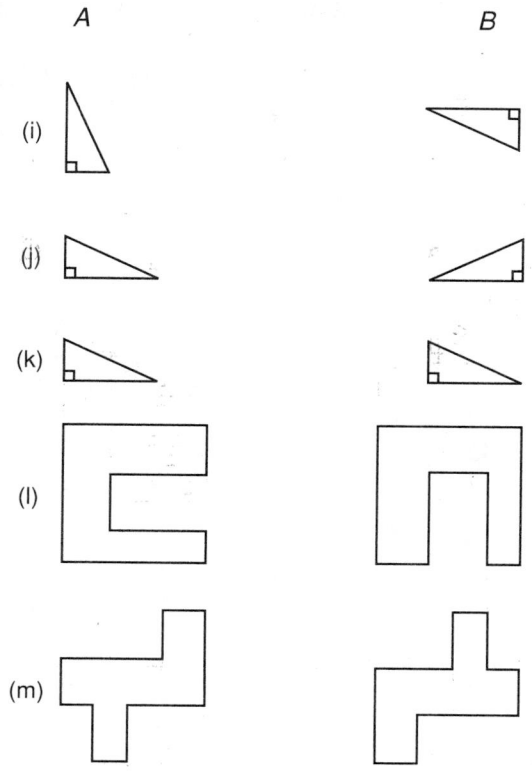

B

(i)

(j)

(k)

(l)

(m)

problem set 123

1. The first urn contains 6 blue marbles and 1 red marble. The second urn contains 9 blue marbles and 3 red marbles. A marble is picked from each urn. What is the probability that both marbles are red?

2. The ratio of the number of philistines to the number of aesthetes was 21 to 2. If 92 were present, how many were aesthetes?

3. Objectionable behavior increased by 50 percent during the present period. If 3600 incidents of objectionable behavior occurred during the present period, how many incidents of objectionable behavior occurred in the previous period?

4. Seven times a number was 32 greater than the product of the number and −9. What was the number?

5. If 1 card is drawn from each of two standard 52-card decks, what is the probability that two black jacks (2 black jacks in each deck) will be drawn?

6. If $10,000 is deposited in an account at 10 percent interest compounded annually, what will be the amount in the account at the end of 4 years?

7. Use a protractor to draw a 41° angle. Then use a straightedge and a compass to copy the angle.

8. Find sides x and y:

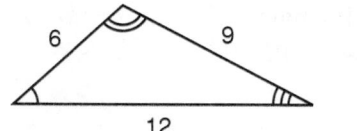 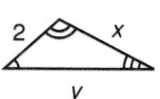

9. Find the measures of angles A, B, and C:

(a) 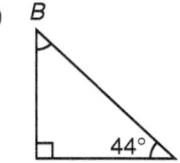   (b)   (c)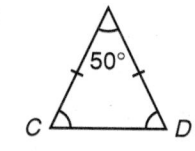

10. Use a metric scale to measure a 6-inch line segment to the nearest millimeter.

11. Use two unit multipliers to convert 6 centimeters to kilometers.

12. Eighty-seven is what percent of 8.7? Draw a diagram of the problem.

13. Sixty-one percent of what number is 183?

14. (a) Tell which rigid movements are necessary to place figure A exactly on top of figure B.

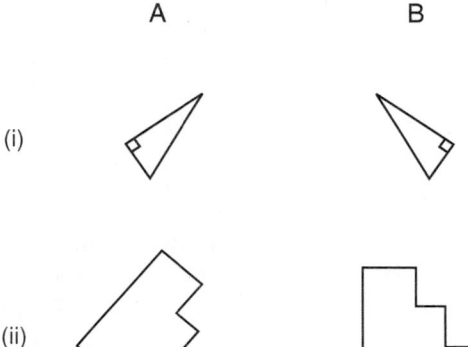

    A   B

(i)

(ii)

(b) Give all the names for the figure shown.

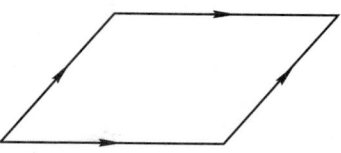

15. Use the distributive property to multiply: $4az\left(z^2 + \dfrac{1}{a} - 3a^2z\right)$

16. Simplify by adding like terms: $7mpm + 2mpm^2 - 4m^2p + 3pm^3 - 9pm^2$

17. Find the lateral surface area in square centimeters of a right circular cylinder whose diameter is 2 centimeters and whose height is 2 centimeters.

18. Find the total surface area in square feet of a cube whose edges measure $2\sqrt{2}$ feet.

19. Write 1999 in Roman numerals.

20. Write $79\frac{1}{8}\%$ as a decimal number.

Solve:

21. $4x - 4 + 5x - 3 = -4x - 46$

22. $-\frac{1}{3}x + \frac{2}{4} = 1\frac{5}{6}$

23. $\dfrac{-\frac{1}{4}}{\frac{7}{11}} = \dfrac{\frac{22}{35}}{x}$

Simplify:

24. $(-2)^3(-1) - \sqrt[6]{64}$

25. $6mp^2m^3p^5m$

26. $4\frac{1}{5} \times 2\frac{1}{2} \div \frac{1}{5} \div \frac{1}{4}$

27. If $9x + 2 = -25$, what is the value of $\frac{1}{9}x + \frac{1}{3}$?

28. If $3x - 15 = 3$, what is the value of $\frac{1}{2}x + 11$?

Evaluate:

29. $-x^y - y^x$ if $x = -2$ and $y = -3$

30. $a^2b + \dfrac{b}{a^2}$ if $a = -4$ and $b = 2$

LESSON 124 *Parallelograms · Trapezoids*

124.A
triangular areas

We have memorized the formulas for the areas of rectangles, triangles, and circles.

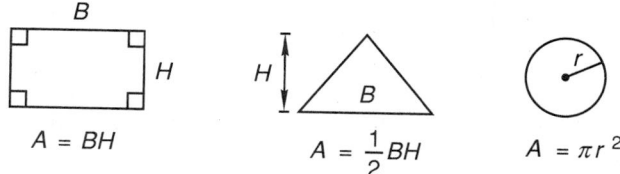

There are formulas for the areas of parallelograms and trapezoids, but formulas are easy to forget. It is poor practice to memorize a formula that can be developed quickly from other formulas that we already know. We can find the area of a parallelogram by dividing it into two triangles. We can find the area of a trapezoid by dividing it into two triangles.

example 124.1 Find the area of this parallelogram. Dimensions are in inches.

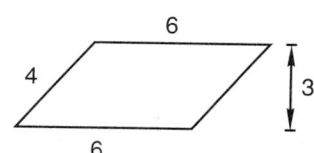

solution   A parallelogram has two diagonals. We can divide the parallelogram into two triangles by drawing either one of the diagonals.

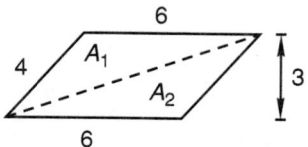

Both triangles in this figure have a base of 6 and a height of 3. The area of each triangle is one-half the base times the height.

$$A_1 = \frac{1}{2}(6)(3) = 9 \text{ in.}^2 \qquad A_2 = \frac{1}{2}(6)(3) = 9 \text{ in.}^2$$

The total area is the sum of the two areas.

$$A_1 + A_2 = 9 \text{ in.}^2 + 9 \text{ in.}^2 = \textbf{18 in.}^2$$

example 124.2   Two equal sides in this parallelogram are B units long. The other two equal sides are not lettered. Divide this parallelogram into two triangles and find the formula for the area of a parallelogram that uses the letters B and H.

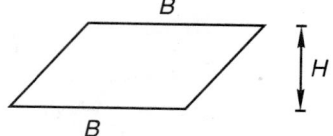

solution   We draw a diagonal to divide the figure into two triangles.

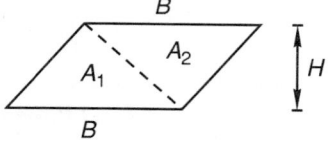

$$A_1 = \frac{1}{2} BH \qquad A_2 = \frac{1}{2} BH$$

$$\text{Total area} = \frac{1}{2} BH + \frac{1}{2} BH = BH$$

The area of a parallelogram equals the base times the height.

$$\text{Area of a parallelogram} = \textbf{BH}$$

This formula is easy to forget because it is used so seldom. If you have to find the area of a parallelogram, use the two-triangle method.

example 124.3   Find the area of this trapezoid. Dimensions are in feet.

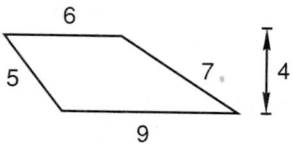

solution   First, we draw a diagonal. Either diagonal will do.

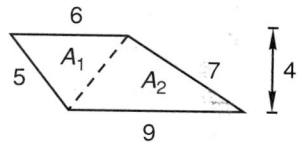

The height of both triangles is 4 ft. The base of one triangle is 6 ft and the base of the other triangle is 9 ft. The area of the trapezoid is:

$$\text{Total area} = \frac{1}{2}B_1H_1 + \frac{1}{2}B_2H_2$$

$$= \frac{1}{2}(6)(4) + \frac{1}{2}(9)(4)$$

$$= 12 \text{ ft}^2 + 18 \text{ ft}^2 = \textbf{30 ft}^2$$

example 124.4   Use two triangles to find a general formula for the area of a trapezoid.

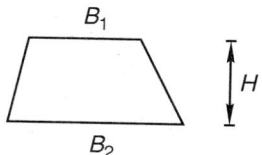

solution   We draw two triangles.

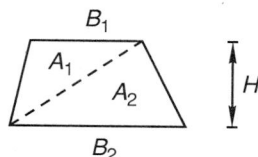

Both triangles have a height of H. One has a base of B_1 and one has a base of B_2.

$$\text{Total area} = \frac{1}{2}B_1H + \frac{1}{2}B_2H$$

practice   Find the area of (a) the trapezoid and (b) the parallelogram. Dimensions are in meters.

a.
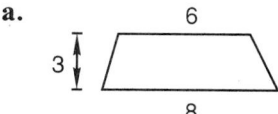

b.
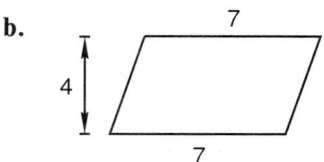

problem set 124

1. The ratio of the number of those who were enthusiastic to the number of those who were blasé was 8 to 7. If 600 were present, how many were blasé?

2. A fair coin is tossed 6 times. What is the probability of getting *HHTTHT* in that order?

3. The importation of tulip bulbs increased 350 percent this year over last year. If 900,000 bulbs were imported this year, how many were imported last year?

4. If the product of a number and -25 is decreased by 108, the result is 16 less than the product of the same number and -10. What is the number?

5. There were $2\frac{3}{5}$ times as many that were amorphous as were not. If there were 6500 that were amorphous, how many were not amorphous?

6. Find sides x and y:

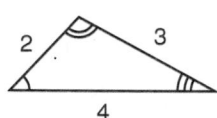 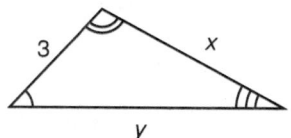

7. Use a protractor to draw a 21° angle. Then use a straightedge and a compass to copy the angle.

8. Find the measures of angles A, B, and C:

(a) 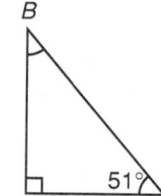   (b) (c)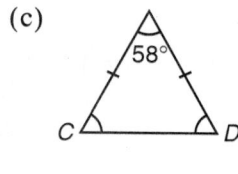

9. Find the area of (a) the trapezoid and (b) the parallelogram. Dimensions are in centimeters.

(a)   (b)

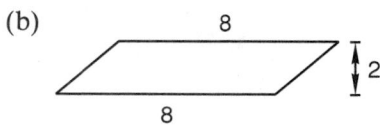

10. Give all the names for each figure:

(a)   (b)   (c)

 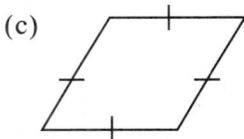

11. Draw a line segment 15 centimeters long. Then use an inch scale to measure the segment to the nearest sixteenth of an inch.

12. Use three unit multipliers to convert 15,000 cubic centimeters to cubic meters.

13. Two hundred six is what percent of 10.3?

14. What percent of 28 is 70?

15. Twenty-three percent of what number is 92?

16. Use the distributive property to multiply: $2xp(x^3 + 4x^2p - 2xp)$

17. Simplify by adding like terms: $3x^2am^4 - xamxmm^2 + 2am^2xmxm$

18. Find the volume in cubic kilometers of a right solid whose base is shown on the left and whose height is 0.6 kilometer. Dimensions are in kilometers.

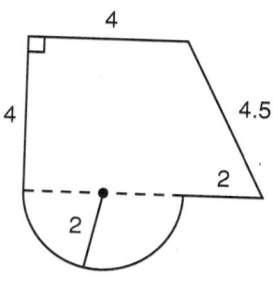 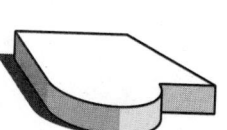

19. Find the lateral surface area of the figure in Problem 18.

Solve:

20. $7x + 4 - 9x = 12x - 48$

21. $-11\frac{1}{2}x - \frac{1}{4} = 5\frac{1}{2}$

22. $\dfrac{-\frac{1}{5}}{3\frac{1}{12}} = \dfrac{x}{\frac{1}{2}}$

23. Graph: $x \not\geq 0$

Simplify:

24. $-[-(-3^3)] + \sqrt[3]{-27}$

25. $-6^2 mp^5 m^2 pm^4 (-1)^5$

26. $\dfrac{-(-4^2) - 2^2 (3 - 4 \cdot 2)}{2^3 (2^3 - 3^2)}$

27. $2\frac{1}{5}\left(3\frac{1}{2} \cdot \frac{1}{4} - \frac{1}{3} \cdot \frac{3}{5}\right)$

28. (a) 3^{-3} (b) $\dfrac{1}{3^{-3}}$

Evaluate:

29. $a^{-3}b^3$ if $a = -2$ and $b = -3$

30. $ab^2 + ab$ if $a = -3$ and $b = 3$

LESSON 125 *Perpendicular bisectors*

The word **bisect** means to divide into two parts. A **perpendicular bisector** of a line segment is a line that is perpendicular to the line segment at the **midpoint** of the line segment. To construct a perpendicular bisector of the line segment $\overline{MN}$, use the compass to draw from point M two arcs of equal radii. The radii should be a little

longer than half of the line segment (a). Then in (b) we draw the same arcs from
point *N*.

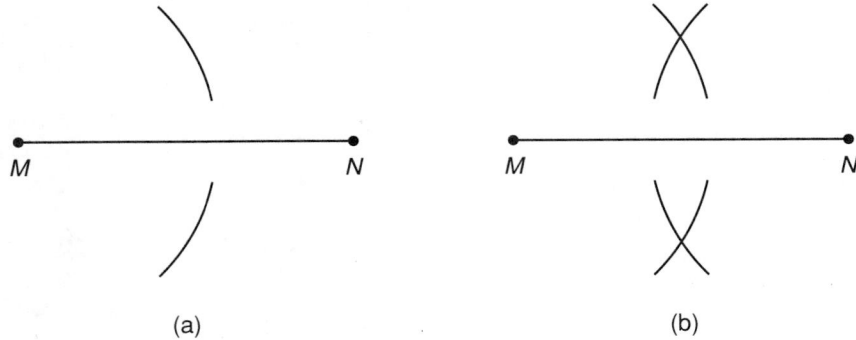

(a)   (b)

The points where the arcs cross are **equidistant** (the same distance) from points *M*
and *N*. **Any point on the line that connects these intersections will also be equidistant
from points *M* and *N*.** A line drawn through these points is the perpendicular
bisector of $\overline{MN}$.

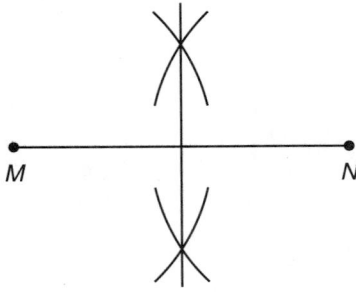

example 125.1   Draw a perpendicular to a line that passes through a designated point that is not on
the line.

solution   We draw a line and a point not on the line. We call the point *P*.

P •

_____

Next we swing a big arc from *P*. We call the points where this arc intersects the line
points *A* and *B*. Then we construct the perpendicular bisector of $\overline{AB}$.

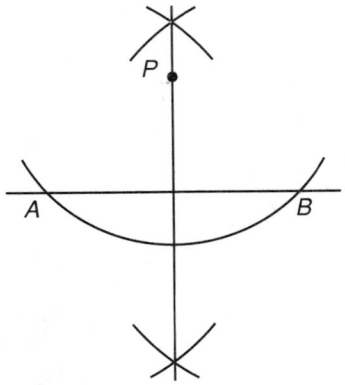

example 125.2 Given the line shown, construct a perpendicular to the line at point A.

A
•———————————————————

solution First we swing an arc in both directions from A. This arc will cut the line at points we call B and C. Then we construct the perpendicular bisector of $\overline{CB}$ by drawing equal arcs that have points B and C as their center. Then we connect the intersection points of the arc.

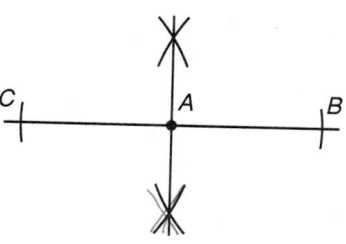

practice **a.** Draw a line segment. Construct the perpendicular bisector of the segment.

b. Draw a line segment. Designate a point not on the segment. Construct a line that is perpendicular to the segment and that passes through the point.

c. Draw a line segment. Designate a point on the segment. Construct a perpendicular at this point.

problem set 125

1. A single die is rolled twice. What is the probability that the first roll will be a 6 and the second roll will be a 2?

2. A pair of dice are rolled. What is the probability of rolling an 11?

3. As the official approached, the number that appeared to be busy increased 150 percent. If 1000 now appeared to be busy, how many appeared busy before the official approached?

4. The ratio of industrious to insouciant was 2 to 5. If 1400 were industrious, how many were insouciant?

5. Use a ruler to draw a line segment 4 centimeters long. Construct a perpendicular to the line at a point 1 centimeter from the left endpoint.

6. Draw a line segment and a point outside the line. Construct a line through the point that is perpendicular to the segment.

7. Use a protractor to draw a 36° angle. Then use a straightedge and a compass to copy the angle.

8. What percent of 70 is 119? Draw a diagram of the problem.

9. Seventy percent of what number is 210? Draw a diagram of the problem.

10. Use three unit multipliers to convert 1620 cubic meters to cubic centimeters.

11. Use one unit multiplier to convert 105 meters to centimeters.

12. Write 2963 in Roman numerals.

13. Write as a decimal number: $98\frac{3}{8}\%$

14. Find the area of (a) the parallelogram and (b) the trapezoid. Dimensions are in feet.

(a)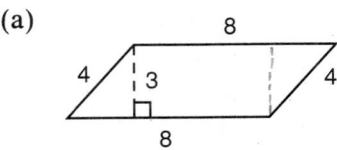

(b)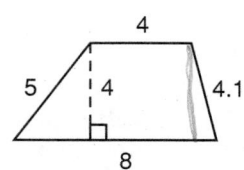

15. (a) Find side *x*:

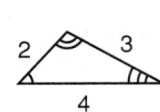 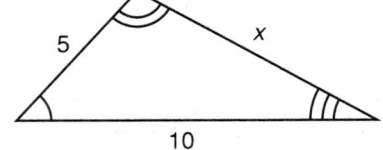

(b) Find angle *Z*:

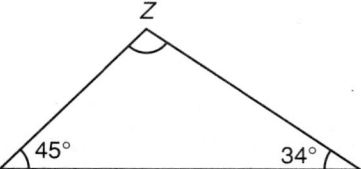

16. Use the distributive property to multiply: $2ab(a + c + b^2 - ac)$

17. Simplify by adding like terms: $2yxxy^2 - 3xy^2x + 3y^2x^2y - 6xyxy$

18. Express in cubic meters the volume of a right solid whose base is shown on the left and whose sides are 4 centimeters high. Dimensions are in centimeters.

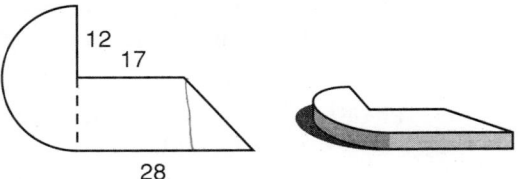

19. Find the surface area of the given prism. Dimensions are in inches.

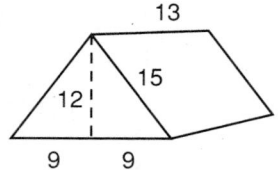

Solve:

20. $4x - 3 + 2x - 6 = -7x + 6$

21. $-1\frac{1}{3}x + \frac{1}{4} = \frac{5}{12}$

22. $\dfrac{-\frac{1}{6}}{\frac{5}{12}} = \dfrac{\frac{1}{2}}{x}$

Simplify:

23. $2(6 - 3 \cdot 4) + 3^2 - (-3)^2$

24. $4ab^2cb^3c^2a^3$

25. $\dfrac{-3^2 + (-2)^3 - 3(2 - 2^2)}{2(3^2 - 2^3)}$

26. $2\frac{1}{4} \times 1\frac{1}{3} \div \frac{1}{5} \div 1\frac{1}{3}$

Evaluate:

27. $xy^2 + 2xy$ if $x = -1$ and $y = 2$

28. $xy^3 + x^2y$ if $x = -2$ and $y = 3$

29. If $3x - 5 = 13$, what is $\frac{1}{3}x - 2$?

30. Write 13621451.4 in words.

LESSON 126 Equations with x^2 · Pythagorean theorem

126.A
equations with
x^2

If we consider the equation

$$x^2 = 4$$

we see that there are two solutions that will satisfy the equation. These are $+2$ and -2.

$$(2)^2 \; = 4 \qquad \text{check}$$
$$(-2)^2 = 4 \qquad \text{check}$$

From this we conclude that we can solve an equation such as $x^2 = 4$ by taking the square root of both sides.

example 126.1   Solve: $p^2 = 16$

solution   We take the square root of p^2 and we get p. Then we take the square root of 16.

$$p^2 = 16 \qquad \text{equation}$$
$$\boldsymbol{p = \pm 4} \qquad \text{square root of both sides}$$

We get ± 4 because $(+4)^2 = 16$ and $(-4)^2 = 16$, so the problem has two solutions.

example 126.2   Solve: $p^2 = 42$

solution   We take the square root of both sides.

$$p^2 = 42 \qquad \text{equation}$$
$$\boldsymbol{p = \pm\sqrt{42}} \qquad \text{solution}$$

126.B
Pythagorean theorem

A right triangle is a triangle that has a right angle. The side opposite the right angle is called the **hypotenuse.** The other two sides are called **legs.**

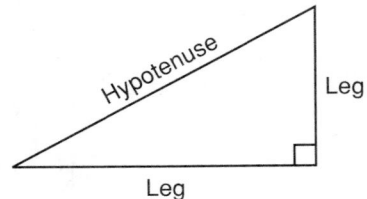

A right triangle has a property that is especially useful. This property was known to the ancient Egyptians before 2000 B.C., but it is named after a Greek mathematician who was born on the Greek island of Samos in the sixth century B.C. Later he lived in southern Italy in a town called Crotona. His name was Pythagoras. The theorem that bears his name is so simple that it is difficult to appreciate how important it is.

> PYTHAGOREAN THEOREM
>
> The area of a square drawn on the hypotenuse of a right triangle equals the sum of the areas of the squares drawn on the other two sides.

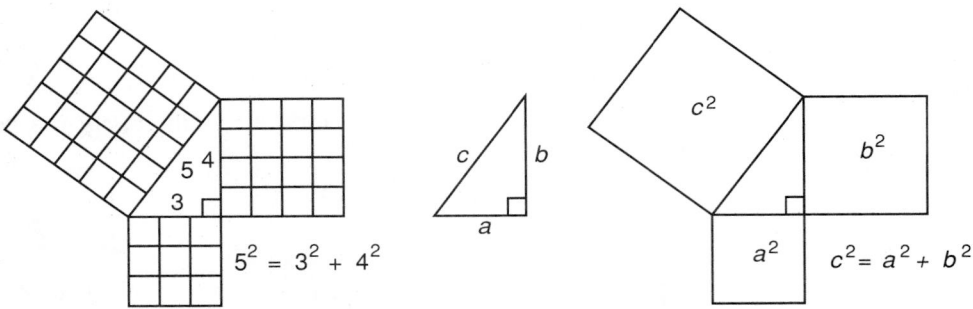

In the triangle on the left the area of the square on the hypotenuse is 25. The areas of the other two squares are 9 and 16, which add to 25. The area of the square on the hypotenuse in the other triangle is c^2. The areas of the other two squares are a^2 and b^2. An algebraic statement of the theorem is

$$c^2 = a^2 + b^2$$

if c is the length of the hypotenuse.

example 126.3   Given the triangle with the lengths of the sides as shown, use the Pythagorean theorem to find the length of side a.

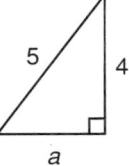

solution   The square of the hypotenuse equals the sum of the squares of the other two sides. Thus,

$$5^2 = 4^2 + a^2 \longrightarrow 25 = 16 + a^2 \longrightarrow 9 = a^2$$

We use the difference of two squares theorem to finish the solution.

$$a^2 = 9 \quad \text{which leads to} \quad a = +3 \quad \text{or} \quad a = -3$$

While -3 is a solution to the equation $a^2 = 9$, it is not a solution to the problem at hand because physical lengths are designated by positive numbers. Thus we reject this solution and say that

$$a = +3$$

example 126.4 Find side p in the triangle shown.

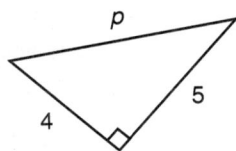

solution The square of the hypotenuse equals the sum of the squares of the other two sides.

$$p^2 = 5^2 + 4^2 \qquad \text{Pythagorean theorem}$$
$$p^2 = 25 + 16 \qquad \text{squared}$$
$$p^2 = 41 \qquad \text{simplified}$$
$$p = \pm\sqrt{41} \qquad \text{square root of both sides}$$

This equation has two answers. They are $+\sqrt{41}$ and $-\sqrt{41}$. But only one of these answers is the length of the hypotenuse because in the real world we say that every length is positive.

$$p = \sqrt{41}$$

We can get a numerical answer by using the square root key on a calculator.

$$p = \sqrt{41} \approx 6.403$$

example 126.5 Find side m.

solution The hypotenuse squared equals the sum of the squares of the other two sides.

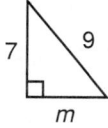

$$9^2 = 49 + m^2 \qquad \text{Pythagorean theorem}$$
$$81 = 49 + m^2 \qquad \text{squared}$$
$$32 = m^2 \qquad \text{simplified}$$
$$\sqrt{32} = m \qquad \text{square root of both sides}$$

practice **a.** Find side M.

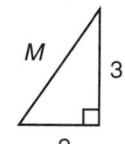

b. Find side p.

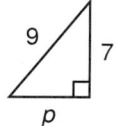

problem set 126 **1.** If one card is drawn from a standard 52-card deck and then replaced and then a second card is drawn, what is the probability of drawing a red 8 and then a black 10 in that order?

2. The ratio of the number of disgruntled customers to the number of satisfied customers was 3 to 2. If 25,000 customers were surveyed, how many were disgruntled?

3. Nine times a number, increased by 5, was 16 less than the product of the number and 2. What was the number?

 4. Aberrations decreased by 92 percent from the first to the second field sample. If only 6 aberrations were noted in the second field sample, how many aberrations were noted in the first field sample? What percent is left?

5. Property values appreciated $9\frac{1}{2}$ percent. If 50 acres were originally appraised at $30,000, what was their value after they appreciated?

6. Draw a line segment. Construct a perpendicular bisector of the segment.

7. Draw a line segment and a point outside the line. Construct a line through the point that is perpendicular to the segment.

8. What percent of 90 is 162? Draw a diagram of the problem.

9. Thirty-five percent of what number is 2450? Draw a diagram of the problem.

10. Use three unit multipliers to convert 1,000,000 cubic meters to cubic kilometers.

11. Write MMMCMXCIV in Arabic numerals.

Find the length of side m in each triangle:

12.
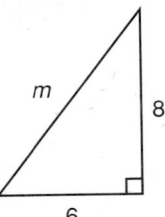

13.
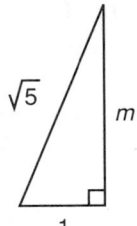

14. Find the area of (a) the parallelogram and (b) the trapezoid. Dimensions are in meters.

(a)
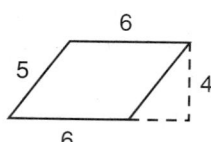

(b)
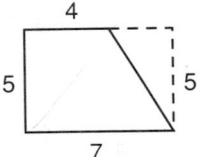

15. (a) Find the length of side x: (b) Find angle z:

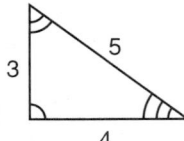 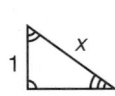 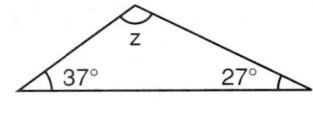

16. Use the distributive property to multiply: $am^2\left(2am + a + \dfrac{m}{b^2}\right)$

17. Simplify by adding like terms: $4a^3pm^2 + a^2pm^{-1} - 2apma^2m$

127.A   *constructing triangles*

18. Express in cubic inches the volume of a right solid whose base is shown on the left and whose sides are 1 foot high. Dimensions are in inches. (There is more than one way to find this volume.)

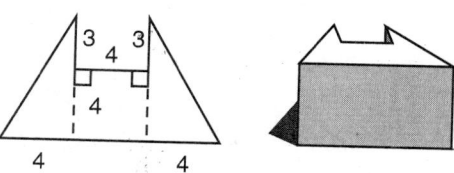

19. Find the perimeter of the figure on the left in Problem 18.

Solve:

20. $16x - 3 + 3x - 5 = -3x - 63$

21. $-2\frac{1}{2}x + 1\frac{1}{16} = \frac{1}{8}$

22. $\dfrac{\frac{1}{5}}{-\frac{4}{9}} = \dfrac{\frac{3}{10}}{x}$

Simplify:

23. $4(7 - 2 \cdot 6) + \frac{1}{3^{-2}} - (-3^2)$

24. $3mn^3pn^2p^3m$

25. $(-4)^3 + \sqrt[3]{-8}$

26. $\dfrac{-(-5^2) - 2^2(3 - 2 \cdot 3)}{-2(\sqrt[3]{-125})}$

27. $4\frac{1}{2}\left(2\frac{1}{3} \cdot 1\frac{1}{2} - \frac{1}{4} \cdot \frac{4}{7}\right)$

28. (a) 1^{-11} (b) $\dfrac{2}{2^{-2}}$ (c) $\dfrac{a^2}{a^{-3}}$

Evaluate:

29. $ab^2 + a^b$ if $a = 2$ and $b = -3$

30. $\sqrt[a]{-b} - (ab)^{-1}$ if $a = 3$ and $b = 8$

LESSON 127 *Constructing triangles · Angle bisectors*

127.A
constructing triangles

We can construct a triangle if we are given the lengths of all three sides.

example 127.1   Construct a triangle whose sides are 4 cm, 3 cm, and 2 cm.

solution   We draw a line segment 4 cm long, and from one end of the segment we draw an arc whose radius is 3 cm.

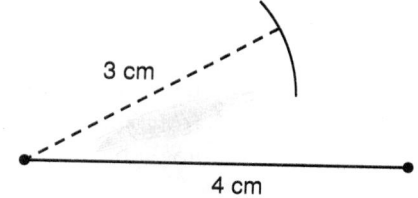

Now from the other end of the line segment, we draw an arc with a 2-cm radius. The intersection of the arcs is the other vertex of the triangle.

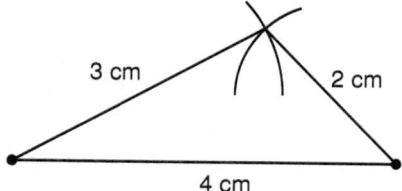

example 127.2   Construct a right triangle whose base is 4 centimeters and whose height is 3 centimeters.

solution   The only way to construct a perpendicular is to bisect a line segment. Thus, we draw a line segment 8 centimeters long and bisect it.

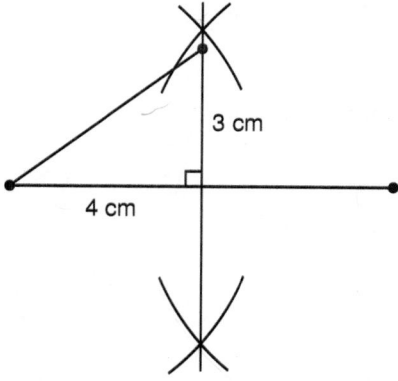

The last step is to measure 3 centimeters on the bisector and then draw the triangle as shown.

127.B

angle bisectors   An angle bisector divides an angle into two equal parts. A similar technique is used to bisect an angle as is used to bisect a line segment. To bisect the angle shown in (a),

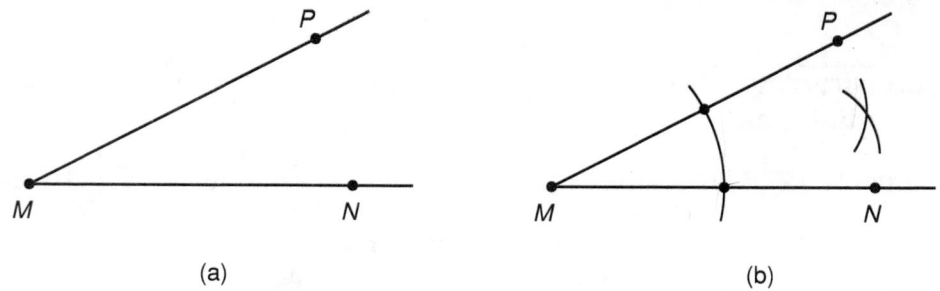

we first draw an arc that intersects both rays (b). Then arcs of equal radii are drawn from each of the points of intersection. The point where these arcs cross lies on the angle bisector.

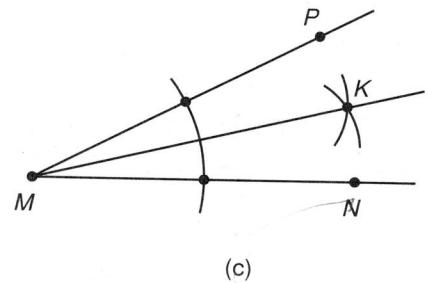

(c)

$$\angle NMK = \frac{1}{2}\angle NMP \quad \text{and} \quad \angle PMK = \frac{1}{2}\angle PMN$$

practice
a. Construct a triangle whose sides are 4 cm, 5 cm, and 2 cm.

b. Construct a right triangle whose base is 6 cm and whose height is 4 cm.

c. Use a protractor to draw a 60° angle. Then construct a bisector of this angle.

problem set 127

1. A fair coin comes up heads twice in a row. If it is flipped again, what is the probability that it will be heads again?

2. An urn contains 2 white marbles, 3 green marbles, and 5 red marbles. A marble is drawn and then replaced. Then a second marble is drawn. What is the probability that the first marble will be white and the second will be green?

3. Forty percent of the items in the showcase were covered with mildew. If there were 2000 items in the showcase, how many were not covered with mildew?

4. Zorba asked for 340 percent more drachmas than the man was willing to give. If the man was willing to give 400 drachmas, how many drachmas did Zorba ask for?

5. Construct a triangle whose sides are 3 cm, 4 cm, and 5 cm.

6. Use a ruler to draw a line segment 5 centimeters long. Construct a perpendicular to the line at a point 2 centimeters from the left endpoint.

7. Draw a line segment and a point outside the line. Construct a line through the point that is perpendicular to the segment.

8. Use a protractor to draw a 110° angle. Then use a straightedge and a compass to construct the bisector of the angle.

9. What percent of 60 is 69? Draw a diagram of the problem.

10. If 120 is increased by 25 percent, what is the resulting number? Draw a diagram of the problem.

11. What number is $6\frac{1}{2}$ percent of 600?

12. Use three unit multipliers to convert 1425 cubic meters to cubic kilometers.

13. Use three unit multipliers to convert 26 cubic centimeters to cubic meters.

14. Find side m:

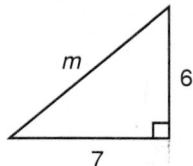

15. (a) Find side x:

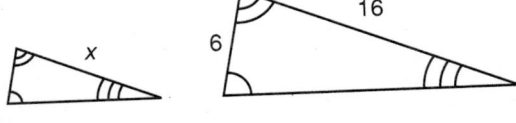

     (b) Find angle z:

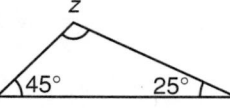

16. Use the distributive property to multiply: $3ab(ab + b - cab)$

17. Simplify by adding like terms: $3xyx^2 - 2xy^2 + 4x^3y - 3xyy$

18. Express in cubic centimeters the volume of a right solid whose base is shown on the left and whose sides are 1 meter high. Dimensions are in meters.

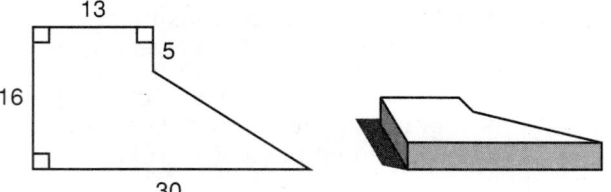

Solve:

19, $3x - 3 + 6x = -5x + 6$

20. $-1\frac{2}{3}x - \frac{1}{6} = 1\frac{2}{3}$ **21.** $\dfrac{-\frac{3}{4}}{\frac{9}{16}} = \dfrac{-\frac{1}{2}}{x}$

Simplify:

22. (a) $\frac{1}{3^{-3}}$ (b) 2^{-3} (c) $\frac{1}{2^{-3}}$ (d) 3^{-3}

23. $-(-2)^2 - 3(3 - 2 \cdot 2^2)$ **24.** $5ab^2a^2ba^3b$

25. $\dfrac{-2^3 + (-3)^2 - 4(2 - 3^2)}{3(2^3 - 3)}$ **26.** $\frac{1}{3}\left(1\frac{1}{3} \cdot \frac{3}{4} - \frac{1}{6} \cdot \frac{3}{2}\right)$

Evaluate:

27. $xy^2 - xy$ if $x = -2$ and $y = -3$

28. $a^2b - a^3b^2$ if $a = -1$ and $b = -3$

29. If $6x - 3 = 21$, what is $\frac{1}{2}x - 12$?

30. Graph: $x \not\geq 3$

LESSON *128* *English volume conversions*

There are quite a few cubic inches in a cubic foot. We can see this clearly by looking at this drawing of a cubic foot.

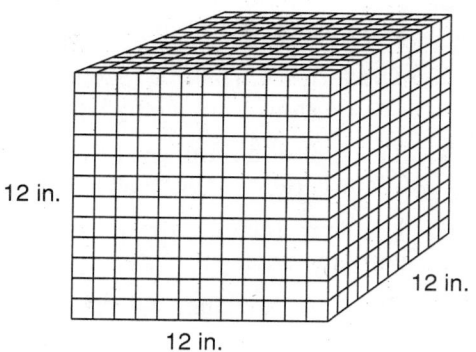

We see that there are $12 \times 12 = 144$ cubes in the front face. The cubes are 12 deep, so there are a total of

$$144 \times 12 = 1728 \text{ cubes}$$

We will use three unit multipliers to make volume conversions. We will concentrate on the process rather than on numerical answers. Thus the answers will not be worked out all the way.

example 128.1   Use three unit multipliers to convert 14 cubic feet to cubic inches.

solution   We will write cubic feet (ft³) as ft · ft · ft so we can see how we use the three unit multipliers.

$$14 \, \cancel{ft} \cdot \cancel{ft} \cdot \cancel{ft} \times \frac{12 \text{ in.}}{1 \, \cancel{ft}} \times \frac{12 \text{ in.}}{1 \, \cancel{ft}} \times \frac{12 \text{ in.}}{1 \, \cancel{ft}} = \textbf{14(12)(12)(12) in.}^3$$

A pocket calculator can be used to do the multiplication if a numerical answer is necessary.

example 128.2   Use six unit multipliers to convert 140 cubic yards (yd³) to cubic inches (in.³).

solution   We will go from cubic yards to cubic feet to cubic inches.

$$140 \, \cancel{yd^3} \times \frac{3 \, \cancel{ft}}{1 \, \cancel{yd}} \times \frac{3 \, \cancel{ft}}{1 \, \cancel{yd}} \times \frac{3 \, \cancel{ft}}{1 \, \cancel{yd}} \times \frac{12 \text{ in.}}{1 \, \cancel{ft}} \times \frac{12 \text{ in.}}{1 \, \cancel{ft}} \times \frac{12 \text{ in.}}{1 \, \cancel{ft}}$$

$$= \textbf{140(3)(3)(3)(12)(12)(12) in.}^3$$

example 128.3   Use six unit multipliers to convert 50,163,529 cubic inches to cubic miles (mi³).

solution   We will go from cubic inches to cubic feet to cubic miles.

$$50{,}163{,}529 \, \cancel{in.^3} \times \frac{1 \, \cancel{ft}}{12 \, \cancel{in.}} \times \frac{1 \, \cancel{ft}}{12 \, \cancel{in.}} \times \frac{1 \, \cancel{ft}}{12 \, \cancel{in.}} \times \frac{1 \text{ mi}}{5280 \, \cancel{ft}} \times \frac{1 \text{ mi}}{5280 \, \cancel{ft}} \times \frac{1 \text{ mi}}{5280 \, \cancel{ft}}$$

$$= \frac{50{,}163{,}529}{(12)(12)(12)(5280)(5280)(5280)} \textbf{ mi}^3$$

practice   Use six unit multipliers to convert 1,000,000 cubic inches to cubic miles.

problem set   1.   Bucolic themes dominated the landscapes, as 70 percent of the paintings were
128   bucolic in nature. If 2000 paintings were exhibited, how many were not
                      bucolic?

2.   Ninety-six percent of the newcomers caviled every time an event was
     announced. If 140 did not cavil, how many newcomers were there?

3.   The ratio of artists to spectators at the art show was 3 to 14. If 1700 attended
     the art show, how many were spectators?

4.   The sum of a number and 40 is 5 less than the product of the number and 6.
     What is the number?

5.   Use a protractor to draw a 30° angle. Then use a compass and a straightedge to
     copy the angle.

6.   Construct a triangle whose sides are 5 cm, 7 cm, and 9 cm.

7.   Draw a line segment and a point outside the line. Construct a line through the
     point that is perpendicular to the segment.

8.   Use a protractor to draw a 135° angle. Then use a straightedge and a compass
     to construct the bisector of the angle.

9.   If 125 is increased by 60 percent, what is the resulting number? Draw a
     diagram of the problem.

10.   If 200 is increased by $12\frac{1}{4}$ percent, what is the resulting number? Draw a
      diagram of the problem.

11.   Use six unit multipliers to convert 10 cubic miles to cubic inches.

12.   If $3x - 5 = 22$, what is the value of $\frac{1}{3}x + \frac{1}{2}$?

13.   If $16x + 21 = 53$, what is the value of $\frac{1}{3}x + \frac{1}{2}$?

14.   Use the distributive property to multiply: $3ac(ab + ac + bc)$

15.   Use the cut-and-try method to estimate $\sqrt[4]{700}$ to the nearest whole number.

Simplify by adding like terms:

16. $ab^2 + abb + 2a^2ab + 2abb - baa^2$

17. $-xy^3 + x^2xyy^2 + x^3y^3 - 3xyy^2$

Solve:

18. $3x + 4 - 2x - x = 4x - 5$ 19. $3\frac{1}{2}x - 1\frac{3}{4} = -\frac{3}{2}$

20. $6 + x - 2x + 3x - 4x = 3x + 4$ 21. $\dfrac{-\frac{1}{4}}{\frac{1}{3}} = \dfrac{\frac{16}{9}}{x}$

Simplify:

22. $(-3)(-2) + (-3)^2$ 23. $aba^2b^2a^3b$

24. $(-3)^2(2) - 3$ 25. $xy^3x^2xxy^3$

26. $\dfrac{-(-3)^2 - 2^3(2^3 - 2 \cdot 6) + 3}{2^2(2^3 - 4)}$

27. $\dfrac{1}{4}\left(2\dfrac{1}{3} \cdot \dfrac{1}{4} - \dfrac{7}{12}\right)$

Evaluate:

28. $-b^2 + a + b$ if $a = -2$ and $b = -1$

29. $-b^3 + 2ab$ if $a = -1$ and $b = -2$

30. Write 6213825.32 in words.

LESSON 129 *Base 2*

129.A

bases, digits, and place value

We remember that the system of numbers that we use is called the Hindu-Arabic system. This system is a base 10 system and uses the 10 digits

$$0, 1, 2, 3, 4, 5, 6, 7, 8, \text{ and } 9$$

Whole numbers in this system have values equal to the sums of the products of the digits and their place values. We remember that the place values in this system are as shown here:

BASE 10 PLACE VALUES

etc.	10,000	1000	100	10	1

Thus, the number 2001 has the value of

etc.	10,000	1000	100	10	1
		2	0	0	1

2 times 1000, plus 0 times 100, plus 0 times 10, plus 1 times 1

and the number 4327 has the value of

etc.	10,000	1000	100	10	1
		4	3	2	7

4 times 1000, plus 3 times 100, plus 2 times 10, plus 7 times 1

Not all systems use 10 digits. Some systems use fewer than 10 digits, and some use more than 10 digits. Each system uses the same number of digits as the base of the system. The base 9 system uses nine digits, the base 6 system uses six digits, the base 4 system uses four digits, etc.

Any whole number greater than 1 can be used as a base of a number system. In this lesson, we will investigate the base 2 number system, which uses the two digits

$$0 \quad \text{and} \quad 1$$

The addition register in a computer consists of a string of electronic devices called bistable multivibrators that have only two states, *on* or *off*. If a multivibrator is on, it represents a 1. If it is off, it represents a 0. All computers are designed to add or subtract using base 2 numbers.

129.B
base 2 numbers

The values of the places in the base 2 system are as shown here. We use base 10 numbers to write the place values.

BASE 2 PLACE VALUES

| etc. | 128 | 64 | 32 | 16 | 8 | 4 | 2 | 1 | . |

The first place to the left of the decimal point has a value of 1. To get the next value, we multiply 1 by 2 and get 2. To get the next value, we multiply 2 by 2 and get 4. To get the next value, we multiply 4 by 2 and get 8, etc. Each place has a value twice the value of the place to its right.

example 129.1 What is the base 10 value of 10101 (base 2)?

solution We begin by making a table of base 2 place values. We use base 10 numerals to write the place values.

BASE 2 PLACE VALUES

64	32	16	8	4	2	1	.
		1	0	1	0	1

We see that we have one 16, one 4, and one 1.

$$16 + 4 + 1 = 21$$

Since the sum of these numbers is 21, the base 10 value of 10101 (base 2) is 21 (base 10).

10101 (base 2)   equals   **21 (base 10)**

example 129.2 What base 10 number does 110101 (base 2) represent?

solution We make a table of place values and write each digit in its proper place.

BASE 2 PLACE VALUES

64	32	16	8	4	2	1	.
1	1	0	1	0	1

We see that we have one 32, one 16, no 8s, one 4, no 2s, and one 1.

$$32 + 16 + 4 + 1 = 53$$

so   110101 (base 2)   equals   **53 (base 10)**

example 129.3 Write 43 (base 10), using base 2 numerals.

solution We always begin by making a table of place values.

BASE 2 PLACE VALUES

| 64 | 32 | 16 | 8 | 4 | 2 | 1 | . |

There are no 64s in 43. There is one 32 in 43. So we put a 1 in the thirty-twos' place and subtract 32 from 43.

64	32	16	8	4	2	1

1

$$\begin{array}{r} 43 \\ -\ 32 \\ \hline 11 \end{array}$$

There are no 16s in 11, so we put a zero in the sixteens' place and subtract zero from 11.

64	32	16	8	4	2	1

1   0

$$\begin{array}{r} 11 \\ -\ 0 \\ \hline 11 \end{array}$$

There is one 8 in 11. We put a 1 in the eights' place and subtract 8 from 11.

64	32	16	8	4	2	1

1   0   1

$$\begin{array}{r} 11 \\ -\ 8 \\ \hline 3 \end{array}$$

There are no 4s in 3. We put a zero in the fours' place and subtract zero from 3.

64	32	16	8	4	2	1

1   0   1   0

$$\begin{array}{r} 3 \\ -\ 0 \\ \hline 3 \end{array}$$

There is one 2 and one 1 in 3. We put a 1 in the twos' place and a 1 in the ones' place.

64	32	16	8	4	2	1

1   0   1   0   1   1

The number 43 has one 32, one 8, one 2, and one 1, so

43 (base 10) = **101011 (base 2)**

example 129.4 Write 102 (base 10), using base 2 numerals.

solution We always begin by writing a table of place values.

BASE 2 PLACE VALUES

64	32	16	8	4	2	1

It is not necessary to go through all the steps used in the preceding example. We can do simple conversions like this one mentally.

The number 102 can be made from one 64, one 32, one 4, and one 2. There are no 16s, no 8s, and no 1s, so we put zeros in these places.

BASE 2 PLACE VALUES

64	32	16	8	4	2	1

1   1   0   0   1   1   0

Thus we see that we can write

102 (base 10)   as   **1100110 (base 2)**

practice   **a.** What is the base 10 value of 110101 (base 2)?

b. What is the base 2 value of 107 (base 10)?

problem set   **1.** The racers surged across the bridge at the start of the marathon. Mary
129   estimated that the ratio of male racers to female racers was 17 to 2. If there
             were 4750 racers in all, how many were males?

2. When the brouhaha commenced, the ratio of natives to intruders was 9 to 5. If
700 were involved, how many were natives?

3. Only one-fourth of those present really believed what the speaker was saying. If
1744 people were present, how many did not believe what the speaker was
saying?

4. Five times a certain number was 6 less than 3 times the same number. What
was the number?

5. What base 10 number does 101001 (base 2) represent?

6. What base 10 number does 11111 (base 2) represent?

7. Write 74 (base 10) using base 2 numerals.

8. What number is 160 percent of 300? Draw a diagram of the problem.

9. If 60 is increased by 75 percent, what is the resulting number? Draw a diagram
of the problem.

10. What percent of 80 is 152? Draw a diagram of the problem.

11. Find the volume in cubic inches of a right solid whose base is shown and whose
height is 1 yard. Dimensions are in inches.

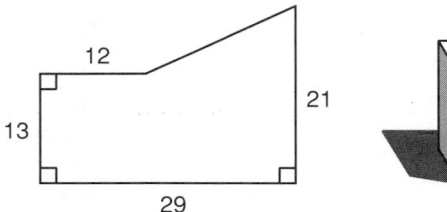

12. If $3x - 7 = 11$, what is the value of $\frac{1}{3}x + 3$?

13. If $2x - 6 = 5$, what is the value of $3x - \frac{1}{2}$?

Use the distributive property to multiply:

14. $2b(3a + b + ab)$ **15.** $3ab(a + b - 4)$

Simplify by adding like terms:

16. $a^2 + b^2 + 2aa + 3bb - 3ab + 2ab$ **17.** $a^3 + 3a^3b + b^3 + 2a^3 + 3aa^2b$

Solve:

18. $2x - 5 = 3x + 10$ **19.** $-6x - 3 = 2x + 6$

20. $-3x - 2 = 5x + 6$ **21.** $\dfrac{-\frac{4}{5}}{2\frac{1}{2}} = \dfrac{\frac{1}{3}}{x}$

Simplify:

22. $(-2)^2 + 2^2$

23. $-[-(-2)] + (-2)^3$

24. $2^3 - 2[2^2(2^3 - 3^2) + 3^2(2 - 1)]$

25. aba^2b^2ab

26. $xy^3x^2yx^2$

27. $m^3m^2n^2n^3m^3$

28. $\dfrac{2^4 - 3(2 - 2^2) + (3^2 - 2 \cdot 4)}{5(2 \cdot 3 - 3)}$

29. Use the method of cut and try to estimate $\sqrt[4]{562}$ to the nearest whole number.

30. Evaluate: $ab + bc$ if $a = -2$, $b = 3$, and $c = 1$

LESSON 130 *Metric volume*

A cubic meter is a relatively large unit of volume which equals approximately 264.17 gallons. A cubic centimeter is a relatively small unit of volume. It takes about 16.39 cubic centimeters to equal a cubic inch. The metric system has a unit of volume that is about the size of a quart—the **liter,** which equals 1000 cubic centimeters. A liter equals about 1.06 quarts. So when we see the word **liter,** we can think **"big quart."**

$$1 \text{ liter} \approx 1 \text{ big quart}$$

Because 1000 cubic centimeters makes a liter, we say that a cubic centimeter is a milliliter, which is one-thousandth of a liter.

$$1 \text{ milliliter} = 1 \text{ cubic centimeter}$$

Only one unit multiplier is required to convert between cubic centimeters and liters. At least three unit multipliers are normally required for any other volume conversions.

example 130.1   Convert 1,451,600 milliliters (ml) to liters.

solution   Only one unit multiplier is required to make this conversion.

$$1{,}451{,}600 \text{ ml} \times \frac{1 \text{ liter}}{1000 \text{ ml}} = \textbf{1451.6 liters}$$

example 130.2   Convert 40 cubic meters (m^3) to cubic centimeters (cm^3).

solution   We need three unit multipliers.

$$40 \text{ m}^3 \times \frac{100 \text{ cm}}{\text{m}} \times \frac{100 \text{ cm}}{\text{m}} \times \frac{100 \text{ cm}}{\text{m}} = 40(100)(100)(100) \text{ cm}^3$$
$$= \textbf{40,000,000 cm}^3$$

example 130.3   Convert 1400 cubic meters to liters.

solution   We will go from m^3 to cm^3 to liters.

$$1400 \text{ m}^3 \times \frac{100 \text{ cm}}{\text{m}} \times \frac{100 \text{ cm}}{\text{m}} \times \frac{100 \text{ cm}}{\text{m}} \times \frac{1 \text{ liter}}{1000 \text{ cm}^3}$$

$$= \frac{1400(100)(100)(100)}{1000} \text{ liters}$$

$$= \mathbf{1,400,000 \text{ liters}}$$

practice **a.** Use three unit multipliers to convert 500 cubic meters to milliliters.

b. Use four unit multipliers to convert 500 liters to cubic meters.

c. Convert 5,000,000 cubic centimeters to milliliters.

problem set   **1.** A single die is rolled twice. What is the probability of getting a 5 and then a 2?
130
2. The sales tax was $7\frac{1}{2}$ percent. If Moriah purchased $40 worth of merchandise, how much tax did he pay?

3. The ratio of the number of formidable contestants to the number who were lackluster was 7 to 2. If there were 45,000 contestants, how many were formidable?

4. The product of a number and -11 increased by 6 is 30 less than the number times 25. What is the number?

5. Construct a triangle whose sides are 3 cm, 4 cm, and 6 cm.

6. Use a ruler to draw a line segment 7 centimeters long. Construct a perpendicular to the line at a point 4 centimeters from the right endpoint.

7. Use a protractor to draw a 71° angle. Then use a straightedge and a compass to construct the bisector of the angle.

8. What percent of 80 is 108? Draw a diagram of the problem.

9. If 160 is increased by $20\frac{1}{2}$ percent, what is the resulting number? Draw a diagram of the problem.

10. Use six unit multipliers to convert 10,000 cubic centimeters to cubic kilometers.

11. Use four unit multipliers to convert 12 cubic meters to liters.

12. Use six unit multipliers to convert 1000 cubic miles to cubic inches.

13. Find side m:

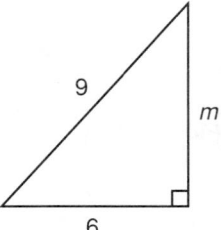

14. (a) Find the length of side k:

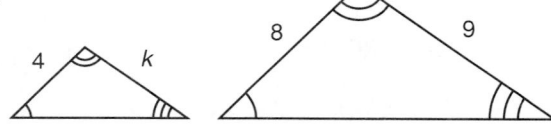

(b) Find the measure of angle Z:

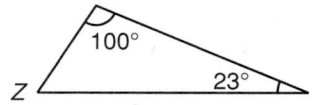

15. Find the area of this trapezoid. Dimensions are in feet.

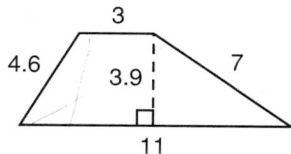

16. Find the volume in cubic centimeters of a right solid whose base is shown on the left and whose height is 3 meters. Dimensions are in centimeters.

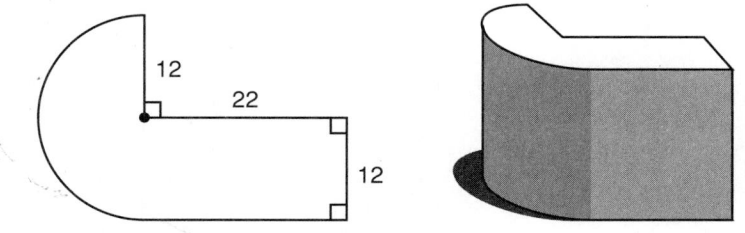

17. Find the lateral surface area in square centimeters of the solid described in Problem 16.

18. Use the distributive property to multiply: $x^2y^3\left(x + yx + \dfrac{2x}{7y^2}\right)$

19. Simplify by adding like terms: $4zmn^3 - 2zm^2 + nzmn^2 - mzm$

Solve:

20. $4x - 9 + x = -3x + 31$

21. $-1\dfrac{1}{2}x - \dfrac{1}{4} = 7\dfrac{1}{2}$

22. $\dfrac{-\dfrac{1}{6}}{\dfrac{4}{3}} = \dfrac{x}{-\dfrac{1}{7}}$

Simplify:

23. (a) $\dfrac{1}{7^{-2}}$ (b) $(-3)^{-3}$

24. $-(-3^2) - 4(3 \cdot 2 - 2^2)$

25. $7mn^4m^2bnb^4$

26. Graph: $x \geq -2$

27. If $9x + 4 = 1$, what is the value of $\dfrac{1}{3}x + \dfrac{1}{9}$?

28. If $\dfrac{1}{4}x + 3 = 2$, what is the value of $2x + \sqrt[5]{-32}$?

Evaluate:

29. $ab^3 + (-ab)^3$ if $a = -2$ and $b = 3$

30. $\sqrt[2]{pz} + \sqrt[3]{z} + p^z$ if $p = -2$ and $z = -8$

LESSON 131 *Rectangular coordinates*

Here we show two number lines. The lines are perpendicular and intersect at the origin of both lines. We call the <u>vertical line the **y axis**</u> and call the horizontal line the <u>**x axis.**</u>

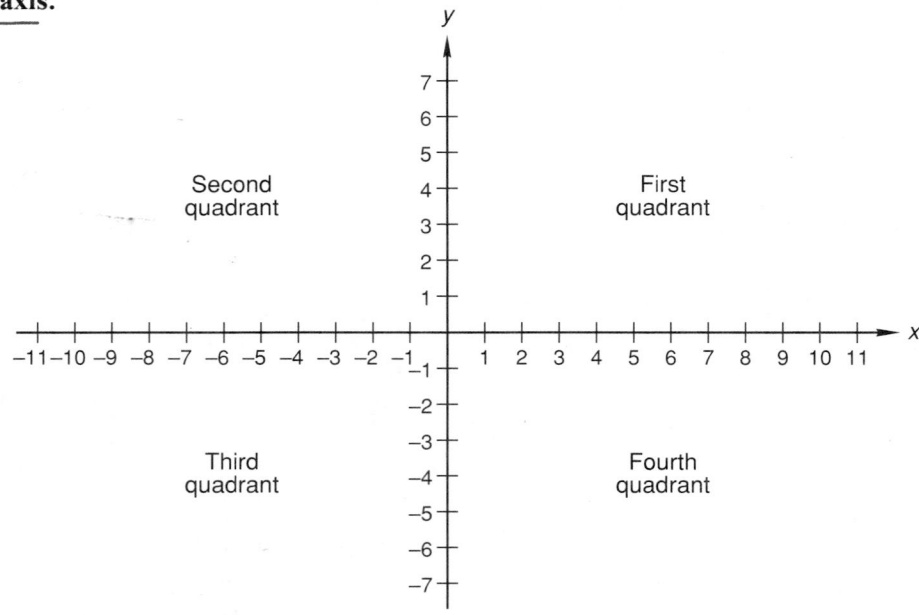

The number lines (axes) divide the plane into four quarters, or **quadrants.** The upper right quadrant is called the first quadrant. The others are numbered in a counter-clockwise direction.

Any point in the plane can be located by telling how far it is to the right or left of the *y* axis and how far up or down it is from the *x* axis. Two numbers are associated with every point in the plane. The <u>first number tells how far to the right</u> (+) or to the left (−) the point is from the *y* axis. This number is called the *x* <u>**coordinate**</u> of the point. The <u>second number tells how far above (+) or below (−) the</u> point is from the *x* axis. This is called the **y coordinate** of the point.

example 131.1   Four points are graphed on this rectangular coordinate system. What are the *x* and *y* coordinates of the points?

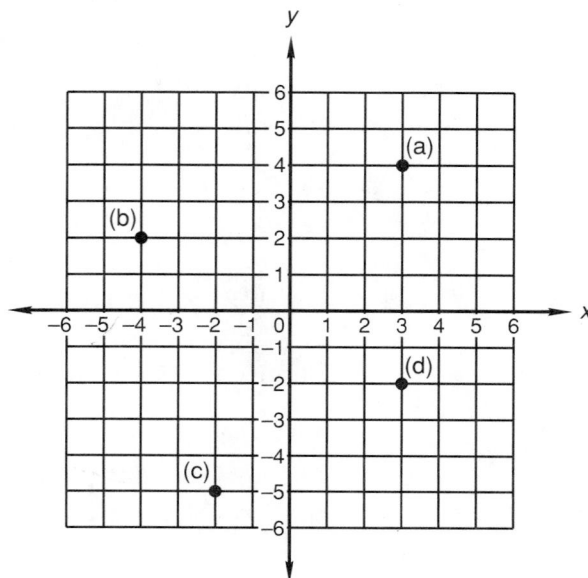

solution   Point (a) is 3 units to the right and 4 units up.

$$\text{(a)} \quad x = 3, y = 4$$

Point (b) is 4 units to the left and 2 units up.

$$\text{(b)} \quad x = -4, y = 2$$

Point (c) is 2 units to the left and 5 units down.

$$\text{(c)} \quad x = -2, y = -5$$

Point (d) is 3 units to the right and 2 units down.

$$\text{(d)} \quad x = 3, y = -2$$

example 131.2   Graph the following points: (a) (4, 2)   (b) (4, −3)   (c) (−4, −3)

solution   The first number is always the x coordinate and the second number is always the y coordinate. Thus,

(a)   is 4 units to the right and 2 units up

(b)   is 4 units to the right and 3 units down

(c)   is 4 units to the left and 3 units down

Now we graph the points.

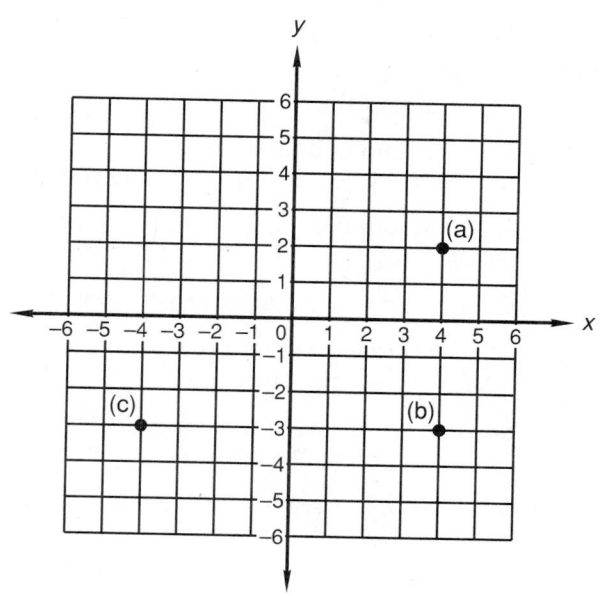

example 131.3   Make a sketch of a rectangular coordinate system and graph the points (−3, 2) and (4, −1).

solution   A quick sketch is all we need. We do not even need to put numbers on the axes.

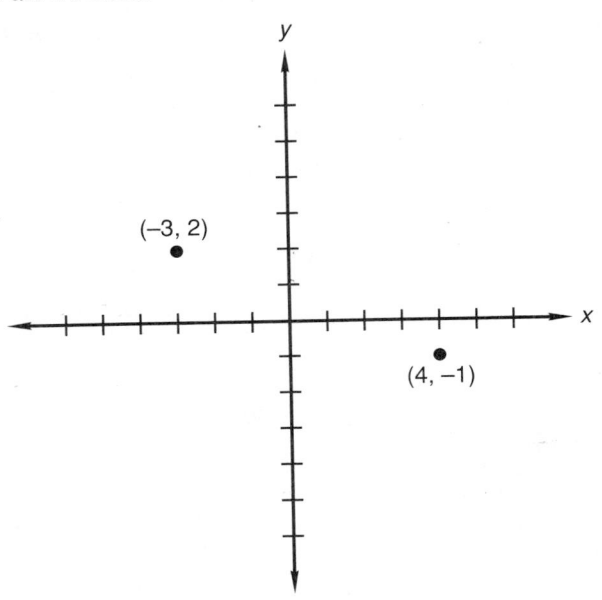

practice   Make a quick sketch of a rectangular coordinate system. Graph these points:
   a. $(-4, -3)$ **b.** $(-4, 2)$ **c.** $(3, -2)$ **d.** $(3, 5)$

problem set   1. A fair coin is tossed 3 times. What is the probability that all 3 tosses will be
131   heads?

   2. A single die is rolled twice. What is the probability of getting a 5 on both rolls?

   3. It had never happened before, but this time the increase was 350 percent. If the
        total this time was 1800, what was the total last time?

   4. The ratio of blue starlings to white egrets was 14 to 3. If 8500 birds had nested
        down by suppertime, how many were white egrets?

   5. Graph the points on a rectangular coordinate system:
        (a) $(4, 3)$ (b) $(-2, 5)$ (c) $(3, 6)$

   6. Construct a triangle whose sides are 4 cm, 3 cm, and 6 cm.

   7. Use a ruler to draw a line segment 7 centimeters long. Construct a perpendicu-
        lar to the line at a point 4 centimeters from the left endpoint.

   8. Use a protractor to draw a 120° angle. Then use a straightedge and a compass
        to construct the bisector of the angle.

   9. Draw a line segment and a point outside the line. Construct through the point a
        line that is perpendicular to the segment.

   10. What base 10 number does 10111 (base 2) represent?

   11. What number is $9\frac{1}{4}$ percent of 1,000,000? Draw a diagram of the problem.

   12. What percent of 80 is 108? Draw a diagram of the problem.

   13. Ninety percent of what number is 261? Draw a diagram of the problem.

   14. Convert 68,921,300 milliliters to liters.

15. Use three unit multipliers to convert 2 cubic miles to cubic feet.

16. Use the distributive property to multiply: $2abc(a + b^2 + c - 1)$

17. Simplify by adding like terms: $2x^2y^2 - 3xyxy + 3x^3y - 2xy^2x$

18. Express in cubic inches the volume of a right solid whose base is shown and whose sides are 2 feet high. Dimensions are in inches. Angles that look square are square.

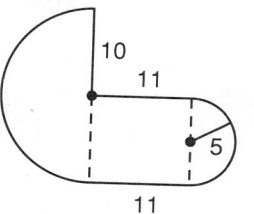 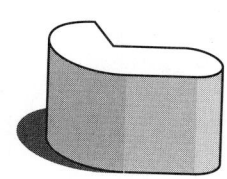

Solve:

19. $2x - 4 + 3x = x + 21$

20. $-2\frac{1}{4}x - \frac{1}{3} = 2\frac{1}{2}$

21. $\dfrac{-\frac{1}{2}}{\frac{3}{4}} = \dfrac{-\frac{1}{3}}{x}$

Simplify:

22. (a) $\dfrac{1}{3^{-2}}$ (b) 2^{-4} (c) $\dfrac{1}{2^{-4}}$ (d) 3^{-2}

23. $(-1)^3 - 2(2^3 - 2 \cdot 3)$

24. $4a^2bab^2a^3b$

25. $\dfrac{-2^3 + (-2)^2 - 3(3 - 3^2)}{2(2^3 - 3^2)}$

26. $\dfrac{1}{2}\left(1\frac{2}{3} \cdot \frac{1}{4} - \frac{1}{2} \cdot \frac{5}{6}\right)$

Evaluate:

27. $-2xy + 2x^2$ if $x = -1$ and $y = 2$

28. $mn - mn^2$ if $m = -2$ and $n = -1$

29. If $4x - 21 = 3$, what is $\frac{1}{4}x - 3$?

30. Find the least common multiple of 6, 9, and 16.

LESSON 132 *Numerals and numbers · Larger than and greater than · Pyramids, cones, and spheres*

132.A
numerals and numbers

A number is an idea. A numeral is the symbol we write to make us think of the idea. The three circles below have the quality of threeness, as do the three squares. The diagram on the right has this same quality of threeness even though the shapes shown are all different.

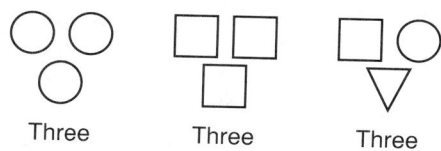

Three   Three   Three

When we want to think of the **idea** of three, we often write the following symbol.

3

This symbol is not a number but is a numeral that makes us think of the number 3. The ancient Romans used the numeral

III

when they wanted to think of the number 3. There are many numerals for 3. All the following numerals have the value of 3.

$$\frac{30}{10} \qquad 2 + 1 \qquad \frac{15}{5} \qquad 6 \div 2 \qquad 14 - 11 \qquad 1 + 1 + 1$$

Each of these groups of numerals can also be called a **numerical expression. We see that the value of a numerical expression is the number that the expression represents. Since all these expressions represent the number 3, we say that each of them has a value of 3.** This distinction between a numeral and a number is often helpful. For instance, we cannot add the fractions

$$\frac{1}{4} + \frac{1}{2}$$

in their present form because the denominators are not the same. But we can add the fractions if we change the numeral for the second number to a numeral whose denominator is 4.

$$\frac{1}{4} + \frac{2}{4} = \frac{3}{4}$$

We did not change the number. We changed the numeral used to represent the number. We could do this because both

$$\frac{1}{2} \qquad \text{and} \qquad \frac{2}{4}$$

have the same value because they represent the same number.

From this we see that there is a very real difference between a number and a numeral. If we pay close attention to this difference, however, it often confuses rather than clarifies. So we will often use the word number when we really should say numeral. It is not a bad error, especially if we remember that there is a difference between the two words and remember that

A number is an idea.

132.B
larger than and greater than

Here we show the numbers 4 and 2.

4   2

We have written the number 4 larger than the number 2. Now we write the number 2 larger than the number 4.

4   2

We have been graphing numbers on the number line, and we have said that 4 is always greater than 2 because its graph is farther to the right on the number line.

Note that we carefully use the words **greater than** and **less than** when we compare numbers. We do this because we want to avoid using the words **bigger than** and **larger than** because some people may think we are talking about the size of the numeral.

13   7

The number 13 is always greater than the number 7, in spite of the fact that here we have shown a numeral for 7 that is "larger than" the numerals used for 13.

132.C
pyramids, cones, and spheres

There are formulas for the volumes of cones, pyramids, and spheres, but formulas are hard to learn and easy to forget.

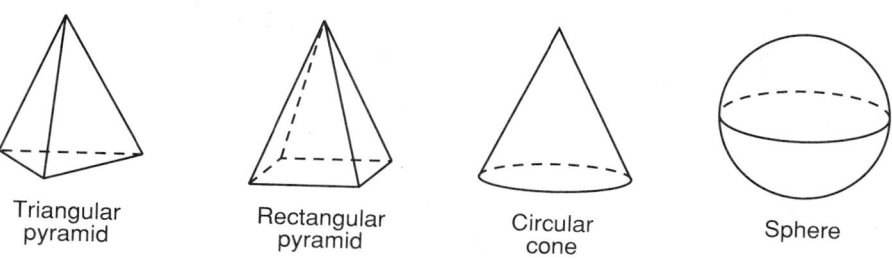

There is an easy way to find volumes. The volume of a pyramid or a cone is exactly one-third the volume of the right solid that has the same base and the same height. The volume of a sphere is exactly two-thirds the volume of the right circular cylinder that has the same radius as the sphere and is just as tall as the sphere.

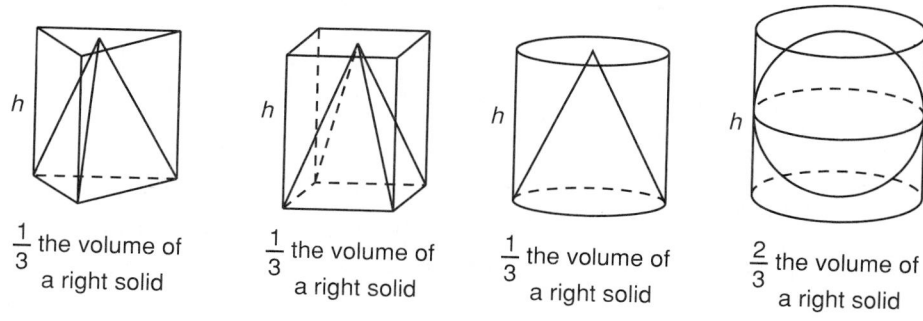

$\frac{1}{3}$ the volume of a right solid $\frac{1}{3}$ the volume of a right solid $\frac{1}{3}$ the volume of a right solid $\frac{2}{3}$ the volume of a right solid

example 132.1 The base of a triangular pyramid 4 meters high is the triangle shown. Find the volume of the pyramid. Dimensions are in meters.

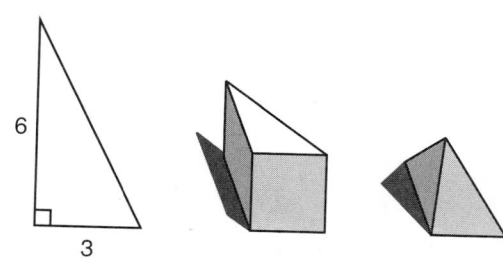

solution   The area of the base of the right solid is the area of the triangle.

$$\text{Area of base} = \frac{1}{2}\, BH = \frac{1}{2}(3)(6) = 9 \text{ m}^2$$

The volume of the right solid is the area of the base times the height.

$$\text{Volume of right solid} = AH = 9(4) = 36 \text{ m}^3$$

The volume of the pyramid is $\frac{1}{3}$ the volume of the solid.

$$\text{Volume of pyramid} = \frac{1}{3}(36) = \mathbf{12 \text{ m}^3}$$

example 132.2   Find the volume of a circular cone 4 cm high whose radius is 10 cm.

solution   First we find the area of the base.

$$\text{Area of base} = \pi r^2 \approx (3.14)(10)^2 \approx 314 \text{ cm}^2$$

The volume of the right solid is the area of the base times the height.

$$\text{Volume of right solid} = AH = 314(4) = 1256 \text{ cm}^3$$

The volume of the cone is $\frac{1}{3}$ the volume of the solid.

$$\text{Volume of cone} = \frac{1}{3}(1256) = \mathbf{418\frac{2}{3} \text{ cm}^3}$$

example 132.3   Find the volume of a sphere whose radius is 10 cm.

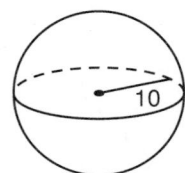

solution   A sphere is perfectly round. Every point on the sphere is the same distance from the center of the sphere. We could fit this sphere into a circular cylinder 20 cm tall whose radius is 10 cm.

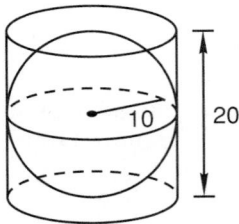

The volume of the cylinder is the area of the base times the height.

$$\text{Volume of cylinder} = \text{area of base} \times \text{height}$$

$$\approx 3.14(10^2)(20)$$

$$\approx 6280 \text{ cm}^3$$

The volume of the sphere is two-thirds the volume of the cylinder.

$$\text{Volume of sphere} = \frac{2}{3}(6280) \approx \textbf{4186.67 cm}^3$$

132.D
surface area

We remember that the lateral surface area of a right solid is the perimeter of the base times its height. The base of a pyramid may be a triangle, a square, or a rectangle or have the shape of any polygon. Because a pyramid is pointed, all the faces are triangular. The surface area of a pyramid equals the area of the base plus the sum of the areas of the triangular faces. The surface area of a cone equals the area of the base plus the lateral surface area. In this book we will consider only the surface area of right circular cones.

$$\text{Lateral surface area of a right circular cone} = \pi r s$$

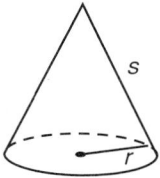

The *s* in this formula is the slant height of the cone.

example 132.4   The base of a pyramid is a square whose sides are 4 meters long. The altitude of each face of the pyramid is 3 meters. Find the surface area.

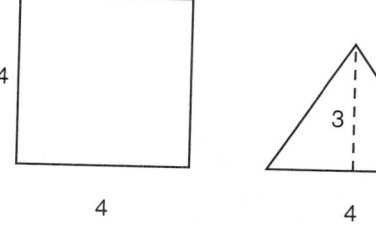

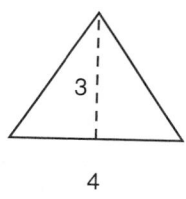

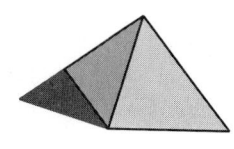

Base   Face

solution   The area of the base is $4 \times 4 = 16$ m². The area of each face is

$$\text{Area of face} = \frac{1}{2}bh = \frac{1}{2}(4)(3) = 6 \text{ m}^2$$

There are four faces, so the total surface area is

$$16 + 6 + 6 + 6 + 6 = \textbf{40 m}^2$$

example 132.5   Find the surface area of a right circular cone whose slant height is 6 meters and whose radius is 5 meters.

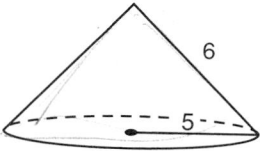

solution   The area of the base is

$$\text{Base area} = \pi r^2 \approx (3.14)(5)^2 = 78.5 \text{ m}^2$$
$$\text{Lateral surface area} = \pi rs \approx (3.14)(5)(6) = 94.2 \text{ m}^2$$

The total surface area is

$$78.5 \text{ m}^2 + 94.2 \text{ m}^2 = \textbf{172.7 m}^2$$

practice   **a.** Find the volume of a pyramid 4 meters high if the base is a square whose sides are 6 meters long.

b. Find the volume of a right circular cone 6 meters high if the radius of the base is 2 meters.

c. Find the volume of a sphere whose radius is 6 centimeters.

d. Find the surface area of a pyramid whose base is a 5 ft × 5 ft square and whose triangular faces have an altitude of 10 ft.

problem set 132

1. One die is red and the other die is green. Both are rolled. What is the probability of getting each of the following sums?
(a) an 8   (b) a 9   (c) a 10

2. The first three cards dealt from a well-shuffled deck of cards are spades. What is the probability that the next card dealt will be a spade?

3. Twenty percent of the flowers in the show were red. If 1400 flowers were not red, how many were red?

4. Among the ballplayers the ratio of ectomorphs to mesomorphs was 3 to 14. If there were 3400 ballplayers in both categories, how many were ectomorphs?

5. What is the difference between a numeral and a number?

6. Draw an 8-centimeter line segment with a ruler and construct the perpendicular bisector of the line segment.

7. Use a protractor to draw a 56° angle. Then use a straightedge and a compass to construct the bisector of the angle.

8. Use a protractor to draw a 55° angle. Then use a straightedge and a compass to copy the angle.

9. Write 96 (base 10) using base 2 numerals.

10. Convert 111110 (base 2) to base 10.

11. (a) Find the volume of a right circular cone whose base has a radius of $\sqrt{2}$ inches and whose height is 6 inches.
(b) Find the volume of a sphere whose radius is $\sqrt{2}$ inches.

12. (a) Find the volume of a pyramid 5 feet high if the base is a square whose sides are 4 feet long.
(b) Find the surface area of a pyramid whose base is a 6 ft × 6 ft square and whose triangular faces have an altitude of 5 ft.

13. What percent of 70 is 112? Draw a diagram of the problem.

14. If 130 is increased by 40 percent, what is the resulting number? Draw a diagram of the problem.

15. Use 4 unit multipliers to convert 1320 cubic meters to liters.

16. Convert 15,321,156 milliliters to liters.

17. Use the distributive property to multiply: $2ab(a + b - ab^2 + a^2b)$

18. Simplify by adding like terms: $2xyy^2 + 3x^2xy - 6yxy^2 + 7xyy^2$

19. Express in cubic meters the volume of a right solid whose base is shown on the left and whose sides are 3 meters high. Dimensions are in centimeters.

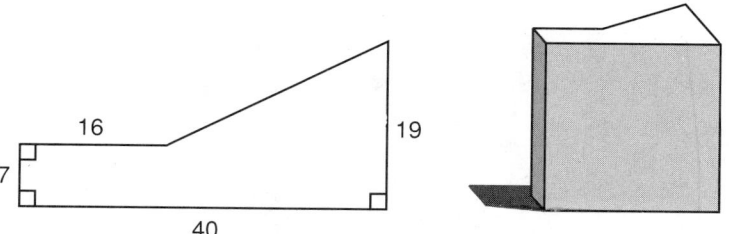

Solve:

20. $5x - 5 = -6x + 2x - 22$

21. $-\dfrac{2}{3}x - \dfrac{1}{5} = 1\dfrac{5}{6}$

22. $\dfrac{-\dfrac{1}{3}}{\dfrac{16}{9}} = \dfrac{-\dfrac{1}{3}}{x}$

Simplify:

23. $-[-(-2)^3] + (-2)$

24. $3a^2ba^3b^2b^3$

25. $\dfrac{-2^3 - 2^2(2^2 - 3) + 3}{3(2^3 - 1)}$

26. $\dfrac{1}{3}\left(2\dfrac{1}{2} \cdot \dfrac{1}{3} - \dfrac{1}{6} \cdot \dfrac{1}{2}\right)$

27. Evaluate: $xy^2 + 2x^2y$ if $x = 1$ and $y = -2$

28. Use the cut-and-try method to estimate $\sqrt{46}$ to one decimal place.

29. Write $0.002\dfrac{3}{5}$ as a decimal number.

30. Write 621321131.2 in words.

LESSON *133 Advanced equations*

The equations that we have discussed thus far have required adding like terms on only one side of the equation. Often we must simplify the expressions on both sides of the equals sign as the first step.

example 133.1   Solve: $6x - 5 + x - 2 = 2x - 3 + x$

solution   We begin by adding like terms to simplify both sides of the equation.

$$7x - 7 = 3x - 3$$

Now we eliminate the $3x$ by adding $-3x$ to both sides.

$$
\begin{array}{lll}
7x - 7 = & 3x - 3 & \text{added like terms} \\
\underline{-3x \qquad -3x} & & \text{add } -3x \text{ to both sides} \\
4x - 7 = & -3 & \\
\underline{\quad +7 \quad +7} & & \text{add } +7 \text{ to both sides} \\
4x \quad = & 4 & \\
x = & 1 & \text{divided both sides by 4}
\end{array}
$$

Check:

$$6(1) - 5 + (1) - 2 = 2(1) - 3 + (1) \longrightarrow 0 = 0 \quad \text{check}$$

example 133.2   Solve: $3x - 2 - x = 4 + x + 3$

solution   Again we begin by adding like terms on both sides. Then we solve.

$$
\begin{array}{lll}
2x - 2 = & x + 7 & \text{added like terms} \\
\underline{-x \qquad -x} & & \text{add } -x \text{ to both sides} \\
x - 2 = & 7 & \\
\underline{\quad +2 \quad +2} & & \text{add } +2 \text{ to both sides} \\
x = & 9 &
\end{array}
$$

Check:

$$3(9) - 2 - (9) = 4 + (9) + 3 \longrightarrow 27 - 2 - 9 = 4 + 9 + 3$$

$$\longrightarrow \quad 16 = 16 \quad \text{check}$$

practice   Solve:

a. $x + 2x + 4 = x - 6 + 8$

b. $2 + 4 + x = 3x + 2 + 5$

problem set 133

1. The urn contained 20 marbles: 10 were black, 5 were red, and 5 were green. One marble was drawn and then put back. Then a second marble was drawn. What is the probability that both marbles were black?

2. In Problem 1 what is the probability that the first marble was black and the second marble was red?

3. Ten percent of those present were able to concentrate on what the speaker was saying. If 600 were concentrating, how many were not concentrating?

4. The ratio of the weight of the canned goods to the weight of the bulk commodities at the warehouse was 4 to 11. If there were 30,000 tons stored in the warehouse, what was the weight of the bulk commodities?

5. Graph the points on a rectangular coordinate system:
   (a) $(3, -2)$ (b) $(-1, 4)$ (c) $(-2, -3)$

6. Construct a triangle whose sides are 5 cm, 6 cm, and 4 cm.

7. Draw a line segment and a point outside the line. Construct through the point a line that is perpendicular to the segment.

8. Convert 10101 (base 2) to base 10.

9. 58.5 is $19\frac{1}{2}$ percent of what number? Draw a diagram of the problem.

10. What percent of 60 is 99? Draw a diagram of the problem.

11. If 130 is increased by 140 percent, what is the resulting number? Draw a diagram of the problem.

12. Use six unit multipliers to convert 256 cubic kilometers to cubic centimeters.

13. Use six unit multipliers to convert 12 cubic miles to cubic inches.

14. Use the distributive property to multiply: $2ac(ab + 2ac - a)$

15. Simplify by adding like terms: $2x^2y^2 - 3x^2y^3 - 3xyxy + 7x^2y^2y$

16. If $3x - 5 = 13$, what is $\frac{1}{3}x - 7$?

17. Find the volume of a solid whose base is shown on the left and whose sides are 3 feet high. Dimensions are in feet.

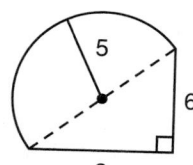 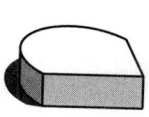

Solve:

18. $3x - 6 + 4x = -3x + 16 - x$

19. $-3\frac{1}{3}x - \frac{1}{6} = \frac{7}{18}$

20. $\dfrac{-\dfrac{1}{4}}{\dfrac{8}{9}} = \dfrac{-\dfrac{1}{4}}{x}$

21. $16x - 15x + x - 4 = 2x - 5 + x$

22. $x - 2 - x = 4 - 2x + 3$

Simplify:

23. (a) $\dfrac{1}{2^{-3}}$ (b) 2^{-3} (c) $\dfrac{1}{2^{-1}}$ (d) 3^{-3}

24. $(-2)^3 - 3(2^3 - 3^2)$

25. $5a^2ba^2ba^3$

26. $\dfrac{-2^3 + (3)^2 - 2^2(2^3 - 3^2)}{3(3 \cdot 2 - 5)}$

27. $\dfrac{1}{3}\left(1\dfrac{1}{4} \cdot \dfrac{1}{3} - \dfrac{1}{6} \cdot \dfrac{1}{2}\right)$

Evaluate:

28. $xy^2 - x^3y^3$ if $x = -1$ and $y = -2$

29. $mn + n^2$ if $m = -1$ and $n = 5$

30. Graph: $x \not\geq -2$

LESSON *134 Forming solids · Symmetry*

134.A

forming solids   Visualizing solids that can be formed by folding patterns is interesting.

example 134.1   What is the volume of the geometric solid formed by folding the shape along the dotted lines? Dimensions are in inches.

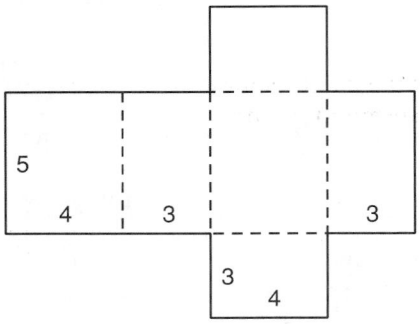

solution   The sides will form the box shown.

Volume = 3 × 4 × 5 = **60 in.³**

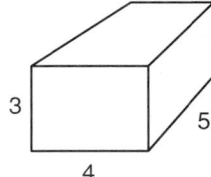

example 134.2   What is the volume of the geometric figure formed by folding this shape along the dotted lines? Dimensions are in centimeters.

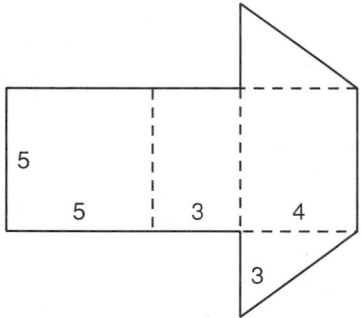

solution   The solid will be a triangular prism. The volume is the area of the triangle times the length.

$$\text{Volume} = \left(\frac{1}{2}BH\right)(L)$$

$$= \frac{1}{2}(4)(3)(5) = \textbf{30 cm}^3$$

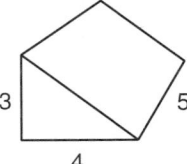

134.B
symmetry

A figure is symmetric about a line if one side of the figure is a mirror image of the other side of the figure. This figure is symmetric about the dotted line.

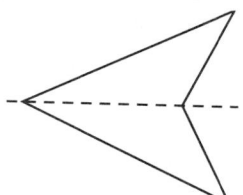

Figures are symmetric about a line if the perpendicular distance from any point on the line to the figure is equal to the perpendicular distance from the same point to a corresponding point in the figure that is on the other side of the line. To demonstrate, we have picked three points on the dotted line.

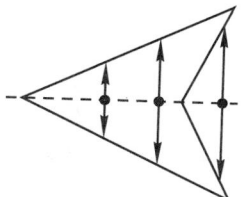

A figure can have more than one line of symmetry. The equilateral triangle has three lines of symmetry, and the square has four lines of symmetry.

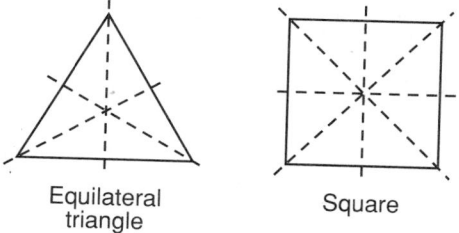

Equilateral triangle

Square

Figures can also be symmetric about a point. This figure is symmetric about the point shown.

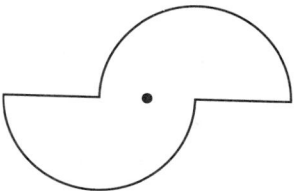

A figure is symmetric about a point if, for any line drawn through the point, it is the same distance in both directions to corresponding points on the figure. We show three such lines.

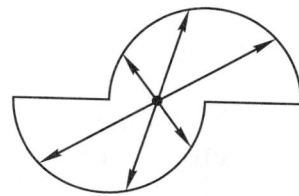

practice Find the volume of the solid formed when each figure is folded along the dotted lines. Dimensions are in inches.

a.

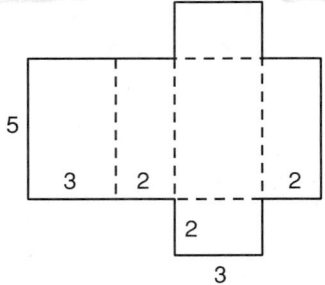

b.

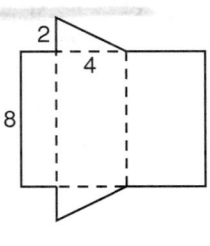

Quickly sketch each figure and draw in the lines of symmetry.

c.

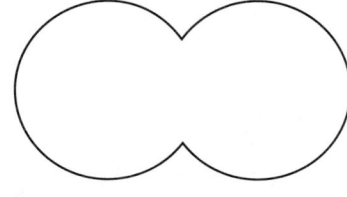

d.

Are these figures symmetric about a point, a line, or both?

e.

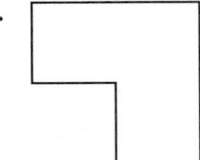

f.

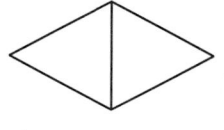

Sometimes we can use symmetry to help us solve problems. Use symmetry to find the area of these symmetric figures. Dimensions are in meters.

g.

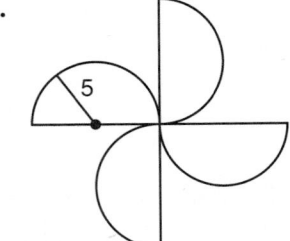

h.

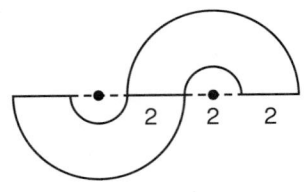

problem set 134

1. A fair coin is tossed 4 times. What is the probability that all 4 tosses will be tails?

2. A red die and a green die are rolled. There are 36 possible combinations. What is the probability that the sum on the dice will equal 8?

3. The number of grant recipients increased 200 percent this year over last year. If 1200 people received grants this year, how many received grants last year?

4. The ratio of azure to teal was 9 to 7. If there were a total of 1600, how many were teal?

5. Graph the points on a rectangular coordinate system:
   (a) (−3, 3)   (b) (1, 1)   (c) (2, −3)

6. Construct a triangle whose sides are 5 cm, 7 cm, and 11 cm.

7. Use a straightedge to draw a line segment 9 centimeters long. Construct a perpendicular to the line at a point 4 centimeters from the right endpoint.

8. Use a protractor to draw a 165° angle. Then use a straightedge and a compass to construct the bisector of the angle.

9. Draw a line segment and a point outside the line. Construct through the point a line that is perpendicular to the segment.

10. Convert 10001 (base 2) to base 10.

11. Find side *m*:

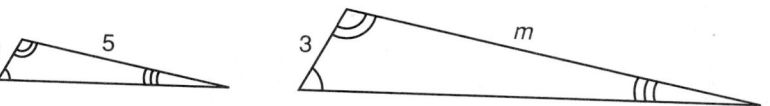

12. What number is $16\frac{1}{2}$ percent of 1100? Draw a diagram of the problem.

13. Sixty percent of what number is 870? Draw a diagram of the problem.

14. Convert 70,000,000 milliliters to cubic meters.

15. Use six unit multipliers to convert 60 cubic inches to cubic miles.

16. Find the volume of the solid formed when this figure is folded along the dotted line. Dimensions are in inches.

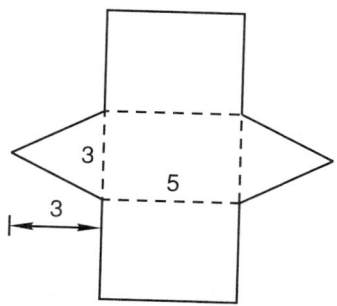

17. Sketch the figures and draw in the lines of symmetry. Which figure is symmetrical about a point?

(a)    (b)

18. Use symmetry to find the area of this symmetrical figure. Dimensions are in feet.

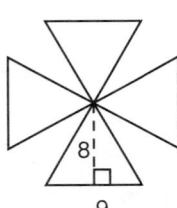

19. Find (a) angle z and (b) side b:

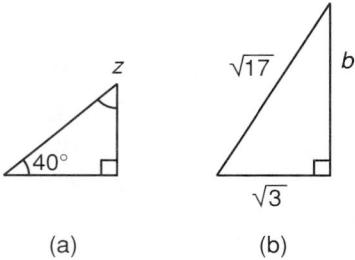

(a)   (b)

Solve:

20. $5x + 4 - x = 3x + 2$

21. $-4\frac{1}{3}x + \frac{1}{4} = -3\frac{1}{12}$

22. $\dfrac{-\frac{1}{6}}{\frac{2}{5}} = \dfrac{x}{\frac{1}{3}}$

Simplify:

23. (a) $\dfrac{1}{9^{-2}}$ (b) 4^{-3}

24. $\dfrac{(-2)^3 - 3(-1^5 - 3 \cdot 2)}{2(3^3 - \sqrt[5]{-243})}$

25. $6b^3ab^2a^5b$

26. $\dfrac{1}{4}\left(2\dfrac{1}{2} \cdot \dfrac{1}{3} - \dfrac{1}{6} \cdot \dfrac{1}{5}\right)$

Evaluate:

27. $ab^3 - b$ if $a = 3$ and $b = -3$

28. $cx^2 - c^5x$ if $c = -1$ and $x = -2$

29. If $14x - 29 = 1$, what is the value of $8x + 14$?

30. If $7x + 20 = -2$, what is the value of $\dfrac{-2}{3}x - \dfrac{1}{3}$?

LESSON 135   *Permutations*

There are 6 ways we can arrange the letters A, B, and C in order. If we put A first, there are 2 ways:

A	B	C

and

A	C	B

If we put B first, there are 2 ways:

B	A	C

and

B	C	A

If we put C first, there are 2 ways:

$$\boxed{C \mid A \mid B} \quad \text{and} \quad \boxed{C \mid B \mid A}$$

We could put any one of the **three** letters in the first box. Then there are **two** letters left that could be put in the second box. Then the last letter is forced into the last box because it is the only letter left. There are

$$3 \cdot 2 \cdot 1 = 6$$

ways these three letters could be arranged in order. Each of these ways is called a **permutation**. If there had been 5 objects, any of the 5 could go in the first place.

$$\boxed{5 \mid \mid \mid }$$

Now any one of the 4 remaining could go in the second place

$$\boxed{5 \mid 4 \mid \mid }$$

And any one of 3 in the next place and 2 in the next place.

$$\boxed{5 \mid 4 \mid 3 \mid 2 \mid 1}$$

$$5 \times 4 \times 3 \times 2 \times 1 = 120$$

There are 120 ways 5 objects can be arranged in order. There is an easier way to write

$$5 \times 4 \times 3 \times 2 \times 1$$

All we do is write 5 and follow it with an exclamation point.

$$5! \quad \text{means} \quad 5 \times 4 \times 3 \times 2 \times 1$$

The notation $5!$ is read as "5 factorial."

If we have 10 objects and want to find how many ways we can arrange 4 of them in a row, we think literally:

Any one of 10 could be in first place.

Any one of the 9 remaining could be in second place.

Any one of the 8 remaining could be in third place.

Any one of the 7 remaining could be in fourth place.

There are $10 \cdot 9 \cdot 8 \cdot 7$ ways 10 objects can be arranged 4 at a time:

$$10 \cdot 9 \cdot 8 \cdot 7 = 5040$$

There are 5040 permutations of 10 objects taken 4 at a time.

example 135.1 Mary has 8 different objects. How many permutations are possible?

solution There are $8!$ permutations.

$$8! = 8 \cdot 7 \cdot 6 \cdot 5 \cdot 4 \cdot 3 \cdot 2 \cdot 1 = \mathbf{40{,}320 \text{ ways}}$$

example 135.2   If Mary chose 3 objects at a time from her 8, how many ways could they be arranged in a row?

solution   Eight can go into the first box, etc.

$$\boxed{8 \mid 7 \mid 6} = 8 \times 7 \times 6 = 336$$

There are **336 ways** 8 objects can be arranged 3 at a time.

practice   **a.**   How many ways can 9 objects be arranged in a row?

               b.   How many ways can 5 of the 9 objects be arranged in a row?

problem set 135

1. The urn contained 34 marbles: 18 were black, 9 were red, and 7 were green. One marble was drawn and then put back. Then a second marble was drawn. What is the probability that both marbles were red?

2. If the product of a number and -17 is decreased by 5, the result is 3 greater than the product of the number and -9. What is the number?

3. Thirty percent of those present were not able to produce a likeness of the subject. If 1260 could produce a likeness, how many were present?

4. The ratio of the number of audiophiles to the number of nonaudiophiles was 2 to 3. If there were 325 in all, what was the number of audiophiles?

5. Graph the points on a rectangular coordinate system:
   (a) $(2, 3)$ (b) $(-1, 4)$ (c) $(2, -2)$

6. Construct a triangle whose sides are 4 cm, 3 cm, and 6 cm.

7. How many ways can 5 objects be arranged in a row?

8. How many ways can 3 of 5 objects be arranged in a row?

9. Draw a line segment and a point outside the line. Construct through the point a line that is perpendicular to the segment.

10. Use a protractor to draw a 33° angle. Then use a straightedge and a compass to construct the bisector at the angle.

11. Find (a) angle z and (b) side m:

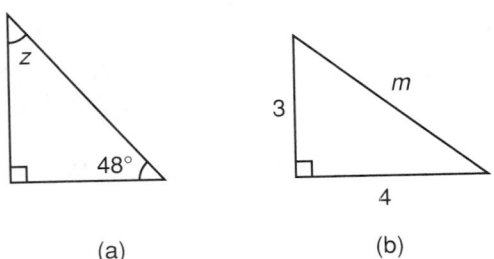

                                      (a)   (b)

12. Find the side m:

13. Sketch the figures and draw in the lines of symmetry. Which figure is symmetrical about a point?

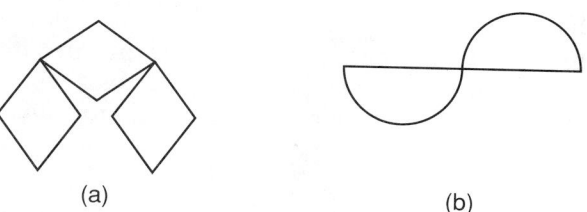

(a)   (b)

14. Use symmetry to find the area of the shaded portion of this symmetrical figure. Dimensions are in inches.

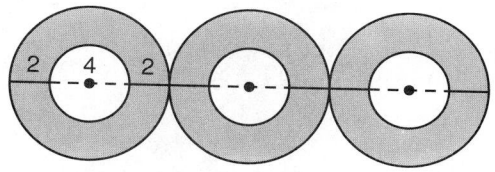

15. What percent of 40 is 54.4? Draw a diagram of the problem.

16. If 180 is increased by 90 percent, what is the resulting number? Draw a diagram of the problem.

17. Nine and one-half percent of what number is 9500? Draw a diagram of the problem.

18. Use six unit multipliers to convert 100,000 cubic centimeters to cubic kilometers.

19. Use six unit multipliers to convert 10 cubic miles to cubic inches.

20. Use the distributive property to multiply: $ab^2(ac + 3ab^2 - b)$

21. Simplify by adding like terms: $2mp^2 - 3m^2p - 4pmp + 5mpm$

Solve:

22. $4x - 6x - 6 = 3x + 14$

23. $\dfrac{-\dfrac{1}{5}}{\dfrac{7}{8}} = \dfrac{-\dfrac{1}{20}}{x}$

Simplify:

24. (a) $(-4)^{-3}$ (b) $\dfrac{1}{-4^{-2}}$

25. $-3^3 + (2)^3 - 3(2^3 - 3^2)$

26. $6m^3n^2bm^4n$

27. $\dfrac{1}{6}\left(2\dfrac{1}{2} \cdot \dfrac{2}{5} - \dfrac{1}{5} \cdot \dfrac{1}{4}\right)$

Evaluate:

28. $x^y - y^x$ if $x = -2$ and $y = -3$

29. $\sqrt[a]{b} + \sqrt{b} + a^2$ if $a = 3$ and $b = 64$

30. Graph: $x \not\le -4$

LESSON *136* *Adding in base 10 and base 2*

136.A
adding in base 10

We review our procedure for adding in base 10 by noting that we split the sum of a column into two parts. One part is a whole number times the base, and the other part is the remainder. We record the remainder and carry the whole number to the next column.

$$
\begin{array}{r}
53\ 9 \\
84\ 4 \\
+\ 63\ 8 \\
\hline
\textcircled{21} = 2(10) + 1
\end{array}
\qquad
\begin{array}{r}
{}^{2} \\
539 \\
844 \\
+\ 638 \\
\hline
1
\end{array}
$$

The sum of the first column is 21. This is 2 times the base (10), with 1 left over. We record the 1 and carry the 2 to the second column.

$$
\begin{array}{r}
{}^{2} \\
5\ 39 \\
8\ 44 \\
+\ 6\ 38 \\
\hline
\textcircled{12}1
\end{array}
\qquad 12 = 1(10) + 2 \qquad \longrightarrow \qquad
\begin{array}{r}
{}^{1\,2} \\
539 \\
844 \\
+\ 638 \\
\hline
21
\end{array}
$$

When we sum the second column, we get 12. This is 1 times the base (10), with 2 left over. We record the 2 and carry the 1 to the next column.

$$
\begin{array}{r}
{}^{1\ 2} \\
5\ 39 \\
8\ 44 \\
+\ \ 6\ 38 \\
\hline
\textcircled{20}21
\end{array}
\qquad 20 = 2(10) + 0 \qquad \longrightarrow \qquad
\begin{array}{r}
{}^{2\,1\,2} \\
539 \\
844 \\
+\ 638 \\
\hline
2021
\end{array}
$$

The total of the third column is 20. This is 2 times the base, with 0 left over. We record the 0 and carry the 2.

136.B
adding in base 2

We will use the same procedure to add in base 2. We will split the sum of a column into two parts. One part is a whole number times the base, and the other part is the remainder. We record the remainder and carry the whole number.

example 136.1 Add: 1111 (base 2) + 1011 (base 2) + 1101 (base 2) + 1101 (base 2)

solution First we write the numbers in base 10 and add.

$$15 + 11 + 13 + 13 = 52 \text{ (base 10)}$$

Now let's add in base 2. We record the numbers vertically and add the first column. **We will always think in base 10.** First we add the right column. We will write the sum in base 10 and circle it.

$$
\begin{array}{r}
1111 \\
1011 \\
1101 \\
+\ 1101 \\
\hline
\textcircled{4} = 2(2) + 0
\end{array}
$$

The sum of the first column in base 10 is 4. This is 2 times the base (2), with a remainder of 0. We record the 0 and carry 2 to the next column and add this column.

$$
\begin{array}{r}
2 \\
1\ 1\ 1\ 1 \\
1\ 0\ 1\ 1 \\
1\ 1\ 0\ 1 \\
+\ 1\ 1\ 0\ 1 \\
\hline
④0
\end{array}
\qquad 4 = 2(2) + 0
$$

This sum is 2 times the base, with a remainder of 0. We record 0 and carry 2. Then we add the third column.

$$
\begin{array}{r}
2\ 2 \\
1\ 1\ 1\ 1 \\
1\ 0\ 1\ 1 \\
1\ 1\ 0\ 1 \\
+\ 1\ 1\ 0\ 1 \\
\hline
⑤0\ 0
\end{array}
\qquad 5 = 2(2) + 1
$$

We get 5, which is 2 times the base, with a remainder of 1. We record 1 and carry the 2 and add the next column.

$$
\begin{array}{r}
2\ 2\ 2 \\
1\ 1\ 1\ 1 \\
1\ 0\ 1\ 1 \\
1\ 1\ 0\ 1 \\
+\ 1\ 1\ 0\ 1 \\
\hline
⑥1\ 0\ 0
\end{array}
\qquad 6 = 3(2) + 0
$$

The number 6 is 3 times the base, with a remainder of 0.

$$
\begin{array}{r}
3\ 2\ 2\ 2 \\
1\ 1\ 1\ 1 \\
1\ 0\ 1\ 1 \\
1\ 1\ 0\ 1 \\
+\quad 1\ 1\ 0\ 1 \\
\hline
③0\ 1\ 0\ 0
\end{array}
$$

But 3 is 1 times the base, with a remainder of 1. So our final sum is

$$① 1 0 1 0 0 \qquad 3 = ① (2) + 1$$

If we write a place-value table for base 2, we can find the number in base 10.

32	16	8	4	2	1

   1   1   0   1   0   0   (base 2) = 32 + 16 + 4 = 52 (base 10)

Our sum was 52 when we added the numbers using base 10 numerals. This is our check.

example 136.2 Convert the base 2 numerals 1101 and 1011 to base 10 numerals and add in base 10. Then add the numbers in base 2 and convert the answer to base 10 to check.

solution We will use a place-value table for base 2 to find the number in base 10.

8	4	2	1

$$
\begin{array}{cccc}
1 & 1 & 0 & 1 = 8 + 4 + 1 = 13 \\
1 & 0 & 1 & 1 = 8 + 2 + 1 = \underline{11} \\
& & & 24 \text{ (base 10)}
\end{array}
$$

Now we will add the numbers in base 2.

$$
\begin{array}{r}
1\;1\;1 \\
1\;1\;0\;1 \\
+\;1\;0\;1\;1 \\
\hline
1\;1\;0\;0\;0
\end{array}
$$

We use a place-value table to write this in base 10.

16	8	4	2	1

$$
1 \quad 1 \quad 0 \quad 0 \quad 0 = 16 + 8 = 24 \text{ (base 10)} \qquad \text{check}
$$

practice   **a.** Add these base 2 numbers. Then write the answer in base 10.

$$
\begin{array}{r}
1011 \\
1111 \\
1101 \\
+\;1110 \\
\end{array}
$$

b. Write 1101 and 1011 as base 10 numbers and add the base 10 numbers. Then add 1101 and 1011 in base 2. Write this answer in base 10. The two answers should check.

problem set 136

1. One card is pulled from a standard deck of 52 cards and then replaced. Then a second card is drawn. What is the probability of getting a black 2 and then a red 9?

2. Two dice are rolled. What is the probability that the sum of their faces will be 12?

3. If $10,000 is deposited in an account that receives $9\frac{1}{2}$ percent interest compounded annually, how much money will be in the account in 3 years?

4. The ratio of philatelists to archivalists was 2 to 8. If there were 1110 total, how many were philatelists and how many were archivalists?

5. Graph each point on a rectangular coordinate system:
   (a) (4, 2)   (b) (0, −1)   (c) (−1, 0)

6. Construct a triangle whose sides are 2 cm, 7 cm, and 8 cm.

7. Use a protractor to draw a 97° angle. Then use a straightedge and a compass to bisect the angle.

8. Convert 101010100 (base 2) to base 10.

9. Add: 1111 (base 2) + 100001 (base 2)

10. What percent of 60 is 84? Draw a diagram of the problem.

11. Seventy percent of what number is 651? Draw a diagram of the problem.

12. Use six unit multipliers to convert 1,000,000 cubic centimeters to cubic kilometers.

13. Use four unit multipliers to convert 10,500 cubic meters to milliliters.

14. (a) How many ways can 7 objects be arranged in a row?
(b) How many ways can 3 of the 7 objects be arranged in a row?

15. Find the volume in cubic feet of the solid formed when this figure is folded along the dotted lines. Dimensions are in feet. (*Note:* The height of the right solid is not 5 feet.)

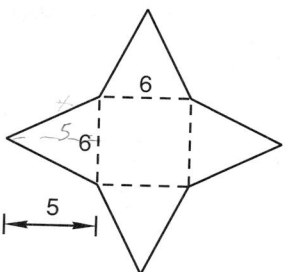

16. Sketch the figures and draw in the lines of symmetry. Which figure is symmetrical about a point?

(a)

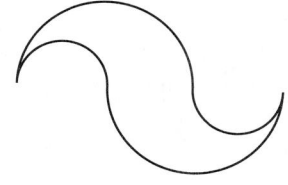

(b)

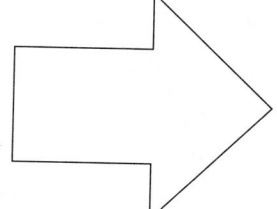

17. Find angle z and side m:

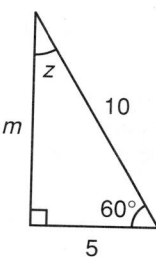

18. Find side m:

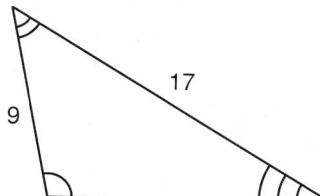

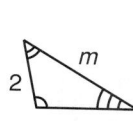

19. Find the area of each figure. Dimensions are in meters.

(a)

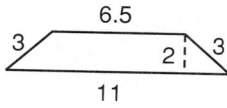

(b)

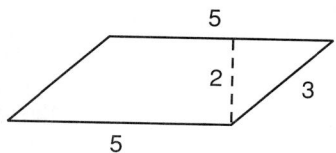

Solve:

20. $7x + 3 + 2x = -4 - 2x + 3 - x$

21. $-1\frac{4}{5}x + 2\frac{1}{2} = -\frac{1}{15}$

22. $\dfrac{\frac{1}{9}}{\frac{4}{7}} = \dfrac{x}{\frac{3}{14}}$

Simplify:

23. (a) $\dfrac{1}{-2^{-6}}$ (b) $(-2)^{-6}$

24. $\dfrac{(-1)^7 - 2(3^2 - 2 \cdot 4)}{3(\sqrt[3]{-216} + 2^3)}$

25. $m^5 p^6 m^2 pmz$

26. $\dfrac{2}{5}\left(3\frac{1}{4} \cdot \frac{1}{6} - \frac{1}{2} \cdot \frac{1}{4}\right)$

Evaluate:

27. $a^2 z - z^2$ if $a = -5$ and $z = -3$

28. $\sqrt[p]{m} + p^z$ if $m = -8$, $p = 3$, and $z = -2$

29. If $16x + 5 = -3$, what is the value of $8x - 1$?

30. If $3x + 21 = -3$, what is the value of $-\frac{3}{8}x + \frac{1}{9}$?

LESSON 137 *Equation of a line · Graphing a line*

137.A
equation of a line

If we have an equation with one unknown, the solution to the equation is the value of x that will make the equation a true equation. The solution to the equation

$$x + 2 = 5$$

is 3 because if we use 3 for x we get a true equation.

$$3 + 2 = 5 \qquad \text{true}$$

If an equation has <u>two variables</u>, a solution to the equation is the values of both variables that make the equation a true equation. The equation

$$y = x + 2$$

is a true equation if we use 5 for y and 3 for x.

$$(5) = (3) + 2 \qquad \text{true}$$

It is also a true equation if we use 9 for y and 7 for x.

$$(9) = (7) + 2 \qquad \text{true}$$

There are many pairs of x and y that make this equation a true equation. If we plot these values of x and y on a rectangular coordinate plane, we find that the points they describe form a straight line. Where we use points to draw a line, we say we are **graphing** the line.

137.B
graphing a line

It takes <u>only two points to determine a line</u>, but since we are beginners we will always find three points.

example 137.1   Graph the line $y = x + 2$.

solution   To find points that lie on the line we pick values of x and use the equation to find the paired values of y. First we make a box.

x		
y		

Then we choose values for x and put them in the box. The numbers 0, 2, and -2 often work well. We will use these numbers.

x	0	2	-2
y			

Then we use these values for x in the equation and solve for y.

Using 0 for x: $y = \ \ (0) + 2 \ \longrightarrow \ y = 2$

Using 2 for x: $y = \ \ (2) + 2 \ \longrightarrow \ y = 4$

Using -2 for x: $y = (-2) + 2 \ \longrightarrow \ y = 0$

Now we put these numbers in our box

x	0	2	-2
y	2	4	0

Now we graph the points (0, 2) (2, 4) and $(-2, 0)$ and draw a line through the points.

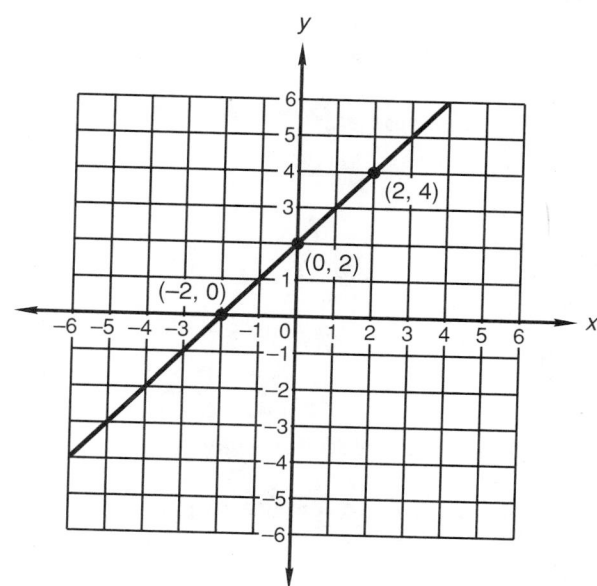

This is the graph of $y = x + 2$.

practice   Use the procedure shown in Example 137.1 to graph each line.

    a. $y = x - 3$

                                       b. $y = 2x + 1$

**problem set
137**

1. A fair coin is tossed 7 times. What is the probability that all tosses will be tails?

2. Reminiscences increased 110 percent. If the previous number of reminiscences was 840, what was the new number of reminiscences?

3. The ratio of the number of ratio problems to the number of nonratio problems was 1 to 29. If there are 3930 problems, how many are ratio problems?

4. The product of a number and 21 is decreased by 102. The result is equal to the product of the number and 11 increased by 8. What is the number?

5. The caravan covered the first 90 miles in 15 hours. Then they doubled their pace for the next 180 miles. How long did it take the caravan to travel the total distance of 270 miles?

6. (a) How many ways can 10 objects be arranged in a row?
   (b) How many ways can 4 of the 10 objects be arranged in a row?

7. Add 101101 (base 2) to 100011 (base 2).

8. Convert the answer to Problem 7 to base 10.

9. Construct a triangle whose sides are 3 cm, 6 cm, and 6 cm.

10. Use a protractor to draw a 91° angle. Then use a straightedge and a compass to bisect the angle.

11. Draw a line segment and a point outside the line. Construct through the point a line that is perpendicular to the segment.

12. Find (a) angle z and (b) side m:

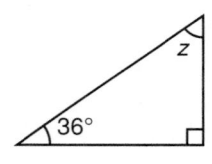
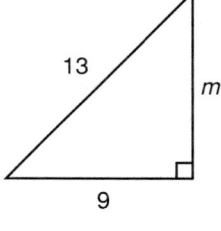

                                (a)   (b)

13. Find side m:

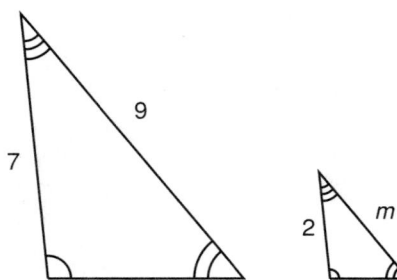

14. Graph the line indicated by the equation $y = 3x - 2$.

15. Eleven and one-fourth percent of what number is 562.5? Draw a diagram of the problem.

16. What percent of 90 is 148.5? Draw a diagram of the problem.

17. Use four unit multipliers to convert 1000 milliliters to cubic meters.

18. Use six unit multipliers to convert 11,000,000 cubic miles to cubic inches.

19. Use the distributive property to multiply: $m^2b(m^3 + b^2 + cm)$

20. Simplify by adding like terms: $2abab - 4ba^2b + 7ab^2a + 3a^2b$

21. Find the volume in cubic meters of a right circular cylinder with a diameter of $4\sqrt{2}$ meters and whose height is 2 meters.

22. Find the surface area in square meters of a cone whose slant height is 5 meters and whose radius is 2 meters.

23. Find the volume in cubic meters of a right cone whose radius is $2\sqrt{2}$ meters and whose height is 2 meters.

24. Find the volume in cubic meters of a right pyramid whose square base has sides 6 meters long and whose height is 4 meters.

25. Find the surface area in square meters of the pyramid in Problem 24 if the faces have a height of 5 meters.

Solve:

26. $5x - 4 + 2x - 1 = 3x + 6 + 2x + 1$

27. $\dfrac{-\dfrac{1}{4}}{\dfrac{2}{7}} = \dfrac{x}{\dfrac{1}{3}}$

28. Simplify: $(-3)^{-2}$

29. Evaluate: $-a^{-2} + b^{-3}$ if $a = 3$ and $b = -5$

30. Graph: $x \geq 0$

APPENDIX A *Additional practice sets*

practice set 1

1. If $8000 is deposited in an account that pays 10 percent interest compounded annually, how much money will be in the account in 4 years?

2. Two dice are rolled. What is the probability that the sum of their faces will be 2?

3. The ratio of greens to blues is 4 to 11. If greens and blues total 300, how many are blue?

4. Forty percent of the crowd ambled. If 3000 did not amble, how many ambled?

5. Graph the points on a rectangular coordinate system:
   (a) $(-1, 2)$ (b) $(2, -1)$

6. Add in base 2 and then write the answer in base 10.
$$\begin{array}{r} 1001 \\ 1000 \\ 1100 \\ + \ 1111 \end{array}$$

7. Eighty percent of what number is 640? Draw a diagram of the problem.

8. Two hundred ninety is what percent of 1450? Draw a diagram of the problem.

9. Use two unit multipliers to convert 14 square meters to square centimeters.

10. How many ways can 5 objects be arranged in a row?

Graph the line indicated by each equation:

11. $y = -2x + 1$

12. $y = -4x + 6$

13. Use the distributive property to multiply: $ab^3\left(b^3 + ab - \dfrac{a}{b}\right)$

14. Simplify by adding like terms: $3abca - 2a^2cb + 4bca^2 - 3bcba$

15. Find the volume in cubic meters of a right circular cylinder whose diameter is $4\sqrt{3}$ meters and whose height is 1 meter.

16. Find the surface area in square feet of a cone whose slant height is 8 feet and whose radius is $\sqrt{3}$ feet.

17. Find angle z and side m:

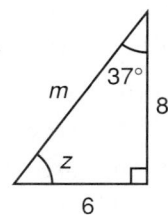

407

18. Find side m:

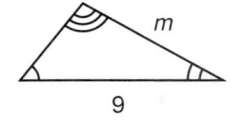 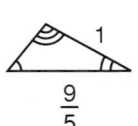

19. Construct a triangle whose sides measure 5 cm, 7 cm, and 9 cm.

20. Use a protractor to draw a 121° angle. Then use a straightedge and a compass to construct the bisector of the angle.

21. Find the volume in cubic inches of a right cone whose radius is $\sqrt{3}$ inches and whose height is 3 inches.

22. Find the volume in cubic centimeters of a right pyramid whose square base has sides 4 centimeters long and whose height is 6 centimeters.

Solve:

23. $4x - 9 + 3x = x + 7 - 2x - 2$

24. $5x + 4 - x = 3x + 4 - 6x - 21$

25. $\dfrac{-\dfrac{1}{8}}{\dfrac{3}{7}} = \dfrac{\dfrac{1}{6}}{x}$

Simplify:

26. $(-2)^{-4}$ **27.** $\dfrac{1}{8^{-2}}$ **28.** $\sqrt[5]{-32}$

29. Evaluate: $ab^2 - \dfrac{1}{a^{-b}}$ if $a = -3$ and $b = 2$

30. Graph: $x \not\leq -2$

practice set 2

1. Romantic themes dominated the literature, as 40 percent of the writings were romantic in nature. If there were 9000 writings, how many were not romantic?

2. Seventy-eight percent of the dryads capered every time a star came out. If 110 did not caper, how many dryads were there?

3. The ratio of prizes to dividends was 5 to 11. If 1600 prizes were awarded, how many dividends were awarded?

4. The sum of a number and 25 is 7 less than the product of the number and 9. What is the number?

5. Use a protractor to draw a 17° angle. Then use a compass and a straightedge to copy the angle.

6. Construct a triangle whose sides are 4 cm, 9 cm, and 11 cm.

7. Draw a line segment and a point outside the line. Construct through the point a line that is perpendicular to the line segment.

8. Use a protractor to draw a 160° angle. Then use a straightedge and a compass to construct the bisector of the angle.

9. If 185 is increased by 30 percent, what is the resulting number? Draw a diagram of the problem.

10. If 400 is increased by 15 percent, what is the resulting number? Draw a diagram of the problem.

11. Use six unit multipliers to convert 100 cubic meters to cubic millimeters.

12. If $8x - 4 = 28$, what is the value of $\frac{1}{4}x + \frac{1}{2}$?

13. If $10x + 21 = 39$, what is the value of $\frac{1}{5}x + \frac{1}{8}$?

14. Use the distributive property to multiply: $mn(4m + mn^2 + m)$

15. Use the cut-and-try method to estimate $\sqrt[4]{980}$ to the nearest whole number.

Simplify by adding like terms:

16. $m^3p + mpm^2 - mp^3 + 3pmp^2$

17. $-a^3b^2 - a^2b - ba^2ba + 2ab^2$

Solve:

18. $8x + 4 - 2x - 3x = 4x + 12$

19. $4\frac{1}{5}x - 2\frac{1}{2} = -\frac{3}{8}$

20. $9 + 3x - 6x + x - 3x = x + 12 - x$

21. $\dfrac{-\frac{1}{9}}{\frac{1}{4}} = \dfrac{\frac{13}{12}}{x}$

Simplify:

22. $(-8)(-2) + (-8)^2$

23. $m^3p^2mpm^2$

24. $(-4)^2(-4) - 4$

25. $x^3y^3x^3y^2x$

26. $\dfrac{-(3)^2 - 2^3(2^3 - 2 \cdot 6) + 3}{2^2(2^3 - 4)}$

27. $\frac{1}{5}\left(3\frac{1}{2} \cdot \frac{1}{6} - \frac{2}{3}\right)$

Evaluate:

28. $a^b - b^a + ab$ if $a = -2$ and $b = -1$

29. $\dfrac{-b}{a} + \dfrac{2a}{b}$ if $a = -1$ and $b = -2$

30. Write 6213825.32 in words.

practice set 3

1. The race cars surged across the ramp at the start of the race. Estelle estimated that the ratio of male racers to female racers was 9 to 2. If there were 4730 racers in all, how many were female?

2. When the brouhaha commenced, the ratio of revelers to debauchers was 8 to 5. If 6500 were involved, how many were revelers?

3. Only one-seventh of those present sampled the cuisine. If 1995 people were present, how many did not sample the cuisine?

4. Five times a certain number was 8 less than 4 times the same number. What was the number?

5. What base 10 number does 111101 (base 2) represent?

6. What base 10 number does 1000011 (base 2) represent?

7. Write 96 (base 10) using base 2 numerals.

8. What number is 225 percent of 500? Draw a diagram of the problem.

9. If 90 is increased by 45 percent, what is the resulting number? Draw a diagram of the problem.

10. What percent of 70 is 133? Draw a diagram of the problem.

11. Find the volume in cubic inches of a right solid whose base is shown and whose height is 1 yard. Dimensions are in inches.

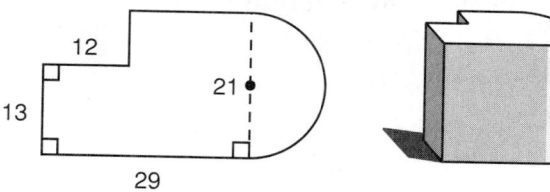

12. If $7x - 7 = 21$, what is the value of $\frac{3}{4}x + 3$?

13. If $9x + 6 = -12$, what is the value of $3x - \frac{1}{2}$?

Use the distributive property to multiply:

14. $2b(3b^2 + ab^2 + a^4)$
15. $4mn(m + n^2 - 4)$

Simplify by adding like terms:

16. $mn^2 + mnm + 7pap - 5n^2m + 3ab$

17. $p^3 + 8p^3b + 2b^3 + 2pbp^2 + 3aa^2b$

Solve:

18. $8x - 8 = 3x + 32$
19. $-16x - 5 = 2x + 31$

20. $-10x - 4 = 3x + 22$
21. $\dfrac{\frac{3}{5}}{1\frac{1}{4}} = \dfrac{\frac{1}{20}}{x}$

Simplify:

22. $(-3)^3 + 3^3$
23. $-[-(-4)] + (2)^2$

24. $5^3 - 2[1^5(-2^3 - 3^2) + 5(2^2 - 5)]$

25. $x^2n^2xn^3x^5$
26. $(-3)^{-3}$
27. $\dfrac{3}{3^{-3}}$

28. $\sqrt[5]{-243}$

29. Graph: $x \not> 4$

30. Evaluate: $a^b + b^a$ if $a = -2$ and $b = -3$

practice set 4

1. A single die is rolled twice. What is the probability of getting a 4 and then a 6?

2. The sales tax was $4\frac{1}{2}$ percent. If Mary purchased $90 worth of merchandise, how much tax did she pay?

3. The ratio of querulous to compliant was 4 to 3. If 30,000 were compliant, how many were querulous?

4. The product of a number and -14 increased by 13 is 7 less than the number times 6. What is the number?

5. Construct a triangle whose sides are 2 cm, 6 cm, and 7 cm.

6. Use a ruler to draw a line segment 9 centimeters long. Construct a perpendicular to the line at a point 3 centimeters from the right endpoint.

7. Use a protractor to draw a 63° angle. Then use a straightedge and a compass to construct the bisector of the angle.

8. What percent of 34 is 85? Draw a diagram of the problem.

9. If 130 is increased by $15\frac{1}{2}$ percent, what is the resulting number? Draw a diagram of the problem.

10. Use six unit multipliers to convert 10,000 cubic inches to cubic miles.

11. Use 4 unit multipliers to convert 300 cubic meters to liters.

12. Use six unit multipliers to convert 10,000 cubic kilometers to cubic centimeters.

13. Find side M:

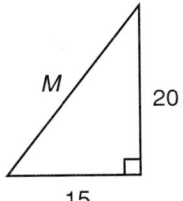

14. (a) Find side x:

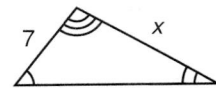

(b) Find angle z:

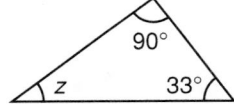

15. Find the area of each figure. Dimensions are in meters.

(a)

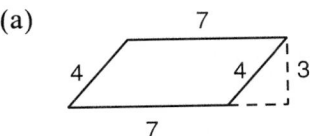

(b)

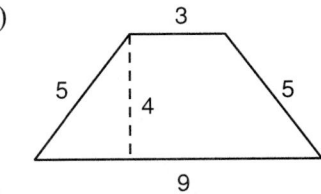

16. Find the volume in cubic centimeters of a right solid whose base is shown and whose height is 4 meters. Dimensions are in centimeters.

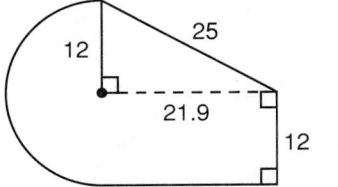

17. Find the lateral surface area in square centimeters for the figure in Problem 16.

18. Use the distributive property to multiply: $x^2y^3(xx + mx^2 + y^m)$

19. Simplify by adding like terms: $-mam + 5ap^3 - 2m^2a + 3pap^2$

Solve:

20. $14x - 6 + x = -3x + 30$

21. $-3\frac{1}{5}x - \frac{1}{12} = 5\frac{1}{3}$

22. $\dfrac{-\dfrac{1}{11}}{\dfrac{5}{2}} = \dfrac{x}{-\dfrac{1}{4}}$

Simplify:

23. (a) $\dfrac{1}{2^{-5}}$ (b) $(-4)^{-3}$

24. $-(-5^2) - 4(8 \cdot 2 - 3^2)$

25. $9x^5p^3c^4xx^2pc^3p^5$

26. Graph: $x \not\geq -9$

27. If $10x + 4 = 26$, what is the value of $\frac{1}{3}x + \frac{1}{8}$?

28. If $\frac{1}{4}x + 13 = 2$, what is the value of $x - \sqrt[5]{-32}$?

Evaluate:

29. $ab^3 + (-ab)^3$ if $a = -3$ and $b = -1$

30. $\sqrt[2]{p} + \sqrt[3]{z} + p^{-1}$ if $p = 121$ and $z = -64$

practice set 5

1. A fair coin is tossed 4 times. What is the probability that the tosses will result in *HTHT* in that order?

2. A single die is rolled twice. What is the probability of getting a 4 on both rolls?

3. It had never happened before, but this time the increase was 460 percent. If the total this time was 3640, what was the total last time?

4. The ratio of sapphires to rubies was 13 to 3. If 4000 gems had been discovered, how many were rubies?

5. Graph the points on a rectangular coordinate system:
(a) (5, 1) (b) (−3, 4) (c) (0, 2)

6. Construct a triangle whose sides are 2 cm, 6 cm, and 7 cm.

7. Use a ruler to draw a line segment 12 centimeters long. Construct a perpendicular to the line at a point 7 centimeters from the left endpoint.

8. Use a protractor to draw a 155° angle. Then use a straightedge and a compass to construct the bisector of the angle.

9. Draw a line segment and a point outside the line. Construct through the point a line that is perpendicular to the segment.

10. Add: 11000 (base 2) + 10000 (base 2)

11. What number is $11\frac{1}{2}$ percent of 1,000,000? Draw a diagram of the problem.

12. What percent of 40 is 176? Draw a diagram of the problem.

13. Thirty percent of what number is 474? Draw a diagram of the problem.

14. Convert 77,463,000 milliliters to liters.

15. Use six unit multipliers to convert 8 cubic miles to cubic inches.

16. Use the distributive property to multiply: $2mnp(n^2 + pm + 3m^2p + n^4)$

17. Simplify by adding like terms: $4x^4m^4 + 3x^2m^2x^2 + 3x^2m^3 - 5xm^3x$

18. Express in cubic inches the volume of a right solid whose base is shown and whose sides are 2 feet high. Dimensions are in inches.

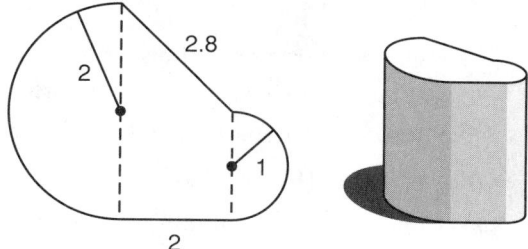

Solve:

19. $12x - 9 + 3x = 4x + 24$

20. $-3\dfrac{1}{6}x - \dfrac{1}{5} = 1\dfrac{7}{30}$

21. $\dfrac{-\dfrac{1}{8}}{\dfrac{3}{4}} = \dfrac{-\dfrac{1}{5}}{x}$

Simplify:

22.   (a) $\dfrac{1}{6^{-2}}$ (b) 3^{-4}

23. $(-1) - 3(3^3 - 2 \cdot 3)$

24. $5m^3nmn^2$

25.   Find side m:

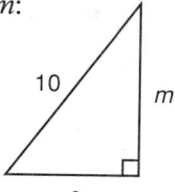

26.   Find side x:

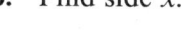 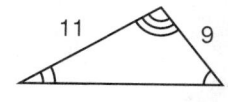

Evaluate:

27. $xy + 3x^2y^2$ if $x = -4$ and $y = 2$

28. $m^{-5}n - mn^2$ if $m = -2$ and $n = -1$

29.   If $14x + 21 = 7$, what is $\dfrac{1}{4}x - 3$?

30.   Find the least common multiple of 16, 15, and 21.

**practice set
6**

1. One die is red and the other die is green. Both are rolled. What is the probability of getting (a) a 4 and (b) a 7?

2. The first three cards dealt from a well-shuffled deck of cards are spades. What is the probability that the next card dealt will be a heart?

3. Sixty percent of the flowers in the show were purple. If 1600 flowers were not purple, how many were purple?

4. The ratio of endomorphs to ectomorphs among the ballplayers was 8 to 12. If there were 3300 ballplayers in all, how many were ectomorphs?

5. What is the difference between a numeral and a number?

6. Draw an 11-centimeter line segment with a ruler and construct the perpendicular bisector of the line segment.

7. Use a protractor to draw a 76° angle. Then use a straightedge and a compass to construct the bisector of the angle.

8. Use a protractor to draw a 35° angle. Then use a straightedge and a compass to copy the angle.

9. Add: 1001 (base 2) + 1111 (base 2) + 1001 (base 2) + 1000 (base 2)

10. Convert 101100 (base 2) to base 10.

11. (a) Find the volume of a right circular cone whose base has a radius of 2 inches and whose height is 6 inches.
     (b) Find the volume of a sphere whose radius is 2 inches.

12. (a) Find the volume of a pyramid 6 feet high if the base is a square whose sides are 2 feet long.
     (b) Find the surface area of a pyramid whose base is a 3 ft × 3 ft square and whose triangular faces have an altitude of 4 ft.

13. What percent of 30 is 78? Draw a diagram of the problem.

14. If 180 is increased by 70 percent, what is the resulting number? Draw a diagram of the problem.

15. Use 4 unit multipliers to convert 3000 cubic meters to liters.

16. Convert 20,430,264 milliliters to liters.

17. Use the distributive property to multiply: $mn(m + n - mn^2 + m^2n)$

18. Simplify by adding like terms: $3a^2b^5 + ab^4ab - ab^4a - 3a^2bb^3$

19. Express in cubic meters the volume of a right solid whose base is shown and whose sides are 3 meters high. Dimensions are in centimeters.

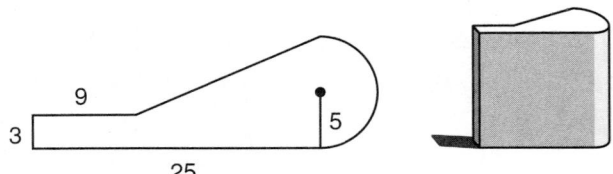

Solve:

20. $15x - 4 = -6x - 3x - 44$

21. $-\dfrac{2}{3}x - \dfrac{1}{9} = 1\dfrac{5}{18}$

22. $\dfrac{-\dfrac{1}{6}}{\dfrac{8}{5}} = \dfrac{-\dfrac{1}{6}}{x}$

Simplify:

23. $-[-(-1)^8] + (-2)$

24. $a^2 bab^2 b^2 b^5 a^4$

25. $\dfrac{-2^3 - 2^2(3 - 2^2) + 9}{3(2^3 - 4)}$

26. $\dfrac{1}{3}\left(\dfrac{1}{2} \cdot \dfrac{1}{8} - 1\dfrac{1}{4} \cdot \dfrac{1}{2}\right)$

Evaluate:

27. $xy + 2x^2y$ \quad if $x = -3$ and $y = 4$

28. Use the cut-and-try method to estimate $\sqrt{67}$ to one decimal place.

29. Write $0.008\dfrac{3}{4}$ as a decimal number.

30. Write 9374223.8 in words.

practice set 7

1. A fair coin is tossed 5 times. What is the probability that all 5 tosses will be tails?

2. A pair of dice is rolled. There are 36 possible combinations. What is the probability that the sum on the dice will equal 9? (Use simple probability.)

3. Laudatory behaviors increased 225 percent this year over last year. If 3328 behaved in a laudatory fashion this year, how many behaved in a laudatory fashion last year?

4. Among the colors, the ratio of puce to fuchsia was 11 to 7. If there was a total of 3798 colors, how many were puce?

5. Graph the points on a rectangular coordinate system:
(a) $(-4, -4)$ \quad (b) $(2, 2)$ \quad (c) $(0, -5)$

6. Construct a triangle whose sides are 5 cm, 9 cm, and 12 cm.

7. Use a ruler to draw a line segment 6 centimeters long. Construct a perpendicular to the line at a point 2 centimeters from the right endpoint.

8. Use a protractor to draw a 145° angle. Then use a straightedge and a compass to construct the bisector of the angle.

9. Draw a line segment and a point outside the line. Construct through the point a line that is perpendicular to the segment.

10. Convert 10101 (base 2) to base 10.

11. Find side m:

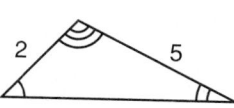
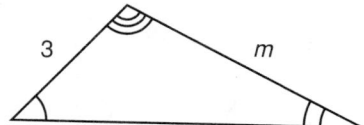

12. What number is $8\dfrac{1}{2}$ percent of 2200? Draw a diagram of the problem.

13. Seventy percent of what number is 1225? Draw a diagram of the problem.

14. Convert 100,000,000 milliliters to cubic meters.

15. Use six unit multipliers to convert 60 cubic millimeters to cubic meters.

16. Find the volume of the solid formed when this figure is folded along the dotted line. Dimensions are in meters.

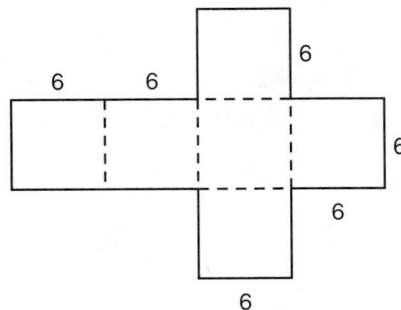

17. Sketch the figures and draw in the line of symmetry. Indicate which figure is symmetrical about a point.

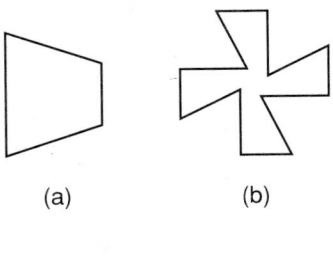

(a)   (b)

18. Use symmetry to find the area of this symmetrical figure. Dimensions are in feet.

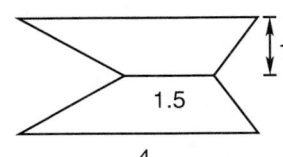

19. Find (a) angle z and (b) side b:

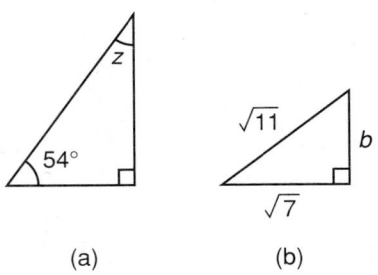

(a)   (b)

Solve:

20. $15x + 9 - 3x = 3x + 24$

21. $-5\frac{1}{3}x + \frac{1}{6} = -3\frac{5}{12}$

22. $\dfrac{-\frac{1}{8}}{\frac{2}{7}} = \dfrac{x}{-\frac{1}{4}}$

Simplify:

23. (a) $\dfrac{1}{10^{-2}}$ (b) 10^{-3}

24. $\dfrac{(-3)^3 - 3(-1^{13} - 3 \cdot 4)}{3(2^3 - \sqrt[7]{-128})}$

25. $6b^3 ab^2 a^5 b$

26. $\dfrac{1}{5}\left(2\frac{1}{3} \cdot \frac{1}{5} - \frac{1}{4} \cdot \frac{1}{5}\right)$

Evaluate:

27. $-3b^a + \dfrac{a^3}{b}$ if $a = 3$ and $b = -2$

28. $cx^2 - c^x x$ if $c = -4$ and $x = 2$

29. If $16x - 39 = -7$, what is the value of $\frac{5}{2}x + 20$?

30. If $9x + 20 = -7$, what is the value of $-\frac{2}{3}x - \frac{1}{3}$?

practice set 8

1. Eleven thousand dollars is deposited in an account which receives 11 percent interest compounded annually. How much money will be in the account in 4 years?

2. Two dice are rolled. What is the probability that the sum of their faces will be 3?

3. The ratio of calm to anxious is 6 to 11. If the calm and anxious total 306, how many are anxious?

4. Forty percent of the choir emoted. If 600 did not emote, how many emoted?

5. Graph the points on a rectangular coordinate system:
(a) $(-2, -3)$ (b) $(4, -1)$

6. Add in base 2 and then write the answer in base 10:

$$\begin{array}{r} 1110 \\ 1101 \\ 1000 \\ 1001 \\ + \ 1111 \\ \hline \end{array}$$

7. Eighty percent of what number is 720? Draw a diagram of the problem.

8. Three hundred thirty is what percent of 1100? Draw a diagram of the problem.

9. Use two unit multipliers to convert 100 square meters to square centimeters.

10. How many ways can 6 objects be arranged in a row?

Graph the line indicated by the equations:

11. $y = -\dfrac{1}{2}x + 2$

12. $y = -3x + 1$

13. Use the distributive property to multiply: $ab^3\left(a^3 + ab^2 - \dfrac{b}{a}\right)$

14. Simplify by adding like terms:

$$abc - 4bca + 2abca - 5caba$$

15. Find the volume in cubic feet of a right circular cylinder whose diameter is $\sqrt{5}$ inches and whose height is 1 foot. (Be careful, 1 ft = 12 in.)

16. Find the surface area in square meters of a cone whose slant height is 4 meters and whose radius is $\sqrt{2}$ meters.

17. Find angle z and side m:

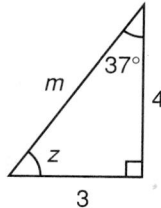

18. Find side m:

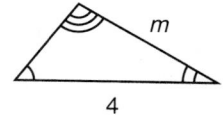
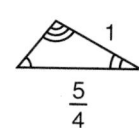

19. Construct a triangle whose sides measure 4 cm, 5 cm, and 8 cm.

20. Use a protractor to draw a 105° angle. Then use a straightedge and a compass to construct the bisector of the angle.

21. Find the volume in cubic inches of a right cone whose radius is $\sqrt{7}$ inches and whose height is 5 inches.

22. Find the volume in cubic centimeters of a right pyramid whose square base has sides 9 centimeters long and whose height is 10 centimeters.

Solve:

23. $4x - 6 + 2x = x + 5 - 2x - 4$

24. $-5x + 9 - x = 8x - 8 - 6x - 23$

25. $\dfrac{-\dfrac{1}{8}}{\dfrac{3}{5}} = \dfrac{\dfrac{1}{2}}{x}$

Simplify:

26. $\dfrac{1}{(-2)^{-5}}$ **27.** 8^{-3} **28.** $\sqrt[5]{-243}$

29. Evaluate: $2ab - \dfrac{1}{a^b}$ if $a = -3$ and $b = 2$

30. Graph: $x \not\geq -3$

practice set 9

1. The urn contained 44 marbles: 28 were black, 9 were red, and 7 were green. One marble was drawn and then put back. Then a second marble was drawn. What is the probability that both marbles were green?

2. If the product of a number and -11 is decreased by 9, the result is 5 greater than the product of the number and 3. What is the number?

3. Forty percent of those present were able to respond. If 1620 could not respond, how many could respond?

4. The ratio of shakes to shimmies was 6 to 5. If there were 4400 shimmies, what was the number of shakes?

5. Graph the points on a rectangular coordinate system:
(a) $(3, 2)$ (b) $(-1, 0)$ (c) $(2, -3)$

6. Construct a triangle whose sides are 3 cm, 6 cm, and 8 cm.

7. How many ways can 11 objects be arranged in a row?

8. How many ways can 5 objects be arranged in a row?

9. Draw a line segment and a point outside the line. Construct through the point a line that is perpendicular to the segment.

10. Use a protractor to draw an 83° angle. Then use a straightedge and a compass to construct the bisector of the angle.

11. Find (a) angle z and (b) side m: **12.** Find side m:

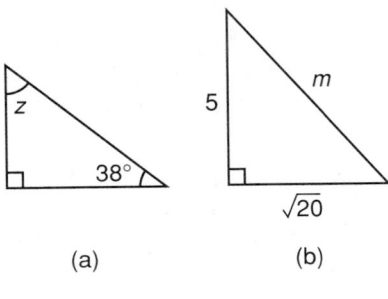

(a)   (b)

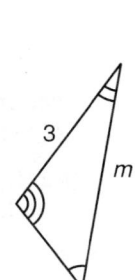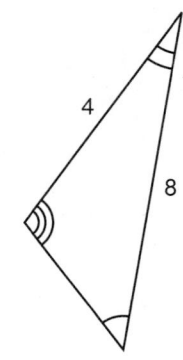

13. Sketch the figures and draw in the lines of symmetry. Which figure is symmetrical about a point?

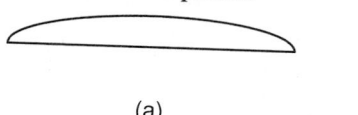

(a)

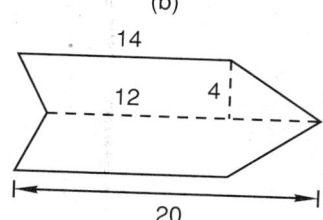

(b)

14. Use symmetry to find the area of the symmetrical figure. Dimensions are in inches.

15. What percent of 50 is 64? Draw a diagram of the problem.

16. If 240 is increased by 65 percent, what is the resulting number? Draw a diagram of the problem.

17. Eight and one-half percent of what number is 0.085? Draw a diagram of the problem.

18. Use six unit multipliers to convert 100,000 cubic inches to cubic miles.

19. Use three unit multipliers to convert 10 cubic meters to cubic centimeters.

20. Use the distributive property to multiply: $z^4p(z^3m + px - zp)$

21. Simplify by adding like terms: $z^4p - 5zpz^3 + z^2p + 5zpz$

Solve:

22. $14x - 6x - 6 = 18x + 14$

23. $\dfrac{-\dfrac{1}{10}}{\dfrac{7}{5}} = \dfrac{-\dfrac{11}{20}}{x}$

Simplify:

24. (a) $(-5)^{-3}$ (b) $\dfrac{1}{-1^{-5}}$

25. $\dfrac{(3)^3 - 3(2^3 - 2^2)}{3(3 \cdot \sqrt[3]{-64} - 4)}$

26. $6p^3n^5cz^2n^2z^4$

27. $\dfrac{1}{9}\left(2\dfrac{1}{3} \cdot \dfrac{2}{3} - \dfrac{1}{5} \cdot \dfrac{1}{6}\right)$

Evaluate:

28. x^yy^x if $x = -2$ and $y = -4$

29. $\sqrt[a]{b} + \sqrt{b} + a^{-2}$ if $a = 6$ and $b = 64$

30. Graph: $x \leq 1$

practice set 10

1. One card is pulled from a standard deck of 52 cards and then replaced. Then a second card is drawn. What is the probability of getting a black jack and then a black 9?

2. Two dice are rolled. What is the probability that the sum of their faces will be 6?

3. Eight thousand dollars is deposited in an account which receives 8 percent interest compounded annually. How much money will be in the account in 6 years?

4. The ratio of detritus to valuables was 2 to 9. If 80 pounds of detritus lay about, how many pounds of valuables were there?

5. Graph the points on a rectangular coordinate system:
   (a) $(-4, 0)$ (b) $(3, -1)$ (c) $(-1, -2)$

6. Construct a triangle whose sides are 6 cm, 7 cm, and 2 cm.

7. Use a protractor to draw a 107° angle. Then use a straightedge and a compass to construct the bisector of the angle.

8. Convert 1000010 (base 2) to base 10.

9. Add: 1110 (base 2) + 111101 (base 2)

10. What percent of 13.8 is 75.9? Draw a diagram of the problem.

11. Draw a line segment and a point outside the line. Construct through the point a line that is perpendicular to the segment.

12. Find (a) angle z and (b) side m:

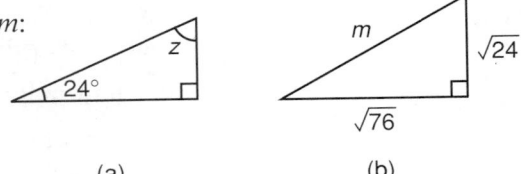

           (a)   (b)

13. Find side m:

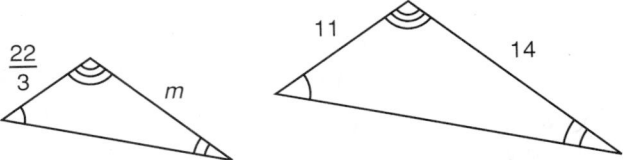

14. Graph the line indicated by the equation $y = -\frac{3}{2}x + 5$.

15. Twelve and one-half percent of what number is 3125? Draw a diagram of the problem.

16. What percent of 40 is 220? Draw a diagram of the problem.

17. Use four multipliers to convert 5234 milliliters to cubic meters.

18. Use six unit multipliers to convert 100,000,000 cubic kilometers to cubic centimeters.

19. Use the distributive property to multiply: $a^4b^2(a^2 + ab - a^2b)$

20. Simplify by adding like terms: $2mcm^2 - 3mcmm + 2m^2c^2 - 4ccm^2$

21. Find the volume in cubic meters of a right circular cylinder with a diameter of 4 meters and whose height is 3 meters.

22. Find the volume in cubic meters of a sphere whose radius is $\sqrt{3}$ meters.

23. Find the volume in cubic meters of a cone whose radius is $\sqrt{2}$ meters and whose height is 3 meters.

24. Find the volume in cubic meters of a pyramid whose square base has sides 4 meters long and whose height is 24 meters.

25. Find the surface area in square meters of the pyramid in Problem 24. (*Note*: the height of the triangular side is 24.1 meters.)

Solve:

26. $15x - 6 + 2x - 4 = 3x + 16 - 5x + 14$

27. $\dfrac{-\dfrac{1}{4}}{\dfrac{2}{7}} = \dfrac{x}{\dfrac{5}{21}}$

28. Simplify: $(-5)^{-3}$

29. Evaluate: $-a^{-2} + b^3$ if $a = -3$ and $b = -4$

30. Graph: $x < 0$

APPENDIX B Glossary

absolute value   The quality of a number that equals the distance of the graph of the number from the origin. Since the graphs of −3 and +3 are both 3 units from the origin, the absolute value of both numbers is +3.

acute angle   An angle whose degree measure is between 0° and 90°.

acute triangle   A triangle in which all three angles are acute angles.

addend   One of two or more numbers that are to be added to find a sum.

adjacent angles   Two angles that have a common side and a common vertex. The angles lie on opposite sides of their common side.

algebraic addition   The combining of positive and/or negative numbers to form a sum.

algebraic subtraction   The sum formed by changing the sign of the subtrahend in a subtraction problem and then adding.

algorithm   A particular process for solving a certain type of problem. Often the process is repetitive, as in the long division algorithm.

altitude of a triangle   The perpendicular distance from the base of a triangle to the opposite vertex. Also called the *height* of the triangle.

angle bisector   A ray from the vertex of an angle that divides the angle into two angles whose measures are equal.

angle   In geometry, the figure formed by two rays that have a common endpoint.

arc   A segment or a piece of a curve.

average   The sum of a group of numbers divided by the number of numbers in the group.

base   (1) A designated side (or a face) of a geometric figure. (2) The lower number in an exponential expression. In the exponential expression 2^5, the number 2 is the base and the number 5 is the exponent.

bisect   To divide or separate into two equal parts.

chord   A segment that connects two points on a circle.

circumference   The perimeter of a circle.

coefficient   A factor of an indicated product. In the product $4x$, 4 is the coefficient of x and x is the coefficient of 4.

common factors   Identical factors of two or more indicated products.

complementary angles   Two angles whose sum is 90°.

composite number   A counting number that is the product of two counting numbers, neither of which is the number 1.

concave polygon   A polygon in which at least one interior angle has a measure that is greater than 180°.

conditional equation   An equation whose truth or falsity depends on the number or numbers used to replace one or more variables.

congruent polygons   Two polygons in which the corresponding sides have equal lengths and the corresponding angles have equal measures.

convex polygon   A polygon in which all interior angles have a measure that is less than 180°.

coordinate   The number associated with a point on a number line.

corresponding sides   Sides of similar polygons that occupy corresponding positions. Corresponding sides are always opposite angles whose measures are equal.

counting numbers   Sometimes called the natural numbers, these numbers are 1, 2, 3, 4, 5….

curve   The path traced by a moving point.

decimal fraction   A decimal number.

decimal point   A dot placed in a decimal number to use as a place value reference point. The place to the left of the decimal point is always the units place.

denominate number   A combination of a number and a descriptor that designates units. Examples are 4 ft, 16 tons, 42 miles per hour.

denominator   The bottom number in a fraction.

diameter   A chord that passes through the center of a circle; the length of the chord.

difference   The answer to a subtraction problem.

digit   In the base 10 system, any of the symbols 0, 1, 2, 3, 4, 5, 6, 7, 8, or 9.

dividend   The number being divided. In the expression $10 \div 2$, the dividend is 10 and the divisor is 2.

divisible   If one whole number is divided by another whole number and the quotient is a whole number (the remainder is zero), we say that the first whole number is divisible by the second whole number.

divisor   One of the numbers in a division problem. In the expression $10 \div 2$, the divisor is 2 and the dividend is 10.

equidistant   The same distance. If points A and B are the same distance from point C, we say that A and B are equidistant from C.

equilateral triangle   A triangle whose sides all have the same length.

equivalent fractions   Fractions that have the same value.

expanded form   A way of writing a number as the sum of the products of the digits and the place values of the digits.

expanded fraction   A fraction of equal value whose denominator is greater than the denominator of the original fraction.

exponential expression   An expression that indicates that one number is to be used as a factor a given number of times. The expression 4^3 tells us that 4 is to used a factor 3 times. The value of 4^3 is 64.

exponential notation   A notation that uses an exponential expression to designate a number.

exponent   The upper number in an exponential expression. In the expression a^x, a is the base and x is the exponent.

factor   (1) Noun. One of two or more numbers that are to be multiplied. In the expression $4xy$, the factors are 4, x, and y. (2) Verb. To write as a product of factors. We can factor the number 6 by writing it as 2×3.

factorial   A counting number followed by an exclamation point. The value of a factorial expression is the product of the counting number and all lesser counting numbers. We write 6 factorial as 6!. The value of 6! is $6 \cdot 5 \cdot 4 \cdot 3 \cdot 2 \cdot 1 = 720$.

fraction line   The line segment that separates the numerator and the denominator of a fraction.

geometric solid   A three-dimensional geometric figure. Spheres, cones, and prisms are examples of geometric solids.

greater than   One number is said to be greater than a second number if the graph of the number on a number line is to the right of the graph of the second number.

height   *See* Altitude.

hypotenuse   The side of a right triangle that is opposite the right angle.

improper fraction   A fraction whose numerator is equal to or greater than the denominator.

independent events   Two events are said to be independent if the outcome of one event does not affect the probability of the other event. If a dime is tossed twice, the outcome (heads or tails) of the first toss does not affect the probability of heads or tails on the second toss.

inverse operation   Two operations are inverse operations if one operation will "undo" the other operation. If we begin with 3 and multiply by 2, the product is 6. If we divide 6 by 2, we will undo the multiplication by 2, and the answer will be 3, the original number.

invert   To turn upside down.

isosceles triangle   A triangle in which at least two sides have equal lengths.

lateral surface area   The word *lateral* comes from the Latin word *latus*, which means side. The lateral surface area of a geometric solid is the sum of the surface areas of its "sides."

least common denominator (LCD)   Of two or more fractions, a denominator that is the least common multiple of the denominators of the fractions.

least common multiple (LCM)   The smallest whole number that every member of a set of whole numbers will divide evenly.

less than   One number is less than a second number if the graph of the number on a number line is to the left of the graph of the second number.

like terms   Terms whose variable components have the same values regardless of the numbers used as replacements for the variables. The terms $14xy$ and $10yx$ are like terms because xy and yx have equal values.

line segment   A part of a line.

line   A straight curve.

liter   The basic unit of volume in the metric system.

lowest terms   In reference to a fraction, when the numerator and denominator contain no common factors.

mathematical point   A location on a line or in space. A mathematical point has no size. The dot we make to indicate the location of the mathematical point is called the graph of the point.

mean   Of a set of numbers, is the average of the set of numbers.

median   The middle number when a set of numbers is arranged in order from the least to the greatest.

midpoint   A point that is equidistant from two designated points.

milliliter   One-thousandth of a liter.

mixed number   A numerical expression composed of a whole number and a fraction.

mode   The number in a set of numbers that appears the most often.

multiple   A product of a selected counting number and any other counting number.

multiplicand   One of two numbers that are to be multiplied. A factor.

multiplier   One of two numbers that are to be multiplied. A factor.

numeral   Symbol or group of symbols used to represent a number.

numerator   The top number of a fraction.

numerical expression   A numeral.

obtuse angle   An angle whose measure is greater than 90° and less than 180°.

obtuse triangle   A triangle that contains an obtuse angle.

opposites   A positive number and a negative number whose absolute values are equal. The numbers −3 and +3 are a pair of opposite numbers.

origin   The point on a number line with which the number zero is associated.

parallelogram   A quadrilateral that has two pairs of parallel sides.

percent   Hundredth. Forty percent is forty hundredths.

perimeter   Of a flat geometric figure, the distance around the figure.

permutation   The number of different ways a set of objects can be arranged in a row.

perpendicular bisector   A line that is perpendicular to a given line segment at the midpoint of the segment.

planar figure   A figure drawn on a flat surface.

plane   A mathematical point has no size. A mathematical line has no ends. A mathematical plane is a "flat surface" that has no boundaries.

polygon   A closed, planar geometric figure whose sides are line segments.

power   The value of an exponential expression. The expression 2^4 is read as 2 to the fourth power and has a value of 16. Thus 16 is the fourth power of 2. The word *power* is also used to describe the exponent.

prime factors   The factors of a number that are prime numbers.

prime number   A whole number greater than 1 whose only whole number divisors are 1 and the number itself.

product   The result obtained when numbers are multiplied.

proper fraction   A fraction whose numerator is a smaller number than the denominator.

proportion   An equation that equates two ratios.

quadrant Any one of the four sectors of a rectangular coordinate system, which is formed by two perpendicular number lines that intersect at the origins of both number lines.

quotient The answer to a division problem.

radical expression An expression that contains radical signs, such as $\sqrt{x}$, $\sqrt[3]{16}$, $\sqrt[4]{xy}$, which indicate roots of a number.

radicand The number under the radical sign.

radius The distance from the center of a circle to a point on the circle.

rate (1) Speed. (2) The decimal equivalent of a percent.

ratio A comparison of two numbers.

ray A part of a line that begins at a point called the origin and continues without end.

reciprocal Of a fraction, the inverted form of the fraction. The reciprocal of $\frac{4}{3}$ is $\frac{3}{4}$.

rectangle A parallelogram that has four right angles.

regular polygon A polygon in which all sides have equal lengths and all angles have equal measures.

rhombus A parallelogram that has four sides whose lengths are equal.

right angle One of the angles formed at the intersection of two perpendicular lines. A right angle has a measure of 90°.

right solid A solid whose sides are perpendicular to both bases.

right triangle A triangle that contains a right angle.

Roman numerals Numerals used by the ancient Romans.

root The solution to an equation. Also the value of a radical expression.

rotation The turning of a figure about a point.

scale factor The number that relates corresponding sides of similar geometric figures.

scalene triangle A triangle that has no two sides whose lengths are equal.

scientific notation A method of writing a number as a product of a decimal number and a power of 10.

semicircle A half circle.

signed numbers Numbers that are either positive numbers or negative numbers.

similar triangles Two triangles that have the same angles.

straight angle An angle whose measure is 180°.

sum The answer we get when we add.

supplementary angles Two angles whose measures sum to 180°.

surface area The total outside area of a geometric solid.

translation The movement of a geometric figure without rotation. A "slide."

unit conversion The process of changing a denominate number to an equivalent denominate number that has different units.

unit multiplier A fraction of denominate numbers whose value is 1.

unity The state of being 1.

variable A letter used to represent a number that has not been designated.

vinculum A bar drawn over two or more algebraic terms. The vinculum in $43.\overline{256}$ means that the digits 256 repeat without end as 43.256256256256.... The vinculum in $\overline{XC}$ means to multiply by 1000, so this is the Roman numeral for 90,000.

volume The space occupied by a geometric solid.

whole number The numbers 0, 1, 2, 3, 4, ...

Answers to odd-numbered problems

practice   **a.** 2000   **c.** $(8 \times 10,000) + (5 \times 1000) + (2 \times 10)$

     e. Thirty-six million, twenty-five thousand, one hundred three

problem set 1

1. (a) 60,000   (b) 900   (c) 3   **3.** 3,666,766   **5.** 41,000,200,520

7. 407,000,090,742,072

9. Five hundred seventeen million, two hundred thirty-six thousand, four hundred twenty-eight

11. Thirty-two billion, six hundred fifty-two   **13.** Six million, forty thousand

15. 304,020   **17.** 9405   **19.** $(5 \times 1000) + (2 \times 100) + (8 \times 10)$

21. $(7 \times 10,000) + (6 \times 100)$ **23.** $(4 \times 1000) + (5 \times 1)$

25. 294   **27.** 22,840   **29.** 3530

practice   **a.** 914,470,000   **c.** Number line with points at 3, 11, 14, 19; scale marked at 0, 5, 10, 15, 20.

problem set 2

1. Number line with points at 2, 5, 6, 11, 19; scale marked at 0, 5, 10, 15, 20.

3. 249,294,924,942   **5.** 83,722,000

7. 777,727,757   **9.** 107,047,020

11. Seven hundred thirty-one million, two hundred eighty-four thousand, six

13. Nine billion, three million, one thousand, two hundred fifty-six

15. 70,654   **17.** 9609

19. $(6 \times 10,000) + (8 \times 1000) + (3 \times 100) + (1 \times 10) + (2 \times 1)$

21. 22,858   **23.** 219,258   **25.** 213,820   **27.** 1036   **29.** 1992

practice   **a.** 349   **c.** 399

problem set 3

1. 225,223   **3.** 4,144,444   **5.** 14,705,052

7. $(6 \times 10,000) + (4 \times 1000) + (3 \times 10)$

9. $(1 \times 100,000) + (2 \times 10,000) + (3 \times 1000) + (4 \times 100) + (1 \times 10) + (9 \times 1)$

11. 569   **13.** 1311   **15.** 364   **17.** 186   **19.** 6606

21. Seven hundred seven million, seventy thousand, seven hundred five

23. 2083   **25.** 2230   **27.** 255,945   **29.** 90,887

practice a. 37,500   c. 7980   e. 409

problem set 1. 7333   3. 47,000,014   5. 720,000,000

4 7. $(7 \times 1000) + (6 \times 100) + (5 \times 10)$ 9. 279   11. 1115 R42   13. 104 R5

15. 91,485   17. 163,840   19. 210,000   21. 669   23. 3182

25. Thirty-nine thousand, two   27. 2,026,416   29. 6781

practice a. 1176

problem set 1. 870   3. 175,000,000   5. 7333   7. 1328 R4   9. 1438 R17

5 11. 139,150   13. 963   15. 412   17. 469   19. 70,430   21. 47,000,014

23. 196, 619, 691, 916, 961

25. Seventy-five billion, four hundred million, seven hundred thousand, two hundred fifteen

27. 720,000,000   29. 1,913,323

practice a. 5000.0742   c. Seven thousand and sixty-five thousandths   e. 43.8742

problem set 1. 18,819,491   3. 773,757,797   5. 372,333   7. 1.2688   9. 14.0168

6 11. 157 R13   13. 811 R1   15. 62 R4   17. 5,144,000

19. Four thousand, one hundred sixty-five and one hundred sixty-two ten-thousandths

21. 63,000.0214   23. 288   25. 952   27. 79.2253   29. 94,579

practice a. 416,200   c. 734,260

problem set 1. 2806   3. 88,838,887   5. 31,621   7. 0.05426   9. 1.3992

7 11. 36.353   13. 12.067836   15. 392   17. 2100 R27   19. 91,600,000

21. 246, 264, 426, 624, 642   23. 702.00942

25. Nine million, fifty-six thousand, two hundred thirteen and fifty-seven thousand, three hundred twenty-eight hundred-millionths

27. 678   29. 4915.524

practice a. 416.04274   c. 2837.0652

problem set 1. 18,014,390 units   3. 413.6268   5. 9.31521   7. 0.0165164

8 9. 0.00200304   11. 388.086   13. 15.5   15. 72.3   17. 1003.21

19. 4526.76   21. 42.123457   23. 47,000,067,000.00417

25. Six thousand, one hundred eighty-four hundred-millionths

27. 0.0164, 0.0426, 0.0461, 0.0614   29. 350.6492

practice a. 74 cm

problem set 1. 15,237   3. 31.64215   5. 417.3652   7. 0.07584   9. 0.0185262

9 11. 1.642   13. 2.8529   15. 212 inches   17. 668   19. 5705.48

21. 61.3737378   23. 742,000,537.010948

25. One hundred twenty-eight thousand, six hundred forty-seven hundred-millionths

27. Fifty-one thousand, seven hundred eighty-six and seven hundred eighty-five hundred-thousandths

29. 11,556.876

practice a. 45,285; 305,961   c. 72,840

problem set 10

1. 1,081,307   **3.** 762,000,442.12792   **5.** 118 meters   **7.** 0.078848
9. 1255.42   **11.** 0.799   **13.** 1.365   **15.** 0.10   **17.** 383.70   **19.** 6392.54
21. 4300   **23.** 47.123   **25.** 0.0000143
27. Five million, one hundred sixty-two and eight ten-thousandths
29. 2986.47

practice   **a.** 1400   **c.** 4 spaces will have 18 kids; 4 spaces will have 19 kids

problem set 11

1. 750   **3.** 43.5369 inches   **5.** 112 inches   **7.** 13.2116   **9.** 0.123631
11. 36.4011   **13.** 49.41   **15.** 61.8088   **17.** 1.079

19. 6.74   **21.** 3.56   **23.** 223.09   **25.** 1,625,000,250,025.123
27. Two hundred twenty-three million, ninety-two thousand, eight hundred seventy
29. 7707.8

practice   **a.** $2 \times 2 \times 2 \times 3 \times 5$ **c.** $2 \times 2 \times 2 \times 3 \times 3 \times 5 \times 7$

problem set 12

1. 10,979.04098 inches   **3.** 148

5. (a) 302; 9172; 3132; 62,120   (b) 3132   (c) 625; 62,120   (d) 62,120
7. 112 feet   **9.** 36,821.1   **11.** 77   **13.** 5651.47
15. (a) 5×19 (b) $2 \times 2 \times 2 \times 2 \times 3 \times 3 \times 5$ (c) $2 \times 3 \times 3 \times 3 \times 53$
17. 0.33   **19.** 3.05   **21.** 251.03   **23.** 4700   **25.** 961,313,000,025
27. Sixteen and five hundred sixty-two ten-thousandths
29. 2898.29

practice   **a.** $22.68

problem set 13

1. 2,076,464   **3.** 13,942,000.000128

5. (a) 1020, 130, 1332, 132   (b) 1020, 125, 130, 185   (c) 1020, 1332, 132
(d) 1020, 130
7. 112 cm   **9.** 3,118,361.52   **11.** 5036.31
13. (a) $2 \times 2 \times 2 \times 3 \times 3 \times 5$ (b) $2 \times 2 \times 2 \times 2 \times 3 \times 3 \times 5$
(c) $2 \times 2 \times 2 \times 2 \times 2 \times 3 \times 3 \times 5$
15. 1.4816   **17.** 385.99   **19.** 19.36   **21.** 259.27   **23.** 1231.626
25. 321,617,212.231
27. Six hundred thirteen and one hundred sixty-two thousandths
29. 728.15

practice   **a.** $\frac{54}{18}$ **c.** $\frac{8}{18}$ **e.** $\frac{1}{2}$

problem set 14

1. 440   **3.** 0.00001197
5. (a) 120, 1620   (b) 120, 135, 1620   (c) 120, 122, 1332, 1620
(d) 120, 135, 1332, 1620
7. 150 km

9. (a) $\frac{10}{20}$ (b) $\frac{4}{20}$ (c) $\frac{140}{20}$ (d) $\frac{5}{20}$ **11.** 11,361.21   **13.** 144.64

15. (a) $2 \times 2 \times 2 \times 3 \times 3 \times 5 \times 5$ (b) $2 \times 2 \times 3 \times 3 \times 5 \times 5$
(c) $2 \times 3 \times 3 \times 5 \times 5$
17. 175.482   **19.** 184.99   **21.** 36,544.8   **23.** 509.34   **25.** 20,100,000,000

27. Eleven thousand, one hundred twenty-three and one hundred twenty-one thousandths

29. 2121.2516

practice   **a.** 4.613614   **c.** $0.\overline{3}$

problem set 15

1. 14,907,987   **3.** 189,903   **5.** (a) 2133, 312, 630   (b) 212, 312, 610, 630

7. 168 ft   **9.** 0.05   **11.** (a) $\frac{7}{6}$ (b) $\frac{2}{3}$ (c) $\frac{2}{3}$ **13.** 12,361.311   **15.** 70.848

17. 999.15   **19.** 424.51   **21.** 272.34   **23.** 2.51   **25.** 4017.336336

27. One million, eight hundred seventy-six thousand, two hundred eleven and thirty-two hundredths

29. 7842.0

practice   **a.** $\frac{476,325}{100,000}$

problem set 16

1. 272   **3.** 79,027   **5.** (a) 650, 625, 15, 20, 30   (b) 650, 20, 30

7. 162 yd   **9.** 0.06   **11.** $\frac{6}{7}$ **13.** $\frac{81}{243}$ **15.** 686.56   **17.** 167,318.38

19. 193.41   **21.** 151.08   **23.** $2 \times 2 \times 2 \times 3 \times 3 \times 3 \times 5$ **25.** 87,621.321789

27. One hundred seventy-two and three hundred twelve thousandths

29. $\frac{17}{20}$

practice   **a.** 12 ft^2 **c.** 84 ft^2

problem set 17

1. 982   **3.** 103,173

5. 238   **7.** 78 cm^2 **9.** 0.18   **11.** 0.92

13. $2 \times 3 \times 5$ **15.** $2 \times 2 \times 3 \times 3 \times 5 \times 7$ **17.** $\frac{3}{4}$ **19.** 538.5141

21. 62,538.76   **23.** 121.8979   **25.** 90.52   **27.** 2.0707071

29. 0.0091, 0.090109, 0.091, 0.3

practice   **a.** $\frac{14}{15}$

problem set 18

1. 71,266   **3.** 44,102,079   **5.** 29, 31, 37, 41, 43, 47   **7.** 570 cm^2

9. $\frac{12}{25}$ **11.** $\frac{5}{7}$ **13.** 0.94   **15.** 0.85   **17.** $\frac{84}{90}$ **19.** 8826.43   **21.** 181.2783

23. 3.18   **25.** 9.86960   **27.** 7562.47   **29.** $\frac{19}{200}$

practice   **a.** $\frac{1}{9}$ **c.** $\frac{2}{5}$

problem set 19

1. Charles's guess: 975.2157; Mary's guess: 975.0137. Mary's guess was closer.

3. 13,292   **5.** 37   **7.** $\frac{5}{12}$ **9.** $\frac{27}{56}$ **11.** 700 yd^2 **13.** $\frac{2}{3}$ **15.** $\frac{16}{45}$

17. 0.29   **19.** $2 \times 2 \times 2 \times 2 \times 2 \times 3 \times 3 \times 5 \times 7$ **21.** 79.488   **23.** 593.448

25. 19.78   **27.** Nine million, six hundred ninety-nine thousand, six hundred ninety

29. 545.6

practice   **a.** 7, 14, 21, 28, 35, 42, 49

problem set 20

1. 36,825   **3.** 114,012   **5.** 7, 14, 21, 28, 35, 42, 49, 56, 63   **7.** $\frac{5}{24}$

9. $\frac{1}{4}$ **11.** 484 m²   **13.** $\frac{3}{4}$ **15.** $\frac{3}{5}$ **17.** 0.46   **19.** 0.0049, 0.0096, 0.04, 0.1

21. 109,670.4   **23.** 8.4041   **25.** 3.14

27. One hundred eleven million, five hundred forty-six thousand, four hundred thirty-five

29. 197.49631

practice   **a.** $89.\overline{3}$ **c.** $7.63

problem set 21

1. 1,527,474,973.0173   **3.** 1713   **5.** 31, 37, 41, 43, 47   **7.** $1357.55

9. $\frac{3}{8}$ **11.** $\frac{1}{5}$ **13.** 380 cm²   **15.** $\frac{5}{6}$ **17.** 0.65

19. $2 \times 2 \times 2 \times 3 \times 3 \times 5 \times 7$ **21.** 112.179   **23.** 121.89   **25.** 7.18

27. 34.7182

29. 0.0098762

practice   **a.** $\frac{2}{35}$ **c.** $\frac{1}{3}$

problem set 22

1. 3,576,999.998662   **3.** 52.4 seconds   **5.** $1189.\overline{3}$ **7.** $\frac{21}{44}$ **9.** 4

11. 357 ft²   **13.** $2 \times 2 \times 2 \times 2 \times 3 \times 3 \times 5 \times 7$ **15.** 0.27   **17.** $\frac{3}{8}$ **19.** $\frac{6}{7}$

21. 60.547   **23.** 6852.449   **25.** 3.77   **27.** 0.30   **29.** 123,713.6

practice   **a.** $137\frac{1}{3}$ yd   **c.** 4944 inches

problem set 23

1. 27,049.4995 cm   **3.** 79   **5.** (a) 61, 67   (b) 63   **7.** 10 ft

9. 504 inches   **11.** $1\frac{1}{5}$ **13.** 170 yd   **15.** $2 \times 2 \times 2 \times 2 \times 2 \times 2 \times 3 \times 3 \times 5$

17. 0.68   **19.** $\frac{45}{64}$ **21.** $\frac{1}{4}$ **23.** 691.04   **25.** 61,059.377   **27.** 23,937.67

29. Sixty-seven million, two hundred eleven thousand, three hundred sixty-one and seventy-two hundredths

practice   **a.** 5800 cm   **c.** 480 m

problem set 24

1. 6065   **3.** 43   **5.** (a) 23, 29   (b) 24, 28   **7.** 8.59 m   **9.** 17 ft

11. $\frac{2}{3}$ **13.** 1   **15.** 520 in.²   **17.** $2 \times 2 \times 2 \times 2 \times 2 \times 3 \times 5 \times 5$

19. 0.24   **21.** $\frac{1}{3}$ **23.** 7465.7   **25.** 12.829   **27.** (a) $\frac{9}{42}$ (b) $\frac{24}{42}$ (c) $\frac{11}{42}$

29. 3,817,300

practice   **a.** 56 cm²

problem set 25

1. 54,285   **3.** 3284   **5.** (a) 41, 43, 47   (b) 42, 45, 48   **7.** 4631 cm

9. 0.416 km   **11.** 1   **13.** 128 ft   **15.** $2 \times 2 \times 2 \times 2 \times 2 \times 3 \times 3$

17. 0.16   **19.** $\frac{13}{24}$ **21.** $\frac{25}{72}$ **23.** 17.3536   **25.** 11.31   **27.** 3403.5

29. One hundred eleven million, three hundred twenty-one thousand, six hundred fifty-four and seven tenths

practice   **a.** 490 ft

problem set 26

1. 4020 pounds   **3.** 9.625   **5.** 1525   **7.** 1.3615 m   **9.** 1.899 km

11. 1   **13.** 575 m^2 **15.** $2 \times 2 \times 3 \times 3 \times 5 \times 5 \times 7$ **17.** 0.54   **19.** $\frac{6}{7}$

21. $\frac{9}{16}$ **23.** 1.8931   **25.** 711.999   **27.** 181.15

29. Six million, two hundred eleven thousand, three hundred fifty-seven and five tenths

practice   **a.** 6 ft^2 **c.** 40 ft^2

problem set 27

1. 2095   **3.** 296,002   **5.** 1229   **7.** 243 ft   **9.** 489,900 cm   **11.** 2

13. 140 ft   **15.** 400 ft^2 **17.** $2 \times 2 \times 3 \times 3 \times 3 \times 3 \times 3 \times 3$ **19.** 0.54

21. $\frac{3}{5}$ **23.** 1713.88   **25.** 8.50311092   **27.** 15,283   **29.** 781.10564

practice   **a.** $\frac{17}{3}$ **c.** $4\frac{2}{3}$

problem set 28

1. 0.03753   **3.** Thirty-three million, seven hundred forty-five thousand, twenty-six

5. $1\frac{8}{9}$ **7.**   **9.** $\frac{60}{7}$ **11.** 1.9272 m

13. $\frac{1}{6}$ **15.** $1\frac{1}{3}$ **17.** 1000 cm   **19.** 0.70   **21.** $\frac{6}{7}$ **23.** $\frac{5}{7}$ **25.** 116.6508

27. 260,376.67

29. Seventy-eight million, two hundred fifty-six thousand, one hundred thirteen and seven tenths

practice   **a.** 40   **c.** $\frac{15}{7} = 2\frac{1}{7}$

problem set 29

1. 40,000.001078   **3.** 38   **5.** 4300   **7.** 30   **9.** $4\frac{1}{5}$ **11.** $\frac{59}{8}$ **13.** $\frac{62}{11}$

15. 95,040 ft   **17.** $\frac{2}{15}$ **19.** $2 \times 2 \times 3 \times 5 \times 5 \times 7$ **21.** 0.70   **23.** $\frac{61}{90}$

25. 0.0375   **27.** 52.593   **29.** 2831.82113

practice   **a.** 30,000

problem set 30

1. 200   **3.** 21   **5.** 46,710   **7.** 42   **9.** $4\frac{2}{5}$

11. $\frac{23}{3}$ **13.** $\frac{37}{8}$ **15.** 192.62 m   **17.** $\frac{2}{3}$ **19.** 9.6 m   **21.** $\frac{16}{45}$ **23.** 0.53

25. $2 \times 3 \times 5 \times 5 \times 11$ **27.** 45.5275   **29.** 137,840

practice   **a.** 1260

problem set 31

1. $50   **3.** 17 hr   **5.** (a) $\dfrac{20\ \text{kursh}}{1\ \text{riyal}}, \dfrac{1\ \text{riyal}}{20\ \text{kursh}}$ (b) 800 riyal   **7.** 121

9. $3\frac{1}{4}$ **11.** 98,700   **13.** 3400   **15.** 147,800 m   **17.** $\frac{2}{3}$ **19.** 4,400,000 cm^2

21. $\frac{7}{8}$ **23.** 0.83   **25.** 1.25154   **27.** 56.779   **29.** 415.629

practice   a. $\frac{7}{20}$ c. $\frac{3}{8}$

problem set 32

1. $28.09   **3.** 1027   **5.** 68,969   **7.** 1100   **9.** $5\frac{6}{7}$ **11.** 1800

13. 360   **15.** $\frac{2}{35}$ **17.** $\frac{4}{5}$ **19.** 0.0024081   **21.** 481.492

23. 425 ft²   **25.** $\frac{52}{61}$ **27.** 0.83   **29.** 42,062,918,000

practice   a. 16

problem set 33

1. 6475   **3.** 94   **5.** $91.60   **7.** 140   **9.** $5\frac{5}{7}$ **11.** 1080   **13.** $\frac{39}{40}$

15. 29   **17.** 2   **19.** 0.0000368   **21.** $\frac{10}{33}$ **23.** 450 ft²

25. $2 \times 2 \times 2 \times 2 \times 2 \times 2 \times 5 \times 13$ **27.** 4.25   **29.** 99,540,000

practice   a. 16

problem set 34

1. 311   **3.** 150   **5.** 17   **7.** 40   **9.** $84\frac{1}{5}$ **11.** 840   **13.** $\frac{15}{16}$

15. 16   **17.** 14,300   **19.** 1730.99

21. (a) $\dfrac{40 \text{ gallons}}{1 \text{ barrel}}, \dfrac{1 \text{ barrel}}{40 \text{ gallons}}$ (b) 100,000 gallons   **23.** 90   **25.** 345 ft²

27. 0.6   **29.** 41,060,000

problem set 35

1. The second measurement was larger by 0.0033   **3.** $40,000   **5.** 402 lb

7. 20   **9.** $23\frac{1}{2}$ **11.** $1\frac{9}{16}$ **13.** 43   **15.** 0.108378   **17.** 513.0011

19. 3   **21.** 4   **23.** $\frac{100}{9}$ yard $= 11\frac{1}{9}$ yard   **25.** 2,027,520 inches   **27.** 150 cm

29. 5.55556

practice   a. $5\frac{3}{10}$ c. 8 yards per second, $\frac{1}{8}$ second per yard

problem set 36

1. 214 pounds   **3.** $997.80   **5.** 5 skins per lira, $\frac{1}{5}$ lira per skin   **7.** 160

9. $7\frac{9}{11}$ **11.** $\frac{8}{15}$ **13.** 15   **15.** $5\frac{3}{8}$ **17.** $905\frac{2}{5}$ **19.** 89,525   **21.** $2\frac{1}{3}$

23. 24   **25.** 1.876258 km   **27.** 114 cm   **29.** 659,000,000

practice   a. $2\frac{8}{15}$

problem set 37

1. 18 dollars per item, $\frac{1}{18}$ item per dollar   **3.** $10,000,000   **5.** 10   **7.** $6\frac{1}{5}$

9. 2520   **11.** $\frac{5}{8}$ **13.** $10\frac{7}{8}$ **15.** $751\frac{3}{5}$ **17.** $517\frac{11}{16}$ **19.** 575.2782

21. 674.056   **23.** 19   **25.** 633,600 inches   **27.** 112 inches   **29.** 1.052222

practice   a. $\dfrac{40 \text{ cents}}{2 \text{ ounces}}, \dfrac{2 \text{ ounces}}{40 \text{ cents}}$, 6 ounces

problem set 38   **1.** $\dfrac{28 \text{ dollars}}{7 \text{ hours}}, \dfrac{7 \text{ hours}}{28 \text{ dollars}}$, \$160   **3.** 6993 tons   **5.** 10   **7.** $25\dfrac{3}{17}$

9. 28,200   **11.** $1\dfrac{19}{40}$ **13.** 21   **15.** $519\dfrac{1}{10}$ **17.** $491\dfrac{3}{4}$ **19.** 272.4

21. $1\dfrac{1}{2}$ **23.** 116,159,480,000   **25.** 4   **27.** 725 cm²   **29.** 0.74

practice   a. 6   c. 20

problem set 39   **1.** 16   **3.** 5.5 m   **5.** 24   **7.** $24\dfrac{4}{13}$ **9.** 72   **11.** $7\dfrac{9}{10}$ **13.** 29

15. $682\dfrac{5}{8}$ **17.** $193\dfrac{11}{20}$ **19.** 101.7   **21.** $1\dfrac{1}{2}$ **23.** 1   **25.** 9

27. 0.00625611 km   **29.** 0.87

practice   a. 2   c. $\dfrac{1}{4}$

problem set 40   **1.** 473   **3.** 5040   **5.** 15   **7.** $13\dfrac{2}{3}$ **9.** 3780   **11.** $5\dfrac{1}{3}$ **13.** 9

15. $660\dfrac{7}{10}$ **17.** $395\dfrac{7}{8}$ **19.** 28,600   **21.** 4   **23.** 17   **25.** $\dfrac{11}{14}$ **27.** 34

29. (a) 41,43   (b) 42

practice   a. $\dfrac{77}{5}$ c. $\dfrac{3}{8}$ e. 15

problem set 41   **1.** 2.6 inches   **3.** 60   **5.** 9   **7.** $16\dfrac{2}{5}$ **9.** 840   **11.** $1\dfrac{1}{3}$ **13.** 60

15. 35   **17.** 19   **19.** 286   **21.** 6601.91   **23.** 7296   **25.** 44   **27.** 291 m²

29. 0.79

practice   a. \$64.00

problem set 42   **1.** 600 units per hour   **3.** \$37,000   **5.** 40   **7.** 1400   **9.** $24\dfrac{1}{12}$ **11.** $\dfrac{2}{3}$

13. 36   **15.** 78   **17.** 34   **19.** 178   **21.** 9232.51   **23.** 0.73962   **25.** 4

27. 1135 ft²   **29.** 190,000,000

practice   a. 3   c. $\dfrac{500}{(12)(12)}$ ft²

problem set 43   **1.** Second guess, 0.066268   **3.** $\dfrac{40 \text{ dollars}}{1 \text{ bunch}}, \dfrac{1 \text{ bunch}}{40 \text{ dollars}}$, \$4000   **5.** 200

7. 52   **9.** $18\dfrac{1}{15}$ **11.** $4\dfrac{1}{2}$ **13.** 300   **15.** $\dfrac{3}{8}$ **17.** $46\dfrac{2}{15}$ **19.** 46

21. 2172.93   **23.** 1   **25.** 54   **27.** 2   **29.** 140 cm

practice   a. $\dfrac{34}{9}$ c. $\dfrac{65}{44}$

problem set 44   **1.** $\dfrac{78 \text{ pots}}{312 \text{ dollars}}, \dfrac{312 \text{ dollars}}{78 \text{ pots}}$, \$1600   **3.** 7,134,108   **5.** 40   **7.** $\dfrac{117}{16}$

9. $\dfrac{32}{7}$ **11.** 588   **13.** $\dfrac{3}{22}$ **15.** 10   **17.** $\dfrac{24}{5}$ **19.** $\dfrac{7}{4}$ **21.** $4\dfrac{5}{6}$

23. 0.02103   **25.** 34   **27.** 1127.5 ft^2 **29.** (a) 41,43,47   (b) 40,45,50

practice   **a.** 700   **c.** $\frac{10}{9}$

problem set 45   **1.** 176   **3.** $\frac{1\text{ avocado}}{79\text{ cents}}, \frac{79\text{ cents}}{1\text{ avocado}}$, 30 avocados   **5.** 112   **7.** 96   **9.** 18

11. $\frac{41}{100}$ **13.** 59   **15.** $\frac{91}{15}$ **17.** $\frac{49}{69}$ **19.** $\frac{25}{16}$ **21.** 6.579   **23.** 16,105,000

25. 1,700,000 cm   **27.** 8   **29.** 3,345,408,000 ft^2

practice   **a.** 64   **c.** 5   **e.** 9

problem set 46   **1.** $\frac{4\text{ large ones}}{40\text{ dollars}}, \frac{40\text{ dollars}}{4\text{ large ones}}$, \$1200   **3.** 29.6 min   **5.** 350   **7.** $\frac{17}{5}$

9. 12   **11.** 28   **13.** 315   **15.** 5   **17.** $7\frac{5}{6}$ **19.** 74   **21.** 6834.03

23. $\frac{361}{36}$ **25.** $\frac{46}{41}$ **27.** 1.04 m   **29.** $10 \times 10 \times 10 \times 10 \times 10 \times 10 \times 10 \times 10 \times 10$

practice   **a.** 240 cubes   **c.** 70 m^3

problem set 47   **1.** 132   **3.** 420   **5.** 84 in.^{3} **7.** 1400   **9.** 144   **11.** 6   **13.** 47

15. 34   **17.** $476\frac{5}{8}$ **19.** 1.7524   **21.** 63,570   **23.** 12   **25.** $\frac{152}{115}$

27. 589 ft^2 **29.** 621.727

practice   **a.** $\frac{3}{14}$

problem set 48   **1.** 40   **3.** $\frac{14\text{ games}}{70\text{ dollars}}, \frac{70\text{ dollars}}{14\text{ games}}$, 40 games

5. $\frac{164}{3}$ seconds   **7.** $12\frac{16}{25}$ **9.** 21   **11.** 45   **13.** 12   **15.** $\frac{46}{35}$ **17.** 5

19. $105\frac{3}{10}$ **21.** 0.844   **23.** $\frac{38}{7}$ **25.** $2\frac{30}{47}$ **27.** 11,800 cm   **29.** 12,960 inches

practice   **a.** 64   **c.** 3

problem set 49   **1.** $\frac{40\text{ pecks}}{640\text{ dollars}}, \frac{640\text{ dollars}}{40\text{ pecks}}$, \$1600   **3.** 62,640   **5.** 600 cubes   **7.** $26\frac{5}{8}$

9. 16   **11.** $2\frac{7}{40}$ **13.** 27   **15.** $\frac{1}{16}$ **17.** $7\frac{3}{16}$ **19.** 2.5296   **21.** 65,455

23. $\frac{81}{95}$ **25.** 243   **27.** 604.5 m^2

29. Seven thousand, one hundred sixty-four and three thousand, one hundred eighty-six millionths

practice   **a.** $\frac{7}{1}$ **c.** 10

problem set 50   **1.** 1600   **3.** $\frac{40\text{ good ones}}{12\text{ dollars}}, \frac{12\text{ dollars}}{40\text{ good ones}}$, 11   **5.** $\frac{49}{15}$ **7.** 72 cubes

9. $\frac{213}{31}$ **11.** 18   **13.** 32   **15.** $\frac{3}{4}$ **17.** $121\frac{9}{14}$ **19.** 8   **21.** 202.097

23. $1\frac{21}{115}$ **25.** 9   **27.** 6000 ft^3

29. Seven hundred and five hundred sixty-three ten-millionths

practice   **a.** 40 ft^2

problem set 51

1. 85,251   **3.** 250 tons   **5.** 1350 m^2 **7.** 105   **9.** 72   **11.** 24

13. $\frac{29}{10}$ **15.** 46   **17.** $\frac{11}{12}$ **19.** $3\frac{19}{30}$ **21.** 0.26912   **23.** 541.67   **25.** $2\frac{12}{17}$

27. 8   **29.** Forty-one thousand and two ten-millionths

practice   **a.** 13

problem set 52

1. 5643   **3.** $\frac{15}{315\ \text{dollars}}, \frac{315\ \text{dollars}}{15}$, $2940

5. 276 ft^3 **7.** 378   **9.** $\frac{7}{8}$ **11.** $8\frac{11}{15}$ **13.** $2\frac{5}{6}$ **15.** 27   **17.** $4\frac{3}{4}$

19. $3\frac{7}{20}$ **21.** $22\frac{9}{14}$ **23.** 44.079   **25.** $\frac{308}{51}$ **27.** 35   **29.** 0.0216218 km

practice   **a.** 4.76×10^5 **c.** 3.056×10^5 **e.** 406,000

problem set 53

1. 40   **3.** $\frac{1900\ \text{oscillators}}{38,000\ \text{dollars}}, \frac{38,000\ \text{dollars}}{1900\ \text{oscillators}}$, $100,000

5. (a) 4.7×10^7 (b) 4.7×10^{-7} **7.** 4200 yd^3 **9.** 180   **11.** 39

13. 96   **15.** 15   **17.** $\frac{5}{4}$ **19.** 14   **21.** $11\frac{1}{2}$ **23.** 1097.388   **25.** $\frac{203}{48}$

27. 72   **29.** 810,000 in.^{2}

practice   **a.** 0.2   **c.** 0.42

problem set 54

1. 155,982,014   **3.** 7700   **5.** 120   **7.** 0.1 m^2 **9.** 516 m^2 **11.** $\frac{3}{8}$

13. $9\frac{1}{20}$ **15.** $\frac{25}{6}$ **17.** 20   **19.** $\frac{11}{12}$ **21.** 0   **23.** 0.23582   **25.** 2797.639

27. $\frac{34}{45}$ **29.** 5

practice   **a.** $\frac{3}{5}$

problem set 55

1. $\frac{70}{3500\ \text{dollars}}, \frac{3500\ \text{dollars}}{70}$, $36,000   **3.** 698,042   **5.** 120

7. (a) 3.87×10^{-4} (b) 869,000,000,000   **9.** 5200 cm^3 **11.** 84   **13.** $\frac{182}{23}$

15. $\frac{23}{14}$ **17.** 18   **19.** $\frac{19}{9}$ **21.** 2.5649   **23.** 3840.6   **25.** $\frac{103}{168}$ **27.** $\frac{1144}{135}$

29. 10

practice   **a.** 0.64   **c.** 0.05

problem set 56

1. (a) $\frac{780\ \text{dollars}}{10\ \text{tires}}$, $4524   (b) $\frac{10\ \text{tires}}{780\ \text{dollars}}$, 2 tires   **3.** 1052 lb

5. (a) 62   (b) 1488   **7.** 80   **9.** $\frac{15}{19}$ **11.** 44   **13.** 360   **15.** 30

17. 66   **19.** $1\frac{3}{5}$ **21.** 1.5924   **23.** 45,020   **25.** $\frac{1}{6}$ **27.** $\frac{39}{8}$ **29.** 26

practice   **a.** $\frac{8}{7}$ **c.** 16

problem set 57

1. $\frac{80 \text{ trees}}{3520 \text{ dollars}}, \frac{3520 \text{ dollars}}{80 \text{ trees}}$, \$440   **3.** 1072   **5.** (a) 36   (b) 576

7. $\frac{625}{9}$ ft²   **9.** $\frac{4}{5}$ **11.** 1040 ft³   **13.** $27\frac{3}{7}$ **15.** $\frac{10}{3}$ **17.** $\frac{27}{28}$ **19.** 28

21. $4\frac{9}{40}$ **23.** 3.0736   **25.** 86,060   **27.** $\frac{22}{15}$ **29.** (a) 13   (b) 3.92×10^{-2}

practice   **a.** 0.125   **c.** $\frac{11}{50}$

problem set 58

1. 27,000.00666   **3.** $\frac{53 \text{ new ones}}{742 \text{ dollars}}, \frac{742 \text{ dollars}}{53 \text{ new ones}}$, \$350

5. (a) $\frac{1}{4}$ (b) 0.25   (c) 0.5   (d) 50   **7.** 60   **9.** 0.65   **11.** 180

13. $\frac{15}{16}$ **15.** $\frac{108}{5}$ **17.** 30   **19.** $\frac{37}{12}$ **21.** 17   **23.** 1249.829   **25.** $\frac{1}{10}$

27. $\frac{4}{7}$ **29.** 25

practice   **a.** $\frac{34}{9}$

problem set 59

1. 21   **3.** 7245 pounds   **5.** (a) $\frac{4}{25}$ (b) 0.16   (c) 0.125   (d) 12.5

7. 80   **9.** 0.7   **11.** $\frac{1}{12}$ **13.** 360   **15.** $\frac{8}{3}$ **17.** $\frac{15}{4}$ **19.** 47   **21.** $\frac{20}{3}$

23. 1.2012   **25.** $20,383.\overline{6}$ **27.** $\frac{29}{10}$ **29.** $\frac{675}{22}$

practice   **a.** $\frac{175}{8}$ **c.** $\frac{395}{136}$

problem set 60

1. $\frac{560 \text{ red ones}}{7 \text{ pesos}}, \frac{7 \text{ pesos}}{560 \text{ red ones}}$, 4800   **3.** $2\frac{\text{mi}}{\text{hr}}$

5. (a) $\frac{6}{25}$ (b) 24   (c) 0.6   (d) 60   **7.** 700   **9.** $0.58\overline{3}$ **11.** $\frac{21}{4}$

13. 144   **15.** $\frac{5}{2}$ **17.** $\frac{25}{14}$ **19.** 48   **21.** $\frac{85}{12}$ **23.** 11.1924   **25.** 2602.14

27. $\frac{5}{48}$ **29.** $\frac{38}{3}$

practice   **a.** 40 miles per hour, 10 hours

problem set 61

1. $\frac{230 \text{ feet}}{1 \text{ second}}, \frac{1 \text{ second}}{230 \text{ feet}}, \frac{10,000}{23}$ seconds

3. 20 hr   **5.** (a) $\frac{11}{50}$ (b) 22   (c) 0.84   (d) 84

7. 120   **9.** 0.83   **11.** $\frac{147}{40}$ **13.** 152.8 in.²   **15.** $\frac{24}{5}$ **17.** $\frac{203}{12}$ **19.** 68

21. $\frac{27}{4}$ **23.** 7.9443   **25.** 37,522   **27.** $\frac{7}{60}$ **29.** $\frac{5}{3}$

practice   a. $\frac{1}{30}$ c. $\frac{1}{2}$

problem set 62

1. $\dfrac{14 \text{ big ones}}{9 \text{ crowns}}, \dfrac{9 \text{ crowns}}{14 \text{ big ones}}$, 560   3. 8934

5. (a) $\frac{3}{25}$ (b) 12   (c) $0.8\overline{3}$ (d) $83.\overline{3}$ 7. 790   9. 0.875   11. $\frac{77}{20}$

13. 240   15. $\frac{36}{5}$ 17. $\frac{45}{23}$ 19. 31   21. $\frac{89}{12}$ 23. 0.43368   25. $2937.1\overline{6}$

27. $\frac{4}{9}$ 29. $\frac{100}{19}$

practice   a. 200 m   c. 31,400 m²

problem set 63

1. 70 minutes   3. 20 hours   5. (a) $\frac{4}{25}$ (b) 16   (c) 0.4   (d) 40

7. $0.\overline{1}$ 9. $\frac{125}{128}$ 11. (a) 62.8 cm   (b) 314 cm²   13. 780,595,200,000 ft²

15. $\frac{16}{5}$ 17. $\frac{49}{132}$ 19. 35   21. $\frac{93}{20}$ 23. 0.86982   25. 20,575   27. $\frac{91}{188}$

29. 252

practice   a. 1   c. $\frac{56}{75}$

problem set 64

1. 64 miles   3. 124 miles   5. (a) $\frac{3}{5}$ (b) 0.6

7. $0.\overline{6}$ 9. $\frac{39}{68}$ 11. 304 ft³   13. 294 m²   15. $\frac{25}{14}$ 17. $\frac{113}{50}$ 19. 66

21. $\frac{58}{21}$ 23. 0.019602   25. 38,522   27. $\frac{35}{48}$ 29. 46

practice   a. 1150

problem set 65

1. 1050   3. $\frac{11}{20}$ 5. (a) $\frac{9}{50}$ (b) 18   7. $0.8\overline{3}$ 9. $\frac{91}{12}$

11. 6600 cm³   13. $458\frac{5}{7}$ 15. $\frac{8}{3}$ 17. $\frac{25}{38}$ 19. 18   21. $\frac{61}{42}$ 23. 2.3556

25. 250.87   27. $\frac{75}{17}$

29. Six hundred twenty-one million, seven hundred twenty-three thousand, one hundred thirty-one and seventy-two hundredths

practice   a. 2.4 hours

problem set 66

1. 4 days   3. 145   5. (a) $\frac{73}{100}$ (b) 0.73   7. $0.\overline{6}$ 9. $\frac{77}{6}$

11. 270   13. (a) 3.92×10^{-3} (b) 0.00000000603   15. $\frac{64}{27}$ 17. $\frac{39}{40}$

19. 123   21. $\frac{6}{5}$ 23. 4.2336   25. 75,380   27. $\frac{1}{5}$ 29. $6944\frac{4}{9}$ ft²

practice   a. 29.7 cm

problem set 67

1. 250   **3.** $\frac{1}{7}$ **5.** (a) 0.4   (b) 40   **7.** 240   **9.** $0.\overline{3}$ **11.** 5950 cm³

13. 120,000 inches   **15.** 4   **17.** $\frac{11}{26}$ **19.** 34   **21.** $\frac{281}{28}$ **23.** 0.22968

25. 2697   **27.** $\frac{817}{672}$ **29.** (a) 1.39×10^6 (b) 426,000,000,000,000,000

practice   **a.** 2

problem set 68

1. 3200   **3.** 20 miles per hour   **5.** (a) $\frac{3}{50}$ (b) 6   **7.** $\frac{3}{5}$

9. (a) 222.8 ft   (b) 3028 ft²   **11.** 693   **13.** $\frac{125}{34,848}$ mi²   **15.** $\frac{3}{28}$ **17.** $\frac{121}{338}$

19. 2   **21.** 21   **23.** $\frac{57}{4}$ **25.** 6801.864   **27.** $\frac{13}{24}$ **29.** 0.75 km

practice   **a.** 22,000

problem set 69

1. 1750   **3.** 2400   **5.** (a) $\frac{1}{4}$ (b) 0.25   **7.** 410   **9.** $0.8\overline{3}$

11. (a) 91.4 ft   (b) 557 ft²   **13.** 180   **15.** $\frac{124}{11}$ **17.** $\frac{26}{21}$ **19.** $\frac{29}{9}$ **21.** 30

23. $9\frac{3}{4}$ **25.** 10,128.279   **27.** $\frac{67}{100}$ **29.** 40

practice   **a.** 1.44 dollars per dozen

problem set 70

1. 1,381,732   **3.** 3000   **5.** (a) $\frac{37}{100}$ (b) 37   **7.** 540   **9.** 0.9375

11. (a) 102.8 in.   (b) 714 in.²   **13.** 360   **15.** $\frac{35}{12}$ **17.** $\frac{207}{152}$ **19.** $\frac{115}{19}$

21. 82   **23.** $15\frac{17}{36}$ **25.** 10,824.3919   **27.** $1\frac{1}{2}$ **29.** 108

practice   **a.** 40%, 420   **c.** 400

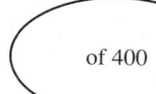

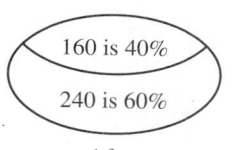

Before, 100%   After

problem set 71

1. 4   **3.** 10.90 dollars per dozen   **5.** 800, 1200, 60%

7. 20%

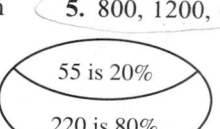

of 275   55 is 20%
             220 is 80%

Before, 100%   After

9. (a) $\frac{3}{10}$ (b) 30   **11.** $\frac{129}{56}$ **13.** 854 ft³   **15.** 360   **17.** $\frac{9}{4}$ **19.** $1\frac{35}{363}$

21. 64   **23.** $8\frac{5}{6}$ **25.** 0.032778   **27.** 40,292.5   **29.** $\frac{1378}{93}$

practice   **a.** −2   **c.** −5

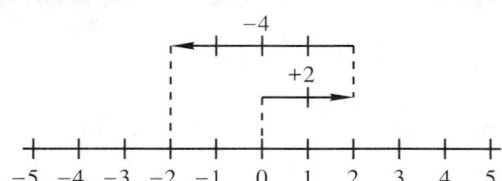
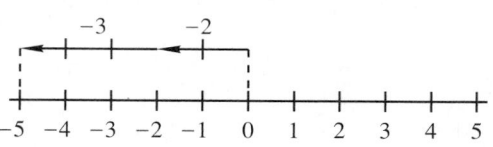

problem set   **1.** 25 cents, $2.50   **3.** 3500
72   **5.** 762 dollars per year, 616.67 dollars per year, 600 dollars per year. Five years of coverage for $3000 is the least expensive insurance plan.

7. 20%, 120, 80%   **9.** 294

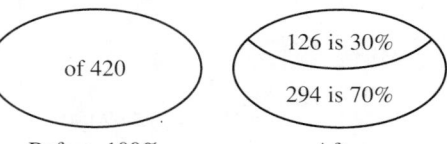

11. 2.16×10^8 **13.** (a) 88.4 in.   (b) 407 in.^{2} **15.** $\frac{443}{12}$ **17.** $\frac{28}{15}$ **19.** $\frac{11}{10}$

21. 111   **23.** $\frac{5}{3}$ **25.** 0.040551   **27.** 36,262   **29.** $\frac{1050}{17}$

practice   **a.** −12   **c.** +33   **e.** −9

problem set   **1.** 2160   **3.** 2200   **5.** (a) −8   (b) −4   **7.** 150, 450, 75%
73   **9.** 40%

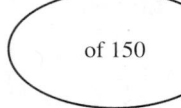
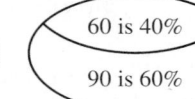

11. 0.00000604   **13.** (a) 100 m   (b) 400 m^2

15. $207\frac{1}{10}$ **17.** $\frac{57}{16}$ **19.** $\frac{41}{32}$ **21.** 76   **23.** $2\frac{15}{16}$ **25.** 0.060384

27. 32,030   **29.** $\frac{960}{169}$

practice   **a.** −10   **c.** −16

problem set   **1.** 5390   **3.** 14,700   **5.** 224, 56, 25%
74   **7.** 410

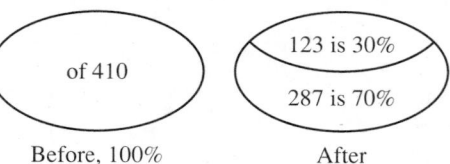

9. (a) 0.37   (b) 37%   **11.** 0.25   **13.** 89.25 m^3 **15.** 300   **17.** $\frac{15}{2}$

19. 23   **21.** 0   **23.** −10   **25.** 0.3996   **27.** 42,638

29. Six hundred twenty-five million, three hundred sixty-one thousand, eight hundred eleven and one hundredth

practice   **a.** $\frac{1}{27}$ **c.** $\frac{1}{16}$

problem set 75   **1.** $8666.67   **3.** 900   **5.** 20%

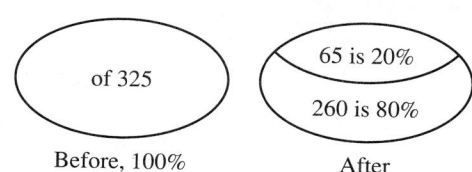

of 325   65 is 20%

Before, 100%   260 is 80%   After

7. (a) $\frac{71}{100}$ (b) 71%   **9.** 0.6   **11.** 150.83 yd³   **13.** $15\frac{155}{198}$ miles   **15.** $\frac{2}{3}$

17. $\frac{1}{2}$ **19.** $\frac{77}{12}$ **21.** 2440   **23.** 4   **25.** 0   **27.** 4.056819   **29.** 547,600

practice   **a.**

problem set 76   **1.** $\frac{3}{29}$ **3.** 3 hours   **5.**

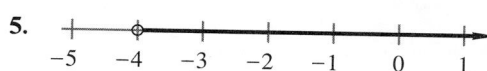

7. 288   **9.** 20%

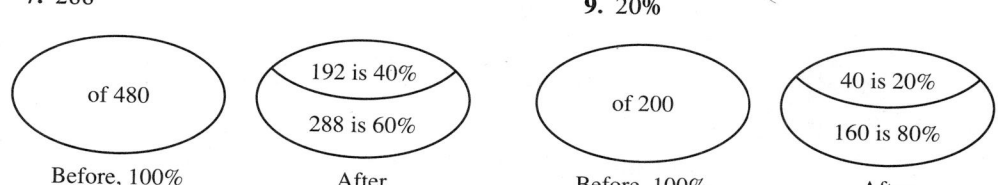

of 480   192 is 40%   of 200   40 is 20%

Before, 100%   288 is 60%   After   Before, 100%   160 is 80%   After

11. (a) $\frac{16}{81}$ (b) $\frac{8}{9}$ **13.** (a) 85.68 ft   (b) 401.04 ft²   **15.** 384 m²

17. $\frac{176}{39}$ **19.** $\frac{15}{52}$ **21.** −10   **23.** −24   **25.** 0.22374   **27.** 36,385

29. 200,000,000

practice   **a.** 9420 ft³

problem set 77   **1.** 105   **3.** 20 miles per hour   **5.**

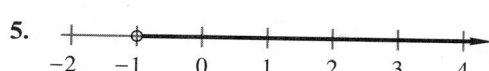

7. 312, 60%   **9.** 500

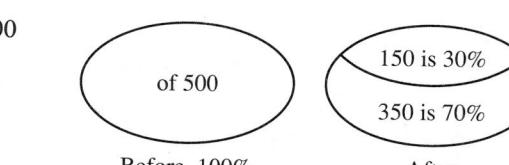

of 500   150 is 30%

Before, 100%   350 is 70%   After

11. (a) $\frac{27}{64}$ (b) $\frac{2}{3}$ **13.** 282.6 m³   **15.** 1.04×10^9

17. 24   **19.** $\frac{25}{81}$ **21.** −7   **23.** −26   **25.** 0.457699   **27.** 104,855

29. 960,000 inches

practice   a. +7

problem set   **1.** 240,000   **3.** 5 hours   **5.**

78

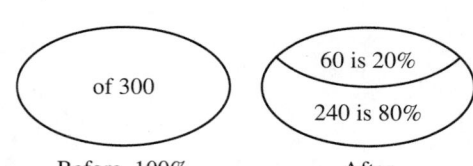

7. 400, 120, 30%   **9.** 20%

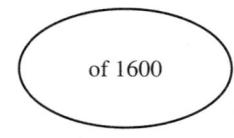 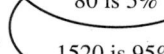

of 300

Before, 100%

60 is 20%

240 is 80%

After

11. 10,300,000,000   **13.** 4368 in.³   **15.** 3140 ft³

17. $\frac{243}{80}$ **19.** $\frac{5}{11}$ **21.** 8   **23.** 0   **25.** $8\frac{17}{20}$ **27.** 3591.694   **29.** $\frac{11}{32}$

practice   a. 44

problem set   **1.** 676 gallons   **3.** 99,000   **5.**

79

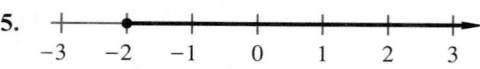

7. 20%, 680, 80%   **9.** 5%

of 1600

Before, 100%

80 is 5%

1520 is 95%

After

11. 1.03×10^{11} **13.** 821.25 m³   **15.** $231\frac{4}{7}$ **17.** $\frac{256}{15}$ **19.** $\frac{536}{495}$

21. −3   **23.** 0.11200255   **25.** 82,965.6685   **27.** $\frac{27}{64}$ **29.** 100,000,000 cm

practice   a. 8×10^{10} c. 4×10^{13}

problem set   **1.** 14 eggs   **3.** 8 hours   **5.**

80

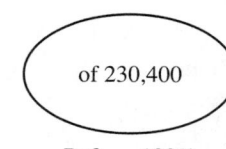

7. 230,400

of 230,400

Before, 100%

110,592 is 48%

119,808 is 52%

After

9. 90, 54, 60%   **11.** 6×10^6 **13.** 6×10^9 **15.** $\frac{9}{64}$

17. 2,509,056,000,000 ft²   **19.** 360 ft³   **21.** $\frac{3}{4}$ **23.** $\frac{26}{81}$ **25.** −30

27. 3.4227   **29.** 11,222.7289

practice   a. 2, 3, 5   c. 260

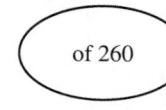 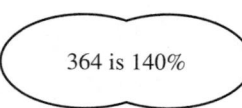

of 260

Before, 100%

364 is 140%

After

problem set 81

1. 990   **3.** 560   **5.**

7. 450%   **9.** 32   **11.** 334,540,800 ft²   **13.** 912 ft²   **15.** 108   **17.** $\frac{72}{5}$

19. 26   **21.** 6   **23.** −24   **25.** 0.0001578   **27.** 4×10^{-47} **29.** 52

practice   **a.** −12   **c.** 12   **e.** $-\frac{1}{2}$ **g.** −2

problem set 82

1. 2500   **3.** 1440   **5.**   **7.** 104

9. 195

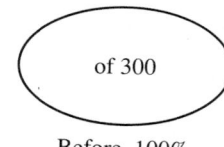

Before, 100%   After

11. (a) 0.29   (b) 29%   **13.** 0.75   **15.** 102.8 ft   **17.** 60   **19.** 2
21. 111   **23.** (a) 12   (b) 12   (c) −12   (d) −12   **25.** −14
27. 7×10^2 **29.** $\frac{19}{36}$ **31.** 134

practice   **a.** −19

problem set 83

1. 600   **3.** 385 miles   **5.**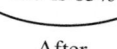

7. 385   **9.** 175%

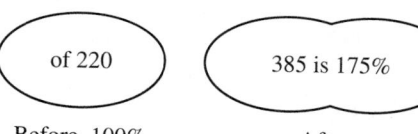   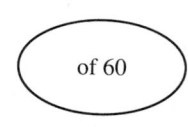

of 220   385 is 175%   of 60   105 is 175%

Before, 100%   After   Before, 100%   After

11. 0.000000000091   **13.** 150,000 cm³   **15.** 6×10^{-27} **17.** $\frac{7}{45}$ **19.** $\frac{31}{20}$

21. (a) −6   (b) −6   (c) 6   **23.** −10   **25.** −12   **27.** $\frac{197}{14}$ **29.** 0.0052

practice   **a.** 18,360

problem set 84

1. 32,500,000   **3.** 147   **5.**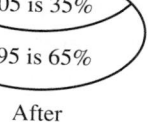

7. 184

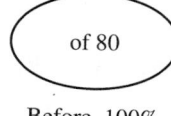   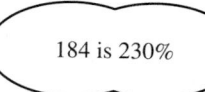

of 80   184 is 230%

Before, 100%   After

9. $\frac{48}{35}$ **11.** 224   **13.** 26.28 ft³   **15.** 3   **17.** $\frac{11}{70}$ **19.** 627

21. (a) −8   (b) −8   (c) 8   **23.** −8   **25.** 1962   **27.** $\frac{25}{64}$ **29.** 0.0067367

practice   **a.** −3   **c.** 16   **e.** Negative

problem set 85

1. 0.69   **3.** 340 miles   **5.**

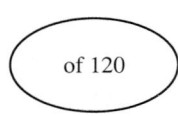

7. 192

of 120   192 is 160%

Before, 100%   After

9. 250%

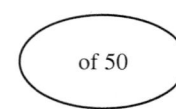

of 50   125 is 250%

Before, 100%   After

11. 9.13×10^{-7} **13.** 52,500 cm³   **15.** $120\frac{11}{15}$ **17.** $\frac{5}{87}$ **19.** $\frac{23}{24}$

21. (a) −18   (b) 18   (c) −18   **23.** 3   **25.** −4   **27.** 0

29. One hundred thirty-six million, one hundred twenty-one thousand, one hundred thirty-four and five-tenths

practice   **a.** −4   **c.** −3

problem set 86

1. 40 miles per hour   **3.** 8000   **5.**

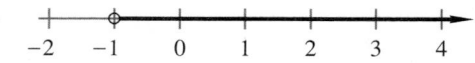

7. 196

of 140   196 is 140%

Before, 100%   After

9. 140%

of 70   98 is 140%

Before, 100%   After

11. 8   **13.** 1371 in.³   **15.** $113\frac{1}{12}$ **17.** $\frac{17}{40}$ **19.** $\frac{8}{7}$

21. (a) −18   (b) 10   (c) −7   **23.** −1   **25.** −8   **27.** 10   **29.** 31

practice   **a.** 49 hours

problem set 87

1. 70 miles per hour   **3.** 3800   **5.** 77 hours

7.

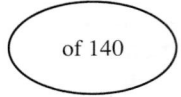

9. 1430

of 650   1430 is 220%

Before, 100%   After

11. 6×10^{-37}

13. 310   **15.** 1000 m²   **17.** 21.195 yd³   **19.** $\frac{15}{16}$ **21.** $-\frac{7}{345}$

23. (a) −28   (b) −28   (c) 28   **25.** −4   **27.** 8   **29.** 11

practice   **a.** −2

problem set 88

1. $\frac{1}{8}$ **3.** 1000   **5.**

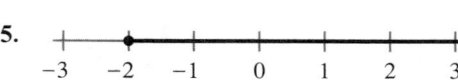

7. 288

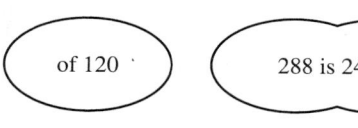

Before, 100%   After

9. 160%

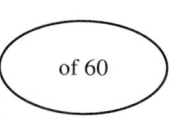

Before, 100%   After

11. 216 m²   **13.** 42.2 m³   **15.** 138,531,000 cm   **17.** −3   **19.** $\frac{9}{40}$ **21.** 38

23. (a) −8   (b) −8   (c) 12   **25.** 51   **27.** 4   **29.** 19

practice

problem set 89

a. $N - 4$ **c.** $-N - 3$ **e.** $5(N - 6)$

1. $2400   **3.** 14,000

5. (a) $2N - 16$ (b) $-3(N + 5)$ (c) $-N - 4$ (d) $5(2N + 6)$

7. 646

9. 160%

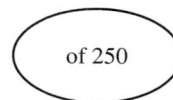

Before, 100%   After

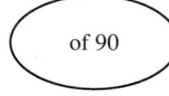

Before, 100%   After

11. 0.375 cm²   **13.** $\frac{17}{36}$ yd²   **15.** $2 \times 2 \times 2 \times 2 \times 3 \times 3 \times 5$ **17.** $-\frac{15}{7}$

19. $-\frac{1}{2}$ **21.** $\frac{33}{4}$ **23.** 6   **25.** 155,302.5   **27.** 0.165494   **29.** −1

practice

problem set 90

a. 720 m²

1. $8325   **3.** $864,000   **5.** (a) $3N - 11$ (b) $-5(N - 2)$ (c) $-3(N - 4)$

7. 675

9. 400%

Before, 100%   After   Before, 100%   After

11. 207.6 ft²   **13.** 5×10^{44} **15.** 13   **17.** −4   **19.** $\frac{4}{11}$ **21.** 89

23. (a) 15   (b) −7   (c) 2   **25.** 15   **27.** 2,633,000   **29.** −30

practice

problem set 91

a. 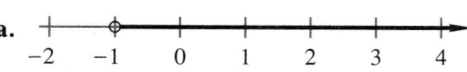   **c.**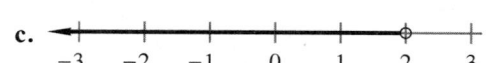

1. 56 inches   **3.** $50,400

5. (a) $4(3N - 6)$ (b) $-2(N + 5)$ (c) $-N - 25$

7.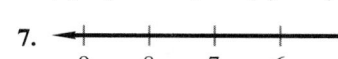

9. 675

(of 250 — Before, 100%)   (675 is 270% — After)

11. 4   **13.** 18,280 cm²   **15.** $-\dfrac{5}{7}$ **17.** $\dfrac{11}{9}$ **19.** 40

21. (a) −6   (b) 4   (c) 6   **23.** −8   **25.** $\dfrac{49}{15}$ **27.** 653.0152   **29.** 11

practice   **a.** 12   **c.** −10

problem set 92   **1.** 42   **3.** −8   **5.** (number line, open circle at 3, shaded right, 2 3 4 5 6 7 8)

7. (number line, closed dot at −2, shaded right, −3 −2 −1 0 1 2 3)

9. 675

(of 300 — Before, 100%)   (675 is 225% — After)

11. (a) $\dfrac{2}{5}$ (b) 0.4   **13.** 360 in.³   **15.** $-\dfrac{2}{3}$ **17.** $\dfrac{7}{19}$ **19.** 16

21. (a) −6   (b) −6   (c) 20   **23.** $-\dfrac{14}{3}$ **25.** $\dfrac{143}{5}$ **27.** 1169.0007   **29.** 1

practice   **a.** 54

problem set 93   **1.** 3   **3.** 3   **5.** (number line, open circle at 1, shaded right, 0 1 2 3 4 5 6)

7. 805   **9.** 135%

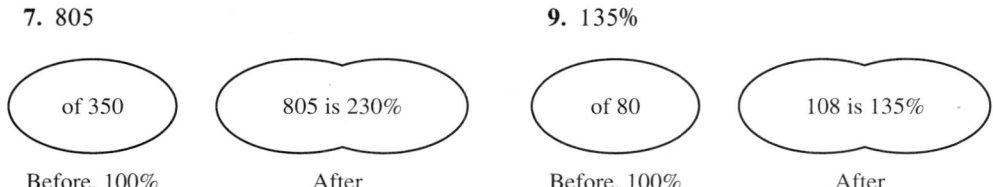

(of 350 — Before, 100%) (805 is 230% — After) (of 80 — Before, 100%) (108 is 135% — After)

11. (a) $\dfrac{9}{25}$ (b) 36%   **13.** 1042 yd³   **15.** 1.2348792 km   **17.** $-\dfrac{1}{3}$ **19.** $-\dfrac{3}{2}$

21. 36   **23.** 32   **25.** $\dfrac{140}{9}$ **27.** $205{,}003.\overline{3}$ **29.** 8

practice   **a.** 2.5

problem set 94   **1.** −6   **3.** 900   **5.** 100   **7.** (number line, open circle at 0, shaded right, −1 0 1 2 3 4 5)

9.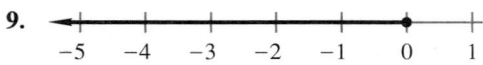
(number line, closed dot at 0, shaded left, −5 −4 −3 −2 −1 0 1)

11. 380%

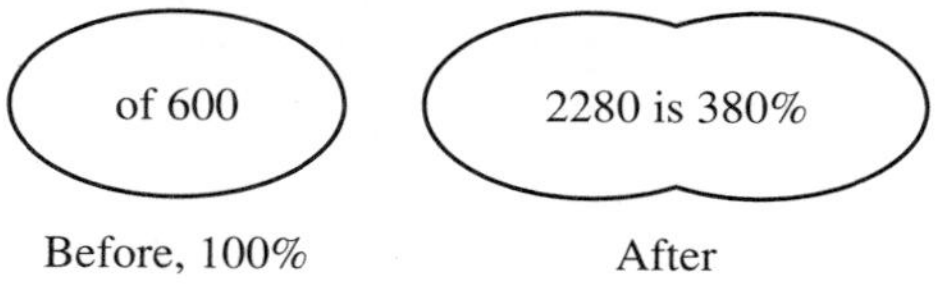

$$\text{of 600}$$
Before, 100%

$$\text{2280 is 380\%}$$
After

13. 12 yd^2 **15.** -1

17. $\frac{1}{4}$ **19.** 9×10^{-36} **21.** $\frac{37}{4}$ **23.** 11   **25.** $\frac{353}{80}$ **27.** 0.000817

29. -30

practice problem set 95

a. -5

1. 5   **3.** -5 **5.**

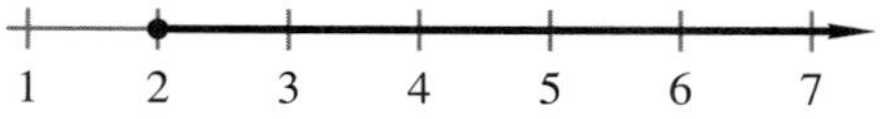

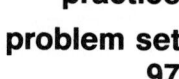
$$\begin{array}{ccccccc} 1 & 2 & 3 & 4 & 5 & 6 & 7 \end{array}$$

7. 286

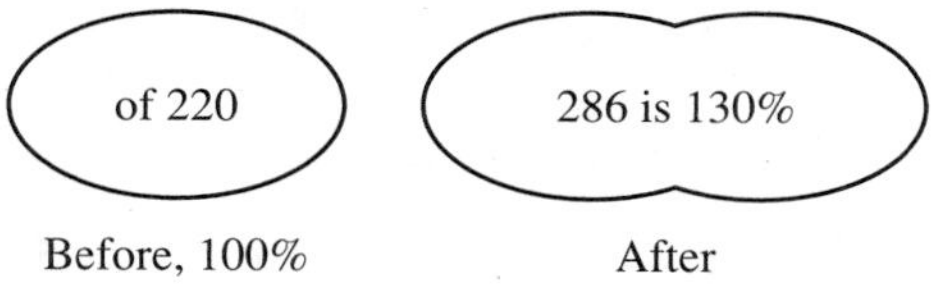

$$\text{of 220}$$
Before, 100%

$$\text{286 is 130\%}$$
After

9. 160%

$$\text{of 70}$$
Before, 100%

$$\text{112 is 160\%}$$
After

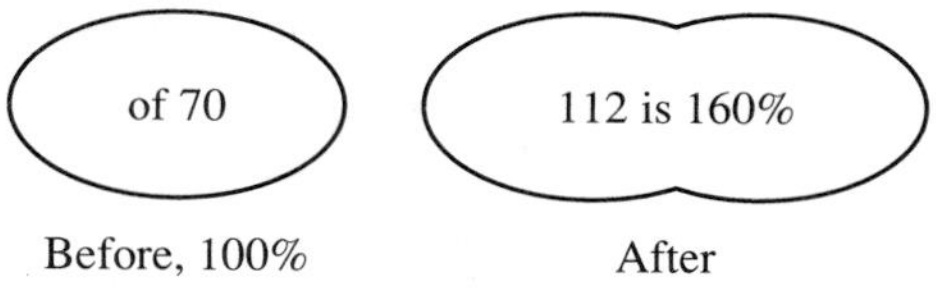

11. (a) $\frac{53}{100}$ (b) 53%   **13.** 14,568 in.^{3} **15.** 1,268,700 cm   **17.** $\frac{2}{3}$

19. $-\frac{51}{56}$ **21.** 26   **23.** $-\frac{21}{5}$ **25.** $\frac{3}{8}$ **27.** 30   **29.** 83

practice problem set 96

a. $2x + 3y - 4$ **c.** $2a + 11b - 8$

1. 5   **3.** 9   **5.** 580   **7.**

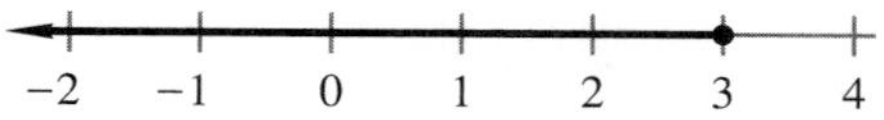

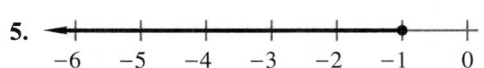
$$\begin{array}{ccccccc} -2 & -1 & 0 & 1 & 2 & 3 & 4 \end{array}$$

9. 560

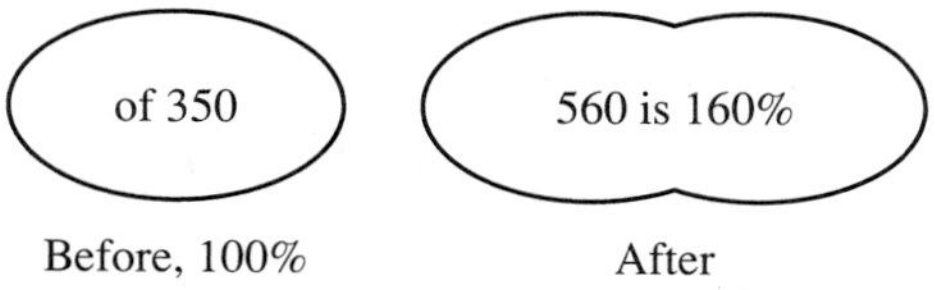

$$\text{of 350}$$
Before, 100%

$$\text{560 is 160\%}$$
After

11. 450%

$$\text{of 80}$$
Before, 100%

$$\text{360 is 450\%}$$
After

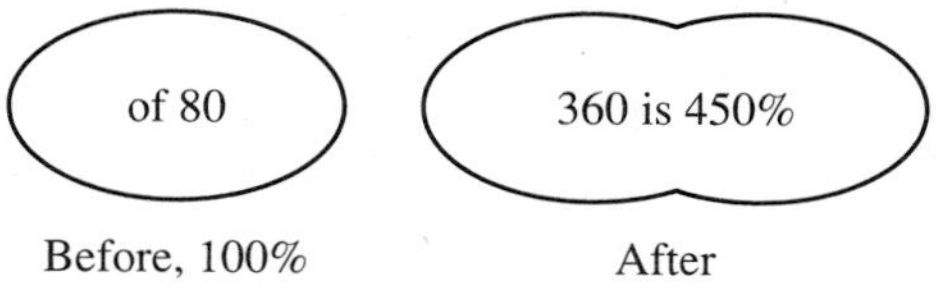

13. 6×10^{-5} **15.** 181.4 ft^3 **17.** 1 km^2 **19.** $-4x - 4$ **21.** $\frac{5}{3}$ **23.** -119

25. $\frac{7}{9}$ **27.** 9   **29.** 4.8×10^{10}

practice problem set 97

a. 1

1. 20   **3.** 5 hours   **5.**

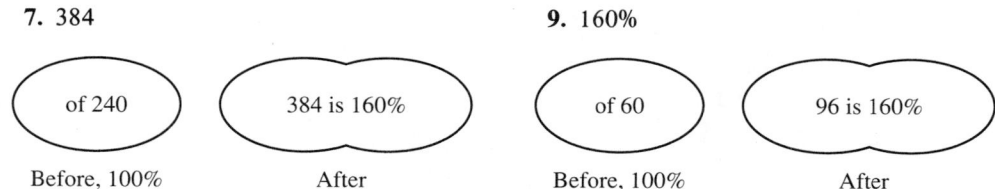
$$\begin{array}{cccccc} -6 & -5 & -4 & -3 & -2 & -1 & 0 \end{array}$$

7. 384

$$\text{of 240}$$
Before, 100%

$$\text{384 is 160\%}$$
After

9. 160%

$$\text{of 60}$$
Before, 100%

$$\text{96 is 160\%}$$
After

11. 6.4×10^{16} **13.** 2.2   **15.** $\frac{1}{2}$ **17.** $\frac{5}{3}$ **19.** $-\frac{14}{39}$ **21.** -24 **23.** $\frac{13}{3}$

25. $\frac{11}{72}$ **27.** $\frac{49}{19}$ **29.** 48

practice   **a.** 17   **c.** $\frac{49}{2}$

problem set 98   **1.** -20 **3.** \$1400   **5.**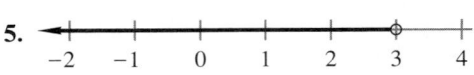

7. 621   **9.** 145%

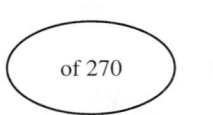 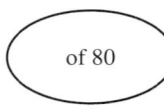

| of 270 | 621 is 230% | of 80 | 116 is 145% |
| Before, 100% | After | Before, 100% | After |

11. (a) 0.9   (b) 90%   **13.** 1062 cm^2 **15.** 0.01287321 km   **17.** 4   **19.** $-\frac{11}{4}$

21. $-\frac{3}{7}$ **23.** -30 **25.** 4   **27.** 69   **29.** -25

practice   **a.** 3

problem set 99   **1.** -24 **3.** 7   **5.**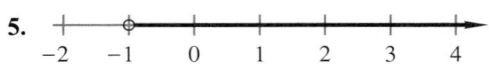

7. 800   **9.** 180%

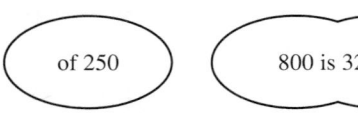 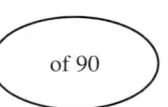

| of 250 | 800 is 320% | of 90 | 162 is 180% |
| Before, 100% | After | Before, 100% | After |

11. 2.1   **13.** 4340 cm^3 **15.** 190,080 inches   **17.** $\frac{7}{6}$ **19.** $\frac{11}{3}$ **21.** $\frac{8}{9}$

23. 4   **25.** 1   **27.** 107   **29.** -1

practice   **a.** 4

problem set 100   **1.** 5   **3.** 720   **5.**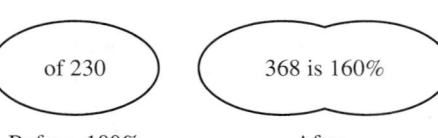

7. 368   **9.** 185%

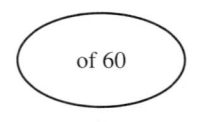 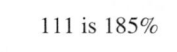

| of 230 | 368 is 160% | of 60 | 111 is 185% |
| Before, 100% | After | Before, 100% | After |

11. 0.0000734   **13.** 6.0   **15.** 0.07892321 km   **17.** $\frac{7}{2}$ **19.** $-\frac{1}{5}$ **21.** $-\frac{9}{20}$

23. -94 **25.** 2   **27.** $\frac{9}{28}$ **29.** -18

practice   **a.** 44, 198
problem set   **1.** 140 acres, 100 acres   **3.** 11   **5.**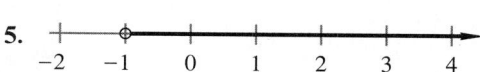
101

7. 195   **9.** 25%

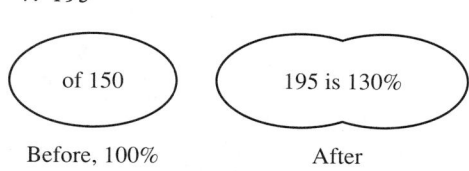   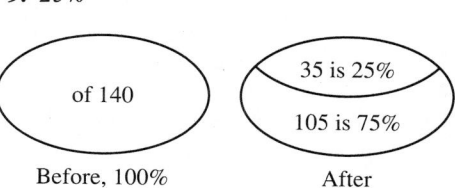

11. 720   **13.** 1177.5 ft^3 **15.** 6,113,112.1 cm   **17.** 3   **19.** 1   **21.** $-\dfrac{6}{5}$

23. 4   **25.** -13 **27.** $\dfrac{17}{9}$ **29.** -14

practice   **a.** 3^8 **c.** a^8b^4
problem set   **1.** 1400   **3.** $700   **5.**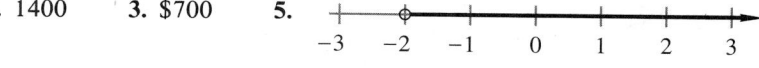
102

7. 224   **9.** 600

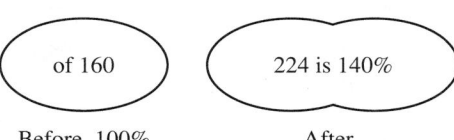   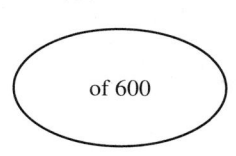

11. 144 inches   **13.** 11   **15.** $\dfrac{13}{4}$ **17.** $\dfrac{7}{5}$ **19.** $\dfrac{1}{5}$ **21.** -9 **23.** $x^{11}y^{21}$

25. a^7m^8 **27.** -1 **29.** 196

practice   **a.** $2x^2y$ **c.** $xy^3 + x^2y^2 - 5x^3y$
problem set   **1.** 250   **3.** -4 **5.**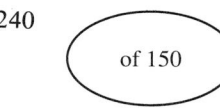
103

7. 240

of 150   240 is 160%

Before, 100%   After

9. 3.1   **11.** 180   **13.** 28   **15.** $\dfrac{3}{2}$ **17.** $-\dfrac{2}{3}$ **19.** $4ab^2 - a + 3$

21. -17 **23.** 22   **25.** x^7y^9 **27.** $\dfrac{7}{24}$ **29.** $\dfrac{280}{3}$

practice   **a.** $12x^2 + 8x$ **c.** $-x^3 - x^2 - 4x$ **e.** 3
problem set   **1.** 42   **3.** 6 hours
104

5. 90   **7.** 155%

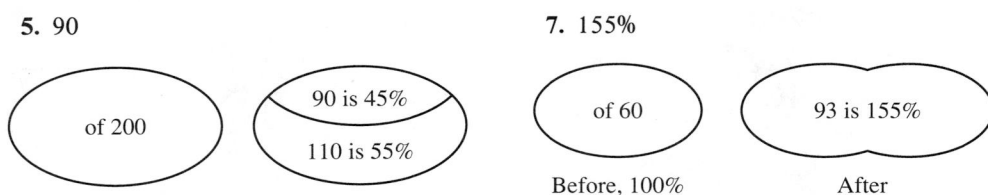

9. 736 cm²   **11.** $\dfrac{37}{3}$ **13.** $2a^2b + 2ab^2$ **15.** $7xy + 9x + 3y + 5$

17. −2   **19.** 0   **21.** −1   **23.** −248   **25.** x^6y^7 **27.** −1   **29.** 7

practice   **a.** −9

problem set   **1.** 2016   **3.** 120   **5.** 2000 m

105   **7.** 204   **9.** 110%

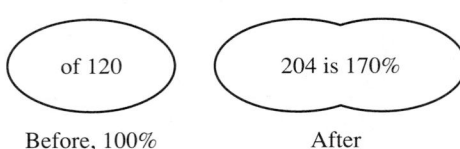
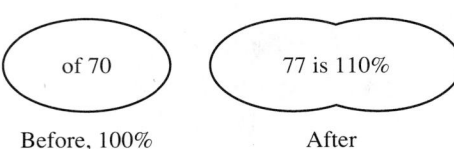

11. 15,120 cm²   **13.** 5   **15.** $a^2b + ab^2 + a^2b^2$ **17.** $2x^2y + 5xy^2$

19. $-\dfrac{9}{4}$ **21.** $-\dfrac{7}{2}$ **23.** 0   **25.** a^6b^3 **27.** $-\dfrac{14}{3}$ **29.** −130

practice   **a.** −3   **c.** $\dfrac{1}{25}$

problem set   **1.** 24   **3.** 800; 6800   **5.** 5 hours

106   **7.** 162   **9.** 158.5 m³   **11.** −2

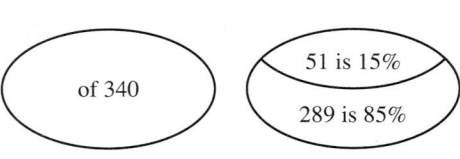

13. (a) $\dfrac{1}{64}$ (b) $\dfrac{1}{9}$ **15.** 50   **17.** $6p^2x^2 + 3p^2x + 3px^2$ **19.** $5m^2n^3$ **21.** 3

23. $\dfrac{275}{52}$ **25.** 45   **27.** 9   **29.** −28

practice   **a.** 210   **c.** 0.2%

problem set   **1.** 24   **3.** −14   **5.** 5 hours

107   **7.** 289   **9.** 165%

11. 8292 cm²   **13.** 1%   **15.** 13   **17.** −97   **19.** $5m + 11mx + 2x + 4$

21. -1 **23.** -2 **25.** -21 **27.** m^5p^7 **29.** 5

practice   **a.** IV   **c.** XXXIV   **e.** XL

g. XV, XVI, XVII, XVIII, XIX, XX, XXI, XXII, XXIII, XXIV, XXV, XXVI, XXVII, XXVIII, XXIX, XXX

problem set 108

1. 4   **3.** 800   **5.** 7 hours

7. 702

9. 180%

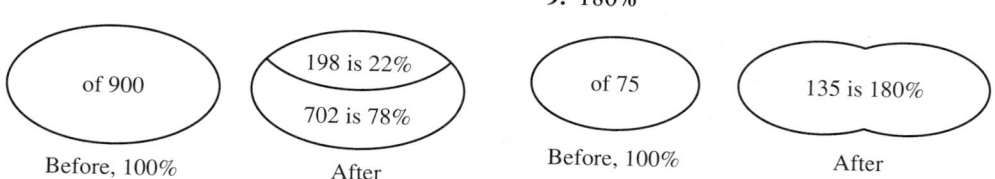

Before, 100%   After   Before, 100%   After

11. 606.96 ft² **13.** XXIX   **15.** 30%   **17.** 19   **19.** -41

21. $-6m^2 + 8c^2 - 4cm$ **23.** $-\dfrac{16}{15}$ **25.** -1 **27.** -8 **29.** 8

practice   **a.** 1400

problem set 109

1. 210   **3.** 10,400   **5.** 5   **7.** $868

9. 112

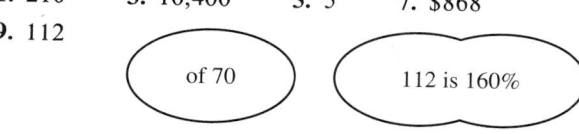

Before, 100%   After

11. 112.8 inches

13. 6   **15.** $3a^2b + 3ab^2 + 3abc + 3abd$ **17.** $a^3 + 4a^2b + 6b^2a$ **19.** $-\dfrac{9}{7}$

21. $-\dfrac{21}{16}$ **23.** -16 **25.** -28 **27.** m^6n^5 **29.** 7

practice   **a.** 0.000462

problem set 110

1. 24,000   **3.** 84,000   **5.** 2000; 2800

7.

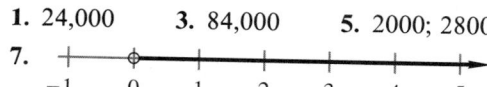

9. 1215

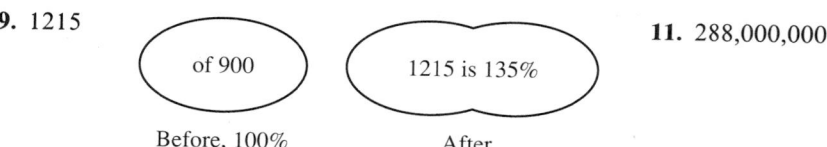

Before, 100%   After

11. 288,000,000 cm³

13. XVII   **15.** 161   **17.** 43   **19.** $14m^2p^2 + 7mpx + 7m^2p + 7mp^2$

21. $4a^2m + p^2 - 2p^3$ **23.** (a) 0.045   (b) 0.06375   **25.** $-\dfrac{7}{10}$ **27.** $-\dfrac{27}{2}$

29. 105

practice   **a.** $8260

problem set 111

1. 1,050,000   **3.** $40,800   **5.** $19,500

7.

9. 240%

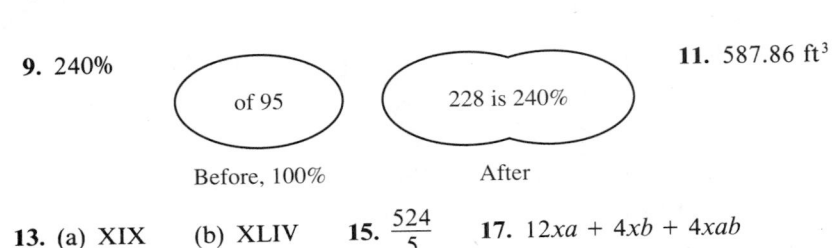

of 95
Before, 100%

228 is 240%
After

11. 587.86 ft³

13. (a) XIX   (b) XLIV   **15.** $\dfrac{524}{5}$ **17.** $12xa + 4xb + 4xab$

19. $mp^2 + m^2p + 3a^2b^2 + mp$ **21.** 0.00725   **23.** 0.092   **25.** -5 **27.** $\dfrac{15}{4}$

29. 8

practice   **a.** $9600

problem set   **1.** 1400   **3.** 36,300,000 bushels   **5.** $5724.50   **7.** $52.50
112   **9.** 171   **11.** 2280

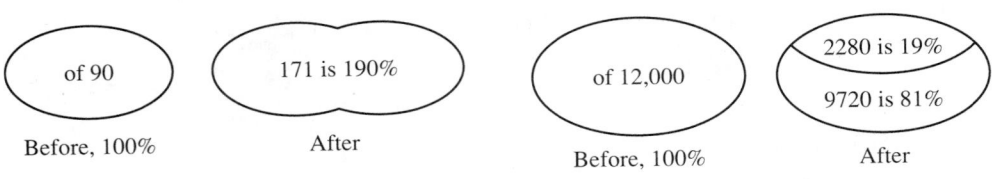

of 90
Before, 100%

171 is 190%
After

of 12,000
Before, 100%

2280 is 19%
9720 is 81%
After

13. 0.0157 ft³, 0.314 ft²   **15.** 33   **17.** $3cm + 3cmn + 3cn$

19. $2a^3p + 2m^2b^2$ **21.** 0.009175   **23.** 0.082   **25.** 12   **27.** 19   **29.** $\dfrac{261}{2}$

practice   **a.** $1674

problem set   **1.** 2000   **3.** -4 **5.** $4850   **7.** 33.8125 ft³   **9.** XXIV
113   **11.** 216

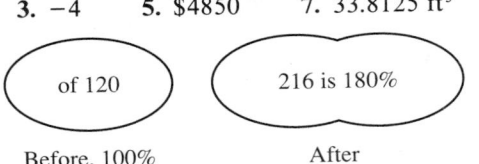

of 120
Before, 100%

216 is 180%
After

13. (a) 0.616   (b) 0.078   **15.** 29   **17.** $-a^3m - am^2 + a^2m^2$ **19.** $\dfrac{1}{2}$

21. $-\dfrac{1}{3}$ **23.** -28 **25.** $-\dfrac{5}{4}$ **27.** 8   **29.**

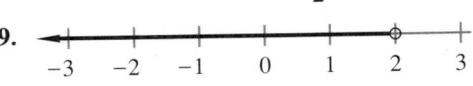

practice   **a.** 140°   **c.** Obtuse angle   **e.** Acute angle   **g.** 26°   **i.** 132°   **k.** 90°
m. 180°

problem set   **1.** 2400   **3.** 60   **5.** (a) 115°   (b) 25°   **7.** (a) 45°   (b) 100°   (c) 135°
114   **9.** 135%

of 620
Before, 100%

837 is 135%
After

11. 55,756,800 ft²   **13.** 3   **15.** $m^2n^2 + mn^2 + m^2n$

17. $2xy^3 + 2x^2y$ **19.** $\dfrac{23}{9}$ **21.** $\dfrac{27}{8}$ **23.** 0   **25.** x^6y^5 **27.** $\dfrac{1}{9}$ **29.** 11

practice   **a.** 56°   **c.** Refer to Lesson 115

problem set   **1.** 200   **3.** 490   **5.** $50,000   **7.** Refer to Lesson 115
115   **9.** (a) 56°   (b) 145°   **11.** (a) 0.052   (b) 0.1775   (c) 0.11375
13. 44,000,000   **15.** 699,840 in.³   **17.** $3a^2c + 3abc - 3ac^2 + 3a^2c^2b$

19. 752,160,000 cm³   **21.** 25   **23.** −18   **25.** $-\frac{3}{7}$ **27.** 2

29.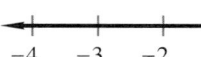

---

practice   **a.** 90°   **c.** 65°

problem set   **1.** 3000   **3.** −4   **5.** $70,800
116   **7.** Refer to Lesson 115   **9.** (a) Obtuse angle   (b) Right angle   (c) Acute angle
11. (a) 0.16875   (b) 0.0425   (c) 0.296
13. 44,000,000 cm³   **15.** $2a^2b^2 + 2a^3b - 2ab^3 + 2ba^2$ **17.** 4771.875 ft³

19. $-\frac{9}{8}$ **21.** 13   **23.** $5a^6b^6$ **25.** 0   **27.** −9

---

practice   **a.** 765   **c.** DCCCLXXXVIII

problem set   **1.** 49,000   **3.** −1   **5.** 6900   **7.** MCDXXXV   **9.** (a) 97°   (b) 7°
117   **11.** 1240

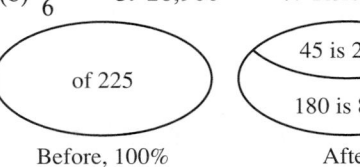

Before, 100%   After

13. 0.069   **15.** 2.3552×10^{13} ft³

17. $4m^2n^2 + 4mnx^2 - 4mnz^2 + 4mnc$ **19.** 288 ft³   **21.** $-\frac{21}{4}$ **23.** $-\frac{19}{126}$

25. 38   **27.** 0   **29.** −23

---

practice   **a.** $\frac{9}{16}$

problem set   **1.** 8000   **3.** (a) $\frac{1}{6}$ (b) $\frac{1}{6}$ **5.** 28,900   **7.** Refer to Lesson 115
118   **9.** 3434   **11.** 20%

of 225

45 is 20%
180 is 80%

Before, 100%   After

13. 52,000,000 cm³   **15.** −5   **17.** $-4xy^2 + 7x^2y$ **19.** −27   **21.** $-\frac{1}{4}$

23. $4a^6b^4$ **25.** $\frac{413}{180}$ **27.** −5   **29.** $2 \times 2 \times 2 \times 2 \times 2 \times 3 \times 3 \times 5 \times 7$

---

practice   **a.** $2\frac{3}{4}$ inches, 7 cm   **c.** 0.6 cm, 1.6 cm, 2.8 cm, 4.7 cm

problem set   **1.** $\frac{6}{11}$ **3.** 990   **5.** −3   **7.** Refer to Lesson 115   **9.** 3015
119

11. 5%

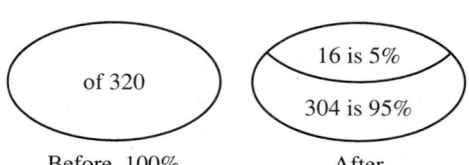

Before, 100%   After

13. 260%

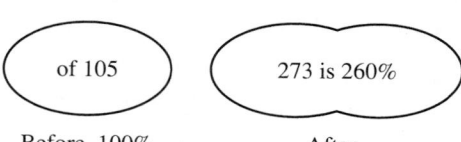

Before, 100%   After

15. $\dfrac{91}{2}$ **17.** $14d^2z + 14dz^2$ **19.** 54 in.^{2} **21.** $-\dfrac{3}{10}$

23. (a) 118°   (b) 64°   (c) 69°   **25.** -38 **27.** $-\dfrac{10}{9}$ **29.** (a) $\dfrac{1}{32}$ (b) 27

31. -196

practice   **a.** $\dfrac{36}{5}, \dfrac{28}{5}$

problem set 120   **1.** $\dfrac{2}{3}$ **3.** 4   **5.** 800, 2800   **7.** Refer to Lesson 115   **9.** MMMCMLXIV

11. 20%

of 265

Before, 100%

53 is 20%

212 is 80%

After

13. 250%

of 94

Before, 100%

235 is 250%

After

15. $\dfrac{97}{26}$ **17.** $6a^3m^{-2} + 2$ **19.** 19.16 m   **21.** $-\dfrac{5}{141}$

23. (a) 93°   (b) 50°   (c) 71°   **25.** $\dfrac{m^3p^{12}}{16}$ **27.** $-\dfrac{8}{15}$ **29.** -114

practice   **a.** $\dfrac{1}{512}$

problem set 121   **1.** $\dfrac{11}{45}$ **3.** $\dfrac{1}{32}$ **5.** 700   **7.** 12, 16   **9.** (a) 81°   (b) 64°   (c) 60°

11. (a) 0.06875   (b) 1.3275   **13.** 120

15. 26,642,829,312,000 ft^3

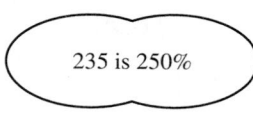

of 120

Before, 100%

48 is 40%

72 is 60%

After

17. $-7x^2y + 8y^3x$ **19.** 136 ft^2 **21.** -11 **23.** 8   **25.** 49   **27.** -9 **29.** 12

practice   **a.** -1

problem set 122   **1.** 140   **3.** 1000   **5.** $\dfrac{1}{16}$ **7.** 104

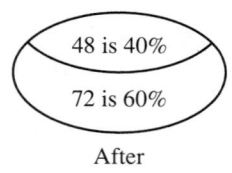

of 65

Before, 100%

104 is 160%

After

9. 0.001275   **11.** 360   **13.** -1 **15.** $a^2b^2c + 2a^2b^2 - 2ab^2c$

17. $2m^2n^3 + 5mn^2$ **19.** $\dfrac{33}{38}$ **21.** 5.7   **23.** a^3b^7 **25.** -27 **27.** $\dfrac{5}{7}$ **29.** 1

practice   **a.** Convex polygon, regular quadrilateral, rectangle, rhombus, square

c. Convex polygon, trapezoid, quadrilateral

e. Convex polygon, rectangle, quadrilateral   **g.** Concave polygon, hexagon

i. Translation, rotation, reflection   **k.** Translation   **m.** Translation, rotation

problem set 123   **1.** $\frac{1}{28}$ **3.** 2400   **5.** $\frac{1}{676}$ **7.** Refer to Lesson 115

9. (a) 141°   (b) 46°   (c) 65°   **11.** 0.00006 km

13. 300

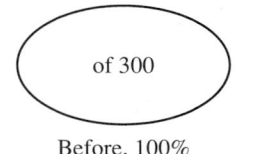 of 300, Before, 100%    117 is 39%, 183 is 61%, After

15. $4az^3 + 4z - 12a^3z^2$

17. 12.56 cm^2 **19.** MCMXCIX   **21.** -3 **23.** $-\frac{8}{5}$ **25.** $6m^5p^7$ **27.** 0

29. $\frac{1}{72}$

practice   **a.** 21 m^2

problem set 124   **1.** 280   **3.** 200,000   **5.** 2500   **7.** Refer to Lesson 115

9. (a) 26 cm^2 (b) 16 cm^2 **11.** $5\frac{5}{16}$ inches   **13.** 2000%   **15.** 400

17. $4x^2am^4$ **19.** 12.468 km^2 **21.** $-\frac{1}{2}$

23. 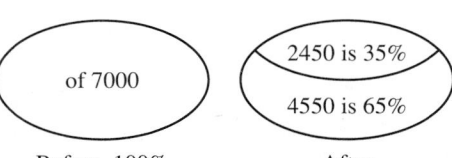   **25.** $36m^7p^6$ **27.** $\frac{297}{200}$ **29.** $\frac{27}{8}$

practice   **a.** Refer to Lesson 125   **c.** Refer to Lesson 125

problem set 125   **1.** $\frac{1}{36}$ **3.** 400   **5.** Refer to Lesson 125   **7.** Refer to Lesson 115

9. 300

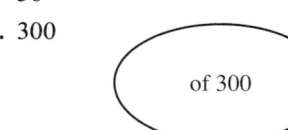 of 300, Before, 100%   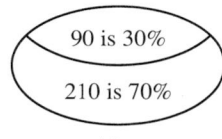 90 is 30%, 210 is 70%, After

11. 10,500 cm   **13.** 0.98375   **15.** (a) 7.5   (b) 101°

17. $5x^2y^3 - 9x^2y^2$ **19.** 840 in.^{2} **21.** $-\frac{1}{8}$ **23.** -12 **25.** $-\frac{11}{2}$ **27.** -8

29. 0

practice   **a.** $\sqrt{13}$

problem set 126   **1.** $\frac{1}{676}$ **3.** -3 **5.** \$32,850   **7.** Refer to Lesson 125

9. 7000

of 7000, Before, 100%   2450 is 35%, 4550 is 65%, After

11. 3994   **13.** 2   **15.** (a) $\frac{5}{3}$ (b) 116°   **17.** $2a^3pm^2 + a^2\,pm^{-1}$

19. $(22 + 2\sqrt{65})$ in.   **21.** $\frac{3}{8}$ **23.** -2 **25.** -66 **27.** $\frac{423}{28}$ **29.** $18\frac{1}{8}$

practice   **a.** Refer to Lesson 127   **c.** Refer to Lesson 127

problem set   **1.** $\frac{1}{2}$ **3.** 1200   **5.** Refer to Lesson 127   **7.** Refer to Lesson 125
127   **9.** 115%

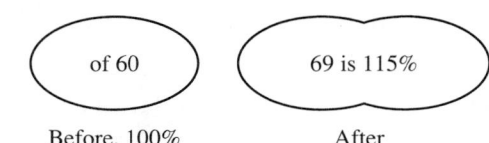

Before, 100%   After

11. 39   **13.** $\frac{26}{(100)(100)(100)}$ m^3 **15.** (a) $\frac{16}{3}$ (b) 110°   **17.** $7x^3y - 5xy^2$

19. $\frac{9}{14}$ **21.** $\frac{3}{8}$ **23.** 11   **25.** $\frac{29}{15}$ **27.** -24 **29.** -10

practice   **a.** $\frac{1,000,000}{(12)(12)(12)(5280)(5280)(5280)}$ mi^3

problem set   **1.** 600   **3.** 1400   **5.** Refer to Lesson 115   **7.** Refer to Lesson 125
128   **9.** 200

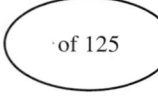 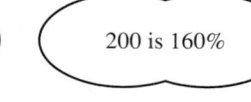

Before, 100%   After

11. $10(5280)(5280)(5280)(12)(12)(12)$ in.^{3} **13.** $\frac{7}{6}$ **15.** 5   **17.** $2x^3y^3 - 4xy^3$

19. $\frac{1}{14}$ **21.** $-\frac{64}{27}$ **23.** a^6b^4 **25.** x^5y^6 **27.** 0   **29.** 12

practice   **a.** 53

problem set   **1.** 4250   **3.** 1308   **5.** 41   **7.** 1001010 (base 2)
129   **9.** 105

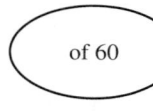

Before, 100%   After

11. $445(3)(12)$ in.^{3} **13.** 16   **15.** $3a^2b + 3ab^2 - 12ab$ **17.** $3a^3 + 6a^3b + b^3$

19. $-\frac{9}{8}$ **21.** $-\frac{25}{24}$ **23.** -10 **25.** a^4b^4 **27.** m^8n^5 **29.** 5

practice   **a.** $500(100)(100)(100)$ ml   **c.** 5,000,000 ml

problem set   **1.** $\frac{1}{36}$ **3.** 35,000   **5.** Refer to Lesson 127   **7.** Refer to Lesson 127
130   **9.** 192.8

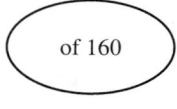

Before, 100%   After

11. 12,000 liters   **13.** $\sqrt{45}$ **15.** 27.3 ft^2

17. 31,704 cm^2 **19.** $5zmn^3 - 3zm^2$ **21.** $-\dfrac{31}{6}$ **23.** (a) 49   (b) $-\dfrac{1}{27}$

25. $7m^3n^5b^5$ **27.** 0   **29.** 162

practice

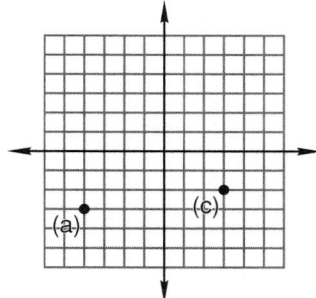

problem set 131   **1.** $\dfrac{1}{8}$ **3.** 400   **5.**

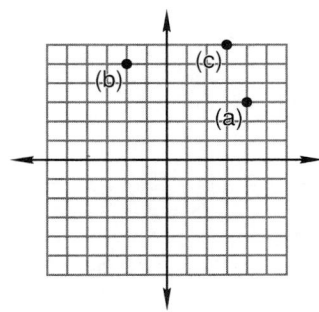

7. Refer to Lesson 125   **9.** Refer to Lesson 125

11. 92,500   **13.** 290

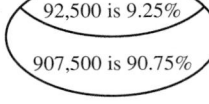
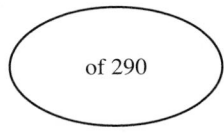
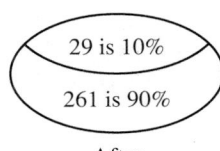

15. 2(5280)(5280)(5280) ft^3 **17.** $-3x^2y^2 + 3x^3y$ **19.** $\dfrac{25}{4}$ **21.** $\dfrac{1}{2}$ **23.** -5

25. -7 **27.** 6   **29.** $-\dfrac{3}{2}$

practice   **a.** 48 m^3 **c.** 904.32 cm^3

problem set 132   **1.** (a) $\dfrac{5}{36}$ (b) $\dfrac{1}{9}$ (c) $\dfrac{1}{12}$ **3.** 350

5. A number is an idea. A numeral is a symbol to make us think of the idea.

7. Refer to Lesson 127   **9.** 1100000 (base 2)   **11.** (a) 12.56 in.^{3} (b) 11.84 in.^{3}

13. 160%

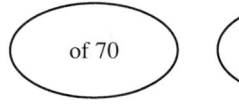

15. $\dfrac{(1320)(100)(100)(100)}{1000}$ liters   **17.** $2a^2b + 2ab^2 - 2a^2b^3 + 2a^3b^2$ **19.** 0.1272 m^3

21. $-\dfrac{61}{20}$ **23.** -10 **25.** $-\dfrac{3}{7}$ **27.** 0 **29.** 0.0026

practice   **a.** -1

problem set 133   **1.** $\dfrac{1}{4}$ **3.** 5400 **5.**

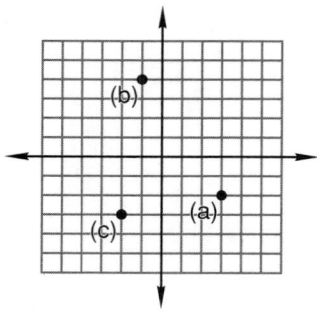

7. Refer to Lesson 125

9. 300 **11.** 312

Before, 100%   After   Before, 100%   After

13. $12(5280)(5280)(5280)(12)(12)(12) \text{ in.}^3$ **15.** $4x^2y^3 - x^2y^2$ **17.** 189.75 ft^3

19. $-\dfrac{1}{6}$ **21.** 1 **23.** (a) 8 (b) $\dfrac{1}{8}$ (c) 2 (d) $\dfrac{1}{27}$ **25.** $5a^7b^2$ **27.** $\dfrac{1}{9}$

29. 20

practice   **a.** 30 in.^3 **c.**   **e.** Line   **g.** 157 m^2

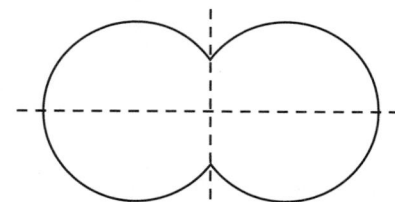

problem set 134   **1.** $\dfrac{1}{16}$ **3.** 400 **5.**

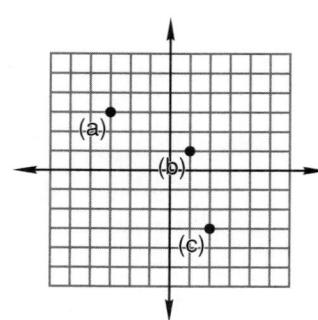

7. Refer to Lesson 125   **9.** Refer to Lesson 125   **11.** 15

13. 1450

Before, 100%   After

15. $\dfrac{60}{(12)(12)(12)(5280)(5280)(5280)}$ mi³   **17.** Figure (b) is symmetrical about a point.

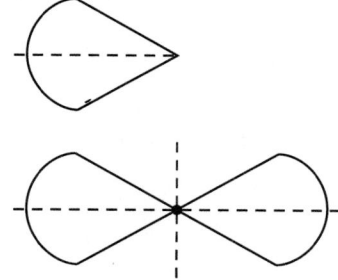

19. (a) 50° (b) $\sqrt{14}$ **21.** $\dfrac{10}{13}$ **23.** (a) 81 (b) $\dfrac{1}{64}$ **25.** $6a^6b^6$ **27.** -78

29. $31\dfrac{1}{7}$

practice   **a.** 362,880

problem set   **1.** $\dfrac{81}{1156}$ **3.** 1800   **5.**   **7.** 120
135

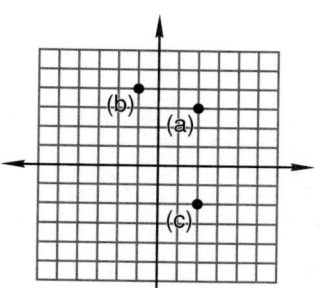

9. Refer to Lesson 125   **11.** (a) 42° (b) 5
13. Figure (b) is symmetrical about a point.

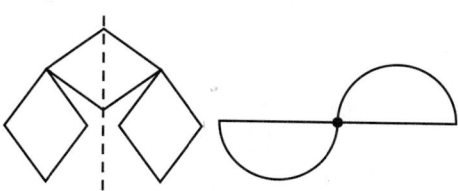

15. 136%

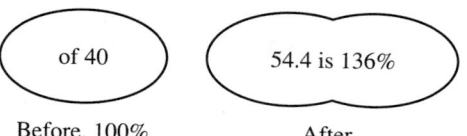

Before, 100%   After

17. 100,000

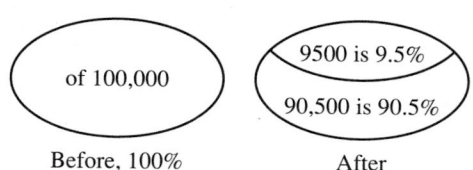

Before, 100%   After

19. 10(5280)(5280)(5280)(12)(12)(12) in.³

21. $-2mp^2 + 2m^2p$ **23.** $\dfrac{7}{32}$ **25.** -16 **27.** $\dfrac{19}{120}$ **29.** 21

practice   **a.** 53

problem set   **1.** $\frac{1}{676}$ **3.** $13,129.32   **5.**
136

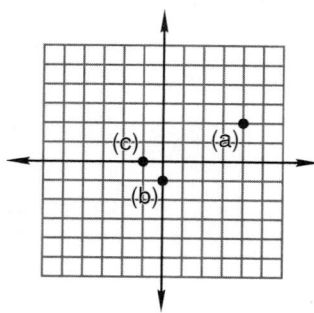

7. Refer to Lesson 127   **9.** 110000 (base 2)

11. 930

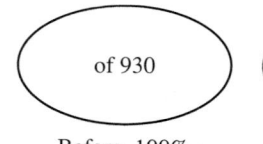

of 930

Before, 100%

279 is 30%

651 is 70%

After

13. (10,500)(100)(100)(100)(1) milliliters   **15.** 48 ft³   **17.** 30°, $\sqrt{75}$

19. (a) 17.5 m²   (b) 10 m²   **21.** $\frac{77}{54}$ **23.** (a) -64 (b) $\frac{1}{64}$ **25.** $m^8 p^7 z$

27. -84 **29.** -5

practice

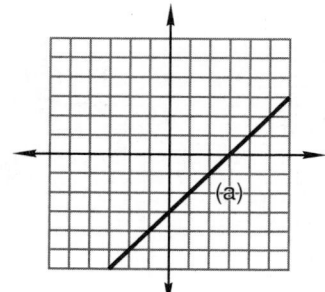

problem set   **1.** $\frac{1}{128}$ **3.** 131   **5.** 30 hours   **7.** 1010000 (base 2)
137

9. Refer to Lesson 127   **11.** Refer to Lesson 125   **13.** $\frac{18}{7}$

15. 5000

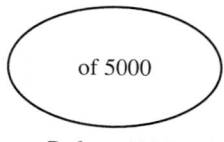

of 5000

Before, 100%

562.5 is 11.25%

4437.5 is 88.75%

After

17. $\frac{(1000)(1)}{(100)(100)(100)}$ m³   **19.** $m^5 b + m^2 b^3 + cm^3 b$ **21.** 50.24 m³

23. 16.75 m³   **25.** 96 m²   **27.** $-\frac{7}{24}$ **29.** $-\frac{134}{1125}$

practice set 1

1. $11,712.80   **3.** 220   **5.**

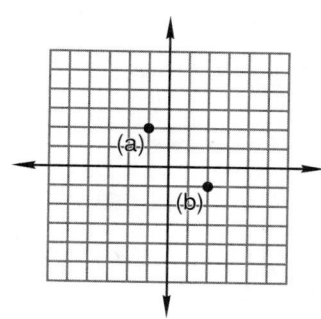

7. 800

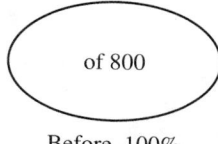

of 800	160 is 20%
Before, 100%	640 is 80%
	After

9. 140,000 cm^2 **11.**

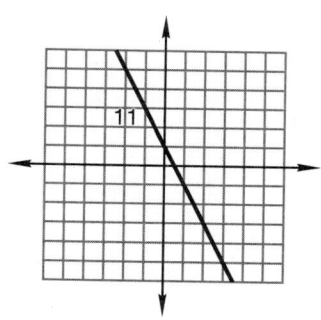

13. $ab^6 + a^2b^4 - a^2b^2$ **15.** 37.68 m^3 **17.** 53°, 10   **19.** Refer to Lesson 127

21. 9.42 in.^{3} **23.** $\frac{7}{4}$ **25.** $-\frac{4}{7}$ **27.** 64   **29.** -21

practice set 2

1. 5400   **3.** 3520   **5.** Refer to Lesson 115   **7.** Refer to Lesson 125

9. 240.5

| of 185 | 240.5 is 130% |
| Before, 100% | After |

11. 100(100)(100)(100)(10)(10)(10) mm^3 **13.** $\frac{97}{200}$ **15.** 6

17. $-2a^3b^2 - a^2b + 2ab^2$ **19.** $\frac{85}{168}$ **21.** $-\frac{39}{16}$ **23.** m^6p^3 **25.** x^7y^5

27. $-\frac{1}{60}$ **29.** -1

practice set 3

1. 860   **3.** 1710   **5.** 61   **7.** 1100000 (base 2)

9. 130.5   **11.** 24,699.33 in.^{3}

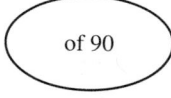

| of 90 | 130.5 is 145% |
| Before, 100% | After |

13. $-6\frac{1}{2}$ **15.** $4m^2n + 4mn^3 - 16mn$ **17.** $p^3 + 10p^3b + 2b^3 + 3a^3b$ **19.** -2

21. $\frac{5}{48}$ **23.** 0   **25.** x^8n^5 **27.** 81   **29.**

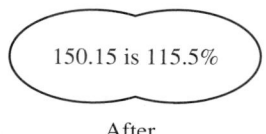

practice set 4

1. $\frac{1}{36}$ **3.** 40,000   **5.** Refer to Lesson 127   **7.** Refer to Lesson 127

9. 150.15

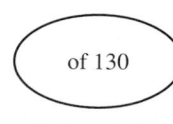

of 130   150.15 is 115.5%

Before, 100%   After

11. $\dfrac{(300)(100)(100)(100)}{1000}$ liters   **13.** 25   **15.** (a) 21 m^2 (b) 24 m^2

17. 38,632 cm^2 **19.** $-3m^2a + 8ap^3$ **21.** $-\dfrac{325}{192}$ **23.** (a) 32   (b) $-\dfrac{1}{64}$

25. $9c^7p^9x^8$ **27.** $\dfrac{103}{120}$ **29.** -24

practice set 5

1. $\frac{1}{16}$ **3.** 650   **5.**

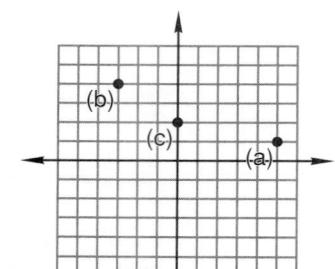

7. Refer to Lesson 125   **9.** Refer to Lesson 125

11. 115,000   **13.** 1580

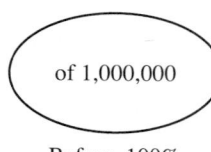 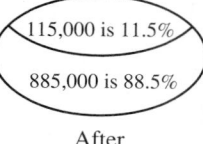 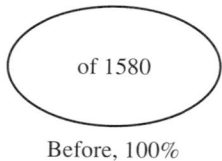 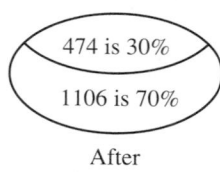

of 1,000,000   115,000 is 11.5%   of 1580   474 is 30%

885,000 is 88.5%   1106 is 70%

Before, 100%   After   Before, 100%   After

15. 8(5280)(5280)(5280)(12)(12)(12) in.^{3} **17.** $4x^4m^4 + 3x^4m^2 - 2x^2m^3$ **19.** 3

21. $\frac{6}{5}$ **23.** -64 **25.** 8   **27.** 184   **29.** $-\dfrac{13}{4}$

practice set 6

1. (a) $\frac{1}{12}$ (b) $\frac{1}{6}$ **3.** 2400

5. A number is an idea. A numeral is the symbol we write to make us think of the idea.

7. Refer to Lesson 127   **9.** 101001 (base 2)   **11.** (a) 25.12 in.^{3} (b) 33.49 in.^{3}

13. 260%

Before, 100%   After

15. 3,000,000 liters   **17.** $m^2n + mn^2 - m^2n^3 + m^3n^2$ **19.** 0.051075 m³   **21.** $-\frac{25}{12}$

23. -1 **25.** $\frac{5}{12}$ **27.** 60   **29.** 0.00875

practice set 7

1. $\frac{1}{32}$ **3.** 1024   **5.**

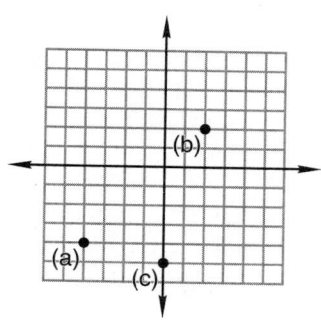

7. Refer to Lesson 125   **9.** Refer to Lesson 125   **11.** 7.5

13. 1750

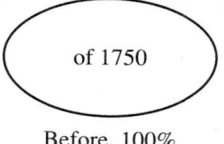 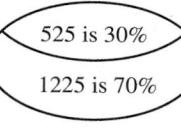

Before, 100%   After

15. $\dfrac{60}{(10)(10)(10)(100)(100)(100)}$ m³   **17.** Figure (b) is symmetrical about a point.

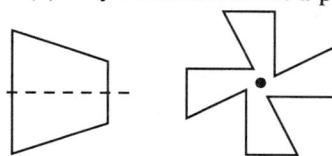

19. (a) 36° (b) 2   **21.** $\frac{43}{64}$ **23.** (a) 100 (b) $\frac{1}{1000}$ **25.** $6a^6b^6$ **27.** $\frac{21}{2}$

29. 25

practice set 8

1. $16,698.77   **3.** 198   **5.**

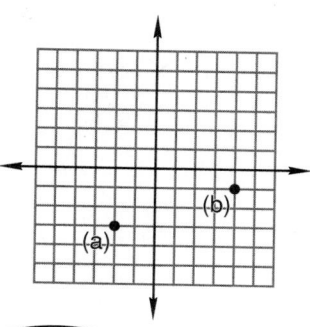

7. 900

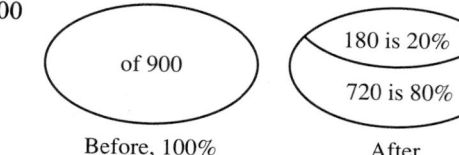

Before, 100%   After

9. $100(100)(100)$ cm^2 **11.**

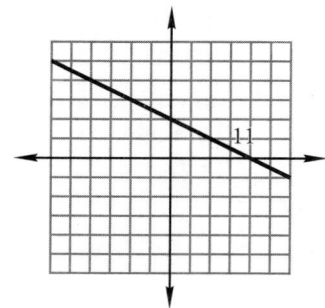

13. $a^4b^3 + a^2b^5 - b^4$ **15.** $\dfrac{47.1}{12(12)(12)}$ ft^3 **17.** $53°$, 5 **19.** Refer to Lesson 127

21. 36.63 in.^{3} **23.** 1 **25.** $-\dfrac{12}{5}$ **27.** $\dfrac{1}{512}$ **29.** $-\dfrac{109}{9}$

practice set 9 **1.** $\dfrac{49}{1936}$ **3.** 1080 **5.**

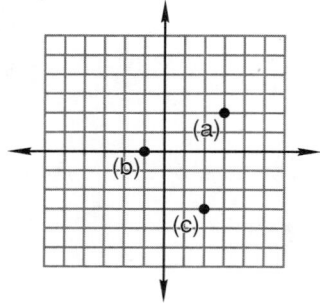

7. 39,916,800 **9.** Refer to Lesson 125 **11.** (a) $52°$ (b) $3\sqrt{5}$

13. Figure (b) is symmetrical about a point.

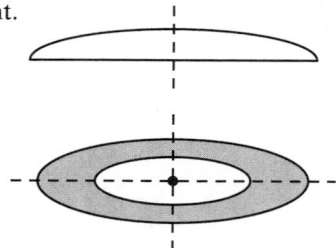

15. 128%

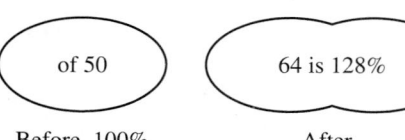

Before, 100%   After

17. 1

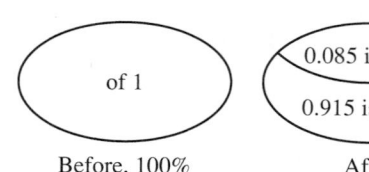

Before, 100%   After

19. 10,000,000 cm^3

21. $6z^2p - 4z^4p$ **23.** 7.7 **25.** $-\dfrac{5}{16}$ **27.** $\dfrac{137}{810}$ **29.** $10\dfrac{1}{36}$

**practice set
10**

1. $\dfrac{1}{676}$ **3.** \$12,695   **5.**

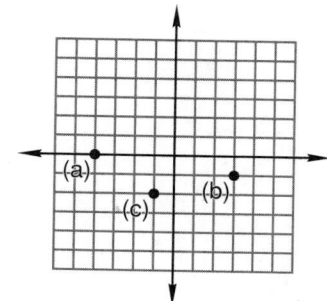

7. Refer to Lesson 127   **9.** 1001011 (base 2)   **11.** Refer to Lesson 125

13. $\dfrac{28}{3}$ **15.** 25,000

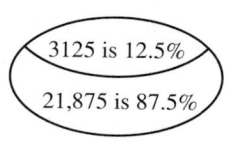

of 25,000

Before, 100%

3125 is 12.5%

21,875 is 87.5%

After

17. $\dfrac{5234}{(100)(100)(100)}$ m³   **19.** $a^6b^2 + a^5b^3 - a^6b^3$ **21.** 37.68 m³   **23.** 6.28 m³

25. 208.8 m²   **27.** $-\dfrac{5}{24}$ **29.** $-\dfrac{577}{9}$

Index

Geometric Formulas

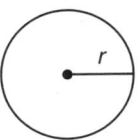

CIRCLE
$$\text{Area} = \pi r^2$$
$$\text{Circumference} = 2\pi r$$

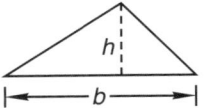

TRIANGLE
$$\text{Area} = \frac{1}{2}bh$$

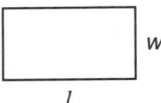

RECTANGLE
$$\text{Area} = l \times w$$

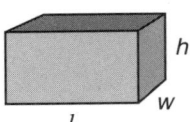

RECTANGULAR SOLID
$$\text{Volume} = l \times w \times h$$

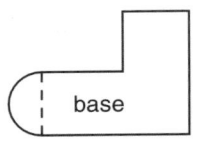 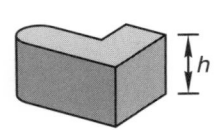

ANY RIGHT SOLID
$$\text{Volume} = \text{area of base} \times \text{height}$$
$(p.190)$

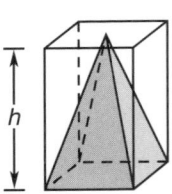

PYRAMID
$$\text{Volume} = \frac{1}{3}\left(\begin{array}{l}\text{the volume of a right} \\ \text{solid of the same height}\end{array}\right)$$

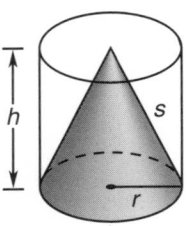

CONE
$$\text{Volume} = \frac{1}{3}\left(\begin{array}{l}\text{the volume of a right} \\ \text{solid of the same height}\end{array}\right)$$
$$\text{Lateral Surface Area} = \pi rs$$

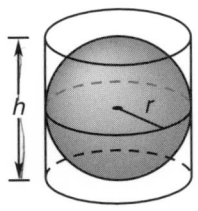

SPHERE
$$\text{Volume} = \frac{2}{3}\left(\begin{array}{l}\text{the volume of a right} \\ \text{cylinder of the same height}\end{array}\right)$$
$$= \frac{4}{3}\pi r^3$$

Lateral S.A. – p. 253
Surface area (calculating) – p. 147